Advances in the Synthesis of Heterocyclic Compounds and Their Applications

Advances in the Synthesis of Heterocyclic Compounds and Their Applications

Guest Editors

Mircea Darabantu
Ionel Mangalagiu

MDPI

Basel • Beijing • Wuhan • Barcelona • Belgrade • Novi Sad • Cluj • Manchester

Guest Editors

Mircea Darabantu
Department of Chemistry
University Babes-Bolyai
Cluj-Napoca
Cluj-Napoca
Romania

Ionel Mangalagiu
Chemistry Department
Alexandru Ioan Cuza
University of Iasi
Iasi
Romania

Editorial Office
MDPI AG
Grosspeteranlage 5
4052 Basel, Switzerland

This is a reprint of the Special Issue, published open access by the journal *Molecules* (ISSN 1420-3049), freely accessible at: https://www.mdpi.com/journal/molecules/special_issues/9H9L15X18P.

For citation purposes, cite each article independently as indicated on the article page online and as indicated below:

Lastname, A.A.; Lastname, B.B. Article Title. *Journal Name* **Year**, *Volume Number*, Page Range.

ISBN 978-3-7258-6996-1 (Hbk)
ISBN 978-3-7258-6997-8 (PDF)
https://doi.org/10.3390/books978-3-7258-6997-8

Contents

molecules

MDPI

Editorial

Advances in the Synthesis of Heterocyclic Compounds and Their Applications

Ionel I. Mangalagiu [1,*] and Mircea Darabantu [2,*]

[1] Faculty of Chemistry, Alexandru Ioan Cuza University of Iasi, 11 Carol 1st Bvd, 700506 Iasi, Romania
[2] Department of Chemistry, Faculty of Chemistry and Chemical Engineering, University Babes-Bolyai Cluj-Napoca, 11 Aranyi Janos str., 400028 Cluj-Napoca, Romania
* Correspondence: ionelm@uaic.ro (I.I.M.); mircea.darabantu@ubbcluj.ro (M.D.)

The synthesis of heterocyclic compounds remains a major goal in organic synthesis, especially because of their numerous applications in medicine, pharmacy, opto-electronics, agriculture, and related fields [1–5]. Subject to all evolutionary aspects of heterocyclic chemistry, modern approaches to heterocycles (reagents, methodologies, strategies, reaction mechanisms, chemical and instrumental auxiliaries, catalysts, etc.) nevertheless obey two crucial aspects, "selectivity" and "specificity" (including typical prefixes such as chemo-, regio-, or stereo-), as compulsory criteria [6–11]. Targeting heteromolecules and their applications in human life and in the environment, this Special Issue (S.I.) provides a current overview of the latest progress in synthetic methodologies and chemical technologies by updating relevant and important original contributions and reviews.

Contribution 1 is related to the synthesis and bioactivity of new boronic derivatives. Their antimicrobial activities (against *M. tuberculosis* and the fungal strains *C. albicans*, *T. mentagrophytes*, and *T. rubrum*) and anticancer properties (against oral squamous cell carcinoma cell lines) were determined, and some of the compounds were found to possess promising activities.

Contribution 2 focuses on the synthesis and spectral characterization of 3,5-bis-aminated pyrazolo[1, 5-*a*]pyrimidines using an innovative catalyzed Ullmann coupling methodology (CuI, microwave heating). Its advantages include very good yields, short reaction times, no toxic reagents, and a broad substrate scope.

Contribution 3 highlights the synthesis and spectral characterization of new tetracyclic heterosteroidal compounds, namely 14-aza-12-oxasteroids. The reaction pathway involves the Bucherer conversion of 2-naphthols to 2-naphthylamines, their subsequent regioselective C-acetylation (via the Sugasawa reaction), and borohydride reduction of the acetyl groups. Key intermediates of the resulting naphthalene amino-alcohols undergo double dehydration and subsequent double intramolecular cyclization in reaction with oxo-acids to afford the desired 14-aza-12-oxasteroids.

The electrochemical synthesis of 2-aminoxazole-based polycyclic compounds is another novel finding presented in the SI. Contribution 4 describes the electrochemical synthesis of *N*-arylnaphtho- and *N*-arylanthra[2, 3-*d*]oxazol-2-amines by reaction of 3-amino-2-naphthol or 3-amino-2-anthracenol with isothiocyanates in the presence of potassium iodide, using a graphite electrode as an anode and a platinum electrode as a cathode. The resultant naphthalene/anthracene-fused tricyclic and tetracyclic oxazoles exhibit extended π-conjugated skeletons and fluoresce in the 340–430 nm region.

Contribution 5 discusses the synthesis and bioactivity of some hybrid indole and 8-hydroxyquinoline derivatives linked to a di- or triaryl methane moiety. A direct synthetic protocol was also the most effective, and the isolated hybrid compounds were proven to have significant anticancer activity against the resistant colon adenocarcinoma cell line Colo320 and non-tumor fibroblast cells.

This SI has also welcomed theoretical research papers, such as Contribution 6, centered on calculating the premise of the regioselective synthesis of 3-nitro-substituted 2-isoxazolines via a molecular cycloaddition mechanism of the [3 + 2] type, involving nitro-substituted formonitrile *N*-oxide and electron-rich alkenes.

In Contribution 7, the authors report the synthesis and utility of some new 3,7-diheteroaryl-substituted 10-(3-(trimethylammonium)propyl)-10H-phenothiazine derivatives. They and their precursors exhibited reversible redox behavior with tunable potentials and blue to green-blue emissive aptitude. Some of the compounds demonstrated good antimicrobial activity against strains of *M. tuberculosis*, *A. baumannii*, *E. coli*, *S. aureus*, and *K. pneumoniae*.

The synthesis and bioactivity of some new 6-iodo-substituted carboxy-quinolines is discussed in Contribution 8. The study describes their one-pot, three-component reaction catalyzed by trifluoroacetic acid, which was efficient thanks to cost-effective catalysts, rapid response times, expeditious purification procedures, and high product yields. Their antimicrobial activity against strains of *K. pneumonie*, *S. epidermidis*, and *C. parapsilosis* was tested, with very promising results.

Contribution 9 concerns the synthesis and spectral characterization of new fused isoxazolidine/isoquinolinone and isoxazole/isoquinolinone hybrids. The reported strategy was facile and efficient, with general applicability, consisting of a three-step reaction sequence: a 1,3-dipolar cycloaddition, a Schmidt reaction, and an Ullmann-type cyclization.

Contribution 10 describes the synthesis and spectral characterization of a new library of C^3-difluoromethyl carbinol-containing imidazo[1, 2-*a*]pyridines. The HFIP-assisted Friedel–Crafts reaction of difluoroacetaldehyde ethyl hemiacetal and imidazo[1, 2-a]pyridines revealed facile and efficient access to the desired compounds. The green synthetic protocol had a wide substrate scope and had several advantages: it was highly efficient, was carried out at room temperature, and involved no transition metals or oxidants.

Finally, the SI also includes two extensive reviews related to the synthesis of heterocyclic compounds and their applications.

Contribution 11 presents advances in the synthetic knowledge of various pathways leading to quinoline-4-ones, together with an overview of their structures, evolutionary development, and structure–activity relationships. Contribution 12 offers an interesting perspective with respect to some classes of quinoline derivatives and related structures (including 4-aminoquinolines, quinoline-4-ols, and fused and spiro-quinolines). The selected examples are mostly discussed in terms of 2-azidobenzaldehyde-based [4 + 2] annulations. In addition, evidence supporting the aptitude of 2-azidobenzaldehyde-initiated synthesis for one-pot stepwise synthesis or multicomponent reactions is presented. Some of the synthetic approaches included in this Special Issue provide novel pathways for quinoline synthesis and could also be used to obtain other heterocyclic compounds.

Author Contributions: The authors have equal contributions. All authors have read and agreed to the published version of the manuscript.

Conflicts of Interest: The authors declare no conflicts of interest.

List of Contributions:

1. Prates, J.L.B.; de Matos Silva, S.; Medina-Alarcón, K.P.; dos Santos, K.S.; Belizario, J.A.; Lopes, J.R.; Marin-Dett, F.H.; Campos, D.L.; Mendes Giannini, M.J.S.; Fusco-Almeida, A.M.; et al. Synthesis and Evaluation of Boron-Containing Heterocyclic Compounds with Antimicrobial and Anticancer Activities. *Molecules* **2025**, *30*, 1117. https://doi.org/10.3390/molecules30051117.

2. Iorkula, T.H.; Tolman, B.A.; Ganiyu, L.O.; Peterson, M.A. An Efficient Synthesis of 3,5-Bis-Aminated Pyrazolo[1,5-a]Pyrimidines: Microwave-Assisted Copper Catalyzed C-3 Amination of 5-Amino-3-Bromo-Substituted Precursors. *Molecules* **2025**, *30*, 458. https://doi.org/10.3390/molecules30030458.
3. Srivastava, S.; Luo, J.; Whalen, D.; Robertson, K.N.; Jha, A. Concise Synthesis of Naphthalene-Based 14-Aza-12-Oxasteroids. *Molecules* **2025**, *30*, 415. https://doi.org/10.3390/molecules30020415.
4. Murata, Y.; Kawakubo, M.; Maruyama, A.; Matsumura, M.; Yasuike, S. Synthesis and Optical Properties of N-Arylnaphtho- and Anthra[2,3-d]oxazol-2-amines. *Molecules* **2025**, *30*, 319. https://doi.org/10.3390/molecules30020319.
5. Hegedűs, D.; Szemerédi, N.; Petrinca, K.; Berkecz, R.; Spengler, G.; Szatmári, I. Synthesis of Tumor Selective Indole and 8-Hydroxyquinoline Skeleton Containing Di-, or Triarylmethanes with Improved Cytotoxic Activity. *Molecules* **2024**, *29*, 4176. https://doi.org/10.3390/molecules29174176.
6. Dresler, E.; Wróblewska, A.; Jasiński, R. Energetic Aspects and Molecular Mechanism of 3-Nitro-substituted 2-Isoxazolines Formation via Nitrile N-Oxide [3+2] Cycloaddition: An MEDT Computational Study. *Molecules* **2024**, *29*, 3042. https://doi.org/10.3390/molecules29133042.
7. Khelwati, H.; van Geelen, L.; Kalscheuer, R.; Müller, T.J.J. Synthesis, Electronic, and Antibacterial Properties of 3,7-Di(hetero)aryl-substituted Phenothiazinyl N-Propyl Trimethylammonium Salts. *Molecules* **2024**, *29*, 2126. https://doi.org/10.3390/molecules29092126.
8. Al-Matarneh, C.M.; Nicolescu, A.; Marinaş, I.C.; Găboreanu, M.D.; Shova, S.; Dascălu, A.; Silion, M.; Pinteală, M. New Library of Iodo-Quinoline Derivatives Obtained by an Alternative Synthetic Pathway and Their Antimicrobial Activity. *Molecules* **2024**, *29*, 772. https://doi.org/10.3390/molecules29040772.
9. Ouzounthanasis, K.A.; Rizos, S.R.; Koumbis, A.E. A Convenient Synthesis of Novel Isoxazolidine and Isoxazole Isoquinolinones Fused Hybrids. *Molecules* **2024**, *29*, 91. https://doi.org/10.3390/molecules29010091.
10. Gao, J.; Liu, Z.; Guo, X.; Wu, L.; Chen, Z.; Yang, K. 1,1,1,3,3,3-Hexafluoro-2-Propanol-Promoted Friedel–Crafts Reaction: Metal-Free Synthesis of C3-Difluoromethyl Carbinol-Containing Imidazo[1,2-a]pyridines at Room Temperature. *Molecules* **2023**, *28*, 7522. https://doi.org/10.3390/molecules28227522.
11. Gach-Janczak, K.; Piekielna-Ciesielska, J.; Waśkiewicz, J.; Krakowiak, K.; Wtorek, K.; Janecka, A. Quinolin-4-ones: Methods of Synthesis and Application in Medicine. *Molecules* **2025**, *30*, 163. https://doi.org/10.3390/molecules30010163.
12. Zhang, X.; Liu, M.; Qiu, W.; Zhang, W. 2-Azidobenzaldehyde-Based [4+2] Annulation for the Synthesis of Quinoline Derivatives. *Molecules* **2024**, *29*, 1241. https://doi.org/10.3390/molecules29061241.

References

1. Naik, A.; Juvale, K. A comprehensive review on the role of nitrogen containing heterocycles in overcoming ABC transporter mediated multidrug resistance in cancer. *Bioorg. Chem.* **2025**, *163*, 108709. [CrossRef] [PubMed]
2. Rusu, A.; Moga, I.-M.; Uncu, L.; Hancu, G. The Role of Five-Membered Heterocycles in the Molecular Structure of Antibacterial Drugs Used in Therapy. *Pharmaceutics* **2023**, *15*, 2554. [CrossRef] [PubMed]
3. Boretti, A.; Banik, B. A Narrative Review of Four-Membered Heterocycles in Next-Generation Energy Conversion and Storage. *Energy Storage* **2025**, *7*, e70233. [CrossRef]
4. Moldoveanu, C.; Mangalagiu, I.I.; Zbancioc, G. Fluorescent Azasteroids through Ultrasound Assisted Cycloaddition Reactions. *Molecules* **2021**, *26*, 5098. [CrossRef] [PubMed]
5. Hu, Y.; Wu, D.; Peng, W.; Li, X.; Hu, F.; Huang, B.; Zhu, J.; Wu, Y. Heterocyclic compound, application thereof and pharmaceutical composition comprising same. *PCT Int. Appl.* **2019**, *2019*, 158107.
6. Pellissier, H. Synthesis of Chiral Heterocycles Through Enantioselective Silver-Catalyzed Transformations. *Eur. J. Org. Chem.* **2025**, *28*, e202401284. [CrossRef]
7. Chaudhary, A.; Srivastava, R. Synthetic Approaches of Six Membered N-Heterocycles Using Bridged N-Based Ionic Liquids: A Review. *Asian J. Org. Chem.* **2025**, *14*, e202400785. [CrossRef]

8. Takasu, K.; Shindo, N. Synthesis of Azaheterocycles and Related Molecules by Tf_2NH-Catalyzed Cycloadditions. *Heterocycles* **2018**, *96*, 195–218. [CrossRef]
9. Singh, M.K.; Lakshman, M.K. Recent developments in the utility of saturated azaheterocycles in peptidomimetics. *Org. Biomol. Chem.* **2022**, *20*, 963–979. [CrossRef] [PubMed]
10. Maji, M.; Panja, D.; Borthakur, I.; Kundu, S. Recent advances in sustainable synthesis of N-heterocycles following acceptorless dehydrogenative coupling protocol using alcohols. *Org. Chem. Front.* **2021**, *8*, 2673–2709. [CrossRef]
11. Wiesenfeldt, M.P.; Nairoukh, Z.; Dalton, T.; Glorius, F. Selective Arene Hydrogenation for Direct Access to Saturated Carbo- and Heterocycles. *Angew. Chem. Int. Ed. Engl.* **2019**, *58*, 10460–10476. [CrossRef] [PubMed]

molecules

MDPI

Article

Synthesis and Evaluation of Boron-Containing Heterocyclic Compounds with Antimicrobial and Anticancer Activities

João Lucas Bruno Prates [1,2,*], Samanta de Matos Silva [1], Kaila Petrolina Medina-Alarcón [1], Kelvin Sousa dos Santos [1], Jenyffie Araujo Belizario [1], Juliana Romano Lopes [1], Freddy Humberto Marin-Dett [1], Debora Leite Campos [1], Maria José Soares Mendes Giannini [1], Ana Marisa Fusco-Almeida [1], Paula Aboud Barbugli [1,3], Fernando Rogério Pavan [1] and Jean Leandro Dos Santos [1,2,*]

1 School of Pharmaceutical Sciences, São Paulo State University (UNESP), Araraquara 14800-903, SP, Brazil; samanta.matos@unesp.br (S.d.M.S.); kaila.medina@unesp.br (K.P.M.-A.); k.santos@unesp.br (K.S.d.S.); jenyffie.belizario@unesp.br (J.A.B.); romano.lopes@unesp.br (J.R.L.); freddy.m.dett@unesp.br (F.H.M.-D.); leite.debora26@gmail.com (D.L.C.); maria.giannini@unesp.br (M.J.S.M.G.); ana.marisa@unesp.br (A.M.F.-A.); paula.barbugli@unesp.br (P.A.B.); fernando.pavan@unesp.br (F.R.P.)

2 Institute of Chemistry, São Paulo State University (UNESP), Araraquara 14800-900, SP, Brazil

3 School of Dentistry, São Paulo State University (UNESP), Araraquara 14801-385, SP, Brazil

* Correspondence: joao.prates@unesp.br (J.L.B.P.); jean.santos@unesp.br (J.L.D.S.); Tel.: +55-16-3301-6962 (J.L.D.S.)

Abstract: Organoboron compounds, especially those containing boronic acid and benzoxaborole in their structure, have been gaining prominence in medicinal chemistry, following the FDA approval of tavaborole for the treatment of onychomycosis and bortezomib for multiple myeloma. The antimicrobial and anticancer effects of organoboron compounds motivate the investigation of the effects of the novel derivatives described here. A total of fourteen new boronic derivatives were synthesized and characterized using analytical methods. The antimicrobial activities were evaluated against *M. tuberculosis* (*Mtb*) H37Rv strains and fungal dermatophytes (*C. albicans*, ATCC 90028; *T. rubrum*, ATCC 28189; and *T. mentagrophytes*, ATCC 11481), while the anticancer effect was evaluated against oral squamous cell carcinoma (SCC) cell lines. Several promising boron-containing prototypes were identified, providing a foundation for further molecular optimization in the development of new antimicrobial and anticancer compounds.

Keywords: organoboron; boronic acid; benzoxaborole; boron-containing compounds; *Mycobacterium tuberculosis*; antimicrobial; fungi; cancer

1. Introduction

Organic molecules containing boron in their structure have garnered significant attention in chemistry due to their diverse applications, in both synthetic and biological contexts [1,2]. Boronic acid derivatives possess the ability to form reversible covalent bonds with nucleophilic amino acid residues [3–6], enabling their use in drug design for a range of therapeutic applications, including anticancer, anti-inflammatory, antibacterial, and antifungal activities (Figure 1) [7–10].

The therapeutic potential of boronic acid and its derivatives was first acknowledged in the mid-2000s, when the Food and Drug Administration (FDA) approved bortezomib (Velcade®) in 2003, establishing it as the first drug of its class (Figure 1). Bortezomib, an anticancer drug utilized in the treatment of multiple myeloma and mantle cell lymphoma, works by reversibly inhibiting the 26S proteasome, a protein complex responsible for degrading ubiquitinated proteins. This inhibition is achieved through the covalent binding of its boron atom to the threonine residue (Thr1) at the catalytic site of the proteasome, demonstrating a notable affinity with a Ki of 0.62 nM [11,12].

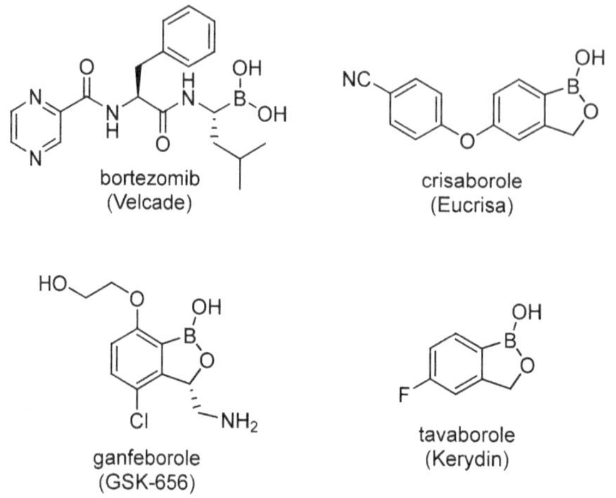

Figure 1. Chemical structure of boronic acid, bortezomibe (Velcade®), crisaborole (Eucrisa®) ganfeborole (GSK-656), and tavaborole (Kerydin®).

Crisaborole (Eucrisa®) (Figure 1), approved by the FDA in 2016 for the treatment of atopic dermatitis, serves as a noteworthy example of a boron-containing anti-inflammatory drug that emerged as an alternative to topical calcineurin inhibitors and corticosteroids. Acting as a reversible and competitive inhibitor, the boron atom in crisaborole binds to the zinc and magnesium ions at the active site of phosphodiesterase 4 (PDE4), leading to increased intracellular levels of cyclic adenosine monophosphate (cAMP) and ultimately reducing pro-inflammatory cytokine production [13].

In the pursuit of new therapeutic agents against tuberculosis, researchers developed the compound (*S*)-3-(aminomethyl)-4-chloro-7-(2-hydroxyethoxy)benzo[*c*][1,2]oxaborol-1(*3H*)-ol, commonly known as GSK-3036656 (ganfeborole) (Figure 1). This benzoxaborole derivative acts as an inhibitor of leucyl-tRNA synthetase (LeuRS) in *Mycobacterium tuberculosis*, disrupting the sequential transport of amino acids by forming a covalent bond between the boron atom in its structure and the adenosine monophosphate (AMP) Ade76 of tRNALeu. By inhibiting protein synthesis, a vital process for the survival of *M. tuberculosis*, this interaction effectively disrupts mycobacterial function. In vitro assays showed that ganfeborole exhibited an IC_{50} of 0.20 μM for LeuRS and an MIC of 0.08 μM against *M. tuberculosis* H37Rv strains. Currently, it is being evaluated in a phase 2a open-label randomized clinical trial [14,15].

Tavaborole, also known as AN2690, received FDA approval in 2014 for the treatment of onychomycosis and is commercially available under the trade name Kerydin® (Figure 1). This antifungal agent exerts its effect by inhibiting tRNA synthetase, a mechanism of action comparable to that of GSK-656. By targeting tRNA synthetase, tavaborole disrupts protein synthesis in the fungus, ultimately compromising its viability [16,17]. Tavaborole exhibits over 1000-fold selectivity for fungal leucyl-tRNA synthetase compared to its mammalian counterpart. Furthermore, it demonstrates potent antifungal activity, with MIC values of 2.0 μg/mL against dermatophyte strains, such as *T. rubrum* (ATCC 1890, ATCC 18759), *T. mentagrophytes* (ATCC 28185), *M. audouinii* (ATCC 42558), and *M. gypseum* (ATCC 24103), and 1.0 μg/mL against *E. floccosum* (ATCC 525066) [18].

In a constant effort to search for new antimicrobial and anticancer agents, our laboratory has carried out molecular optimization focusing on boron-containing derivatives.

Previously, we synthesized and evaluated a series of benzofuroxan derivatives (1–6) with promising anti-*Mycobacterium tuberculosis* with MIC_{90} values ranging from 0.09 to 19.20 μM, including multi-drug-resistant clinical isolates (MDR-TB) (Figure 2) [19]. Preliminary studies on the mode of action suggested that protein synthesis inhibition could be a potential mechanism [19]. Therefore, by exploring the bioisosteric replacement of benzofuroxan with benzoxaboroles and boronic acids, we synthesized and evaluated a total of fourteen novel compounds (Figure 2). Moreover, we explored their antifungal activity against dermatophytes and *Candida albicans*, drawing inspiration from the well-established antifungal properties of tavaborole.

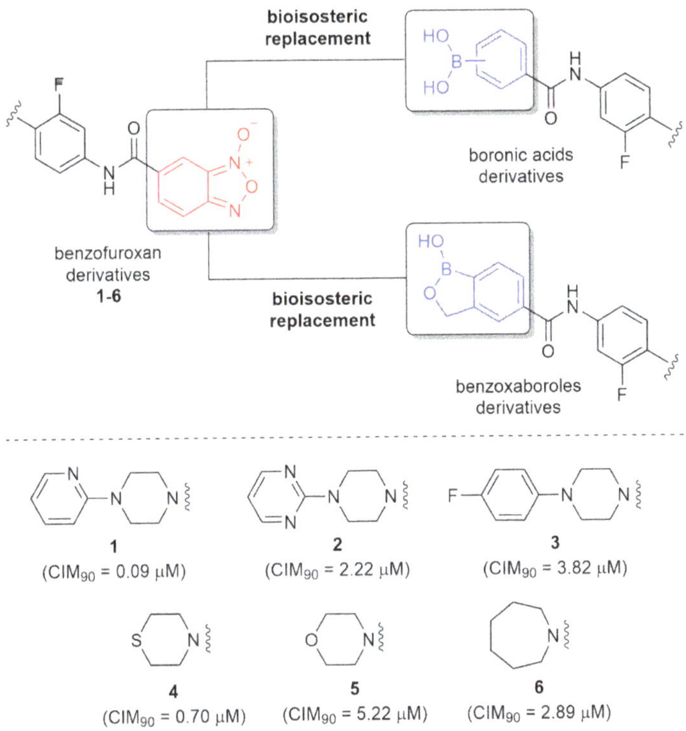

Figure 2. Design of boronic acid and benzoxaborole derivatives.

Furthermore, this screening was enhanced by a comprehensive evaluation of its anticancer potential. Although the anticancer effects of boron-containing compounds have been reported in the literature [7,20], there is limited information regarding their effects on head and neck cancer. Thus, to better understand their potential application, we further evaluated their activity against SCC-25 and NOK-si cell lines.

2. Results

2.1. Chemistry

2.1.1. Synthesis of Boronic Acid Derivatives

The synthesis of the target series **29–40**, **44**, and **45** started with the preparation of intermediates **13–18** (Scheme 1), implying, firstly, the expected regioselective *p*-S_N2Ar amination of 3,4-difluoronitrobenzene carried out with a variety of cyclamines.

The second step consisted of reductive conversion of nitro derivatives **13–18** into the corresponding amines **19–24** in classical Béchamp conditions.

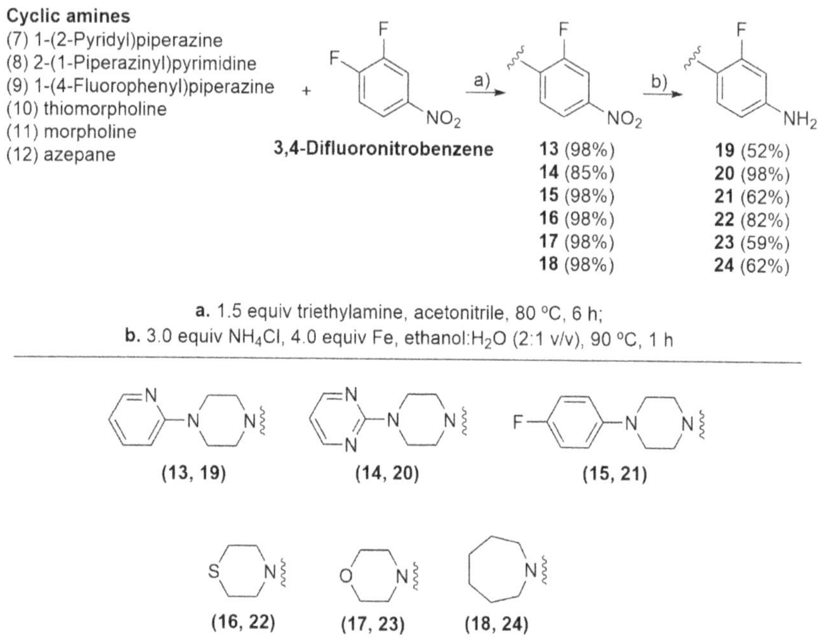

Scheme 1. Synthesis of nitro (**13–18**) and amine aromatics (**19–24**).

In the final step, direct amidation coupling between 3-carboxyphenylboronic acid **25** or 4-carboxyphenylboronic acid **26** and arylamines **19–24** was performed. *N*-(3-dimethylaminopropyl)-*N′*-ethylcarbodiimide (EDC) was used as the coupling agent, yielding the boronic derivatives **29–40** [21–23] (Scheme 2).

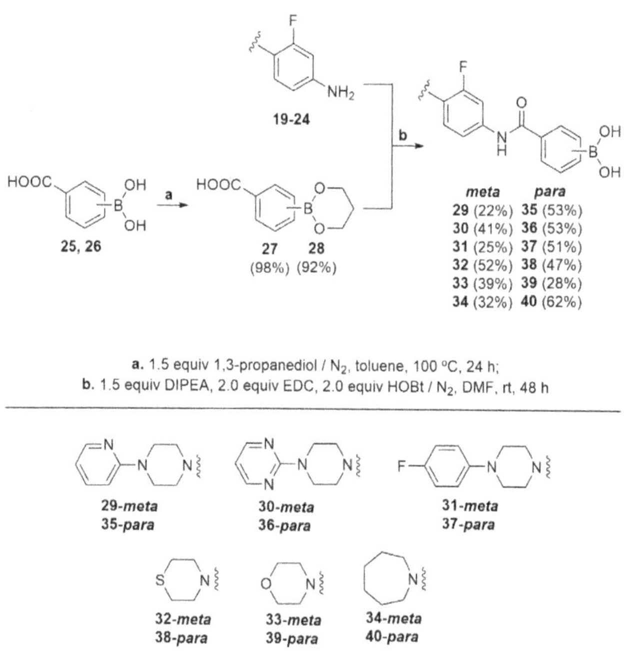

Scheme 2. Synthesis of boronic acid derivatives (**29–40**).

2.1.2. Synthesis of Benzoxaboroles Derivatives

First (a), the bromine in the commercial 4-bromo-3-methylbenzoate was replaced by the pinacol boronic acid ester group, applying the Miyaura borylation protocol [24–26] to afford the arylboronic acid **41** as a white solid (98% yield). The second step (b) involved the Wohl–Ziegler benzylic bromination of **41** with the use of the tandem NBS/AIBN to obtain compound **42** (yellow oil, 62% yield). The third stage (c) occurred in a one-pot two-step manipulation (**c-i** and **c-ii**), the first being the alkaline SN hydroxylating replacement of bromine (**c-i**); this was followed by the intramolecular transesterification/acidolysis of the pinacol boronic ester (**c-ii**), thus leading to a five-membered oxaborole ring closure [27]. In addition, the same acidolysis conditions also favored formation of the free carboxyl termini group, producing the benzoxaborole acid **43** (slightly yellow solid, 65% yield). The last step (**d**) consisted of the direct amidation coupling between the carboxylic benzoxaborole acid **43** and fluorinated phenyl amines **19** and **22**, carried out in the presence of azabenzotriazole tetramethyl uranium (HATU) [23], producing benzoxaborole phenyl amides **44** and **45** (31% and 15% yields, respectively, solid materials) (Scheme 3).

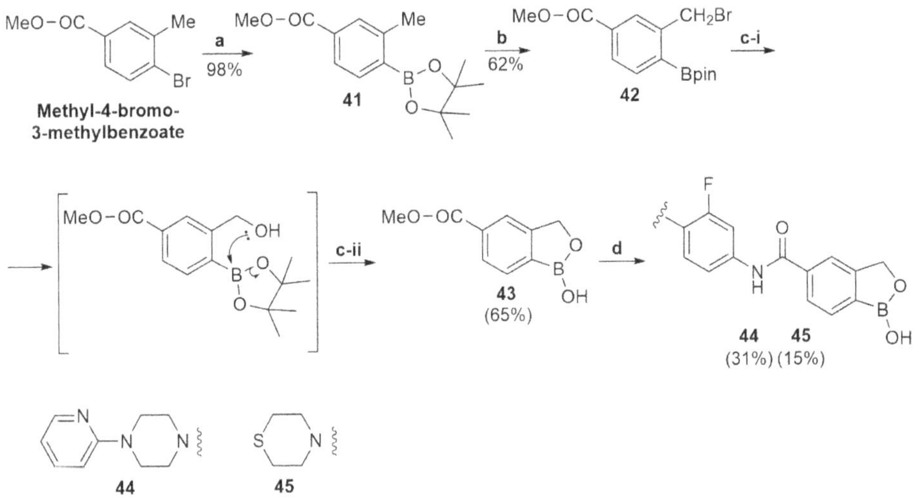

a. 1,5 equiv B_2pin_2, 3.0 equiv KOAc, (10% mol) $PdCl_2$(dppf) / N_2, 1,4-dioxane, 80 °C, 12 h; **b.** 1.5 equiv NBS, (25% mol) AIBN / ACN, 80 °C, 24 h; **c-i.** 10.0 equiv NaOH / H_2O:THF (2:1 v/v), 50 °C, 2 h; **c-ii.** HCl 6 M, pH~2,0 / 2 h; **d.** 1.0 equiv **19** (or 1.0 equiv **22**), 1.0 equiv DIPEA, 1.1 equiv HATU / N_2, DMF, rt, 24 h

Scheme 3. Synthesis of benzoxaborole phenyl amide **44** and **45**.

2.2. Biological Assay

2.2.1. In Vitro Antimicrobial Activity Against *Mycobacterium tuberculosis* and Fungi

All the compounds were evaluated against the inhibition activity of the *M. tuberculosis* H37RV ATCC 27294 strain, using the resazurin microtiter assay (REMA) method. The results were determined using the minimum inhibitory concentration (MIC_{90}), which is the minimum concentration required to inhibit 90% of the mycobacterium's activity [28]. As in-house protocol, we established a cut-off for the MIC_{90} as being less than 25 μM [29]. All the compounds, **29–40**, **44**, and **45**, were considered inactive, with a MIC_{90} value above 25 μM (Table 1).

Table 1. Antimicrobial and cLogD *.

Compounds	*T. rubrum* ATCC 28189 MIC_{90} (μM)	*T. mentagrophytes* ATCC 28189 MIC_{90} (μM)	*C. albicans* ATCC 90028 MIC_{90} (μM)	*M. tuberculosis* $H_{37}R_v$ MIC_{90} (μM)	cLogD *
29	297	297	297	>25	4.44
30	297	297	297	>25	3.60
31	286	286	286	>25	5.19
32	347	347	347	>25	3.34
33	363	363	363	>25	2.76
34	42.0	21.0	351	>25	4.22
35	297	297	297	>25	4.44
36	297	297	297	>25	3.60
37	286	286	286	>25	5.19
38	347	347	347	>25	3.34
39	363	363	363	>25	2.76
40	88.0	42.0	351	>25	4.22
44	289	146	289	>25	4.90
45	336	336	336	>25	3.80
Terbinafine	0.10 nM	0.025 nM	–	–	–
Itraconazole	–	–	0.34 nM	–	–
Isoniazide	–	–	–	0.11	–

* cLogD calculated using ChemAxon's calculator.

Compounds **29–40**, **44**, and **45** were evaluated for their antifungal activity against *C. albicans* (ATCC 90028), *T. rubrum* (ATCC 28189), and *T. mentagrophytes* (ATCC 11481). The MIC_{90} values were determined, along with the minimum fungicidal concentration (MFC), which represents the lowest concentration required to kill the fungi. Compound **34** demonstrated promising inhibition activity against *T. rubrum* and *T. mentagrophytes* with MIC_{90} values of 42.0 μM and 21.0 μM, respectively. Compound **40** did not exhibit inhibition activity comparable to **34**, but it did demonstrate activity with MIC_{90} values of 88.0 μM and 42.0 μM against *T. rubrum* and *T. mentagrophytes*, respectively. Compound **44** exhibits inhibition activity against *T. mentagrophytes* with an MIC_{90} of 146 μM (Table 1). None of the compounds exhibited activity against *C. albicans*.

2.2.2. Cellular Viability

For the cytotoxicity assay, the HaCaT cell line, which is characterized as immortalized normal human keratinocytes, was used. The cytotoxic effects of the most active antifungal compounds **34**, **40**, **44**, and **45** were evaluated. It was observed that the boronic acid derivatives **34** and **40** exhibited higher cytotoxic effects compared to the benzoboroxole derivatives **44** and **45**.

For compound **34**, no cytotoxic effects were observed at concentrations ranging from 0.2 to 1.1 μg/mL. Similarly, for compound **40**, the absence of cytotoxic effects was noted at concentrations between 0.2 and 9.3 μg/mL. In contrast, the benzoboroxole derivatives **44** and **47** did not exhibit cytotoxic effects at concentrations ranging from 0.2 to 75 μg/mL (Figure 3).

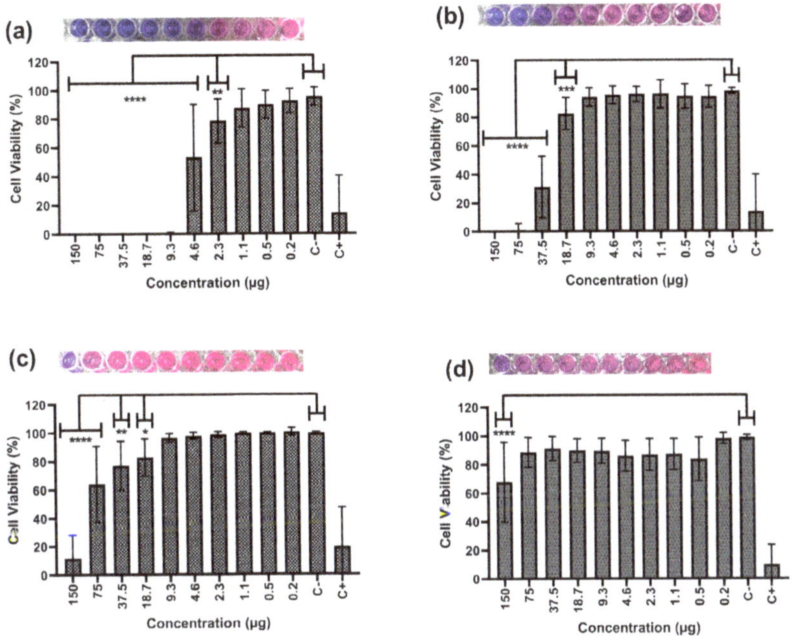

Figure 3. (**a**) Resazurin cytotoxicity assay using HaCaT cells for compound **24**, (**b**) for compound **30**, (**c**) for compound **36**, and (**d**) for compound **37**. **** $p < 0.0001$ *** $p < 0.001$ ** $p < 0.01$ * $p < 0.1$.

The distribution coefficient (cLog D) of the boronic derivatives was calculated using ChemAxon's calculator (Table 1) to assess the impact of this physicochemical property on their biological evaluation. All the compounds exhibited cLogD values ranging from 2.76 to 5.19. For the most active antifungal compounds **34**, **40**, **44**, and **45**, the calculated cLogD values were 4.22, 4.22, 4.9, and 3.8, respectively. These compounds displayed high lipophilicity, which is consistent with the characteristics of antifungal drugs effective against dermatophytes.

2.2.3. In Vitro Anticancer Activity

Boronic acid and benzoboroxole derivatives have been identified as potential anti-cancer agents [3,9,20,30]. To investigate their efficacy against head and neck cancer cell lines, SCC-25 cells were utilized in this study, as a model for oral squamous cell carcinoma. Additionally, cytotoxicity was performed in normal oral keratinocytes NOK-si cells to evaluate the selective index of the compounds. A compound is considered promising when it demonstrates high cytotoxicity against tumor cells while exhibiting minimal toxicity toward normal cells. The selectivity of the compounds is quantified by the selectivity index (SI), with those with an SI greater than 3.0 being considered promising for further evaluation, in accordance with internal protocols. Among the fourteen boronic acid derivatives tested, compound **35** exhibited the highest potency against SCC-25 cells (IC_{50}: 45.01 μM) with a selectivity index of 3.37. It displayed minimal toxicity to NOK-si cells (IC_{50}: 220.7 μM), demonstrating its potential for further investigation (Table 2). In contrast, its regioisomer **29** showed low inhibition activity (SCC-25 IC_{50}: 79.06 μM), but non-selectivity (SI < 1.0) (Table 2). The benzoboroxol derivative **44** IC_{50} values were not determined (N/A) for either SCC-25 or NOK-si cells due to its absent killing effect in SCC-25 cells (Table 2).

Table 2. Anticancer activity.

Compounds	SCC-25 ** IC_{50} (μM)	NOK-si *** IC_{50} (μM)	SI
29	117.4	110.0	<1
30	N/D *	N/D *	<1
31	59.07	220.7	3.73
32	N/D *	N/D *	<1
33	N/D *	N/D *	<1
34	67.57	129.3	1.91
35	45.61	153.7	3.37
36	133.4	133.5	1.00
37	79.06	123.0	1.55
38	N/D *	149.0	<1
39	74.09	77.09	1.04
40	85.09	127.2	1.49
44	N/D *	N/D *	<1
45	82.98	65.77	<1
5-fluorouracil	1800	–	–

* N/D: not determined. ** SCC-25: squamous cell carcinoma. *** NOK-si: normal oral keratinocytes–spontaneously immortalized.

Like compound **35**, compound **31** decreased SCC-25 cell viability (IC_{50}: 59.07 μM) with excellent selectivity (SI: 3.73) and no cytotoxicity effect on NOK-si cells (IC_{50} > 220.7 μM). However, its derivative **27** showed moderate inhibition (IC_{50}: 79.06 μM) and lacked significant selectivity (SI: 1.55) (Table 2). The compounds **34**, **39**, and **40** exhibited moderate inhibitory effects against SCC-25 cells (IC_{50}: 67.57, 74.09, and 85.09 μM, respectively), but displayed low selectivity, as evidenced by their selectivity index (SI) values below 3.0 (SI: 1.91, 1.04, and 1.49, respectively). Derivative **36** demonstrated weak inhibitory activity against SCC-25 cells (IC_{50}: 133.4 μM) and was also cytotoxic to NOK-si cells (IC_{50}: 133.5 μM), whereas compound **45** exhibited the highest cytotoxicity towards NOK-si cells within the series (IC_{50}: 65.77 μM). It was not feasible to ascertain the IC_{50} values for compounds **30**, **32**, and **33** against either cell line (Table 2). These data suggested the potential of compounds **31** and **35** as prototypes for further optimization, aiming towards new anticancer agents for the treatment of oral cancers.

3. Discussion

In recent years, there has been growing interest in the application of boron derivatives in medicinal chemistry, following the approval of new drugs, such as crisaborole, bortezomib, and tavaborole. The distinctive ability of boronic acid and benzoboroxole derivatives to form reversible covalent bonds with amino acid residues has inspired the exploration of new therapeutic agents for the treatment of cancer, inflammation, and microbial infections [31,32].

In previous work conducted by our research group, we identified potent benzofuroxan derivatives with activity against *Mycobacterium tuberculosis*, exhibiting MIC_{90} values of 0.09 μM [19]. Transcriptomic studies indicated that the mechanism of action of these compounds involved the inhibition of protein synthesis. A similar mechanism of action has been reported for ganfeborole (GSK-3036656, MIC_{90}: 0.08 μM), which inhibits the leucyl-tRNA synthetase (LeuRS) enzyme [14]. These findings led us to hypothesize that the biosynthetic substitution of the benzofuroxan scaffold with benzoboroxole and boronic acid derivatives could further optimize anti-*Mtb* activity. To test this hypothesis, fourteen new derivatives were synthesized and evaluated as antimicrobial agents, yielding overall synthesis rates between 15% and 62%.

The presence of boronic acids has been investigated in the development of new anti-*MtB* drugs. Umerisi and colleagues synthesized and tested a series of derivatives containing this functional group as inhibitors of the enoyl acyl carrier protein reductase (InhA) enzyme, which plays a key role in the biosynthesis of fatty acids, including mycolic acid. The MIC_{90} values ranged from 0.083 to 0.125 mg/mL, highlighting the potential of this chemical group in anti-*Mtb* activity [33]. In another study, Chatterjee and collaborators explored the ability of boronic acids to inhibit the LexA/RecA system in mycobacteria. This system is critical for the "SOS" response, and its inhibition disrupts the mycobacteria's ability to repair DNA damage under stress conditions. Molecular modeling studies indicated that boronic acid can interact with the catalytic residues (S160 and K197) of the *Mtb* LexA enzyme [34]. The reactivity of boronic acids with glycans in the mycobacterial membrane structure has prompted the synthesis of dimeric boronic acid derivatives, aimed at forming reversible covalent bonds with the 1,2- and 1,3-diols found in the carbohydrate structure of the *Mtb* membrane. It is well understood that the glycan layer of mycobacteria is vital for its survival and pathogenesis, and disruption of this structure can result in the bacillus's death. In this study, the researchers identified boronic acid derivatives with MIC_{90} values ranging from 0.78 to 12.5 mM. Additionally, the authors proposed optimal linker distances, ranging from 1.59 to 9.68 nm, to facilitate binding with the glycans in the *Mtb* cell wall [35,36]. Boronic acid transition state inhibitors (BATSIs) are recognized as an important class of beta-lactamase inhibitors. The BlaC enzyme is known to be involved in the inactivation of beta-lactam antibiotics, and the combination of meropenem with clavulanate has been shown to have sterilizing effects on *Mtb* cultures. In this regard, and given the ability of boronic acids to form reversible covalent bonds, Kurz and colleagues described new boronic acid derivatives that mimic the transition state and inhibit the BlaC enzyme in mycobacteria, offering a promising new approach in the search for antimycobacterial agents [37]. In light of the potential of boronic acids and benzoboroxol, and leveraging a bioisosterism strategy, this study investigates the capacity of these compounds to inhibit mycobacterial growth. However, exploratory studies assessing their activity against *M. tuberculosis* did not confirm the anticipated bioisosteric relationship, as the compounds showed no significant activity against the bacillus, with MIC_{90} values exceeding 25 μM. One possible reason for the lack of effect is that benzofuroxan compounds may exert additional influences, such as by increasing oxidative stress and generating free radicals, which could enhance their anti-*Mtb* activity [19,29].

Alongside the antimycobacterial assays, we investigated the effects of the compounds on fungal cells to assess their activity in eukaryotic systems. This investigation aimed to explore the spectrum of action of the compounds and conduct preliminary screening, given that tavaborole had been previously reported as an antifungal agent against dermatophytes. Tavaborole has been reported to exhibit fungistatic effects, with MIC values of 8 μg/mL against *T. mentagrophytes* and *T. rubrum* [38]. The rates of mycological cure and complete cure have been observed to reach up to 36% and 10%, respectively [39]. Furthermore, resistance to tavaborole has already been documented, highlighting the need for the discovery of new antifungal agents.

The antifungal properties of organoboron compounds have been extensively documented [40]. Notably, these compounds have been shown to interfere with carbohydrate metabolism and other metabolic pathways essential for fungal growth and reproduction [41]. For instance, boric acid disrupts hyphal transformation and the development of fungal biofilms [42]. Campbell and colleagues synthesized a series of 3-substituted-2(5*H*)-oxaboroles with antifungal activity against several species, including *T. mentagrophytes*, *P. chrysogenum*, and *A. flavus*, with MIC values ranging from 5.2 to 6.25 μg/mL. The planarity of the system was found to enhance the antifungal effects, and none of the compounds exhibited hemolytic effects or cytotoxicity against rat myoblast cells (H9c2) [43]. Borys and colleagues have reported the antifungal properties of a series of 2-formylphenylboronic acids, with MIC values ranging from <1 to 125 µg/mL against *Aspergillus*, *Fusarium*, *Penicillium*, and *Candida* species. Notably, these derivatives exhibit cyclic tautomers in solution, leading to the formation of 3-hydroxybenzoxaborole [44]. A similar approach was reported by Adamczyk-Woźniak and colleagues, who identified 5-trifluoromethyl-2-formyl phenylboronic acid as an inhibitor of leucyl-tRNA synthetase (LeuRS). The authors observed MIC values ranging from 8 to 250 µg/mL against *Candida albicans*, *Aspergillus niger*, *Escherichia coli*, and *Bacillus cereus* [45].

In our studies, we evaluated the antifungal activity of the compounds against the dermatophytes *T. rubrum* (ATCC 28189) and *T. mentagrophytes* (ATCC 11481), as well as against *C. albicans* (ATCC 90028). Compounds **34**, **40**, **44**, and **45** demonstrated moderate antifungal activity, with compound 34 showing MIC_{90} values of 42.0 µM and 21.0 µM against *T. rubrum* and *T. mentagrophytes*, respectively. Although the boronic acid derivatives exhibited stronger antifungal activity than the benzoboroxole derivatives, cytotoxicity studies using HaCaT cells revealed that the latter showed lower toxicity. The benzoboroxole derivatives did not exhibit cytotoxic effects at concentrations ranging from 0.2 to 75 µg/mL. No clear structure–activity relationship was identified for this series; however, compounds containing 3- or 4-(azepan-1-yl) (**34** and **40**, respectively) exhibited promising antifungal effects, suggesting their potential for further molecular optimization.

To explore potential secondary targets involved in the action of the compounds, and in light of the observed cytotoxic effects, we proceeded to screen for anticancer activity. The anticancer activity of boron-containing compounds has been well documented in the literature [20], which inspired the evaluation of these compounds against tumor cells. Head and neck cancers and oral squamous cell carcinoma rank among the fifth most common cancer types and can present challenges in treatment [46]. Boronic acid has been shown to inhibit the proliferation of head and neck cancer cell lines, with GI_{50} values ranging from 800 to 1600 µg/mL [47]. Additionally, boric acid was evaluated for its effects on medullary thyroid cancer, revealing an IC_{50} of 35 µM after 48 h of treatment, through the modulation of apoptotic pathways [48]. In addition to bortezomib, other organoboron drugs, such as ixazomib, have been shown to treat cancers with greater efficacy and safety compared to bortezomib [32]. Furthermore, non-dipeptide boron derivatives have demonstrated potent anticancer effects against head and neck cancer cell lines. For instance, Kong and collaborators synthesized novel tubulin polymerization derived from combretastatin, where a hydroxyl group was replaced by a boronic acid. These compounds showed activity across 16 human cancer cell lines, with MIC values ranging from 10 to 200 nM. In the case of SNC, they tested the compounds on cell lines SF-268, SF-295, SF-539, SNB-19, and U251, reporting GI50 values ranging from <0.01 to 2.73 µM [49]. These findings prompted us to explore the anticancer effects of the synthesized boronic acid and benzoboroxole derivatives. During the screening, two promising prototype compounds, **31** and **35**, demonstrated potent activity against SCC-25 cells, with IC_{50} values of 59.07 µM (**31**) and 45.61 µM (**35**), and SI values of 3.73 (**31**) and 3.37 (**35**), while showing no cytotoxicity toward non-tumoral NOK-si cells (IC_{50} > 220.7 µM). Like the findings in antifungal activity, no clear structure–activity relationship was identified for the anticancer effects.

Thus, in this screening study, we identified the benzoxaborole compounds **44** and **45** as promising prototypes for the development of new antifungal agents, while boronic acid derivatives emerged as potential prototypes for anticancer agents. The identification of these prototype compounds represents a key paradigm in medicinal chemistry, enabling continuous optimization in the pursuit of new therapeutic agents for the treatment of infectious diseases and cancer.

4. Experimental

A. General

All the reagents were of analytical purity and purchased from Acros, Alfa Aesar, Oakwood Chemical, Sigma-Aldrich, and Synth, with no need for additional purification, except for 1,4-dioxane, acetonitrile, and toluene which were dried over molecular sieves porosities 5 Å and 1.6 Å, (Sigma-Aldrich, St Louis, MO, USA) prior to use. Sensitive reactions were carried out under an inert atmosphere (N_2).

Melting points were determined using Stuart Scientific® Melting Point Apparatus SMP3 capillary equipment (Bibby Stuart Scientific, Cole-Parmer, Staffordshire, UK) and were not corrected. TLC was carried out on aluminum sheets (plates) with silica gel 90 Å Merck® (visualization with fluorescence indicator under UV at λ = 254 nm). Column chromatography was conducted on Sigma-Aldrich® 60 Å silica gel (40–60 μm, 230–400 mesh) (Sigma-Aldrich, St Louis, MO, USA) using ethyl acetate, n-hexane, and methanol as eluents (mobile phase).

IR spectra (KBr) were recorded using Prestige-21 Shimadzu® equipment (Shimadzu, Kyoto, Japan) operating in the range absorption between 400 and 4000 cm^{-1}. Only relevant absorption maxima (ν_{max}) are listed as s (strong), m (medium), and w (weak). 1D-^{1}H- and -^{13}C-NMR spectra were carried out using a Bruker Fourier 300 spectroscope (Bruker Corporation, Billerica, MA, USA) operating at 300 and 75 MHz for ^{1}H and ^{13}C nuclei, respectively. 1D-ATP and 2D-^{1}H, ^{13}C-HSQS, -HMBC spectra/charts were obtained using Bruker Avance III 600 (Bruker Corporation, Billerica, Massachusetts, USA) equipment operating at 600 and 150 MHz for ^{1}H and ^{13}C nuclei, respectively. 1D-ATP and 2D-^{1}H, ^{13}C-HSQS, -HMBC spectra/charts were obtained using Bruker Avance III 600 equipment operating at 600 and 150 MHz for ^{1}H and ^{13}C nuclei, respectively. All chemical shifts (δ_X values, X = ^{1}H and ^{1}) are given in parts per million (ppm); all homo- and heterocoupling patterns ($^nJ_{C,F}$ values) are given in Hertz. No TMS was added; the chemical shifts were measured against the solvent peak: Acetone-d_6 (δ_H = 2.09 and δ_C = 205.87 ppm), $CDCl_3$ (δ_H = 7.26 and δ_C = 77.36 ppm) and DMSO-d_6, (δ_H = 2.50 and δ_C = 39.52 ppm). The multiplicity of ^{1}H and ^{13}C signals was assigned as: singlet (s), doublet (d), triplet (t), quadruplet (q), double doublet (dd), double double of doublet (ddd), and multiplet (m).High-resolution mass spectrometry (ESI+) spectra were obtained by using the Bruker Maxis Impact quadrupole time-of-flight tandem mass spectrometer (Q-TOF MS/MS) (Bruker Corporation, Billerica, MA, USA).

For further details, please refer to the Supplementary Materials, where the spectra for all analytical methods are described.

B. General Procedure for the Synthesis of Nitroaromatic Intermediates **13–18** [17,19]

At room temperature, under inert atmosphere (N_2) and after 15 min. of vigorous stirring, 3,4-difluoronitrobenzene (7.03 mmol, 1.0 equiv.) was added to an anhydrous acetonitrile (8 mL) solution containing (7.03 mmol, 1 equiv.) of phenyl amine **7–12** and triethylamine (10.55 mmol, 1.5 equiv.). The reaction mixtures were heated at reflux for 6 h. After cooling at room temperature, the solvent and the excess triethylamine were removed to dryness under reduced pressure. The resulting residues were purified by flash column chromatography (eluent n-hexane/AcOEt 1:1 v/v) [17,19].

1-(2-Fluoro-4-nitrophenyl)-4-(pyridin-2-yl)piperazine **13** yellow solid; yield 98%; mp: 152–154 °C (n-hexane/AcOEt, 5:5); ^{1}H-NMR (600 MHz, DMSO-d_6) δ_H 8.15–8.17 (ddd, J = 4.9, 2.0, 0.8 Hz, H-6, pyridine), 8.06–8.03 (m, H-5, nitrobenzene), 8.03–8.02 (m, H-3, nitrobenzene), 7.58–7.55 (ddd, J = 8.8, 7.1, 2.0 Hz, H-4, pyridine), 7.23–7.20 (m, H-6, nitrobenzene), 6.89–6.88 (d, J = 8.8 Hz, H-3, pyridine), 6.69–6.67 (dd, J = 7.1, 4.9, 0.7 Hz, H-5, pyridine), 3.67–3.65 (dd, J = 6.2, 4.2 Hz, H-2, -6, piperazine), 3.42–3.40 (dd, J = 6.2, 4.9 Hz, H-3, -5, piperazine) ppm. ^{13}C-NMR (150 MHz, DMSO-d_6) δ_C 158.8 (C-2, pyridine), 152.9 (d, $^1J_{C\text{–}F}$ = 247.4 Hz, C-2, nitrobenzene), 147.6 (C-6, pyridine), 145.3 (d, $^2J_{C–F}$ = 7.6 Hz, C-1, nitrobenzene), 139.4 (d, $^3J_{C–F}$ = 9.2 Hz, C-4, nitrobenzene), 137.7 (C-4, pyridine), 121.3 (d, $^4J_{C–F}$ = 2.9 Hz, C-5, nitrobenzene), 118.0 (d, $^3J_{C–F}$ = 4.3 Hz, C-6, nitrobenzene), 113.4 (C-5, pyridine), 112.3 (d, $^2J_{C–F}$ = 26.2 Hz, C-3, nitrobenzene), 107.2 (C-3, pyridine), 48.8 (C-2, -6, piperazine), 44.3 (C-3, -5, piperazine) ppm.

2-[4-(2-Fluoro-4-nitrophenyl)piperazin-1-yl]pyrimidine **14** yellow solid; yield 85%; mp: 191–192 °C (n-hexane/AcOEt, 5:5); ^{1}H-NMR (600 MHz, DMSO-d_6) δ_H 8.41–8.40 (d, J = 4.7 Hz, H-4, -6, pyrimidine), 8.06–8.04 (m, H-5, nitrobenzene), 8.04–8.02 (m, H-3, nitrobenzene), 7.23–7.20 (m, H-6, nitrobenzene), 6.69–6.67 (t, J = 4.7 Hz, H-5, pyrimidine), 3.91–3.89 (m, H-3, -5, piperazine), 3.39–3.37 (m, H-2, -6, piperazine) ppm. ^{13}C-NMR (150 MHz, DMSO-d_6) δ_C 161.1 (C-2, pyrimidine), 158.0 (C-4, -6, pyrimidine), 152.9 (d, $^1J_{C–F}$ = 247.8 Hz, C-2, nitrobenzene), 145.4 (d, $^2J_{C–F}$ = 8.0 Hz, C-1, nitrobenzene), 139.5 (d, $^3J_{C–F}$ = 9.4 Hz, C-4, nitrobenzene), 121.3 (d, $^4J_{C–F}$ = 3.1 Hz, C-5, nitrobenzene), 118.1 (d, $^3J_{C–F}$ = 4.4 Hz, C-6, nitrobenzene), 112.4 (d, $^2J_{C–F}$ = 26.4 Hz, C-3, nitrobenzene), 110.4 (C-5, pyrimidine), 48.9 (C-3, -5, piperazine), 43.0 (C-2, -6, piperazine) ppm.

1-(2-Fluoro-4-nitrophenyl)-4-(4-fluorophenyl)piperazine **15** yellow solid; yield 98%; mp: 135–137 °C (n-hexane/AcOEt, 5:5); ^{1}H-NMR (600 MHz, DMSO- d_6) δ_H 8.06–8.04 (dd, J = 6.4, 2.0 Hz, H-5, fluorobenzene), 8.04–8.03 (m, H-3, nitrobenzene), 7.23–7.22 (m, H-6, nitrobenzene), 7.10–7.06 (m, H-3, -5, fluorobenzene), 7.03–7.01 (m, H-2, -6, fluorobenzene), 3.45–3.43 (dd, J = 4.7, 2.4 Hz, H-2, -6, piperazine), 3.25–3.24 (dd, J = 4.7, 2.4 Hz, H-3, -5, piperazine) ppm. ^{13}C-NMR (150 MHz, DMSO-d_6) δ_C 157.1 (d, $^1J_{C–F}$ = 236.1 Hz, C-4, fluorobenzene), 152.9 (d, $^1J_{C–F}$ = 247.5 Hz, C-2, nitrobenzene), 147.6 (d, $^4J_{C–F}$ = 2.3 Hz, C-1, fluorobenzene), 145.2 (d, $^2J_{C–F}$ = 7.7 Hz, C-1, nitrobenzene), 139.6 (d, $^3J_{C–F}$ = 9.4 Hz, C-4, nitrobenzene), 121.3 (d, $^4J_{C–F}$ = 2.9 Hz, C-5, nitrobenzene), 118.1 (d, $^2J_{C–F}$ = 4.4 Hz, C-3, -5, fluorobenzene), 117.5 (d, $^3J_{C–F}$ = 7.9 Hz, C-6, nitrobenzene), 115.4 (d, $^3J_{C–F}$ = 22.0 Hz, C-2, -6, fluorobenzene), 112.3 (d, $^2J_{C–F}$ = 26.5 Hz, C-3, nitrobenzene), 49.1 (C-2, -6, piperazine), 48.9 (C-3, -5, piperazine) ppm.

4-(2-Fluoro-4-nitrophenyl)thiomorpholine **16** orange solid; yield 98%; mp: 54–56 °C (n-hexane/AcOEt, 5:5); ^{1}H-NMR (600 MHz, $CDCl_3$) δ_H 7.97 (dd, J = 9.0, 1.9 Hz, H-5, nitrobenzene), 7.89 (dd, J = 12.9, 2.6 Hz, H-3, nitrobenzene), 6.92 (t, J = 8.8 Hz, H-6, nitrobenzene), 3.59–3.56 (m, H-3, -5, thiomorpholine), 2.81–2.78 (m, H-2, -6, thiomorpholine) ppm. ^{13}C-NMR (150 MHz, $CDCl_3$) δ_C 153.9 (d, $^1J_{C–F}$ = 249.0 Hz, C-2, nitrobenzene), 146.0 (d, $^2J_{C–F}$ = 7.6 Hz, C-1, nitrobenzene), 140.7 (d, $^3J_{C–F}$ = 8.7 Hz, C-4, nitrobenzene), 121.1 (d, $^4J_{C–F}$ = 2.6 Hz, C-5, nitrobenzene), 117.9 (d, $^3J_{C–F}$ = 3.5 Hz, C-6, nitrobenzene), 112.9 (d, $^2J_{C–F}$ = 26.5 Hz, C-3, nitrobenzene), 52.5 (C-3, -5, thiomorpholine), 27.5 (C-2, -6, thiomorpholine) ppm [19].

4-(2-Fluoro-4-nitrophenyl)morpholine **17** orange solid; yield 98%; mp: 109–111 °C (n-hexane/AcOEt, 5:5); ^{1}H-NMR (600 MHz, $CDCl_3$) δ_H 7.99 (ddd, J = 9.0, 2.6, 0.9 Hz, H-5, nitrobenzene), 7.91 (dd, J = 13.1, 2.6 Hz, H-3, nitrobenzene), 6.92 (t, J = 8.8 Hz, H-6, nitrobenzene), 3.89–3.86 (m, H-2, -6, morpholine), 3.29–3.26 (m, H-3, -5, morpholine) ppm. ^{13}C-NMR (150 MHz, $CDCl_3$) δ_C 154.1 (d, $^1J_{C–F}$ = 249.5 Hz, C-2, nitrobenzene), 145.6 (d, $^2J_{C–F}$ = 7.4 Hz, C-1, nitrobenzene), 141.0 (d, $^3J_{C–F}$ = 8.7 Hz, C-4, nitrobenzene), 121.1 (d, $^4J_{C–F}$ = 2.7 Hz, C-5, nitrobenzene), 117.0 (d, $^3J_{C–F}$ = 3.9 Hz, C-6, nitrobenzene), 112.8 (d, $^2J_{C–F}$ = 26.3 Hz, C-3, nitrobenzene), 66.7 (C-3, -5, morpholine), 50.0 (C-2, -6 morpholine) ppm [19].

1-(2-Fluoro-4-nitrophenyl)azepane **18** yellow solid; yield 98%; mp: 32–33 °C (n-hexane/AcOEt, 5:5); ^{1}H-NMR (600 MHz, $CDCl_3$) δ_H 7.91 (dd, J = 9.3, 2.7 Hz, H-5, nitrobenzene), 7.86 (dd, J = 15.0, 2.6 Hz, H-3, nitrobenzene), 6.71 (t, J = 9.2 Hz, H-6, nitrobenzene), 3.57 (td, J = 6.3; 1.5 Hz, H-2, -7, azepane), 1.87–1.82 (m, H-3, -5, azepane), 1.63–1.56 (m, H-4, -5, azepane) ppm. ^{13}C-NMR (150 MHz, $CDCl_3$) δ_C 150.6 (d, $^{1}J_{C-F}$ = 245.1 Hz, C-2, nitrobenzene), 144.4 (d, $^{2}J_{C-F}$ = 7.0 Hz, C-1, nitrobenzene), 136.7 (d, $^{3}J_{C-F}$ = 8.5 Hz, C-4, nitrobenzene), 121.7 (d, $^{4}J_{C-F}$ = 1.4 Hz, C-5, nitrobenzene), 113.8 (d, $^{3}J_{C-F}$ = 5.5 Hz, C-6, nitrobenzene), 113.5 (d, $^{2}J_{C-F}$ = 27.8 Hz, C-3, nitrobenzene), 52.3 (C-2, -7, azepane), 28.5 (C-3, -6, azepane), 26.9 (C-4, -5, azepane) ppm [19].

C. General Procedure for the Synthesis of Amine Aromatic Intermediates **19–24** [17,19]

In a mixture of EtOH (15.0 mL)/H_2O (7.5 mL), (3.31 mmol, 1.0 equiv.) of the corresponding nitro aromatic intermediates **13–18** were added, followed by the addition of (9.93 mmol, 3.0 equiv.) of ammonium chloride and (13.24 mmol, 4.0 equiv.) of metallic iron, in a 50.0 mL round-bottom flask, with stirring for 1 h and refluxing. Then, the reaction mixture was vacuum filtered with the aid of celite 545 and washed in methanol; the resulting solvent was removed under reduced pressure. Soon afterwards, the resulting mixture was solubilized in 50.0 mL of dichloromethane and washed (3 × 50.0 mL) with saturated sodium chloride solution and (3 × 50.0 mL) saturated sodium bicarbonate solution. The organic phase was collected and the solvent removed under reduced pressure. The crude product was purified by column chromatography (eluent n-hexane/AcOEt, 6:4) [17,19].

3-Fluoro-4-[4-(pyridin-2-yl)piperazin-1-yl]aniline **19** brown solid; yield 52%; mp: 92–95 °C (n-hexane/AcOEt, 6:4); ^{1}H-NMR (600 MHz, DMSO-d_6) δ_H 8.13–8.11 (ddd, J = 4.9, 2.0, 0.8 Hz, H-6, pyridine), 7.55–7.52 (ddd, J = 8.7, 7.1, 2.0 Hz, H-4, pyridine), 6.86–6.84 (d, J = 8.7 Hz, H-3, pyridine), 6.82–6.78 (dd, J = 9.9, 8.6 Hz, H-5, aniline), 6.66–6.64 (ddd, J = 7.1, 4.9, 0.8 Hz, H-5, pyridine), 6.37–6.34 (dd, J = 14.4, 2.5 Hz, H-2, aniline), 6.32–6.30 (m, H-6, aniline), 5.01 (s, 2H, NH_2), 3.59–3.57 (m, H-2, -6, piperazine), 2.91–2.90 (m, H-3, -5, piperazine) ppm. ^{13}C-NMR (150 MHz, DMSO-d_6) δ_C 159.0 (C-2, pyridine), 157.1 (d, $^{1}J_{C-F}$ = 241.6 Hz, C-3, aniline), 147.6 (C-6, pyridine), 145.6 (d, $^{3}J_{C-F}$ = 11.0 Hz, C-1, aniline), 137.7 (C-4, pyridine), 129.0 (d, $^{2}J_{C-F}$ = 10.3 Hz, C-4, aniline), 120.8 (d, $^{4}J_{C-F}$ = 4.9 Hz, C-5, aniline), 113.1 (C-5, pyridine), 109.3 (d, $^{3}J_{C-F}$ = 2.9 Hz, C-6, aniline), 107.1 (C-3, pyridine), 101.9 (d, $^{2}J_{C-F}$ = 23.5 Hz, C-2, aniline), 51.1 (C-3, -5, piperazine), 49.0 (C-2, -6, piperazine) ppm.

3-Fluoro-4-[4-(pyrimidin-2-yl)piperazin-1-yl]aniline **20** brown solid; yield 98%; mp: 81–82 °C (n-hexane/AcOEt, 6:4); ^{1}H-NMR (600 MHz, DMSO-d_6) δ_H 8.37–8.36 (d, J = 4.7 Hz, H-6, -4, pyrimidine), 6.81–6.77 (dd, J = 9.9, 8.6 Hz, H-5, aniline), 6.64–6.63 (t, J = 4.7 Hz, H-5, pyrimidine), 6.37 (dd, J = 14.4, 2.5 Hz, H-2, aniline), 6.31–6.29 (m, H-6, aniline), 5.01 (s, 2H, NH_2), 3.84–3.83 (m, H-2, -6, piperazine), 2.87–2.85 (m, H-3, -5, piperazine) ppm. ^{13}C-NMR (150 MHz, DMSO-d_6) δ_C 161.2 (C-2, pyrimidine), 157.9 (C-4, -6, pyrimidine), 157.2 (d, $^{1}J_{C-F}$ = 241.3 Hz, C-3, aniline), 145.7 (d, $^{1}J_{C-F}$ = 11.2 Hz, C-1, aniline), 129.1 (d, $^{1}J_{C-F}$ = 10.0 Hz, C-4, aniline), 120.9 (d, $^{1}J_{C-F}$ = 5.0 Hz, C-5, aniline), 110.2 (C-5, pyrimidine), 109.5 (d, $^{1}J_{C-F}$ = 3.0 Hz, C-6, aniline), 101.8 (d, $^{1}J_{C-F}$ = 25.2 Hz, C-2, aniline), 51.2 (C-3, -5, piperazine), 43.6 (C-2, -6, piperazine) ppm.

3-Fluoro-4-[4-(4-fluorophenyl)piperazin-1-yl]aniline **21** brown solid; yield 62%; mp: 161–164 °C (n-hexane/AcOEt, 6:4); ^{1}H-NMR (600 MHz, DMSO-d_6) δ_H 7.07–7.04 (m, H-3, -5, fluorobenzene), 6.99–6.97 (dt, J = 4.5, 2.3 Hz, H-2, -6, fluorobenzene), 6.82–6.79 (dd, J = 9.9, 8.5 Hz, H-5, aniline), 6.37–6.34 (dd, J = 14.4, 2.5 Hz, H-2, aniline), 6.33–6.31 (dd, J = 8.4, 2.5 Hz, H-6, aniline), 5.01 (s, 2H, NH_2), 3.19–3.17 (m, H-2, -6, piperazine), 2.98–2.95 (m, H-3, -5, piperazine) ppm. ^{13}C NMR (150 MHz, DMSO-d_6) δ_C 157.1 (d, $^{1}J_{C-F}$ = 242.1 Hz, C-4, fluorobenzene), 156.9 (d, $^{1}J_{C\text{–}F}$ = 235.3 Hz, C-3, aniline), 147.9 (d, $^{4}J_{C\text{–}F}$ = 2.1 Hz, C-1, fluorobenzene), 145.5 (d, $^{2}J_{C\text{–}F}$ = 11.0 Hz, C-1, aniline), 128.9 (d, $^{3}J_{C\text{–}F}$ = 10.1 Hz, C-4, aniline), 120.6 (d, $^{4}J_{C\text{–}F}$ = 5.0 Hz, C-5, aniline), 117.3 (d, $^{3}J_{C\text{–}F}$ = 8.2 Hz, C-2, -6, fluorobenzene), 115.3 (d, $^{2}J_{C\text{–}F}$ = 22.0 Hz, C-3, -5, fluorobenzene), 109.5 (d, $^{4}J_{C\text{–}F}$ = 2.8 Hz, C-6, aniline), 101.9 (d, $^{2}J_{C\text{–}F}$ = 23.7 Hz, C-2, aniline), 51.2 (C-3, -5, piperazine), 49.4 (C-2, -6, piperazine) ppm.

3-Fluoro-4-thiomorpholinoaniline **22** brown solid; yield 82%; mp: 63–66 °C (n-hexane/AcOEt, 6:4); ^{1}H-NMR (600 MHz, $CDCl_3$) δ_H 6.81 (t, J = 8.8 Hz, H-5, aniline), 6.41 (dd, J = 13.3, 2.3 Hz, H-2, aniline), 6.38 (dd, J = 8.3 Hz, H-6, aniline), 3.56 (s, 2H, NH_2), 3.24–3.16 (m, H-3, -5, thiomorpholine), 2.84–2.76 (m, H-2, -6, thiomorpholine) ppm. ^{13}C-NMR (150 MHz, $CDCl_3$) δ_C 157.9 (d, $^{1}J_{C\text{–}F}$ = 245.3 Hz, C-3, aniline), 143.3 (d, $^{3}J_{C\text{–}F}$ = 10.4 Hz, C-1, aniline), 133.3 (d, $^{2}J_{C\text{–}F}$ = 9.9 Hz, C-4, aniline), 122.0 (d, $^{3}J_{C\text{–}F}$ = 4.2 Hz, C-5, aniline), 110.7 (d, $^{4}J_{C\text{–}F}$ = 3.1 Hz, C-6, aniline), 103.9 (d, $^{2}J_{C\text{–}F}$ = 23.7 Hz, C-2, aniline), 54.1 (C-3, -5, thiomorpholine), 28.4 (C-2, -6, thiomorpholine) ppm [19].

3-Fluoro-4-morpholinoaniline **23** brown solid; yield 59%; mp: 119–121 °C (n-hexane/AcOEt, 6:4); ^{1}H-NMR (600 MHz, $CDCl_3$) δ_H 6.81–6.77 (m, H-5, aniline), 6.43 (dd, J = 13.5, 2.5 Hz, H-2, aniline), 6.40 (dd, J = 8.4, 3.1 Hz, H-6, aniline), 3.87–3.83 (m, H-2, -6, morpholine), 3.56 (s, 2H, NH_2), 2.98–2.94 (m, H-3, -5, morpholine) ppm. ^{13}C-NMR (150 MHz, $CDCl_3$) δ_C 157.6 (d, $^{1}J_{C\text{–}F}$ = 245.2 Hz, C-3, aniline), 142.9 (d, $^{3}J_{C\text{–}F}$ = 10.4 Hz, C-1, aniline), 131.8 (d, $^{2}J_{C\text{–}F}$ = 9.7 Hz, C-4, aniline), 120.3 (d, $^{3}J_{C\text{–}F}$ = 4.2 Hz, C-5, aniline), 110.7 (d, $^{4}J_{C\text{–}F}$ = 3.1 Hz, C-6, aniline), 103.9 (d, $^{2}J_{C\text{–}F}$ = 23.9 Hz, C-2, aniline), 67.3 (C-3, -5, morpholine), 51.8 (C-2, -6, morpholine) ppm [19].

4-(Azepan-1-yl)-3-fluoroaniline **24** black oil; yield 62%; (n-hexane/AcOEt, 6:4); ^{1}H-NMR (600 MHz, DMSO-d_6) δ_H 6.71 (dd, J = 10.2, 8.6 Hz, H-5, aniline), 6.30 (dd, J = 14.9, 2.5 Hz, H-2, aniline), 6.25 (dd, J = 8.5, 2.5 Hz, H-6, aniline), 4.81 (s, 2H, NH_2), 3.07–3.03 (m, H-2, -7, azepane), 1.70–1.66 (m, H-3, -6, azepane), 1.58–1.55 (m, H-4, -5, azepane), ^{13}C-NMR (150 MHz, DMSO d_6) δ_C 156.1 (d, $^{1}J_{C\text{–}F}$ = 243.6 Hz, C-3, aniline) ppm. 143.5 (d, $^{3}J_{C\text{–}F}$ = 10.2 Hz, C-1, aniline), 133.8 (d, $^{2}J_{C\text{–}F}$ = 9.4 Hz, C-4, aniline), 120.4 (d, $^{3}J_{C\text{–}F}$ = 4.9 Hz, C-5, aniline), 109.6 (d, $^{4}J_{C\text{–}F}$ = 3.0 Hz, C-6, aniline), 100.2 (d, $^{2}J_{C\text{–}F}$ = 24.4 Hz, C-2, aniline), 53.1 (C-4, -5, azepane), 28.8 (C-3, -7, azepane), 26.8 (C-3, -6, azepane) ppm [19].

D. General Procedure for the Synthesis of Boronic Acid Derivatives **29–40** [19,24]

At room temperature and under nitrogen atmosphere, (0.49 mmol, 1.0 equiv.) of boronic acids **27** and **28**, (0.98, 2.0 equiv.) of the coupling agent *N*-(3-dimethylaminopropyl)-*N′*-Ethylcarbodiimide hydrochloride, and (0.98 mmol, 2.0 equiv.) of 1-hydroxybenzotriazole reagent were added to 10.0 mL of anhydrous *N′N*-dimethylformamide, in which the reaction medium was stirred for 2 h. Then, simultaneously, (0.74 mmol, 1.5 equiv.) of the aromatic amines **19–24** and *N,N*-diisopropylethylamine (DIPEA) were added, and they remained under stirring for 48 h, in N_2, at room temperature. After this time, 50.0 mL of water was added to the reaction medium, and the resulting solution was washed with ethyl acetate (3 × 50.0 mL) in a separatory funnel. The organic phase was collected and concentrated under reduced pressure. The resulting products were purified by column chromatography (eluent, n-hexane/AcOEt/MeOH, gradient mode, starting with Hex/AcOEt 80:20 increasing the percentage of AcOEt by 10%, up to AcOEt/MeOH 90:10) [18,24].

{3-({3-Fluoro-4-[4-(pyridin-2-yl)piperazin-1-yl]phenyl}carbamoyl)phenyl}boronic acid **29** white solid; yield 22%; mp: 235–239 °C (n-hexane/AcOEt/MeOH, gradient mode, starting with Hex/AcOEt 80:20 increasing the percentage of AcOEt by 10%, up to AcOEt/MeOH 90:10); IR (KBr) ν_{max} 3263 (B–OH boronic acid; N–H amide), 2848 (B–O boronic acid), 1643 (C=O

amide), 1600 (C=C aromatic), 1440 (C=C aromatic), 1315 (C–F allyl floride), 1242 (C–H methylene) cm^{-1}. ^{1}H-NMR (600 MHz, DMSO-d_6) δ_H 10.30 (s, 1H, NH), 8.32 (m, H-2, carbamoyphenylboronic acid), 8.24 (s, 2H, OH), 8.15–8.13 (ddd, *J* = 4.9, 2.0 Hz, H-6, pyridine), 7.98–7.96 (dd, *J* = 7.4, 1.2 Hz, H-6, carbamoyphenylboronic acid), 7.94–7.93 (m, H-4, carbamoyphenylboronic acid), 7.75–7.72 (dd, *J* = 14.9, 2.4 Hz, H-2, fluorophenyl), 7.57–7.55 (ddd, *J* = 9.0, 7.1, 2.0 Hz, H-4, pyridine), 7.50–7.49 (m, H-5, carbamoyphenylboronic acid), 7.49–7.48 (m, H-6, fluorophenyl), 7.10–7.07 (m, H-5, flhorophenyl), 6.90–6.88 (ddd, *J* = 7.1, 4.9 Hz, H-3, pyridine), 3.65–3.63 (m, H- 2, -6, piperazine), 3.09–3.07 (m, H-3, -5, piperazine) ppm. ^{13}C-NMR (150 MHz, DMSO-d_6) δ_C 166.0 (1C, C=O), 158.9 (C-2, pyridine), 155.2 (d, $^{1}J_{C\text{–}F}$ = 247.2 Hz, C-3, fluorophenyl), 147.6 (C-6, pyridine), 137.6 (C-4, pyridine), 137.1 (C-6, carbamoyphenylboronic acid), 135.6 (d, $^{1}J_{C\text{–}F}$ = 9.4 Hz, C-4, fluorophenyl), 134.6 (d, $^{3}J_{C\text{–}F}$ = 11.0 Hz, C-1, fluorophenyl), 134.1 (C-3, carbamoyphenylboronic acid), 133.4 (C-2, carbamoyphenylboronic acid), 129.2 (C-4, carbamoyphenylboronic acid), 127.4 (C-5, carbamoyphenylboronic acid), 119.4 (d, $^{3}J_{C\text{–}F}$ = 4.6 Hz, C-5, fluorophenyl), 116.2 (d, $^{4}J_{C\text{–}F}$ = 3.7 Hz, C-6, fluorophenyl), 113.3 (C-5, pyridine), 108.4 (d, $^{2}J_{C\text{–}F}$ = 25.7 Hz, C-2, fluorophenyl), 107.2 (C-3, pyridine), 50.3 (C-3, -5, piperazine), 44.8 (C-2, -6, piperazine) ppm. HMRS (ESI$^+$, methanol) $C_{22}H_{23}BFN_4O_3$ $[M+H]^+$ = 421.1843.

[3-({3-Fluoro-4-[4-(pyrimidin-2-yl)piperazin-1-yl]phenyl}carbamoyl)phenyl]boronic acid **30** white solid; yield 41%; mp: 262–266 °C (n-hexane/AcOEt/MeOH, gradient mode, starting with Hex/AcOEt 80:20 increasing the percentage of AcOEt by 10%, up to AcOEt/MeOH 90:10); IR (KBr) ν_{max} 3360 (N–H amide), 3265 (B–OH boronic acid), 2908 (C–H methylene), 1647 (C=O amide), 1591 (C=C aromatic), 1442 (C=C aromatic), 1311 (B–O boronic acid), 1247 (C–F allyl floride) cm^{-1}. ^{1}H-NMR (600 MHz, DMSO-d_6) δ_H 10.30 (s, 1H, NH), 8.40–8.39 (d, *J* = 4.7 Hz, H-4, -6, pyrimidine), 8.32 (m, H-2, carbamoyphenylboronic acid), 8.24 (s, 2H, OH), 7.98–7.97 (m, H-6, carbamoyphenylboronic acid), 7.94–7.93 (m, H-4, carbamoyphenylboronic acid), 7.75–7.72 (dd, *J* = 14.9, 2.3 Hz, H-2, fluorophenyl), 7.50–7.49 (m, H-5, carbamoyphenylboronic acid), 7.49–7.48 (m, H-6, fluorophenyl), 7.09–7.06 (m, H-5, fluorophenyl), 6.67–6.65 (t, *J* = 4.7 Hz, H-5, pyrimidine), 3.90–3.89 (m, H-2, -6, piperazine), 3.05–3.03 (m, H-3, -5, piperazine) ppm. ^{13}C-NMR (151 MHz, DMSO-d_6) δ_C 166.0 (1C, C=O), 161.2 (C-2, pyrimidine), 158.0 (C-4, -6, pyrimidine), 155.2 (d, $^{1}J_{C\text{–}F}$ = 243.4 Hz, C-3, fluorophenyl), 137.2 (C-6, carbamoyphenylboronic acid), 135.5 (d, $^{2}J_{C\text{–}F}$ = 9.4 Hz, C-4, fluorophenyl), 134.6 (d, $^{3}J_{C\text{–}F}$ = 11.6 Hz, C-1, fluorophenyl), 134.1 (C-3, carbamoyphenylboronic acid), 133.4 (C-2, carbamoyphenylboronic acid), 129.2 (C-4, carbamoyphenylboronic acid), 127.4 (C-5, carbamoyphenylboronic acid), 119.4 (d, $^{3}J_{C\text{–}F}$ = 4.5 Hz, C-5, fluorophenyl), 116.2 (d, $^{4}J_{C\text{–}F}$ = 3.5 Hz, C-6, fluorophenyl), 110.4 (C-5, pyrimidine), 108.4 (d, $^{2}J_{C\text{–}F}$ = 25.0 Hz, C-2, fluorophenyl), 50.4 (C-3, -5, piperazine), 49.4 (C-2, -6, piperazine) ppm. HMRS (ESI$^+$, methanol) $C_{21}H_{22}BFN_5O_3$ $[M+H]^+$ = 422.1798.

[3-({3-Fluoro-4-[4-(4-fluorophenyl)piperazin-1-yl]phenyl}carbamoyl)phenyl]boronic acid **31** white solid; yield 25%; mp: 241–246 °C (n-hexane/AcOEt/MeOH, gradient mode, starting with Hex/AcOEt 80:20 increasing the percentage of AcOEt by 10%, up to AcOEt/MeOH 90:10); IR (KBr) ν_{max} 3304 (B–OH boronic acid; N–H amide), 2825 (C–H methylene), 1649 (C=O amide), 1589 (C=C aromatic), 1423 (C=C aromatic), 1365 (B–O boronic acid), 1222 (C–F allyl floride) cm^{-1}. ^{1}H-NMR (600 MHz, DMSO-d_6) δ_H 10.30 (s, 1H, NH), 8.33 (m, H-2, carbamoyphenylboronic acid), 8.24 (s, 2H, OH), 7.98–7.97 (dd, *J* = 7.4, 1.2 Hz, H-6, carbamoyphenylboronic acid), 7.94–7.93 (m, H-4, carbamoyphenylboronic acid), 7.74–7.71 (dd, *J* = 15.0, 2.4 Hz, H-2, fluorophenyl), 7.51–7.50 (m, H-5, carbamoyphenylboronic acid), 7.50–7.48 (m, H-6, fluorophenyl), 7.11–7.09 (m, H-5, fluorophenyl), 7.07–7.05 (m, H-3, -5, fluorobenzene), 7.03–7.00 (dt, *J* = 7.1, 2.3 Hz, H-2, -6, fluorobenzene), 3.65–3.63 (m, H-2, -6, piperazine), 3.09–3.07 (m, H-3, -5, piperazine) ppm. ^{13}C-NMR (150 MHz,

DMSO-d_6) δ_C 166.0 (1C, C=O), 156.9 (d, $^1J_{C-F}$ = 236.6 Hz, C-4, fluorobenzene), 155.2 (d, $^1J_{C-F}$ = 241.8 Hz, C-3, fluorophenyl), 147.9 (d, $^4J_{C-F}$ = 2.1 Hz, C-1, fluorobenzene), 137.2 (C-6, carbamoyphenylboronic acid), 135.4 (d, $^2J_{C-F}$ = 9.4 Hz, C-4, fluorophenyl), 134.5 (d, $^3J_{C-F}$ = 11.8 Hz, C-1, fluorophenyl), 134.1 (C-3, carbamoyphenylboronic acid), 133.4 (C-2, carbamoyphenylboronic acid), 129.2 (C-4, carbamoyphenylboronic acid), 127.4 (C-5, carbamoyphenylboronic acid), 119.3 (d, $^3J_{C-F}$ = 4.5 Hz, C-5, fluorophenyl), 117.4 (d, $^3J_{C-F}$ = 7.9 Hz, C-2, -6, fluorobenzene),116.2 (d, $^4J_{C-F}$ = 3.2 Hz, C-6, fluorophenyl), 115.3 (d, $^2J_{C-F}$ = 21.7 Hz, C-3, -5, fluorobenzene), 108.4 (d, $^2J_{C-F}$ = 24.6 Hz, C-2, fluorophenyl), 50.3 (C-3, -5, piperazine), 44.8 (C-2, -6, piperazine) ppm. HMRS (ESI$^+$, methanol) $C_{23}H_{23}BF_2N_3O_3$ $[M+H]^+$ = 438.1796.

{3-[(3-Fluoro-4-thiomorpholinophenyl)carbamoyl]phenyl}boronic acid **32** white solid; yield 52%; mp: 116–122 °C (n-hexane/AcOEt/MeOH, gradient mode, starting with Hex/AcOEt 80:20 increasing the percentage of AcOEt by 10%, up to AcOEt/MeOH 90:10); IR (KBr) ν_{max} 3327 (N–H amide), 3268 (B–OH boronic acid), 2810 (C–H methylene), 1654 (C=O amide), 1589 (C=C aromatic), 1433 (C=C aromatic), 1319 (B–O boronic acid), 1249 (C–F allyl floride) cm^{-1}. ^{1}H-NMR (600 MHz, DMSO-d_6) δ_H 10.28 (s, 1H, NH), 8.32 (m, H-2, carbamoyphenylboronic acid), 8.22 (s, 2H, OH), 7.98–7.96 (m, H-6, carbamoyphenylboronic acid), 7.94–7.93 (m, H-4, carbamoyphenylboronic acid), 7.71–7.69 (dd, J = 14.6, 2.3 Hz, H-2, fluorophenyl), 7.50–7.49 (m, H-5, carbamoyphenylboronic acid), 7.48–7.47 (m, H-6, fluorophenyl), 7.09–7.05 (t, J = 9.3 Hz, H-5, fluorophenyl), 3.28–3.20 (m, H-3, -5, thiomorpholine), 2.76–2.74 (m, H-2, -6, thiomorpholine) ppm. ^{13}C-NMR (151 MHz, DMSO-d_6) δ_C 166.0 (1C, C=O), 155.3 (d, $^1J_{C-F}$ = 244.8 Hz, C-3, fluorophenyl), 137.1 (C-6, carbamoyphenylboronic acid), 136.4 (d, $^2J_{C-F}$ = 9.9 Hz, C-4, fluorophenyl), 134.8 (d, $^3J_{C-F}$ = 10.8 Hz, C-1, fluorophenyl), 134.1 (C-3, carbamoyphenylboronic acid), 133.4 (C-2, carbamoyphenylboronic acid), 129.1 (C-4, carbamoyphenylboronic acid), 127.4 (C-5, carbamoyphenylboronic acid), 120.4 (d, $^3J_{C-F}$ = 3.8 Hz, C-5, fluorophenyl), 116.2 (d, $^4J_{C-F}$ = 3.5 Hz, C-6, fluorophenyl), 108.4 (d, $^2J_{C-F}$ = 25.8 Hz, C-2, fluorophenyl), 50.4 (C-2, -6, thiomorpholine), 49.4 (C-3, -5, thiomorpholine) ppm. HMRS (ESI$^+$, methanol) $C_{17}H_{19}BFN_2O_3S$ $[M+H]^+$ = 361.1192.

{3-[(3-Fluoro-4-morpholinophenyl)carbamoyl]phenyl}boronic acid **33** white solid; yield 39%; mp: 248–256 °C (n-hexane/AcOEt/MeOH, gradient mode, starting with Hex/AcOEt 80:20 increasing the percentage of AcOEt by 10%, up to AcOEt/MeOH 90:10); IR (KBr) ν_{max} 3533 (N–H amide), 3298 (B–OH boronic acid), 2843 (C–H methylene), 1643 (C=O amide), 1589 (C=C aromatic), 1409 (C=C aromatic), 1315 (B–O boronic acid), 1255 (C–F allyl floride) cm^{-1}. ^{1}H-NMR (600 MHz, DMSO-d_6) δ_H 10.28 (s, 1H, NH), 8.32 (m, H-2, carbamoyphenylboronic acid), 8.23 (s, 2H, OH), 7.98–7.96 (dt, J = 7.4, 1.1 Hz, H-6, carbamoyphenylboronic acid), 7.94–7.92 (m, H-4, carbamoyphenylboronic acid), 7.72–7.69 (dd, J = 15.0, 2.3 Hz, H-2, fluorophenyl), 7.50–7.49 (m, H-5, carbamoyphenylboronic acid), 7.49–7.48 (m, H-6, fluorophenyl), 7.05–7.02 (m, H-5, fluorophenyl), 3.75–3.73 (m, H-3, -5, morpholine), 2.98–2.96 (m, H-2, -6, morpholine) ppm. ^{13}C-NMR (151 MHz, DMSO-d_6) δ_C 166.0 (1C, C=O), 155.2 (d, $^1J_{C-F}$ = 243.0 Hz, C-3, fluorophenyl), 137.2 (C-6, carbamoyphenylboronic acid), 135.5 (d, $^2J_{C-F}$ = 9.5 Hz, C-4, fluorophenyl), 134.4 (d, $^3J_{C-F}$ = 11.5 Hz, C-1, fluorophenyl), 134.1 (C-3, carbamoyphenylboronic acid), 133.4 (C-2, carbamoyphenylboronic acid), 129.1 (C-4, carbamoyphenylboronic acid), 127.4 (C-5, carbamoyphenylboronic acid), 119.0 (d, $^3J_{C-F}$ = 4.2 Hz, C-5, fluorophenyl), 116.2 (d, $^4J_{C-F}$ = 3.2 Hz, C-6, fluorophenyl), 108.4 (d, $^2J_{C-F}$ = 24.9 Hz, C-2, fluorophenyl), 66.2 (C-3, -5, morpholine), 50.8 (C-2, -6, morpholine) ppm. HMRS (ESI$^+$, methanol) $C_{17}H_{19}BFN_2O_4$ $[M+H]^+$ = 345.1420.

(3-{[4-(Azepan-1-yl)-3-fluorophenyl]carbamoyl}phenyl)boronic acid **34** white solid; yield 32%; mp: 107–114 °C (n-hexane/AcOEt/MeOH, gradient mode, starting with Hex/AcOEt 80:20

increasing the percentage of AcOEt by 10%, up to AcOEt/MeOH 90:10); IR (KBr) ν_{max} 3495 (N–H amide), 3321 (B–OH boronic acid), 2924 (C–H methylene), 1643 (C=O amide), 1591 (C=C aromatic), 1425 (C=C aromatic), 1340 (B–O boronic acid), 1271 (C–F allyl floride) cm^{-1}. ^{1}H-NMR (600 MHz, DMSO-d_6) δ_H 10.14 (s, 1H, NH), 8.31 (m, H-2, carbamoyphenylboronic acid), 8.21 (s, 2H, OH), 7.97–7.95 (dt, J = 7.5, 1.1 Hz, H-6, carbamoyphenylboronic acid), 7.93–7.91 (dt, J = 7.8, 1.1 Hz, H-4, carbamoyphenylboronic acid), 7.63–7.60 (dd, J = 16.2, 2.2 Hz, H-2, fluorophenyl), 7.49–7.46 (t, J = 7.5 Hz, H-5, carbamoyphenylboronic acid), 7.39–7.37 (dd, J = 9.0, 2.2 Hz, H-6, fluorophenyl), 6.92–6.89 (t, J = 10.4, 9.0 Hz, H-5, fluorophenyl), 3.20–3.28 (t, J = 5.7 Hz, H-2, -7, azepane), 1.75 (m, H-3, -6, azepane), 1.57–1.55 (t, J = 5.7, 2.6 Hz, H-4, -5, azepane) ppm. ^{13}C-NMR (150 MHz, DMSO-d_6) δ_C 165.7 (1C, C=O), 152.8 (d, $^{1}J_{C\text{–}F}$ = 239.2 Hz, C-3, fluorophenyl), 137.0 (C-7, carbamoyphenylboronic acid), 135.6 (d, $^{2}J_{C\text{–}F}$ = 9.8 Hz, C-4, fluorophenyl), 134.1 (C-3, carbamoyphenylboronic acid), 133.3 (C-2, carbamoyphenylboronic acid), 130.9 (d, $^{3}J_{C\text{–}F}$ = 11.2 Hz, C-1, fluorophenyl), 129.0 (C-4, carbamoyphenylboronic acid), 127.4 (C-5, carbamoyphenylboronic acid), 117.1 (d, $^{3}J_{C\text{–}F}$ = 5.8 Hz, C-5, fluorophenyl), 116.2 (d, $^{4}J_{C\text{–}F}$ = 3.2 Hz, C-6, fluorophenyl), 108.8 (d, $^{2}J_{C\text{–}F}$ = 26.6 Hz, C-2, fluorophenyl), 51.7 (C-2, -7, azepane), 28.5 (C-3, -6, azepane), 26.7 (C-4, -5, azepane). HMRS (ESI$^+$, methanol) $C_{19}H_{23}BFN2O_3$ [M+H]$^+$ = 357.1787.

[4-({3-Fluoro-4-[4-(pyridin-2-yl)piperazin-1-yl]phenyl}carbamoyl)phenyl]boronic acid **35** white solid; yield 53%; mp: 198–204 °C (n-hexane/AcOEt/MeOH, gradient mode, starting with Hex/AcOEt 80:20 increasing the percentage of AcOEt by 10%, up to AcOEt/MeOH 90:10); IR (KBr) ν_{max} 3277 (B–OH boronic acid; N–H amide), 1643 (C=O amide), 1591 (C=C aromatic), 1444 (C=C aromatic), 1317 (B–O boronic acid), 1240 (C–F allyl floride) cm^{-1}. ^{1}H-NMR (600 MHz, DMSO-d_6) δ_H 10.28 (s, 1H, NH), 8.25 (s, 2H, OH), 8.14–8.13 (dd, J = 4.6, 1.6 Hz, H-6, pyridine), 7.93–7.91 (m, H-3, -5, carbamoyphenylboronic acid), 7.90–7.89 (m, H-2, -6, carbamoyphenylboronic acid), 7.75–7.73 (dd, J = 14.9, 2.3 Hz, H-2, fluorophenyl), 7.56–7.54 (ddd, J = 8.6, 7.1, 2.0 Hz, H-4, pyridine), 7.51–7.49 (dd, J = 8.7, 1.8 Hz, H-6, fluorophenyl), 7.10–7.07 (m, H-5, fluorophenyl), 6.89–6.88 (d, J = 8.6 Hz, H-3, pyridine), 6.68–6.66 (m, H-5, pyridine), 3.65–3.66 (m, H-2, -6, piperazine), 3.09–3.07 (m, H-3, -5, piperazine) ppm. ^{13}C-NMR (150 MHz, DMSO-d_6) δ_C 165.5 (1C, C=O), 159.0 (C-2, pyridine), 155.2 (d, $^{1}J_{C\text{–}F}$ = 248.2 Hz, C-3, fluorophenyl), 147.6 (C-6, pyperidine), 137.6 (C-4, pyperidine), 136.0 (C-4, carbamoyphenylboronic acid), 135.8 (d, $^{3}J_{C\text{–}F}$ = 9.3 Hz, C-1, fluorophenyl), 134.4 (d, $^{2}J_{C\text{–}F}$ = 11,0 Hz, C-4, fluorophenyl), 134.0 (C-3, -5, carbamoyphenylboronic acid), 126.5 (C-2, -4, carbamoyphenylboronic acid), 119.3 (d, $^{3}J_{C\text{–}F}$ = 4.8 Hz, C-5, fluorophenyl), 116.3 (d, $^{4}J_{C\text{–}F}$ = 2.8 Hz, C-6, fluorophenyl), 113.3 (C-2, pyridine), 108.4 (d, $^{2}J_{C\text{–}F}$ = 25.2 Hz, C-2, fluorophenyl), 107.2 (C-3, pyridine), 50.3 (C-3, -5, piperazine), 44.8 (C-2, -6, piperazine) ppm. HMRS (ESI$^+$, methanol) $C_{22}H_{23}BFN_4O_3$ [M+H]$^+$ = 421.1840.

[4-({3-Fluoro-4-[4-(pyrimidin-2-yl)piperazin-1-yl]phenyl}carbamoyl)phenyl]boronic acid **36** white solid; yield 53%; mp: 214–220 °C (n-hexane/AcOEt/MeOH, gradient mode, starting with Hex/AcOEt 80:20 increasing the percentage of AcOEt by 10%, up to AcOEt/MeOH 90:10); IR (KBr) ν_{max} 3460 (N–H amide), 3286 (B–OH boronic acid), 2852 (C–H methylene), 1645 (C=O amide), 1591 (C=C aromatic), 1442 (C=C aromatic), 1317 (B–O boronic acid), 1251 (C–F allyl floride) cm^{-1}. ^{1}H-NMR (600 MHz, DMSO-d_6) δ_H 10.27 (s, 1H, NH), 8.39–8.38 (ddd, J = 4.7 Hz, H-4, -6, pyrimidine), 8.24 (s, 2H, OH), 7.92–7.91 (m, H-3, -5, carbamoyphenylboronic acid), 7.90–7.88 (m, H-2, -6, carbamoyphenylboronic acid), 7.75–7.72 (dd, J = 14.9, 2.0 Hz, H-2, fluorophenyl), 7.50–7.48 (m, H-6, fluorophenyl), 7.09–7.07 (m, H-5, fluorophenyl), 6.66–6.65 (t, J = 4.7 Hz, H-5, pyrimidine), 3.90–3.89 (m, H-2, -6, piperazine), 3.05–3.03 (m, H-3, -5, piperazine) ppm. ^{13}C-NMR (150 MHz, DMSO-d_6) δ_H 165.5 (1C, C=O), 161.2 (C-2, pyrimidine), 158.0 (C-4, -6, pyrimidine), 155.2 (d, $^{1}J_{C\text{–}F}$ = 242.8 Hz, C-3, fluorophenyl), 136.0 (C-4, carbamoyphenylboronic acid), 135.6 (d, $^{3}J_{C\text{–}F}$ = 8.9 Hz, C-1,

fluorophenyl), 134.5 (d, $^{2}J_{C\text{–}F}$ = 10.8 Hz, C-4, fluorophenyl), 134.0 (C-3, -5, carbamoyphenylboronic acid), 126.5 (C-2, -6, carbamoyphenylboronic acid), 119.4 (d, $^{3}J_{C\text{–}F}$ = 3.7 Hz, C-5, fluorophenyl), 116.3 (d, $^{4}J_{C\text{–}F}$ = 2.4 Hz, C-6, fluorophenyl), 110.3 (C-5, pyrimidine), 108.4 (d, $^{2}J_{C\text{–}F}$ = 25.5 Hz, C-2, fluorophenyl), 50.3 (C-3, -5, piperazine), 43.4 (C-2, -6, piperazine) ppm. HMRS (ESI$^+$, methanol) $C_{21}H_{22}BFN_5O_3$ [M+H]$^+$ = 422.1801.

[4-({3-Fluoro-4-[4-(4-fluorophenyl)piperazin-1-yl]phenyl}carbamoyl)phenyl]boronic acid **37** white solid; yield 51%; mp: 225–232 °C (n-hexane/AcOEt/MeOH, gradient mode, starting with Hex/AcOEt 80:20 increasing the percentage of AcOEt by 10%, up to AcOEt/MeOH 90:10); IR (KBr) ν_{max} 3304 (B–OH boronic acid; N–H amide), 2850 (C–H methylene), 1643 (C=O amide), 1589 (C=C aromatic), 1222 (C–F allyl floride) cm^{-1}. ^{1}H-NMR (600 MHz, DMSO-d_6) δ_H 10.28 (s, 1H, NH), 8.24 (s, 2H, OH), 7.93–7.91 (m, H-3, -5, carbamoyphenylboronic acid), 7.90–7.88 (m, H-2, -6, carbamoyphenylboronic acid), 7.76–7.71 (dd, *J* = 15.0, 2.3 Hz, H-2, fluorophenyl), 7.52–7.50 (m, H-6, fluorophenyl), 7.12–7.09 (m, H-5, fluorophenyl), 7.08–7.06 (m, H-3, -5, fluorobenzene), 7.03–7.09 (m, H-H-2, -6, fluorobenzene), 3.25–3.23 (m, H-2, -6, piperazine), 3.15–3.12 (m, H-3, -5, piperazine) ppm. ^{13}C-NMR (150 MHz, DMSO-d_6) δ_C 165.5 (1C, C=O), 156.9 (d, $^{1}J_{C\text{–}F}$ = 234.5 Hz, C-4, fluorobenzene), 155.2 (d, $^{1}J_{C\text{–}F}$ = 243.3 Hz, C-3, fluorophenyl), 147.9 (C-1, fluorobenzene), 136.0 (C-4, carbamoyphenylboronic acid), 135.5 (d, $^{3}J_{C\text{–}F}$ = 9.8 Hz, C-1, fluorophenyl), 134.4 (d, $^{2}J_{C\text{–}F}$ = 11.0 Hz, C-4, fluorophenyl), 134.0 (C-3, -5, carbamoyphenylboronic acid), 126.5 (C-2, -6, carbamoyphenylboronic acid), 119.2 (d, $^{3}J_{C\text{–}F}$ = 4.1 Hz, C-5, fluorophenyl), 117.4 (d, $^{3}J_{C\text{–}F}$ = 8.6 Hz, C-2, -6, fluorobenzene), 116.3 (d, $^{4}J_{C\text{–}F}$ = 2.9 Hz, C-6, fluorophenyl), 115.4 (d, $^{2}J_{C\text{–}F}$ = 22.1 Hz, C-3, -5, fluorobenzene), 108.4 (d, $^{2}J_{C\text{–}F}$ = 26.7 Hz, C-2, fluorophenyl), 50.4 (C-3, -5, piperazine), 49.3 (C-2, -6, piperazine) ppm. HMRS (ESI$^+$, methanol) $C_{23}H_{23}BF_2N_3O_3$ [M+H]$^+$ = 438.1795.

{4-[(3-Fluoro-4-thiomorpholinophenyl)carbamoyl]phenyl}boronic acid **38** white solid; yield 47%; mp: 194–198 °C (n-hexane/AcOEt/MeOH, gradient mode, starting with Hex/AcOEt 80:20 increasing the percentage of AcOEt by 10%, up to AcOEt/MeOH 90:10); IR (KBr) ν_{max} 3302 (B–OH boronic acid; N–H amide), 2823 (C–H methylene), 1643 (C=O amide), 1589 (C=C aromatic), 1425 (C=C aromatic), 1327 (B–O boronic acid), 1284 (C–F allyl floride) cm^{-1}. ^{1}H-NMR (600 MHz, DMSO-*d6*) δ_H 10.27 (s, 1H, NH), 8.25 (s, 2H, OH), 7.92–7.91 (m, H-3, -5 carbamoyphenylboronic acid), 7.89–7.88 (m, H-2, -6, carbamoyphenylboronic acid), 7.72–7.69 (dd, *J* = 14.6, 2.2 Hz, H-2, fluorophenyl), 7.49–7.48 (m, H-6, fluorophenyl), 7.09–7.06 (m, H-5, fluorophenyl), 3.22–3.20 (m, H-3, -5, thiomorpholine), 3.15–3.12 (m, H-2, -6, thiomorpholine) ppm. ^{13}C-NMR (150 MHz, DMSO-d_6) δ_C 165.5 (1C, C=O), 155.3 (d, $^{1}J_{C\text{–}F}$ = 242.1 Hz, C-3, fluorophenyl), 136.5 (d, $^{3}J_{C\text{–}F}$ = 10.1 Hz, C-1, fluorophenyl), 135.9 (C-4, carbamoyphenylboronic acid), 134.6 (d, $^{2}J_{C\text{–}F}$ = 11.0 Hz, C-4, fluorophenyl), 134.0 (C-3, -5, carbamoyphenylboronic acid), 126.5 (C-2, -6, carbamoyphenylboronic acid), 120.4 (d, $^{3}J_{C\text{–}F}$ = 4.2 Hz, C-5, fluorophenyl), 116.3 (d, $^{4}J_{C\text{–}F}$ = 2.8 Hz, C-6, fluorophenyl), 108.4 (d, $^{2}J_{C\text{–}F}$ = 26.2 Hz, C-2, fluorophenyl), 50.1 (C-2, -6, thiomorpholine), 50.4 (C-3, -5, thiomorpholine) ppm. HMRS (ESI$^+$, methanol) $C_{17}H_{19}BFN_2O_3S$ [M+H]$^+$ = 361.1184.

{4-[(3-Fluoro-4-morpholinophenyl)carbamoyl]phenyl}boronic acid **39** brown solid; yield 28%; mp: 206–214 °C (n-hexane/AcOEt/MeOH, gradient mode, starting with Hex/AcOEt 80:20 increasing the percentage of AcOEt by 10%, up to AcOEt/MeOH 90:10); IR (KBr) ν_{max} 3273 (B–OH boronic acid; N–H amide), 1647 (C=O amide), 1591 (C=C aromatic), 1440 (C=C aromatic), 1317 (B–O boronic acid), 1257 (C–F allyl floride) cm^{-1}. ^{1}H-NMR (600 MHz, DMSO-d_6) δ_H 10.26 (s, 1H, NH), 8.24 (s, 2H, OH), 7.92–7.91 (m, H-3, -5, carbamoyphenylboronic acid), 7.89–7.88 (m, H-2, -6, carbamoyphenylboronic acid), 7.73–7.70 (dd, *J* = 15.0, 2.0 Hz, H-2, fluorophenyl), 7.50–7.48 (dd, *J* = 9.0, 2.0 Hz, H-6, flhorophenyl), 7.05–7.02 (t, *J* = 9.0 Hz, H-5, fluorophenyl), 3.75–3.73 (m, H-3, -5, morpholine), 2.98–2.96 (m, H-

2, -6, morpholine) ppm. ^{13}C-NMR (151 MHz, DMSO-d_6) δ_C 165.4 (1C, C=O), 155.1 (d, $^1J_{C-F}$ = 242.0 Hz, C-3, fluorophenyl), 136.0 (C-4, carbamoyphenylboronic acid), 135.5 (d, $^3J_{C-F}$ = 10.5 Hz, C-1, fluorophenyl), 134.3 (d, $^2J_{C-F}$ = 11.2 Hz, C-4, fluorophenyl), 134.0 (C-3, -5, carbamoyphenylboronic acid), 126.5 (C-2, -6, carbamoyphenylboronic acid), 118.9 (d, $^3J_{C-F}$ = 4.4 Hz, C-5, fluorophenyl), 116.3 (d, $^4J_{C-F}$ = 2.8 Hz, C-6, fluorophenyl), 108.5 (d, $^2J_{C-F}$ = 26.1 Hz, C-2, fluorophenyl), 66.2 (C-3, -5, morpholine), 50.8 (C-2, -6, morpholine) ppm. HMRS (ESI$^+$, methanol) $C_{17}H_{19}BFN_2O_4$ [M+H]$^+$ = 345.1420.

(4-{[4-(Azepan-1-yl)-3-fluorophenyl]carbamoyl}phenyl)boronic acid **40** white solid; yield 62%; mp: 211–216 °C (n-hexane/AcOEt/MeOH, gradient mode, starting with Hex/AcOEt 80:20 increasing the percentage of AcOEt by 10%, up to AcOEt/MeOH 90:10); IR (KBr) ν_{max} 3321 (B–OH boronic acid; N–H amide), 2924 (C–H methylene), 1643 (C=O amide), 1427 (C=C aromatic), 1340 (B–O boronic acid), 1271 (C–F allyl floride) cm^{-1}. ^{1}H-NMR (600 MHz, DMSO-d_6) δ_H 10.14 (s, 1H, NH), 8.24 (s, 2H, OH), 7.92–7.90 (m, H-3, -5, carbamoyphenylboronic acid), 7.89–7.87 (m, H-2, -6, carbamoyphenylboronic acid), 7.64–7.61 (dd, J = 16.2, 2.1 Hz, H-2, fluorophenyl), 7.40–7.38 (dd, J = 9.0, 2.1 Hz, H-6, fluorophenyl), 6.92–6.89 (t, J = 10.3, 9.0 Hz, H-5, fluorophenyl), 3.30–3.28 (m, H-2, -7, azepane), 1.75 (m, H-3, -6, azepane), 1.57–1.55 (m, H-4, -5, azepane) ppm. ^{13}C-NMR (150 MHz, DMSO-d_6) δ_C 165.1 (1C, C=O), 155.7 (d, $^1J_{C-F}$ = 240.9 Hz, C-3, fluorophenyl), 136.1 (C-4, carbamoyphenylboronic acid), 135.7 (d, $^3J_{C-F}$ = 9.4 Hz, C-1, fluorophenyl), 130.8 (d, $^2J_{C-F}$ = 11.3 Hz, C-2, fluorophenyl), 126.5 (C-2, -6, carbamoyphenylboronic acid), 117.0 (d, $^3J_{C-F}$ = 5.8 Hz, C-5, fluorophenyl), 116.5 (d, $^4J_{C-F}$ = 1.9 Hz, C-6, fluorophenyl), 109.0 (d, $^2J_{C-F}$ = 26.5 Hz, C-2, fluorophenyl), 51.7 (C-2, -7, azepane), 39.5 (C-3, -6, azepane), 26.7 (C-4, -5, azepane) ppm. HMRS (ESI$^+$, methanol) $C_{19}H_{23}BFN_2O_3$ [M+H]$^+$ = 357.1780.

E. General Miyaura borylation synthesis procedure **41** [25–27]

In a 25.0 mL flask, (0.65 mmol, 1.0 equiv.) methyl 4-bromo-3-methylbenzoate, (0.98 mmol, 1.5 equivalent) of bis(pinacolato)diboron (B_2pin_2), (1.95, 3.0 equivalent) of potassium acetate (KOAc), and (0.065 mmol, 10 mol%) of the catalyst were added. Then, [1,1′-bis(diphenylphosphino)ferrocene]palladium(II) dichloride [$PdCl_2$(dppf)] was added to 10.0 mL of previously deaerated 1,4-dioxane, under a nitrogen atmosphere, at a temperature of 80 °C. After 12 h of reaction, the reaction mixture was filtered through celite 545, and the solvent was removed under reduced pressure; then, the resulting product was solubilized in dichloromethane, which was taken to a separatory funnel and washed with saturated sodium chloride solution (3 × 50.0 mL). The collected organic phase and the product were purified by column chromatography (eluent n-hexane/AcOEt, 9:1) [25,27].

Methyl 3-methyl-4-(4,4,5,5-tetramethyl-1,3,2-dioxaborolan-2-yl)benzoate **41** slightly yellow solid; yield 98%; (n-hexane/AcOEt, 9:1). ^{1}H-NMR (300 MHz, DMSO-d_6) δ_H 7.75–7.74 (m, H-2, -5, -6, methyl benzoate), 3.85 (s, 3H, CH_3, methyl benzoate), 2.51 (s, 3H, CH_3, methyl acetate), 1.31 (s, 12H, CH_3, boronic ester) ppm.

F. General procedure of bromination synthesis **42** [27]

In a 25.0 mL flask, (0.54 mmol, 1.0 equiv.) of compound **41**, (0.54 mmol, 1.5 equiv.) of *N*-bromosuccinimide (NBS), and (0.09 mmol, 25 mol%) of 2,2′-azobis(2-methylpropionitrile) (AIBN) were added to 6.0 mL of anhydrous acetonitrile. The reaction was kept under stirring for 24 h at a temperature of 80 °C and a nitrogen atmosphere. After the reaction time, excess solvent was removed under reduced pressure, and the resulting product was solubilized in 50.0 mL of dichloromethane. The organic phase was washed with saturated $NaHCO_3$ solution (3 × 50.0 mL), where the solvent was removed under reduced pressure, without the need to purify the final product [27].

Methyl 3-(bromomethyl)-4-(4,4,5,5-tetramethyl-1,3,2-dioxaborolan-2-yl)benzoate **42** colorless oil; yield 62%; ^{1}H-NMR (300 MHz, DMSO-d_6) δ_H 8.04–8.03 (d, J = 1.5 Hz, H-2, benzoate), 7.91–7.87 (dd, J = 7.8, 1.5 Hz, H-6, benzoate), 7.83–7.80 (d, J = 7.8 Hz, H-5, benzoate), 5.02 (s, 2H, CH_2), 3.87 (s, 3H, CH_3, methyl acetate), 1.34 (s, 12H, CH_3, boronic ester).

G. General procedure for obtaining the benzoxaborole intermediate **43**

To a THF (2 mL)/NaOH aq (5 mL distilled water, 3.60 mmol, 10 equiv. NaOH) solution and under vigorous stirring, compound 42 (0.32 mmol, 1.0 equiv.) was added and the reaction mixture was heated at 50 °C for 2 h. Next, in the same conditions, a 6.0 M hydrochloric acid solution was added dropwise, until pH~2.0; stirring was continued for an additional 2 h at 50 °C. The reaction mixture as a solution was then cooled to room temperature and extracted with ethyl acetate (3 × 50 mL). The combined organic layers were dried over anhyd Na_2SO_4, filtered off, and evaporated under reduced pressure to dryness. The residue was purified by column chromatography in gradient mode, starting the mobile phase n-hexane/AcOEt (1:4 v/v), up to 95% EtOAc and 5% glacial acetic acid. [27].

1-Hydroxy-1,3-dihydrobenzo[c][1,2]oxaborole-5-carboxylic acid **43** white solid; yield 65%; (n-hexane/AcOEt (20:80), up to 95% EtOAc and 5% glacial acetic acid); IR (KBr) ν_{max} 3331 (O–H carboxylic acid), 3068 (B–OH benzoxaborole), 1693 (C=O carboxylic acid), 1489 (C=C aromatic), 1421 (C–H methylene), 1367 (B–O benzoxaborole) cm^{-1}. ^{1}H-NMR (600 MHz, DMSO-d_6) δ_H 13.03 (s, 1H, OH, carboxylic acid), 9.39 (s, 1H, OH, oxaborole), 7.96 (m, H-4, benzoxaborole), 7.93–7.90 (d, J = 7.6 Hz, H-6, benzoxaborole), 7.84–7.81 (d, J = 7.6 Hz, H-7, benzoxaborole), 5.05 (s,H-3, oxaborole) ppm.

H. General procedure for obtaining benzoxaborole derivative **44** and **45**

For the synthesis of the benzoxaborole derivative **44** and **45**, (0.68 mmol, 1.0 equiv.) of the compound **43** and anhydrous *N'N*-diisopropylethylamine (DIPEA) in 5.0 mL of *N'N*-dimethylformamide were added to a 25.0 mL flask. (DMF) and stirred for 10 min at room temperature in a nitrogen atmosphere. After this time, (0.75 mmol, 1.1 equiv.) of the HATU coupling agent was added, and the reaction medium was stirred for another 2 h. Then, (0.68 mmol, 1.0 equiv.) of the compounds **19** and **22** was added, where the reaction remained under stirring for 24 h, at room temperature in a nitrogen atmosphere. The reaction was terminated with the addition of 50.0 mL of distilled water, in a separatory funnel, where it was extracted with ethyl acetate (3 × 50.0 mL); the organic phase was collected, dried with anhydrous sodium sulfate, and filtered, and the resulting solvent was removed under reduced pressure. The crude product was purified by column chromatography (eluent n-hexane/AcOEt 20:80) [50].

N-{3-fluoro-4-[4-(pyridin-2-yl)piperazin-1-yl]phenyl}-1-hydroxy-1,3-dihydrobenzo[c][1,2]oxaborole-5-carboxamide **44** light brown solid; yield 31%; mp: 211–216 °C (n-hexane/AcOEt 20:80) IR (KBr) ν_{max} 3327 (B–OH benzoxaborole), 1643 (C=O amide), 1593 (C=C aromatic), 1525 (N–H amide), 1421 (C=C aromatic), 1319 (C–F aryl fluoride) cm^{-1}. ^{1}H-NMR (600 MHz, DMSO-d_6) δ_H 10.37 (s, 1H, NH), 9.39 (s, 1H, OH), 8.14–8.13 (dd, J = 4.8, 1.2 Hz, H-6, pyridine), 7.94 (m, H-4, benzoxaborole), 7.90–7.88 (d, J = 7.7 Hz, H-6, benzoxaborole), 7.86–7.85 (d, J = 7.7 Hz, H-7, benzoxaborole), 7.75–7.72 (dd, J = 14.9, 2.2 Hz, H-2, fluoropheyl), 7.57–7.54 (ddd, J = 8.9, 7.2, 2.0 Hz, H-4, pyridine), 7.51–7.49 (m, H-6, fluorophenyl), 7.11–7.07 (t, J = 9.3 Hz, H-5, fluorophenyl), 6.89–6.88 (d, J = 8.6 Hz, H-5, pyridine), 6.68–6.66 (dd, J = 7.0, 5.4 Hz, H-6, pyridine), 5.08 (s, H-3, benzoxaborole), 3.65–3.63 (m, H-2, -6, piperazine), 3.09–3.07 (m, H-3, -5, piperazine) ppm. ^{13}C-NMR (150 MHz, DMSO-d_6) δ_C 165.6 (1C, C=O), 159.0 (C-2, pyridine), 155.2 (d, $^{1}J_{C-F}$ = 242.8 Hz, C-3, fluorophenyl), 154.0 (C-3a, benzoxaborole), 147.6

(C-6, pyridine), 137.6 (C-5, benzoxaborole), 136.9 (C-4, pyridine), 135.7 (d, $^{2}J_{C\text{–}F}$ = 9.2 Hz, C-4, fluorophenyl), 134.3 (d, $^{3}J_{C\text{–}F}$ = 10.9 Hz, C-1, fluorophenyl), 130.5 (C-7, benzoxaborole), 126.3 (C-6, benzoxaborole), 120.5 (C-4, benzoxaborole), 119.3 (d, $^{3}J_{C\text{–}F}$ = 4.5 Hz, C-5, fluorophenyl), 116.3 (d, $^{4}J_{C\text{–}F}$ = 3.3 Hz, C-6, fluorophenyl), 113.3 (C-5, pyridine), 108.4 (d, $^{2}J_{C\text{–}F}$ = 25.7 Hz, C-2, fluorophenyl), 107.2 (C-3, pyridine), 69.9 (C-3, benzoxaborole), 50.2 (C-3, -5, piperazine), 44.8 (C-2, -6, piperazine) ppm. HRMS (ESI^{+}, methanol) $C_{23}H_{23}BF_{4}O_{3}$ $[M+H]^{+}$ = 433.1840.

N-(3-fluoro-4-thiomorpholinophenyl)-1-hydroxy-1,3-dihydrobenzo[c][1,2]oxaborole-5-carboxamide **45** brown oil; yield 15%; (n-hexane/AcOEt 20:80) IR (KBr) ν_{max} 3439 (B–OH benzoxaborole), 1647 (C=O amide), 1527 (N–H amide), 1423 (C=C aromatic), 1278 (C–F aryl fluoride) cm^{-1}. ^{1}H-NMR (600 MHz, DMSO-d_6) δ_H 10.36 (s, 1H, NH), 9.37 (s, 1H, OH), 7.93 (m, H-4, benzoxaborole), 7.89–7.88 (d, J = 7.8 Hz, H-6, benzoxaborole), 7.86–7.85 (d, J = 7.8 Hz, H-7, benzoxaborole), 7.72–7.69 (dd, J = 14.6, 1.9 Hz, H-2, fluorophenyl), 7.48–7.46 (m, H-6, fluorophenyl), 7.09–7.06 (t, J = 9.3 Hz, H-5, fluorophenyl), 5.08 (s, H-2, benzoxaborole), 3.22–3.21 (m, H-3, -5, thiomorpholine), 2.76–2.74 (m, H-2, -6, thiomorpholine) ppm. ^{13}C-NMR (150 MHz, DMSO-d^{6}) δ_C 165.6 (1C, C=O), 155.3 (d, $^{1}J_{C\text{–}F}$ = 242.9 Hz, C-3, fluorophenyl), 154.0 (C-3a, benzoxaborole), 136.9 (C-5, benzoxaborole), 136.6 (d, $^{3}J_{C\text{–}F}$ = 9.6 Hz, C-1, fluorophenyl), 134.5 (d, $^{2}J_{C\text{–}F}$ = 11.9 Hz, C-4, fluorophenyl), 130.5 (C-7, benzoxaborole), 126.3 (C-6, benzoxaborole), 120.4 (d, $^{3}J_{C\text{–}F}$ = 4.0 Hz, C-5, fluorophenyl), 116.3 (d, $^{4}J_{C\text{–}F}$ = 2.9 Hz, C-6, fluorophenyl), 108.5 (d, $^{2}J_{C\text{–}F}$ = 26.5 Hz, C-2, fluorophenyl), 69.9 (C-3, benzoxaborole), 53.1 (C-3, -5, thiomorpholine), 27.3 (C-2, -6, thiomorpholine) ppm. HRMS (ESI^{+}, methanol) $C_{23}H_{23}BF_{4}O_{3}$ $[M+H]^{+}$ = 373.1199.

4.1. Biological Evaluation

4.1.1. *Mycobacterium tuberculosis* Biological Assays

The biological tests were carried out in partnership with Prof. Dr. Fernando Rogério Pavan's research group at the Prof. Dr. Hugo David Mycobacteria Laboratory, at the School of Pharmaceutical Sciences, São Paulo State University—Araraquara/Brazil

Determination of the Minimum Inhibitory Concentration (MIC_{90}) Against Strains of *Mycobacterium tuberculosis* $H_{37}R_V$

The minimum inhibitory concentration (MIC_{90}) was determined using the REMA (resazurin microtiter assay) methodology, in which stock solutions of the compounds were prepared in DMSO (10 mg/mL) and then diluted in Middlebrook 7H9 broth (Difco, New Jersey, USA) supplemented with OADC (0.09–25 μg/mL). Rifampicin and isoniazid (Sigma-Aldrich, St Louis, MO, USA) were used as standards, solubilized in DMSO and distilled water, respectively, at concentrations of 0.004–1.0 μg/mL. In the *M. tuberculosis* $H_{37}R_V$ strain suspension, 100 μL of inoculum was added to each well of the 96-well microplate, as well as 100 μL of the compounds at the concentrations described above. The plates were then incubated for seven days in an oven at 37 °C and 5% CO_2. After this time, 30 μL of resazurin (Sigma-Aldrich, St Louis, MO, USA) was added. After the 24 h incubation period, the fluorescence of the wells was measured using Cytation 3 equipment Biotek® (Agilent, Santa Clara, CA, USA). The MIC_{90} was determined as the lowest concentration of the compounds capable of inhibiting 90% of the bacillus' growth [51].

Cytotoxicity Assay (IC_{50}) and Selectivity Index (IS)

The cytotoxicity test was carried out on the most active compounds in the series, i.e., those with the lowest MIC_{90} values. For the test, MRC-5 (ATCC CCL-171) and J774A.1 (ATCC TIB-67) cells were used; they were cultured in a medium containing DNEM (Vitrocell®, Waldkirch, Germany) and RPMI (Vitrocell®, Waldkirch, Germany). Then, 10%

fetal bovine serum, gentamicin sulfate (50 mg/mL) and amphotericin B (2 mg/L) were added. The cells were incubated in an oven at 37 °C with 5% CO_2, and a specific amount of each cell type (2.5×10^5 cells/mL for MRC-5 and 1.0 × 106 for J774A.1) was distributed in a 96-well microplate, with a total volume of 100 µL. They were then incubated for 24 h at 37 °C in an atmosphere containing 5% CO_2. The compounds were then added to the wells, at dilutions of 0.39–100 µg/mL, and incubated again for a further 24 h. After this interval, 50 µL of 0.01% resazurin solution was added, and after 3 h, the fluorescence of the wells was measured using a Synergy H1 reader Biotek® (Agilent, Santa Clara, CA, USA). To determine the IC_{50} values, those concentrations capable of maintaining the viability of 50% of the cells were characterized. The selectivity index (SI) was determined by the ratio between the IC_{50} and MIC_{90} values ($SI = IC_{50}/MIC_{90}$). Compounds with an IS greater than or equal to 10 are considered selective, meaning that the compound is 10 times more toxic to the *M. tuberculosis* strain than to eukaryotic cells [52,53].

4.1.2. Antifungal Assays

Microorganisms and Culture Conditions

This research used strains of *Candida albicans* ATCC 90028, *T. rubrum* ATCC 28189, and *T. mentagrophytes* 11481 clinical isolates. The ATCC strains were obtained from the library of the Mycology Laboratory of the Clinical Analysis Department of the School of Pharmaceutical Sciences, UNESP, Brazil.

Candida albicans were maintained on Sabouraud agar for 24 h at 35 °C, according to the norms stipulated by the Clinical and Laboratory Standards Institute (CLSI) M27-A3 protocol. *T. rubrum* and T mentagrophytes were maintained on Sabouraud agar and incubated at 28 °C for 7 to 15 days. Then, the strain was cultivated on malt extract agar [malt extract (Kasvi) 2%, peptone from animal tissue (Sigma-Aldrich, St Louis, MO, USA) 2%, glucose (Synth, São Paulo, Brazil) 2%, and agar (Kasvi, Paraná, Brazil) 2%], pH 5.7, and kept at 25 °C for 7 days or until sporulation [54,55].

Determination of Minimum Inhibitory Concentration (MIC)

The susceptibility tests were carried out according to the Clinical and Laboratory Standards Institute (CLSI) M27-A3 for *C. albicans* and to the document M38-A2 for *T. rubrum* and *T. mentagrophytes* [56,57].

Briefly, the compounds were solubilized in 100% dimethyl sulfoxide (DMSO) at a stock concentration of 30,000 mg/L and stored at −80 °C. The working solution of the compounds was prepared in Roswell Park Memorial Institute (RPMI-1640) with L-glucose, without sodium bicarbonate, and with phenol red as the pH indicator (Gibco, New York, USA) and buffered with 4-Morpholinepropanesulfonic acid hemisodium salt (MOPS) (Sigma-Aldrich, St Louis, MO, USA), pH = 7, at concentrations ranging from 0.24 to 125 mg/L. Terbinafine (Sigma-Aldrich, St Louis, MO, USA) (0.001 to 0.5 mg/L) and fluconazole) (0.12 to 64 mg/L) were used for quality control.

For the establishment of the MIC, an initial inoculum of *C. albicans* was prepared in saline 0.85% NaCl containing 1 to 5×10^6 cells/mL; from this inoculum, two dilutions of 1:100 and 1:20 were subsequently made in RPMI 1640, so that the final concentration of cells was from 5×10^2 to 2.5×10^3 cells/mL. For dermatophytes, fungal suspensions were prepared in 0.85% NaCl and conidia, adjusted to a final concentration of 2.5×10^4 cells/mL using a hemocytometer, and then added to microdilution plates. The volume of 100 µL of the dilutions of the compounds and antifungal drugs was added to a 96-well microplate following 100 µL of the inoculum. The MIC was stipulated using visual and colorimetric readings by adding 30 µL of 0.03% resazurin [56–58].

Determination of the Minimum Fungicide Concentration (MFC)

Briefly, aliquots of 100 μL from each well from the MIC assay were added to Sabouraud dextrose agar (BD Difco TM, New Jersey, USA) in Petri dishes and incubated for 96 h at 28 °C. The MFC was defined as the lowest concentration of the compound capable of no growth of fungal colonies [56].

Cytotoxicity Assay

For the cytotoxicity assay, the HaCat cell line, characterized as immortalized normal human keratinocytes, was used. The cells were maintained in (DMEM) supplemented with 10% of bovine fetal serum (BFS) (Sigma-Aldrich, St Louis, MO, USA) in cell culture bottles with 5% CO_2 at 37 °C. After a confluence of 75%, the cells were treated with trypsin + EDTA 0.25%, centrifugated at 1500 rpm for 10 min, and 5×10^4 cells/well were seeded in 96-well plates for incubation for 24 h for cell adhesion before treatment. After adhesion, the cells were treated with 100 μL of the compounds **34** and **40**, with fungi activity in concentrations between 3.90 and 125 μg/mL. Regarding the controls, DMSO in the concentration range of (1–4%) was used as a positive control for cytotoxicity; the beyond negative control comprised HaCaT cells only in DMEM medium (Sigma-Aldrich, St Louis, MO, USA). After 24 h of treatment, 30 μL of resazurin 50 μM was added to each well for 4 h, and the plates were read at 560 and 600 nm using microplate reader BioTek (Agilent, Santa Clara, CA, USA), Gen 5 software (version 2.09). The minimum inhibitory concentration (IC_{50}) for each compound was determined based on resazurin reduction. Three independent assays were performed [59].

4.1.3. Anticancer Assays

The SCC-25 was purchased from Rio de Janeiro Cell Bank, and NOK-si was kindly provided by Prof. Carlos Rossa from the School of Dentistry—UNESP. The cell lines were initially thawed and cultured in 25 cm^2 cell culture flasks. Cell growth was monitored using a microscope, and the culture medium was changed periodically. Upon reaching confluency, cell dissociation was achieved by adding 5 mL of trypsin (Sigma-Aldrich, St Louis, MO, USA) and incubating the bottles in a CO_2 incubator for 10 min. Subsequently, the cells were transferred to 15 mL Falcon tubes and centrifuged at 2000 rpm for 10 min. Cell counting was performed using a Countess II FL automated cell counter (ThermoFischer, Waltham, MA, USA). The cells were seeded at a density of 30,000 cells per well in 96-well plates. The plates were then incubated at 37 °C in 5% CO_2 for 24 h. Following this, test compounds were incubated for another 24 h at 10 different concentrations: 50, 25, 12.5, 6.25, 3.12, 1.56, 0.78, 0.39, 0.19, and 0.097 μg/mL. After the specified time, MTT reagent (3 mg/mL) (Sigma-Aldrich, St Louis, MO, USA) dissolved in phenol-free RPMI 1640 was added and incubated for 4 h. The formed MTT crystals were solubilized using 2-propanol P.A (Sigma-Aldrich, St Louis, MO, USA) with agitation. Absorbance was read at 562 nm using an EZ Read 400 microplate reader (Biochrom, Cambridge, UK) and ADAP 2.0 Biochrom software (version 2.0). IC_{50} values and graphs were calculated by Dr. Paula Aboud Barbugli using Excel and Prism software (version 10.4.1). The MTT assay results were evaluated for normality using the Shapiro–Wilk test, followed by one-way ANOVA with Tukey's post hoc test at a significance level of 5%.

5. Conclusions

The objective of this study was to develop and characterize novel boronic acid and benzoboroxole compounds due to their potential as antimycobacterial, antifungal, and antitumor agents. Fourteen new derivatives were synthesized and evaluated for biological activity. While none of the compounds exhibited significant antimycobacterial properties against *Mycobacterium tuberculosis* ($H_{37}R_v$), one derivative, **34**, demonstrated moderate antifungal activity against dermatophyte fungi, *T. rubrum*, and *T. mentagrophytes* (MIC_{90} = 42.0 μM and 21.0 μM, respectively). Furthermore, two compounds, **31** and **35**, exhibited promising antitumor activity against head and neck cancer cell lines, demonstrating selectivity with IC_{50} values of 59.07 μM and 45.61 μM, respectively. Based on these findings, these compounds may represent promising candidates for further exploration as potential therapeutic agents for head and neck cancer. However, additional research is necessary to elucidate their mechanism of action and to evaluate their efficacy in clinical trials.

Supplementary Materials: The following supporting information can be downloaded at: https://www.mdpi.com/article/10.3390/molecules30051117/s1, Figures S1–S12: Spectra of compounds **13–15** and **19–21** (^{1}H, ^{13}C NMR), Figures S13–S100: Spectra of compounds **29–45** (^{1}H NMR, ^{13}C NMR, HRMS/ESI, IR). Detailed ^{1}H NMR spectral data for compounds **16–18** and **22–24** is reported in reference [19].

Author Contributions: J.L.B.P. and J.L.D.S., designed this study and wrote this manuscript. The biological assays in this work were evaluated in vitro as anticancer, J.R.L., F.H.M.-D. and P.A.B.; antifungal, S.d.M.S., K.P.M.-A., K.S.d.S., J.A.B., M.J.S.M.G. and A.M.F.-A.; and anti-*Mycobacterium tuberculosis*, D.L.C. and F.R.P. All authors contributed to revision of the manuscript. All authors have read and agreed to the published version of the manuscript.

Funding: This research was funded by Fundação de Amparo à Pesquisa do Estado de São Paulo (FAPESP), grant numbers: 2023/05739-6 (J.L.D.S), 2023/03556-1 (M.J.S.M.G.), 2023/01664-1 (F.R.P.), 2022/15826-0 (A.M.F.A.) and 2019/07574-9 (P.A.B.). National Council for Scientific and Technological Development (CNPq): Process: 302433/2023-6; Coordenação de Aperfeiçoamento de Pessoal de Nível Superior (CAPES): Finance Code 001. Pro-reitoria de Pós-Graduação UNESP—PROPG edital 01/2025.

Institutional Review Board Statement: Not applicable.

Informed Consent Statement: Not applicable.

Data Availability Statement: Data are contained within the article and Supplementary Materials.

Conflicts of Interest: The authors declare no conflicts of interest.

References

1. Hall, D.G. Strucuture, properties, and preparation of boronic acid derivatives. Overview of their reactions and applications. In *Boronic Acids: Preparation and Applications in Organic Synthesis, Medicine and Materials*, 2nd ed.; Hall, D.G., Ed.; Wiley-VCH Verlag GmbH & Co. KGaA: Weinheim, DE, USA, 2011; pp. 1–133.
2. Defrancesco, H.; Dudley, J.; Coca, A. Boron Chemistry: An Overview. In *Boron Reagents in Synthesis*; Coca, A., Ed.; ACS Symposium Series; American Chemical Society: Washington, DC, USA, 2016; Volume 1236, pp. 1–25.
3. Trippier, P.C.; Mcghigan, C. Boronic acids in medicinal chemistry: Anticancer, antibacterial and antiviral applications. *MedChemComm* **2010**, *3*, 183–198. [CrossRef]
4. Ni, N.; Wang, B. *Applications of boronic acids in chemical biology and medicinal chemistry In Boronic Acids: Preparation and Applications in Organic Synthesis, Medicine and Materials*, 2nd ed.; Hall, D.G., Ed.; Wiley-VCH Verlag GmbH & Co. KGaA: Weinheim, DE, USA, 2011; pp. 591–620.
5. Kahlert, J.; Austin, C.J.D.; Kassiou, M.; Rendina, L.M. The fifth element in drug design: Boron in medicinal chemistry. *Aust. J. Chem.* **2013**, *66*, 1118–1123. [CrossRef]
6. Prates, J.L.B.; Pavan, A.R.; Santos, J.L. Boron in medicinal and organic chemistry. *Curr. Org. Chem.* **2021**, *25*, 1853–1867. [CrossRef]
7. Fernandes, G.F.S.; Denny, W.A.; Santos, J.L. Boron in drug design: Recent advances in the development of new therapeutic agents. *Eur. J. Med. Chem.* **2019**, *179*, 791–804. [CrossRef]
8. Grams, R.J.; Santos, W.L.; Scorei, I.R.; Abad-García, A.; Rosenblum, C.A.; Bita, A.; Cerecetto, H.; Viñas, C.; Soriano-Ursúa, M.A. The rise of boron-containing compounds: Advancements in synthesis, medicinal chemistry, and emerging pharmacology. *Chem. Rev.* **2024**, *124*, 2441–2511. [CrossRef]
9. Silva, M.P.; Saraiva, L.; Pinto, M.; Sousa, M.M. Boronic acids and their derivatives in medicinal chemistry: Synthesis and biological applications. *Molecules* **2020**, *25*, 4323. [CrossRef]

10. Das, B.C.; Nandwana, N.K.; Das, S.; Nandwana, V.; Shareef, M.A.; Das, Y.; Saito, M.; Weiss, L.M.; Almaguel, F.; Hosmane, N.S.; et al. Boron chemicals in drug discovery and development: Synthesis and medicinal perspective. *Molecules* **2022**, *27*, 2615. [CrossRef]
11. Adams, J.; Behnke, M.; Chen, S.; Cruickshank, A.A.; Dick, L.R.; Grenier, L.; Klunder, J.M.; Ma, Y.T.; Plamondon, L.; Stein, R.L. Potent and selective inhibitors of the proteasome: Dipeptidyl boronics acids. *Bioorg. Med. Chem. Lett.* **1998**, *8*, 333–338. [CrossRef]
12. Glynn, S.J.; Gaffney, K.J.; Sainz, M.A.; Louie, S.G.; Petasis, N.A. Molecular characterization of the boron adducts of the proteasome inhibitor Bortezomib with epigallocatechin-3-gallate and related polyphenols. *Org. Biomol. Chem.* **2015**, *13*, 3887–3899. [CrossRef]
13. McDowell, L.; Olin, B. Crisaborole: A novel nonsteroidal topical treatment for atopic dermatitis. *J. Pharm. Technol.* **2019**, *35*, 172–178. [CrossRef]
14. Li, X.; Hernandez, V.; Rock, F.L.; Choi, W.; Mak, Y.S.L.; Mohan, M.; Mao, W.; Zhou, Y.; Easom, E.E.; Plattner, J.J.; et al. Discovery of a potent and specific *M. tuberculosis* Leucyl-tRNA synthetase inhibitor (S)-3-(aminomethyl)-4-chloro-7-(2-hydroxyethoxy)benzo[*c*][1,2]oxaborol-1(3H)-ol. *J. Med. Chem.* **2017**, *60*, 8011–8026. [CrossRef] [PubMed]
15. Diacon, A.H.; Barry, C.E., 3rd; Carlton, A.; Chen, R.Y.; Davies, M.; de Jager, V.; Fletcher, K.; Koh, G.C.K.W.; Kontsevaya, I.; Heyckendorf, J.; et al. A first-in-class leucyl-tRNA synthetase inhibitor, ganfeborole, for rifampicin-susceptible tuberculosis: A phase 2a open label, randomized trial. *Nat. Med.* **2024**, *30*, 896. [CrossRef] [PubMed]
16. Markham, A. Tavaborole: First global approval. *Drugs* **2014**, *74*, 1555–1558. [CrossRef]
17. Baker, S.J.; Zhang, Y.K.; Akama, T.; Lau, A.; Zhou, H.; Hernandez, V.; Mao, W.; Alley, M.R.; Sanders, V.; Plattner, J.J. Discovery of a new boron-containing antifungal agent, 5-fluoro-1,3-dihydro-1-hydroxy-2,1-benzoxaborole (AN2690), for the potential treatment of onychomycosis. *J. Med. Chem.* **2006**, *49*, 4447–4450. [CrossRef]
18. Rock, F.L.; Mao, W.; Yaremchuk, A.; Tukalo, M.; Crépin, T.; Zhou, H.; Zhang, Y.K.; Hernandez, V.; Akama, T.; Baker, S.J.; et al. An antifungal agent inhibits an aminoacyl-tRNA synthetase by trapping tRNA in the editing site. *Science* **2007**, *316*, 1759–1761. [CrossRef]
19. Fernandes, G.F.S.; Campos, D.L.; Da Silva, I.C.; Prates, J.L.B.; Pavan, A.R.; Pavan, F.R.; Dos Santos, J.L. Benzofuroxan derivatives as potent agents against multidrug-resistant *Mycobcterium tuberculosis*. *ChemMedChem* **2021**, *16*, 1268–1282. [CrossRef]
20. Scorei, R.I.; Popa, R., Jr. Boron-containing compounds as preventive and chemotherapeutic agents for cancer. *Anticancer Agents Med. Chem.* **2010**, *10*, 346–351. [CrossRef]
21. Carey, F.A.; Sundberg, R.J. (Eds.) Aromatic Substitution. In *Advanced Organic Chemistry: Part A: Structure and Mechanisms*, 5th ed.; Springer: New York, NY, USA, 2007; pp. 771–831.
22. Wang, Z. Béchamp Reduction. In *Comprehensive Organic Name Reactions and Reagents*, 2nd ed.; Wang, Z., Ed.; John Wiley & Sons: Hoboken, NJ, USA, 2009; Volume 3, pp. 284–287.
23. Valeur, E.; Bradley, M. Amide bond formation: Beyond the myth of coupling reagents. *Chem. Soc. Rev.* **2009**, *2*, 606–631. [CrossRef]
24. Carpino, L.A. 1-Hydroxy-7-azabenzotriazole. An efficient peptide coupling aditive. *J. Am. Chem. Soc.* **1993**, *10*, 4397–4398. [CrossRef]
25. Wuttke, A.; Geyer, A. Self-assembly of peptide boroxoles on cis-dihydroxylated oligoamide templates in water. *J. Pept. Sci.* **2017**, *23*, 549–555. [CrossRef]
26. Carey, F.A.; Sundberg, R.J. (Eds.) Free Radical Reactions. In *Advanced Organic Chemistry: Part A: Structure and Mechanism*, 5th ed.; Springer: New York, NY, USA, 2007; pp. 965–1071.
27. Fuscaldo, R.S.; Vontobel, P.H.V.; Boeira, E.O.; Moro, A.V.; Da Costa, J.S. Synthesis of Amino- and Hydroxymethyl Benzoxaboroles: Prominent Scaffolds for Further Functionalization. *Eur. J. Org. Chem.* **2019**, *2019*, 2050–2055. [CrossRef]
28. Santos, J.L.; Yamasaki, P.R.; Chin, C.M.; Takashi, C.H.; Pavan, F.R.; Leite, C.Q. Synthesis and in vitro anti *Mycobacterium tuberculosis* activity of a series of phthalimide derivatives. *Bioorg. Med. Chem.* **2009**, *17*, 3795–3799. [CrossRef] [PubMed]
29. Fernandes, G.F.D.S.; de Souza, P.C.; Marino, L.B.; Chegaev, K.; Guglielmo, S.; Lazzarato, L.; Fruttero, R.; Chung, M.C.; Pavan, F.R.; Dos Santos, J.L. Synthesis and biological activity of furoxan derivatives against *Mycobacterium tuberculosis*. *Eur. J. Med. Chem.* **2016**, *123*, 523–531. [CrossRef]
30. Plescia, J.; Moitessier, N. Design and discovery of boronic acid drugs. *Eur. J. Med. Chem.* **2020**, *195*, 112270. [CrossRef]
31. Adamczyk-Woźniak, A.; Borys, K.M.; Sporzyński, A. Recent developments in the chemistry and biological applications of benzoxaboroles. *Chem. Rev.* **2015**, *115*, 5224–5247. [CrossRef]
32. Al-Omari, M.K.; Elaarag, M.; Al-Zoubi, R.M.; Al-Qudimat, A.R.; Zarour, A.A.; Al-Hurani, E.A.; Fares, Z.E.; Alkharraz, L.M.; Shkoor, M.; Bani-Yaseen, A.D.; et al. Organoboronic acids/esters as effective drug and prodrug candidates in cancer treatments: Challenge and hope. *J. Enzyme Inhib. Med. Chem.* **2023**, *38*, 2220084. [CrossRef]
33. Umesiri, F.E.; Funk, J.; Lick, A.; Fricke, C.; Nathaniel, T.I. Boronic-aurone derivatives as anti-tubercular agents: Design, synthesis and biological evaluation. *Med. Chem.* **2015**, *5*, 437–441. [CrossRef]
34. Chatterjee, C.; Mohan, G.R.; Hariharan, V.C.; Biswas, B.; Sundaram, V.; Srivastava, A.; Matheshwaran, S. Boronic acid derivative inhibits LexA mediated SOS response in *Mycobacteria*. *bioRxiv* **2023**, *15*, 567215.

35. Guy, C.S.; Gibson, M.I.; Fullam, E. Targeting extracellular glycans: Tuning multimeric boronic acids for pathogen-selective killing of *Mycobacterium tuberculosis*. *Chem. Sci.* **2019**, *10*, 5935–5942. [CrossRef]
36. Guy, C.S.; Tomás, R.M.F.; Tang, Q.; Gibson, M.I.; Fullam, E. Imaging of antitubercular dimeric boronic acids at the mycobacterial cell surface by click-probe capture. *Chem. Commun.* **2022**, *58*, 9361–9364. [CrossRef]
37. Kurz, S.G.; Hazra, S.; Bethel, C.R.; Romagnoli, C.; Caselli, E.; Prati, F.; Blanchard, J.S.; Bonomo, R.A. Inhibiting the β-lactamase of *Mycobacterium tuberculosis* (*Mtb*) with novel boronic acid transition-state inhibitors (BATSIs). *ACS Infect. Dis.* **2015**, *1*, 234–242. [CrossRef] [PubMed]
38. Tachibana, H.; Kumagai, N.; Tatsumi, Y. Fungicidala activity in the presence of keratin as an important factor contributing to in vivo efficacy: A comparison of efinaconazole, tavaborole, and ciclopirox. *J. Fungi* **2017**, *3*, 58. [CrossRef] [PubMed]
39. Jinna, S.; Finch, J. Spotlight on tavaborole for the treatment of onychomycosis. *Drug Des. Devel. Ther.* **2015**, *9*, 6185–6190. [PubMed]
40. Bowen, J.E.; Gauch, H.G. Nonessentiality of boron in fungi and the nature of its toxicity. *Plant. Physiol.* **1996**, *41*, 319–324. [CrossRef]
41. Estevez-Fregoso, E.; Farfán-García, E.D.; García-Coronel, I.H.; Martínez-Herrera, E.; Alatorre, A.; Scorei, R.I.; Soriano-Ursúa, M.A. Effects of boron-containing compounds in the fungal kingdom. *J. Trace Elem. Med. Biol.* **2021**, *65*, 126714. [CrossRef]
42. De Seta, F.; Schmidt, M.; Vu, B.; Essmann, M.; Larsen, B. Antifungal mechanisms supporting boric acid therapy of *Candida vaginitis*. *J. Antimicrob. Chemother.* **2009**, *63*, 325–336. [CrossRef]
43. Campbell, R.; Buchbinder, N.W.; Szwetkowski, C.; Zhu, Y.; Piedl, K.; Truong, M.; Matson, J.B.; Santos, W.L.; Mevers, E. Design, synthesis, and antifungal activity of 3-substituted-2(5*H*)-oxaboroles. *ACS Med. Chem. Lett.* **2024**, *15*, 349–354. [CrossRef]
44. Borys, K.M.; Wieczorek, D.; Pecura, K.; Lipok, J.; Adamczyk-Woźniak, A. Antifungal activity and tautomeric cyclization equilibria of formylphenylboronic acids. *Bioorg. Chem.* **2019**, *91*, 103081. [CrossRef]
45. Adamczyk-Woźniak, A.; Gozdalik, J.T.; Wieczorek, D.; Madura, I.D.; Kaczorowska, E.; Brzezińska, E.; Sporzyński, A.; Lipok, J. Synthesis, properties and antimicrobial activity of 5-Trifluoromethyl-2-formylphenylboronic acid. *Molecules* **2020**, *25*, 799. [CrossRef]
46. Zhang, Y.; Peng, Y.; Lin, B.; Yang, S.; Deng, F.; Yang, X.; Li, A.; Xia, W.; Gao, C.; Lei, S.; et al. Non-coding RNA and drug resistance in head and neck cancer. *Cancer Drug Resist.* **2024**, *7*, 34. [CrossRef]
47. Gunduz, M.; Acar, M.; Fakioglu, K.; Dogan, B.; Öznur, M.; Gunduz, E. Effect of boric acid on head and neck cancer cell lines. *Eur. J. Cancer* **2014**, *50*, S68. [CrossRef]
48. Yıldırım, O.; Seçme, M.; Dodurga, Y.; Mete, G.A.; Fenkci, S.M. In vitro effects of boric acid on cell cycle, apoptosis, and miRNAs in medullary thyroid cancer cells. *Biol. Trace Elem. Res.* **2024**, *203*, 799–809. [CrossRef] [PubMed]
49. Kong, Y.; Wang, K.; Edler, M.C.; Hamel, E.; Mooberry, S.L.; Paige, M.A.; Brown, M.L. A boronic acid chalcone analog of combretastatin A-4 as a potent anti-proliferation agent. *Bioorg. Med. Chem.* **2010**, *18*, 971–977. [CrossRef] [PubMed]
50. Wagner, F.F.; Lundh, M.; Kaya, T.; McCarren, P.; Zhang, Y.L.; Chattopadhyay, S.; Gale, J.P.; Galbo, T.; Fisher, S.L.; Meier, B.C.; et al. An Isochemogenic Set of Inhibitors to Define the Therapeutic Potential of Histone Deacetylases in β-Cell Protection. *ACS Chem. Biol.* **2016**, *11*, 363–374. [CrossRef]
51. Palomino, J.C.; Martin, A.; Camacho, M.; Guerra, H.; Swings, J.; Portaels, F. Resazurin Microtiter Assay Plate: Simple and Inexpensive Method for Detection of Drug Resistance in *Mycobacterium tuberculosis* Resazurin Microtiter Assay Plate: Simple and Inexpensive Method for Detection of Drug Resistance in *Mycobacterium tuberculosis*. *Antimicrob. Agents Chemother.* **2002**, *46*, 2720–2722. [CrossRef]
52. Pavan, F.R.; Maia, P.I.D.S.; Leite, S.R.; Deflon, V.M.; Batista, A.A.; Sato, D.N.; Franzblau, S.G.; Leite, C.Q. Thiosemicarbazones, semicarbazones, dithiocarbazates and hydrazide/hydrazones: Anti-*Mycobacterium tuberculosis* activity and cytotoxicity. *Eur. J. Med. Chem.* **2010**, *45*, 1898–1905. [CrossRef]
53. Campos, D.L.; Machado, I.; Ribeiro, C.M.; Gambino, D.; Pavan, F.R. Bactericidal effect of pyridine-2-thiol 1-oxide sodium salt and its complex with iron against resistant clinical isolates of *Mycobacterium tuberculosis*. *J. Antibiot.* **2020**, *73*, 120–124. [CrossRef]
54. Costa-Orlandi, C.B.; Sardi, J.C.; Santos, C.T.; Fusco-Almeida, A.M.; Mendes-Giannini, M.J. In vitro characterization of Trichophyton rubrum and *T. mentagrophytes* biofilms. *Biofouling* **2014**, *30*, 719–727. [CrossRef]
55. Garcia, L.M.; Costa-Orlandi, C.B.; Bila, N.M.; Vaso, C.O.; Gonçalves, L.N.C.; Fusco-Almeida, A.M.; Mendes-Giannini, M.J.S. A two-way road: Antagonistic interaction between dual-species biofilms formed by *Candida albicans/Candida parapsilosis* and *Trichophyton rubrum*. *Front. Microbiol.* **2020**, *11*, 1980. [CrossRef]
56. Costa-Orlandi, C.B.; Serafim-Pinto, A.; da Silva, P.B.; Bila, N.M.; Bonatti, J.L.C.; Scorzoni, L.; Singulani, J.L.; Dos Santos, C.T.; Nazaré, A.C.; Chorilli, M.; et al. Incorporation of nonyl 3,4-Dihydroxybenzoate into nanostructured lipid systems: Effective alternative for maintaining anti-dermatophytic and antibiofilm activities and reducing toxicity at high concentrations. *Front. Microbiol.* **2020**, *11*, 1154. [CrossRef]

57. Bila, N.M.; Costa-Orlandi, C.B.; Vaso, C.O.; Bonatti, J.L.C.; de Assis, L.R.; Regasini, L.O.; Fontana, C.R.; Fusco-Almeida, A.M.; Mendes-Giannini, M.J.S. 2-Hydroxychalcone as a potent compound and photosensitizer against dermatophyte biofilms. *Front. Cell. Infect. Microbiol.* **2021**, *11*, 679470. [CrossRef] [PubMed]
58. Scorzoni, L.; Benaducci, T.; Almeida, A.M.F.; Silva, D.H.S.; Bolzani, V.D.S.; Gianinni, M.J.S.M. The use of standard methodology for determination of antifungal activity of natural products against medical yeasts *Candida* sp. and *Cryptococcus* sp. *Braz. J. Microbiol.* **2007**, *38*, 391–397. [CrossRef]
59. Assay Guidance Manual. Available online: https://www.ncbi.nlm.nih.gov/books/NBK144065 (accessed on 10 December 2023).

molecules

MDPI

Article

An Efficient Synthesis of 3,5-Bis-Aminated Pyrazolo[1,5-*a*]Pyrimidines: Microwave-Assisted Copper Catalyzed C-3 Amination of 5-Amino-3-Bromo-Substituted Precursors

Terungwa H. Iorkula , Bryce A. Tolman, Latifat O. Ganiyu and Matt A. Peterson *

Department of Chemistry and Biochemistry, Brigham Young University, Provo, UT 84602, USA
* Correspondence: mapeterson@chem.byu.edu; Tel.: +1-801-422-6843

Abstract: An efficient method has been developed for the rapid production of diverse arrays of 3,5-bis-aminated pyrazolo[1,5-*a*]pyrimidines. The method utilizes CuI (5 mol%) and carbazole-based ligand **L-1** (*N*-(9H-carbazol-9-yl)-1H-pyrrole-2-carboxamide) (10 mol%) for efficient Ullmann-type coupling of various amines to 5-amino-3-bromopyrazolo[1,5-*a*]pyrimidine precursors after heating in diethylene glycol (DEG) for only 1 h at 80 °C (microwave heating). 3,5-Bis-aminated products were obtained in good to excellent yields (60–93%, 83% average for 29 examples). 1° and 2° alkylamines, as well as a variety of aryl- or heteroarylamines coupled efficiently, and 1° and 2° alkyl (or aryl) amines at C-5 were well tolerated. The optimized conditions worked well on both the 50 mg and 1.0 g scales and gave products in only two steps from commercially available 3-bromo-5-chloropyrazolo[1,5-*a*]pyrimidine. Advantages provided by this method include short reaction time, excellent yields, broad substrate scope, and avoidance of toxic reagents commonly utilized for reductive aminations of C-3 NH_2 substituted precursors.

Keywords: pyrazolo[1,5-*a*]pyrimidines; copper catalyzed amination; microwave-assisted reactions; ullmann-type arylation

1. Introduction

The pyrazolo[1,5-*a*]pyrimidine scaffold (1) is a versatile template that has been extensively investigated for a number of important medicinal and biologically relevant applications. For example, pyrazolo[1,5-*a*]pyrimidine derivatives have been investigated as potential therapeutics for diabetes [1], cancer [2], neurodegenerative diseases [3], and various infectious diseases, including viral [4], protozoal [5], fungal [6], trypanosomal [7], and bacterial infections [8]. In addition, pyrazolo[1,5-*a*]pyrimidines have been used in dyes [9], fluorophores [10], chemosensors [11], tumor imaging [12], and as inhibitors of protein or lipid kinases [13]. Anti-inflammatory [14], anti-anxiolytic [15], and sedative-hypnotic [16] properties have also been reported. The clinical anticancer drugs Larotrectinib [17] and Reprotrectinib [18], along with several other structurally related clinical [19] or preclinical [13] candidates, as well as other promising disease-related enzyme inhibitors [20], have stimulated increased interest in the pyrazolo[1,5-*a*]pyrimidine scaffold (Figure 1). Of particular interest in recent years has been the synthesis and biological evaluation of C-3 aminated analogs, as underscored by the number of recently reported methodologies for converting C-3 brominated precursors to C-3 aminated derivatives (Figure 2) [21–26]. These methods involve transition-metal-catalyzed Buchwald-Hartwig or Ullmann-type

 https://doi.org/10.3390/molecules30030458

coupling reactions, whereas the originally reported method for preparing Larotrectinib involved the reduction of a C-3 nitro group followed by a two-step formation of the urea (Figure 3) [27,28]. The most commonly reported methods for preparing 3,5-bis-aminated

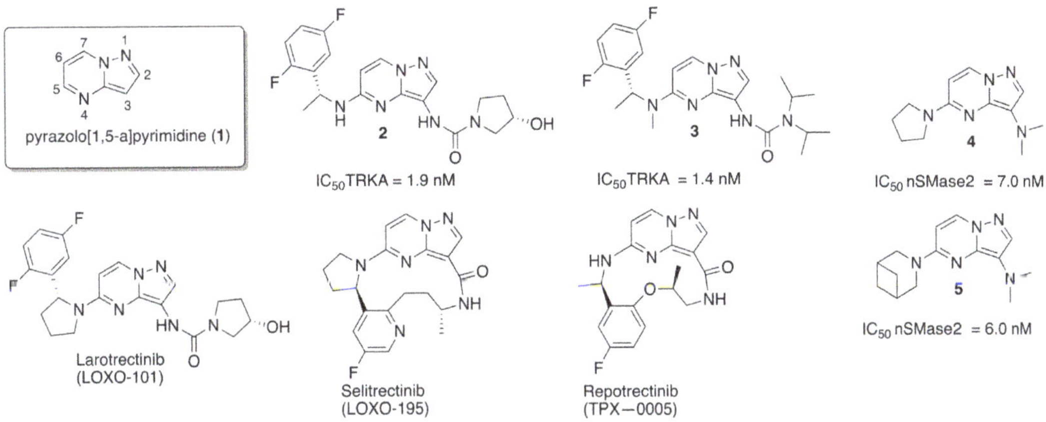

Figure 1. Clinical and preclinical biologically active pyrazolo[1,5-*a*]pyrimidines.

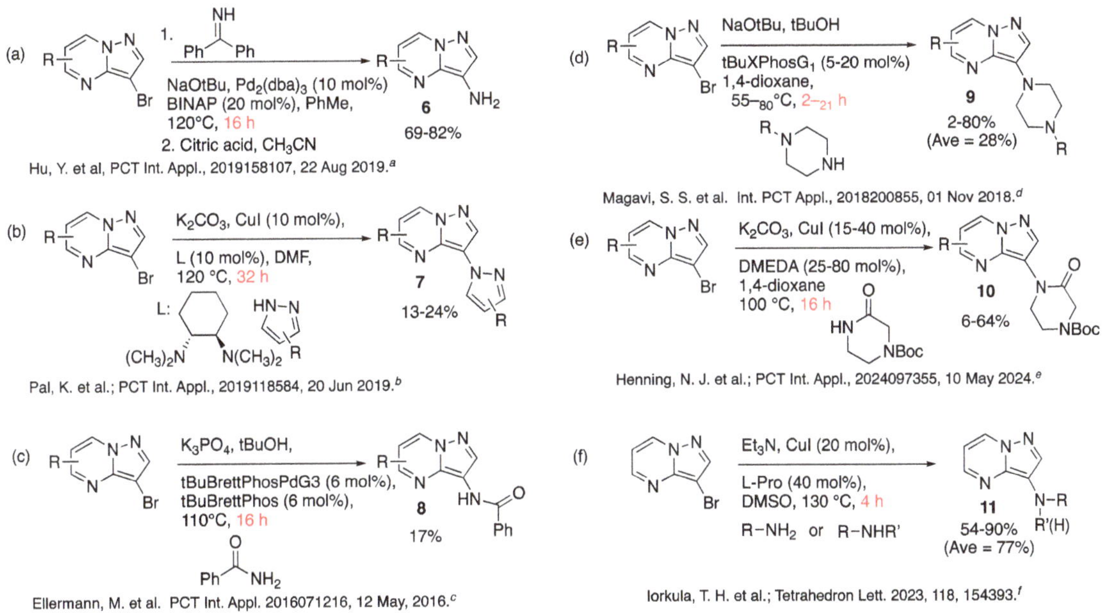

Figure 2. Buchwald-Hartwig or Ullmann-type coupling methods for C-3 amination. [a] [21], [b] [22], [c] [23], [d] [24], [e] [25], [f] [26].

pyrazolo[1,5-*a*]pyrimidine derivatives have either followed the route originally reported for Larotrectinib or related approaches varying essentially only in the order of C-3 nitration, reduction, and/or C-5 Nucleophilic Aromatic Substitution (SNAr) (Figure 3) [13,20,27–29]. Although these methods have provided compound libraries for SAR (Structure Activity Relationship) screening, they suffer from several drawbacks, including low yields, multiple steps, and/or limited synthetic versatility for functionalizing the C-3 NH_2 moiety. Indeed, the C-3 NH_2 of C-5 aminated derivatives has so far only been converted to ureas or thioureas via two-step carbamoylation/amination or to the $N(CH_3)_2$ group via reductive amination utilizing toxic reagents paraformaldehyde and $NaBH_3CN$ (Figure 3). Ideally, a synthetic method that avoids these limitations would provide efficient access to a broader array of analogs in short reaction times, high yields, and fewer steps, and it would avoid the use of toxic reagents. Given the considerable biological potential of such derivatives, methods meeting these criteria are highly desirable. Here, we report a new method that achieves these objectives, giving 3,5-bis-aminated products in 83% yield (average from 29 examples), utilizing 5 mol% of CuI catalyst and 10 mol% of carbazole-based ligand (**L-1**) after heating for only 1 h at 80 °C for C-3 amination. The reaction proceeded in only two steps from commercially available 3-bromo-5-chloropyrazolo[1,5-*a*]pyrimidine (Scheme 1) and provided a broad array of 3,5-bis-aminated products in good to excellent yields (Table 2). Substrates with both 1° alkyl (or aryl) and 2° alkylamines at C-5 were coupled effectively, and efficient C-3 amination was achieved using a variety of 1° or 2° alkylamine as well as aryl- and heteroarylamine C-3 coupling partners. With this method, rapid and efficient access to diverse libraries of 3,5-bis-aminated derivatives consisting of previously unavailable analogs is now readily attainable.

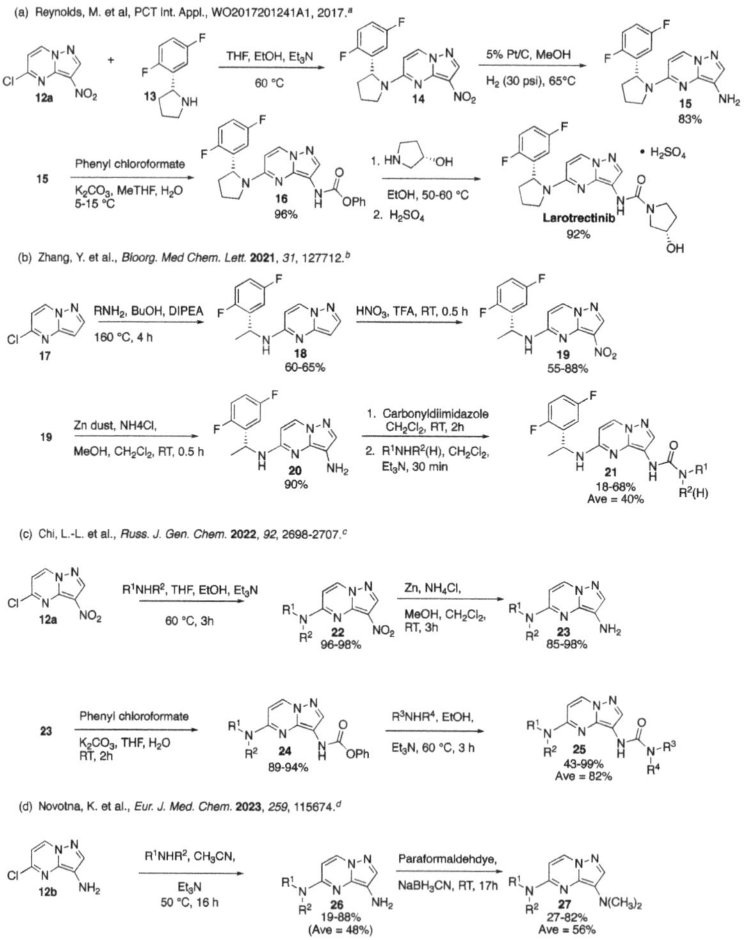

Figure 3. Synthesis of 3,5-bis-aminated pyrazolo[1,5-*a*]pyrimidines. [a] [27], [b] [13], [c] [29], [d] [20].

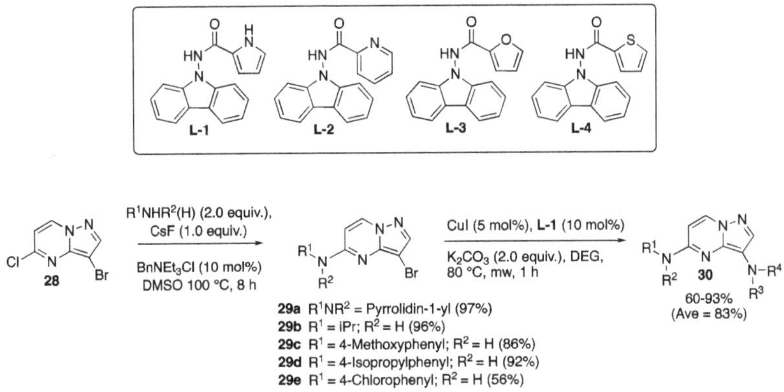

Scheme 1. Two-step preparation of 3,5-bis-aminated pyrazolo[1,5-*a*]pyrimidines **30**.

2. Results

2.1. Optimization

2.1.1. Temperature, Solvent, and Base

From the recent report by Huang and co-workers wherein **L-1** was shown to promote the efficient N-arylation of cyclopropylamine with arylbromides and a CuI catalyst at room temperature [30], we became interested in the possibility that this novel ligand might prove useful for preparing 3,5-bis-aminated pyrazolo[1,5-*a*]pyrimidines that were inaccessible via previously published methods (see Figures 2 and 3). Despite the fact that our previous L-proline/CuI method worked well for preparing C-3 mono-aminated derivatives from substrates that lacked a C-5 amino substituent (see Figure 2f) [26], the yields for converting C-5 amino substituted precursors (e.g., **29a** or **29b**) to 3,5-bis-aminated products **30** were consistently (and unacceptably) low (<35%), with much of the mass balance being attributed to the debrominated byproduct (e.g., **31**). In order to determine if the **L-1**/CuI catalyst system could overcome this limitation, we applied Huang's conditions, utilizing **29a** and morpholine as our model system (Table 1, entry 1). Disappointingly, **L-1**/CuI failed to produce any detectable amount of **30e**, and only unreacted starting material **29a** (with trace amounts of **31**) were detected, even after 7 days at RT (Table 1, entry 1). Heating the reaction at incrementally higher temperatures (50–70 °C, using an oil bath) for 1–2 days gave a slightly better conversion, but substantial quantities of unreacted starting material (70–92%) remained (Table 1, entries 2–3). Heating at 80 °C was required for the complete conversion of **29a**; however, 18% of the debrominated byproduct (**31**) was also obtained (Table 1, entry 4). Next, microwave heating was examined. Encouragingly, our first attempt with microwave heating (Table 1, entry 5) gave 61.4% **30e**, with 1.4% of the debrominated byproduct **31** after heating for only 30 min at 70 °C. Increasing the reaction time (from 30 min to 3h) at 70 °C gave increased conversions, but debromination also increased from 1.8–2.7% (Table 1, entries 5–9). Microwave heating at 80 °C for 1 h gave approximately equivalent results to the reaction run at 70 °C for 3 h (Table 1, entries 9 and 11), while a reaction time of 30 min at 80 °C left 4.4% unreacted **29a** (Table 1, entries 10–11). A longer reaction time at 80 °C (1.5 h) gave product with essentially identical conversion to that obtained after only 1 h (Table 1, entries 11–12). At this point, it seemed reasonable to perform a solvent survey in order to confirm that, in keeping with Huang's observations regarding arylation of cyclopropylamine with **L-1**/CuI, diethylene glycol (DEG) was the most suitable solvent [30]. Accordingly, six common polar solvents were evaluated (Table 1, entries 13–18). Of these solvents, ethylene glycol and 1,2-propanediol gave comparable results (90–94% conversion), but suffered from higher percentages of debromination (5.7–9.9%, Table 1, entries 13 and 18). NMP gave a lower yield (78%) with 19% debromination (Table 1, entry 17). The reaction in 1,4-dioxane failed to give any detectable C-3 aminated product (entry 16), while n-butanol and DMSO gave only 3.6–10.2% conversion, respectively (Table 1, entries 14–15). Next, a survey of four bases commonly used for transition-metal-catalyzed coupling was conducted (Table 1, entries 19–22). The inorganic base K_3PO_4 gave inferior conversion (94% compared to 97%), while tBuOK and TMSONa gave product **30e** in yields comparable to those obtained with K_2CO_3 (Table 1, entries 11 and 19–20). DBU afforded the least effective conversion (48.8%; Table 1, entry 22). Based on these data and the relative cost efficiency of K_2CO_3 compared to TMSONa (which is approx. 55-fold more expensive than K_2CO_3 on a 100 g basis), we opted to use the conditions in entry 11 (80 °C, 1 h, DEG, K_2CO_3) for further optimization and substrate scope determination (*vide infra*).

Table 1. Optimization. [a–e]

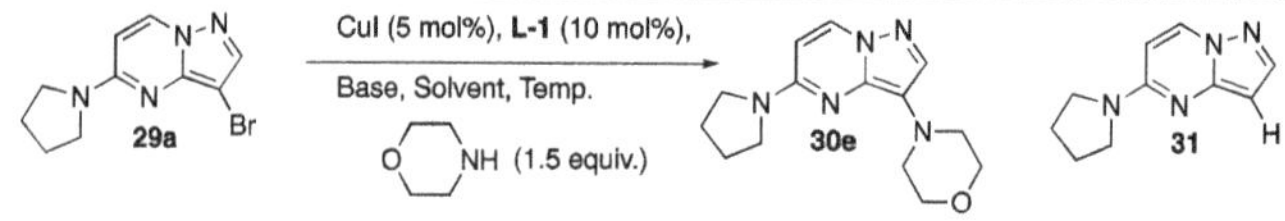

Entry	Temp. (°C)	Ligand	Time	Solvent	Base [c]	Products [d]
1	RT	**L-1**	7 days	DEG	K_2CO_3	**30e** (---) **31** (---) [e] **29a** (100%)
2	50 [a]	**L-1**	2 days	DEG	K_2CO_3	**30e** (7.7%) **31** (---) [e] **29a** (92.3%)
3	70 [a]	**L-1**	24 h	DEG	K_2CO_3	**30e** (27.3%) **31** (2.8%); **29a** (69.9%)
4	80 [a]	**L-1**	2 days	DEG	K_2CO_3	**30e** (82.4%) **31** (17.6%); **29a** (---)
5	70 [b]	**L-1**	30 min	DEG	K_2CO_3	**30e** (61.4%) **31** (1.8%); **29a** (36.8%)
6	70 [b]	**L-1**	1 h	DEG	K_2CO_3	**30e** (78.7%) **31** (2.4%); **29a** (18.9%)
7	70 [b]	**L-1**	1.5 h	DEG	K_2CO_3	**30e** (87.2%) **31** (2.1%); **29a** (10.7%)
8	70 [b]	**L-1**	2 h	DEG	K_2CO_3	**30e** (96.7%) **31** (2.5%); **29a** (0.8%)
9	70 [b]	**L-1**	3 h	DEG	K_2CO_3	**30e** (97.3%) **31** (2.7%); **29a** (---)
10	80 [b]	**L-1**	30 min	DEG	K_2CO_3	**30e** (93.7%) **31** (1.9%); **29a** (4.4%)
11	80 [b]	**L-1**	1 h	DEG	K_2CO_3	**30e** (97.1%) **31** (2.9%); **29a** (---)
12	80 [b]	**L-1**	1.5 h	DEG	K_2CO_3	**30e** (97.0%) **31** (3.0%); **29a** (---)
13	80 [b]	**L-1**	1 h	Ethylene Glycol	K_2CO_3	**30e** (90.1%) **31** (9.9%); **29a** (---)
14	80 [b]	**L-1**	1 h	DMSO	K_2CO_3	**30e** (10.2%) **31** (5.1%); **29a** (84.7%)
15	80 [b]	**L-1**	1 h	^{n} Butanol	K_2CO_3	**30e** (3.6%) **31** (6.3%); **29a** (90.1%)
16	80 [b]	**L-1**	1 h	Dioxane	K_2CO_3	**30e** (---) **31** (7.2%); **29a** (92.8%)
17	80 [b]	**L-1**	1 h	NMP	K_2CO_3	**30e** (78%) **31** (19%); **29a** (3%)

Table 1. *Cont.*

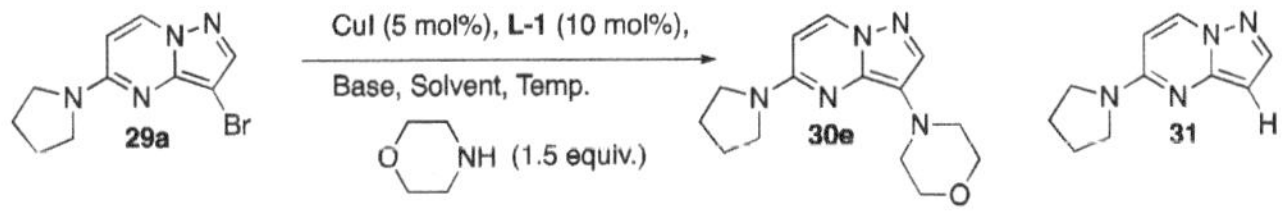

Entry	Temp. (°C)	Ligand	Time	Solvent	Base [c]	Products [d]
18	80 [b]	**L-1**	1 h	1,2-Propanediol	K_2CO_3	**30e** (94.3%) **31** (5.7%); **29a** (---)
19	80 [b]	**L-1**	1 h	DEG	tBuOK	**30e** (95.2%) **31** (4.8%); **29a** (---)
20	80 [b]	**L-1**	1 h	DEG	TMSONa	**30e** (97.1%) **31** (2.9%); **29a** (---)
21	80 [b]	**L-1**	1 h	DEG	K_3PO_4	**30e** (94.3%) **31** (5.7%); **29a** (---)
22	80 [b]	**L-1**	1 h	DEG	DBU	**30e** (48.8%) **31** (1.4%); **29a** (49.8%)
23	80 [b]	**L-2**	1 h	DEG	K_2CO_3	**30e** (21%) **31** (39%); **29a** (40%)
24	80 [b]	**L-3**	1 h	DEG	K_2CO_3	**30e** (17.4%) **31** (0.7%); **29a** (81.9%)
25	80 [b]	**L-4**	1 h	DEG	K_2CO_3	**30e** (12.3%) **31** (---); **29a** (87.7%)

[a] Conventional oil bath heating; [b] Microwave heating; [c] 2.0 equiv.; [d] Percent conversion as determined by ^{1}HNMR; [e] Trace amounts (^{1}H NMR signal detected, but barely distinguishable from baseline noise).

2.1.2. Ligand and C-3 Halogenated Precursor

A brief survey of structurally related ligands (**L-2**, **L-3**, and **L-4**) further confirmed the suitability of the **L-1**/CuI catalyst system (Table 1, entries 23–25). Notably, **L-2** (which was recently shown to promote the efficient CuI-catalyzed arylation of phenols) [31] gave a nearly 2:1 ratio of debrominated **31**:**30e**, while **L-3** and **L-4** gave substantially less efficient conversions (12–17%) under our optimized microwave conditions. In order to confirm the suitability of the C-3 brominated precursors, we also screened 3-chloro and 3-iodo derivatives (**32a–b**) using our optimized conditions (Scheme 2). The reaction for the 3-chloro derivative was markedly inferior (<4% conversion), while the 3-iodo derivative gave excellent yields but suffered from greater dehalogenated byproduct (**31**) compared to the 3-bromo derivative **29a** (5.4 and 2.9%, respectively).

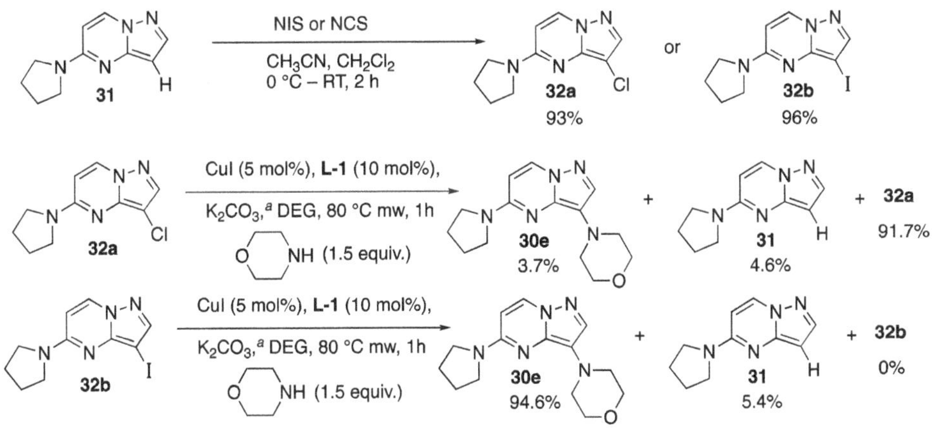

Scheme 2. C-3 amination with 3-chloro or 3-iodo precursors. [a] (2.0 equiv).

3. Discussion

Using the optimized conditions established in our preliminary experiments (Table 1 and Scheme 2), we applied this methodology to the substrates listed in Table 2. We were pleased to note the broad substrate scope and functional group compatibility exemplified by a variety of 1° alkyl and 2° alkylamines and aryl- or heteroarylamines, with key functionalities such as azides or BOC groups being well tolerated (Table 2). However, amides and ureas did not couple efficiently with our method (and the coupling products from those substrates formed in only trace amounts under our optimized conditions). Notwithstanding, the substrate scope presented in Table 2 is the broadest and most comprehensive scope of 3,5-bis-aminated products yet to be reported for the pyrazolo[1,5-*a*]pyrimidine scaffold coupled with alkyl- or arylamines [21–28]. Previous methods have been restricted to only C-3 mono-aminated derivates or an extremely limited number of 3,5-bis-aminated products, as illustrated in Figure 3. (The reactions highlighted in Figure 3 are summarized from an exhaustive Scifinder Scholar search conducted in November 2024). These methods were limited to either C-3 ureas, urethanes, or the $(CH_3)_2N$ group, each of which was obtained via a somewhat circuitous approach requiring C-3 nitration, followed by reduction, with a final step involving either acylation or reductive amination. In our case, C-5 alkylamine- or arylamine-substituted precursors were obtained directly in one step from commercially available 3-bromo-5-chloropyrazolo[1,5-*a*]pyrimidine (Scheme 1). C-3 amination of both C-5 alkylamine and C-5 arylamine substrates worked equally well, and **29a–b** and **29c–e** gave 3,5-bis-aminated products in 82.2% and 83.1% (average yields, respectively). Thus, a broad range of 3,5-bis-aminated products, consisting of both 3,5-bis arylamine, 3,5-bis alkylamine, or 3,5-bis alkyl/aryl amine combinations, could be efficiently obtained in only two steps utilizing our method. The functional group tolerance is noteworthy, given the significant potential that derivatives such as **30d**, **30g**, **30n**, and **30z–c′** have for further elaboration into more complex libraries of compounds. Facile TFA-promoted cleavage of the BOC protecting groups or reduction of the azide would provide the corresponding amine products for further functionalization, and the aryl chloride moieties of **30z–c′** could also be used for Suzuki-Miyaura coupling. An additional advantage of azide **30n** is its potential use in conventional "click" coupling chemistry for conjugating biologically relevant payloads for biological applications.

3,5-Bis-aminated derivatives with other C-5 amino groups could also be envisioned since model substrates **29a–e** establish that a variety of alkyl or aryl amines at this position are well tolerated. Advantages of this method include rapid reaction time (1 h for C-3 amination), high yield (83% average for all 29 substrates examined), avoidance of toxic reagents such as paraformaldehyde and $NaBH_3CN$, and a short reaction sequence (only two steps from commercially available 3-bromo-5-chloropyrazolo[1,5-*a*]pyrimidine). In addition, our method avoids expensive air- and moisture-sensitive palladium catalysts and provides much greater synthetic versatility than any previously reported method (Figures 2 and 3). Perhaps the most striking drawback of the previous methods for 3,5-bis-aminated pyrazolo[1,5-*a*]pyrimidines is their limited scope (limited to C-3 ureas or dimethylamine). Our method opens up the possibility of preparing much broader, diversity-rich compound libraries for screening novel and/or enhanced biological, medicinal, or fluorometric properties.

Table 2. Substrate scope for C-3 amination. [a,b,c]

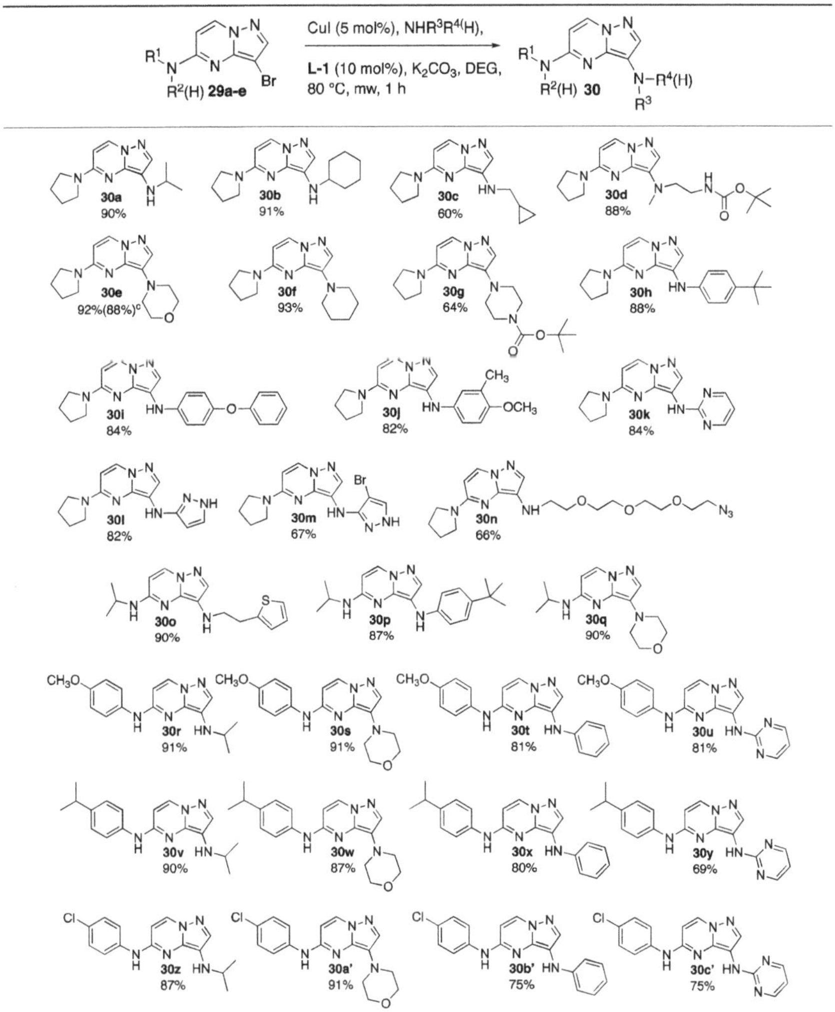

[a] Reaction conditions: **29a–e** (50 mg, 0.15–0.20 mmol), amine (1.5 equiv.), K_2CO_3 (2.0 equiv.), CuI (5 mol%), **L-1** (10 mol%), diethyleneglycol (DEG) (1.0 mL); [b] Isolated yields; [c] Reaction performed on 1.0 g (3.80 mmol) scale.

4. Materials and Methods

4.1. General Experimental

All reactions were performed in either appropriately sized pressure vessels (conventional heating) or in 10 mL microreactor vessels (microwave heating) without previous drying, unless otherwise noted. Conventional heating was performed in an oil bath, and microwave heating was achieved using a CEM Microwave reactor. Reagent-grade solvents and all other reagents were used directly as supplied by commercial suppliers without additional drying or purification. ^{1}H NMR and ^{13}C NMR spectra were determined in $CDCl_3$ or DMSO-d_6 using internal references: δ 7.27 or 2.50 (^{1}H) and δ 77.23 or 39.52 (^{13}C), respectively. Abbreviations such as "d", "t", or "bs" refer to "apparent doublet", "apparent triplet", or "apparent broad singlet", where such apparent multiplicities represent experimentally observed splitting patterns that result from line-broadening caused by conformational isomerism and/or concentration. High-resolution mass spectra were obtained using fast atom bombardment electrospray (ES) ionization techniques and had errors of less than ±5 ppm.

4.2. Ligand Synthesis

The method for preparing ligands **L-1**, **L-2**, **L-3**, and **L-4** is a modification of the method reported by Huang and co-workers [30] (Supplementary Materials, Scheme S1). In our hands, their method for preparing **L-1**, which employed $SOCl_2$ for generating an acyl chloride from pyrrole 2-carboxylic acid, consistently gave yields <35%, even after rigorous drying of glassware, solvents, and reagents. We employed Ph_3P/NIS to generate a reactive acyl iodide intermediate [32,33] and obtained the desired amidation of N-amino-9H-carbazole [34] in higher yields (38–67%; ave = 49%). (Scheme S1).

4.3. 5-Amino-3-Bromopyrazolo[1,5-a]Pyrimidines 29

General Procedure: A solution of **28** (200 mg, 0.86 mmol), pyrrolidine (123 mg, 1.73 mmol, 2.0 equiv), CsF (131 mg, 0.86 mmol, 1.0 equiv), and $BnNEt_3Cl$ (20 mg, 0.088 mmol, 10 mol%), in DMSO (6 mL) was stirred in a sealed pressure vessel (15 mL) and heated at 100 °C for 8 h [32,35]. After cooling to ambient temperature, H_2O (60 mL) was added. The mixture was diluted with ethyl acetate and the organic layer was separated. The aqueous layer was partitioned with EtOAc (2 × 60 mL) and the combined organic layers were washed with saturated sodium bicarbonate solution (2 × 60 mL) and water (2 × 60 mL). The organic layer was dried over anhydrous magnesium sulfate, filtered, and evaporated under reduced pressure. The crude product was purified using flash chromatography (EtOAc/Hexanes) to give pure **29a** (222 mg, 0.83 mmol, 97%).

3-Bromo-5-(pyrrolidin-1-yl)pyrazolo[1,5-a]pyrimidine (29a)

3-Bromo-5-chloropyrazolo[1,5-*a*]pyrimidine (200 mg, 0.86 mmol), pyrrolidine (123 mg, 1.73 mmol, 2.0 equiv), CsF (131 mg, 0.86 mmol, 1.0 equiv), $BnNEt_3Cl$ (20 mg, 0.088 mmol, 10 mol%), DMSO (6 mL); **29a** (222 mg, 0.83 mmol, 97%); ^1H NMR (DMSO-d_6, 500 MHz) δ 8.70 (d, J = 7.7 Hz, 1H), 8.01 (s, 1H), 6.53 (d, J = 7.7 Hz, 1H), 3.62 ("bs", 4H), 2.06 ("bs", 4H); ^{13}C NMR (DMSO-d_6, 125 MHz) δ 154.5, 145.5, 144.0, 136.2, 99.2, 76.9, 47.2, 25.7, 24.9; HRMS Calcd. for $C_{10}H_{12}BrN_4$ [M + H]: 267.0245; Found: 267.0237 (Δ = 3.0 ppm).

3-Bromo-5-(N-(isopropyl)amino)pyrazolo[1,5-a]pyrimidine (29b)

3-Bromo-5-chloropyrazolo[1,5-*a*]pyrimidine (100 mg, 0.43 mmol), isopropylamine (51 mg, 0.86 mmol, 2.0 equiv), CsF (65 mg, 0.43 mmol, 1.0 equiv), $BnNEt_3Cl$ (10 mg, 0.044 mmol,10 mol%), DMSO (3 mL); **29b** (105 mg, 0.41 mmol, 96%); ^1H NMR (DMSO-d_6, 500 MHz) δ 8.42 (d, J = 7.5 Hz, 1H), 7.85 (s, 1H), 7.54 (d, J = 7.4 Hz, 1H), 6.23 (d, J = 7.4 Hz, 1H), 4.15 ("d", J = 5.0 Hz, 1H), 1.19 (d, J = 6.5 Hz, 6H); ^{13}C NMR (DMSO-d_6, 125 MHz) δ 155.8, 145.6, 143.5, 135.6, 101.6, 77.5, 42.1, 22.5; HRMS Calcd. for $C_9H_{12}BrN_4$ [M + H]: 255.0245; Found: 255.0237 (Δ = 3.1 ppm).

3-Bromo-5-(4-(methoxy)anilin-1-yl)pyrazolo[1,5-a]pyrimidine (29c)

3-Bromo-5-chloropyrazolo[1,5-*a*]pyrimidine (400 mg, 1.73 mmol), 4-methoxyaniline (426 mg, 3.46 mmol, 2.0 equiv), CsF (263 mg, 1.73 mmol, 1.0 equiv), $BnNEt_3Cl$ (39 mg, 0.172 mmol,10 mol%), DMSO (12 mL); **29c** (473 mg, 1.49 mmol, 86%); ^1H NMR (DMSO-d_6, 500 MHz) δ 9.73 (s, 1H), 8.62 (d, J = 7.6 Hz, 1H), 7.99 (s, 1H), 7.81 (d, J = 8.8 Hz, 2H), 6.96 (d, J = 9.0 Hz, 2H), 6.50 (d, J = 7.6 Hz, 1H), 3.76 (s, 3H); ^{13}C NMR (DMSO-d_6, 125 MHz) δ 155.3, 153.9, 144.9, 143.9, 136.4, 133.5, 121.2, 114.5, 102.1, 78.8, 55.7; HRMS Calcd. for $C_{13}H_{12}BrN_4O$ [M + H]: 319.0194; Found: 319.0181 (Δ = 4.1 ppm).

3-Bromo-5-(4-(isopropyl)anilin-1-yl)pyrazolo[1,5-a]pyrimidine (29d)

3-Bromo-5-chloropyrazolo[1,5-*a*]pyrimidine (400 mg, 0.43 mmol), 4-isopropylaniline (467 mg, 3.46 mmol, 2.0 equiv), CsF (263 mg, 1.73 mmol, 1.0 equiv), $BnNEt_3Cl$ (39 mg, 0.172 mmol,10 mol%), DMSO (12 mL); **29d** (525 mg, 1.59 mmol, 92%); ^{1}H NMR (DMSO-d_6, 500 MHz) δ 9.82 (s, 1H), 8.61 (d, *J* = 7.6 Hz, 1H), 7.99 (s, 1H), 7.83 (d, *J* = 8.5 Hz, 2H), 7.22 (d, *J* = 8.3 Hz, 2H), 6.53 (d, *J* = 7.6 Hz, 1H), 2.83 (sept, *J* = 6.9 Hz, 1H), 1.18 (d, *J* = 6.9 Hz, 6H); ^{13}C NMR (DMSO-d_6, 125 MHz) δ 153.9, 144.8, 143.9, 143.2, 138.0, 136.4, 127.0, 119.8, 102.3, 79.0, 33.3, 24.4; HRMS Calcd. for $C_{15}H_{16}BrN_4$ [M + H]: 331.0558; Found: 331.0556 (Δ = 0.6 ppm).

3-Bomo-5-(4-(chloro)anilin-1-yl)-pyrazolo[1,5-a]pyrimidine (29e)

3-Bromo-5-chloropyrazolo[1,5-*a*]pyrimidine (400 mg, 1.73 mmol), 4-chloroaniline (439 mg, 3.46 mmol, 2.0 equiv), CsF (263 mg, 1.73 mmol, 1.0 equiv), $BnNEt_3Cl$ (39 mg, 0.172 mmol,10 mol%), DMSO (12 mL); **29e** (312 mg, 0.97 mmol, 56%); δ 10.04 (s, 1H), 8.67 (d, *J* = 7.5 Hz, 1H), 8.03 (s, 1H), 7.93 (d, *J* = 8.9 Hz, 2H), 7.41 (d, *J* = 8.9 Hz, 2H), 6.55 (d, *J* = 7.6 Hz, 1H); ^{13}C NMR (DMSO-d_6, 125 MHz) δ 153.7, 144.5, 144.1, 139.3, 136.8, 129.1, 126.5, 121.0, 102.3, 79.4; HRMS Calcd. for $C_{12}H_9BrClN_4$ [M + H]: 322.9699; Found: 322.9694 (Δ = 1.5 ppm).

4.4. 3,5-Bis-Aminopyrazolo[1,5-a]Pyrimidines 30

General Procedure: A solution of 3-Bromo-5-(pyrrolidin-1-yl)pyrazolo[1,5-*a*]pyrimidine (**29a**) (50 mg, 0.19 mmol), morpholine (25 mg, 0.29 mmol, 1.5 equiv.), **L-1** (6 mg, 0.02 mmol, 10 mol%), CuI (2 mg, 0.01 mmol, 5 mol%), and K_2CO_3 (52 mg, 0.38 mmol, 2 equiv.) in diethyleneglycol (DEG, 1 mL) was stirred in a 10 mL microwave reactor vessel and heated at 80 °C for 1 h. After cooling to ambient temperature, H_2O (5 mL) was added. The mixture was diluted with CH_2Cl_2 (5 mL) and the organic layer was separated. The aqueous layer was partitioned with CH_2Cl_2 (2 × 10 mL), and the combined organic layers were washed with brine (2 × 10 mL), followed by water (2 × 10 mL). The organic layer was dried over anhydrous magnesium sulfate, filtered, and removed under reduced pressure. The crude product was then purified using flash chromatography (EtOAc/Hexanes or MeOH/CH_2Cl_2) to give pure product **30e** (48 mg, 0.176 mmol, 92%).

3-(N-(Isopropyl)amino)-5-(pyrrolidin-1-yl)pyrazolo[1,5-a]pyrimidine (30a)

3-Bromo-5-(pyrrolidin-1-yl)pyrazolo[1,5-*a*]pyrimidine (50 mg, 0.19 mmol), isopropylamine (17 mg, 0.29 mmol, 1.5 equiv.), **L-1** (6 mg, 0.02 mmol, 10 mol%), CuI (2 mg, 0.01 mmol, 5 mol%), and K_2CO_3 (52 mg, 0.38 mmol, 2 equiv.); **30a** (42 mg, 0.171 mmol, 90%); ^{1}H NMR (DMSO-d_6, 500 MHz) δ 8.39 (d, *J* = 7.8 Hz, 1H), 7.51 (s, 1H), 6.23 (d, *J* = 7.7 Hz, 1H), 3.47–3.41 (m, 5H), 1.94 (t, *J* = 6.4 Hz, 4H), 1.07 (d, *J* = 6.3 Hz, 6H); ^{13}C NMR (DMSO-d_6, 125 MHz) δ 152.2, 138.7, 136.6, 135.6, 116.3, 97.7, 47.8, 47.0, 25.4, 23.7; HRMS Calcd. for $C_{13}H_{20}N_5$ [M + H]: 246.1719; Found: 246.1712 (Δ = 2.8 ppm).

3-(N-(Cyclohexyl)amino)-5-(pyrrolidin-1-yl)pyrazolo[1,5-a]pyrimidine (30b)

3-Bromo-5-(pyrrolidin-1-yl)pyrazolo[1,5-*a*]pyrimidine (50 mg, 0.19 mmol), cyclohexylamine (29 mg, 0.29 mmol, 1.5 equiv.), **L-1** (6 mg, 0.02 mmol, 10 mol%), CuI (2 mg, 0.01 mmol, 5 mol%), and K_2CO_3 (52 mg, 0.38 mmol, 2 equiv.); **30b** (49 mg, 0.172 mmol, 91%); ^{1}H NMR (DMSO-d_6, 500 MHz) δ 8.38 (d, *J* = 7.7 Hz, 1H), 7.50 (s, 1H), 6.22 (d, *J* = 7.7 Hz, 1H), 3.47 ('bs', 4H), 3.12 (t, *J* = 9.7 Hz, 1H), 1.95 (t, *J* = 6.3 Hz, 4H), 1.89–1.87 (m, 2H), 1.70– 1.68 (m, 2H), 1.57–1.55 (m, 1H), 1.26–1.09 (m, 5H); ^{13}C NMR (DMSO-d_6, 125 MHz) δ 152.1, 138.5, 136.6, 135.6, 115.9, 97.7, 72.7, 60.7, 55.4, 47.0, 33.8, 26.2, 25.4, 25.1; HRMS Calcd. for $C_{16}H_{24}N_5$ [M + H]: 286.2032; Found: 286.2024 (Δ = 2.8 ppm).

3-(N-((Cyclopropyl)methyl)amino)-5-(pyrrolidin-1-yl)pyrazolo[1,5-a]pyrimidine (30c)

3-Bromo-5-(pyrrolidin-1-yl)pyrazolo[1,5-*a*]pyrimidine (50 mg, 0.19 mmol), cyclopropylamine (21 mg, 0.29 mmol, 1.5 equiv.), **L-1** (6 mg, 0.02 mmol, 10 mol%), CuI (2 mg, 0.01 mmol, 5 mol%), and K_2CO_3 (52 mg, 0.38 mmol, 2 equiv.); **30c** (29 mg, 0.113 mmol, 60%); ^{1}H NMR ($CDCl_3$, 500 MHz) δ 8.11 (d, *J* = 7.7 Hz, 1H), 7.61 (s, 1H), 6.04 (d, *J* = 7.2 Hz, 1H), 3.55 ("bs", 4H), 2.98 (d, *J* = 6.9 Hz, 2H), 2.02 (t, *J* = 6.6 Hz, 4H), 1.17–1.13 (m, 1H), 0.53 (dd, J = 12.6, 5.5 Hz, 2H), 0.23 (dd, J = 10, 4.9 Hz, 2H); ^{13}C NMR ($CDCl_3$, 125 MHz) δ 152.2, 138.7, 135.3, 134.7, 117.5, 97.2, 54.1, 46.9, 25.5, 11.6, 3.4; HRMS Calcd. for $C_{14}H_{20}N_5$ [M + H]: 258.1719; Found: 258.1704 (Δ = 2.7 ppm).

3-[2-(tert-Butyloxycarbonyl)amino)-1-N-methylamino)ethyl]-5-(pyrrolidin-1-yl)pyrazolo pyrimidine (30d)

3-Bromo-5-(pyrrolidin-1-yl)pyrazolo[1,5-*a*]pyrimidine (50 mg, 0.19 mmol), 1-(*tert*-butyloxycarbonylamino)-2-(*N*-methylamino)ethane (51 mg, 0.29 mmol, 1.5 equiv.), **L-1** (6 mg, 0.02 mmol, 10 mol%), CuI (2 mg, 0.01 mmol, 5 mol%), and K_2CO_3 (52 mg, 0.38 mmol, 2 equiv.); **30d** (60 mg, 0.167 mmol, 88%); ^{1}H NMR ($CDCl_3$, 500 MHz) δ 8.07 (d, *J* = 7.7 Hz, 1H), 7.67 (s, 1H), 6.42 (s, 1H), 6.02 (d, *J* = 7.7 Hz, 1H), 3.53 ("bs", 4H), 3.12–3.09 (m, 2H), 2.97–2.95 (m, 2H), 2.77 (s, 3H), 1.97 ("bs", 4H), 1.38 (s, 9H); ^{13}C NMR $CDCl_3$, 125 MHz) δ 156.2, 152.7, 141.5, 137.4, 135.1, 119.6, 97.6, 78.5, 57.6, 47.1, 43.0, 38.8, 28.6, 25.5; HRMS Calcd. for $C_{18}H_{29}N_6O_2$ [M + H]: 361.2352; Found: 361.2344 (Δ = 2.2 ppm).

3-(Morpholin-1-yl)-5-(pyrrolidin-1-yl)pyrazolo[1,5-a]pyrimidine (30e)

(a) 50 mg scale: 3-bromo-5-(pyrrolidin-1-yl)pyrazolo[1,5-*a*]pyrimidine **29a** (50 mg, 0.19 mmol), morpholine (25 mg, 0.29 mmol, 1.5 equiv.), **L-1** (6 mg, 0.02 mmol, 10 mol%), CuI (2 mg, 0.01 mmol, 5 mol%), and K_2CO_3 (52 mg, 0.38 mmol, 2 equiv.); **30e** (48 mg, 0.176 mmol, 92%); (b) 1.0 g scale: **29a** (1.0 g, 3.8 mmol), morpholine (496 mg, 5.7 mmol, 1.5 equiv.), **L-1** (105 mg, 0.38 mmol, 10 mol%), CuI (36 mg, 0.19 mmol, 5 mol%), and K_2CO_3 (1.04 g, 7.6 mmol, 2 equiv.); **30e** (913 mg, 3.34 mmol, 88%); ^{1}H NMR (DMSO-d_6, 500 MHz) δ 8.45 (d, *J* = 7.8 Hz, 1H), 7.58 (s, 1H), 6.30 (d, *J* = 7.7 Hz, 1H), 3.74 (t, *J* = 4.6 Hz, 4H), 3.47 (t, *J* = 6.6 Hz, 4H), 3.09 (t, *J* = 4.4 Hz, 4H), 1.95 ("bs", 4H); ^{13}C NMR ($CDCl_3$, 125 MHz) δ 152.0, 139.4, 134.8, 134.3, 120.6, 97.4, 67.0, 51.5, 46.8, 25.4; HRMS Calcd. for $C_{14}H_{20}N_5O$ [M + H]: 274.1668; Found: 274.1659 (Δ = 3.3 ppm).

3-(Piperidin-1-yl)-5-(pyrrolidin-1-yl)pyrazolo[1,5-a]pyrimidine (30f)

3-Bromo-5-(pyrrolidin-1-yl)pyrazolo[1,5-*a*]pyrimidine (50 mg, 0.19 mmol), piperidine (25 mg, 0.29 mmol, 1.5 equiv.), **L-1** (6 mg, 0.02 mmol, 10 mol%), CuI (2 mg, 0.01 mmol, 5 mol%), and K_2CO_3 (52 mg, 0.38 mmol, 2 equiv.); **30f** (48 mg, 0.177 mmol, 93%); ^{1}H NMR (DMSO-d_6, 500 MHz) δ 8.42 (d, *J* = 7.8 Hz, 1H), 7.56 (s, 1H), 6.28 (d, *J* = 7.7 Hz, 1H), 3.47 (t, *J* = 6.5 Hz, 4H), 3.05 (t, *J* = 5.1 Hz, 4H), 1.95 ("bs", 4H), 1.65–1.61 (m, 4H), 1.51– 1.46 (m, 2H); ^{13}C NMR (DMSO-d_6, 125 MHz) δ 151.9, 138.5, 135.7, 134.5, 121.7, 97.9, 52.2, 46.9, 25.9, 25.4, 24.4; HRMS Calcd. for $C_{15}H_{22}N_5$ [M + H]: 272.1875; Found: 272.1875 (Δ = 2.9 ppm).

tert-Butyl 4-(5-(pyrrolidin-1-yl)pyrazolo[1,5-a]pyrimidin-3-yl)piperazine-1-carboxylate (30g)

3-Bromo-5-(pyrrolidin-1-yl)pyrazolo[1,5-*a*]pyrimidine (50 mg, 0.19 mmol), *tert*-butyl piperazine-1-carboxylate (54 mg, 0.29 mmol, 1.5 equiv.), **L-1** (6 mg, 0.02 mmol, 10 mol%), CuI (2 mg, 0.01 mmol, 5 mol%), and K_2CO_3 (52 mg, 0.38 mmol, 2 equiv.); **30g** (45 mg, 0.120 mmol, 64%); ^{1}H NMR (DMSO-d_6, 500 MHz) δ 8.44 (d, *J* = 7.7 Hz, 1H), 7.59 (s, 1H), 6.29 (d, *J* = 7.7 Hz, 1H), 3.48–3.30 (m, 8H), 3.04 (t, *J* = 4.7 Hz, 4H), 1.94 ("bs", 4H), 1.42 (s, 9H); ^{13}C NMR (DMSO-d_6, 125 MHz) δ 154.4, 152.1, 138.8, 135.8, 134.8, 120.2, 98.1, 79.3, 50.9, 47.0, 28.5, 25.4; HRMS Calcd. for $C_{19}H_{29}N_6O_2$ [M + H]: 373.2352; Found: 373.2347 (Δ = 1.3 ppm).

3-(4-(tert-Butyl)anilin-1-yl)-5-(pyrrolidin-1-yl)pyrazolo[1,5-a]pyrimidine (30h)

3-Bromo-5-(pyrrolidin-1-yl)pyrazolo[1,5-*a*]pyrimidine (50 mg, 0.19 mmol), 4-*tert*-butylaniline (43 mg, 0.29 mmol, 1.5 equiv.), **L-1** (6 mg, 0.02 mmol, 10 mol%), CuI (2 mg, 0.01 mmol, 5 mol%), and K_2CO_3 (52 mg, 0.38 mmol, 2 equiv.); **30h** (56 mg, 0.167 mmol, 88%); ^{1}H NMR (DMSO-d_6, 500 MHz) δ 8.53 (d, *J* = 7.7 Hz, 1H), 7.78 (s, 1H), 7.07 (d, *J* = 8.7 Hz, 2H), 6.87 (s, 1H), 6.55 (dd, *J* = 6.8, 1.9 Hz, 2H), 6.35 (d, *J* = 7.8Hz, 1H), 3.45 (t, *J* = 6.4 Hz, 4H), 1.92 ("bs", 4H), 1.21 (s, 9H); ^{13}C NMR (DMSO-d_6, 125 MHz) δ 153.4, 146.7, 143.3, 142.4, 138.8, 136.0, 125.8, 112.7, 108.4, 98.3, 47.2, 33.9, 32.0, 25.4; HRMS Calcd. for $C_{20}H_{26}N_5$ [M + H]: 336.2188; Found: 336.2196 (Δ = 2.4 ppm).

3-(4-(Phenoxy)anilin-1-yl)-5-(pyrrolidin-1-yl)pyrazolo[1,5-a]pyrimidine (30i)

3-Bromo-5-(pyrrolidin-1-yl)pyrazolo[1,5-*a*]pyrimidine (50 mg, 0.19 mmol), 4-phenoxyaniline (54 mg, 0.29 mmol, 1.5 equiv.), **L-1** (6 mg, 0.02 mmol, 10 mol%), CuI (2 mg, 0.01 mmol, 5 mol%), and K_2CO_3 (52 mg, 0.38 mmol, 2 equiv.); **30i** (59 mg, 0.160 mmol, 84%); ^{1}H NMR (DMSO-d_6, 500 MHz) δ 8.54 (d, *J* = 7.7 Hz, 1H), 7.82 (s, 1H), 7.30 (dd, *J* = 8.4, 7.6 Hz, 2H), 7.10 (s, 1H), 7.00 (t, *J* = 7.4 Hz, 1H), 6.86 (d, *J* = 7.8 Hz, 2H), 6.82 (dd, *J* = 12.2, 3.3 Hz, 2H), 6.66–6.63 (m, 2H), 6.36 (d, *J* = 7.8 Hz, 1H), 3.46 (t, *J* = 6.4 Hz, 4H) 1.93 ("bs", 4H); ^{13}C NMR (DMSO-d_6, 125 MHz) δ 159.4, 153.4, 146.3, 145.9, 143.2, 142.3, 136.1, 130.1, 122.2, 121.3, 116.9, 114.1, 108.3, 98.3, 47.2, 25.3; HRMS Calcd. for $C_{22}H_{22}N_5O$ [M + H]: 372.1824; Found: 372.1830 (Δ = 1.6 ppm).

3-((4-Methoxy-3-methyl)anilin-1-yl)-5-(pyrrolidin-1-yl)pyrazolo[1,5-a]pyrimidine (30j)

3-Bromo-5-(pyrrolidin-1-yl)pyrazolo[1,5-*a*]pyrimidine (50 mg, 0.19 mmol), 4-methoxy-3-methylaniline (40 mg, 0.29 mmol, 1.5 equiv.), **L-1** (6 mg, 0.02 mmol, 10 mol%), CuI (2 mg, 0.01 mmol, 5 mol%), and K_2CO_3 (52 mg, 0.38 mmol, 2 equiv.); **30j** (50 mg, 0.156 mmol, 82%); ^{1}H NMR (DMSO-d_6, 500 MHz) δ 8.52 (d, *J* = 7.7 Hz, 1H), 7.76 (s, 1H), 6.66 (d, *J* = 8.7 Hz, 1H), 6.60 (s, 1H), 6.48 (d, *J* = 2.4 Hz, 1H), 6.40 (dd, *J* = 8.7, 2.7 Hz, 1H), 6.34 (d, *J* = 7.8 Hz, 1H), 3.65 (s, 3H), 3.45 (t, *J* = 6.4Hz, 4H), 2.03 (s, 3H), 1.93 ("bs", 4H); ^{13}C NMR (DMSO-d_6, 125 MHz) δ 153.3, 149.6, 143.0, 142.8, 142.1, 136.0, 126.3, 116.1, 112.0, 110.9, 109.2, 98.2, 56.1, 47.1, 25.4, 16.7; HRMS Calcd. for $C_{18}H_{22}N_5O$ [M + H]: 324.1824; Found: 324.1817 (Δ = 2.2 ppm).

3-(N-(Pyrimidin-2-yl)amino)-5-(pyrrolidin-1-yl)pyrazolo[1,5-a]pyrimidine (30k)

3-Bromo-5-(pyrrolidin-1-yl)pyrazolo[1,5-*a*]pyrimidine (50 mg, 0.19 mmol), 2-aminopy rimidine (28 mg, 0.29 mmol, 1.5 equiv.), **L-1** (6 mg, 0.02 mmol, 10 mol%), CuI (2 mg, 0.01 mmol, 5 mol%), and K_2CO_3 (52 mg, 0.38 mmol, 2 equiv.); **30k** (45 mg, 0.160 mmol, 84%); ^{1}H NMR (DMSO-d_6, 500 MHz) δ 8.53 (d, *J* = 7.7 Hz, 1H), 8.31 (t, *J* = 4.7 Hz, 3H), 7.95 (s, 1H), 6.66 (t, *J* = 4.8 Hz, 1H), 6.35 (d, *J* = 7.7 Hz, 1H), 3.47 (t, *J* = 6.6 Hz, 4H), 1.93 ("bs", 4H); ^{13}C NMR (DMSO-d_6, 125 MHz) δ 162.3, 158.5, 153.4, 142.1, 141.6, 135.8, 111.4, 105.9, 98.3, 47.1, 25.4; HRMS Calcd. for $C_{14}H_{16}N_7$ [M + H]: 282.1467; Found: 282.1459 (Δ = 2.8 ppm).

3-(N-(Pyrazol-3-yl)amino)-5-(pyrrolidin-1-yl)pyrazolo[1,5-a]pyrimidine (30l)

3-Bromo-5-(pyrrolidin-1-yl)pyrazolo[1,5-*a*]pyrimidine (50 mg, 0.19 mmol), 3-aminopyr azole (24 mg, 0.29 mmol, 1.5 equiv.), **L-1** (6 mg, 0.02 mmol, 10 mol%), CuI (2 mg, 0.01 mmol, 5 mol%), and K_2CO_3 (52 mg, 0.38 mmol, 2 equiv.); **30l** (42 mg, 0.156 mmol, 82%); ^{1}H NMR (DMSO-d_6, 500 MHz) δ 8.64 (d, *J* = 7.8 Hz, 1H), 8.00 (s, 1H), 7.26 (d, *J* = 1.8 Hz, 1H), 6.47 (d, *J* = 7.8 Hz, 1H), 5.65 (s, 2H), 5.41 (d, *J* = 1.8 Hz, 1H), 3.50 ("bs", 4H), 1.97 ("bs", 4H); ^{13}C NMR (DMSO, 125 MHz) δ 153.7, 148.0, 140.4, 139.7, 139.4, 136.6, 108.1, 98.8, 88.7, 47.3, 25.8; HRMS Calcd. for $C_{13}H_{16}N_7$ [M + H]: 270.1467; Found: 270.1460 (Δ = 2.6 ppm).

3-(N-(5-Bromopyrazol-3-yl)amino)-5-(pyrrolidin-1-yl)pyrazolo[1,5-a]pyrimidine (30m)

3-Bromo-5-(pyrrolidin-1-yl)pyrazolo[1,5-*a*]pyrimidine (50 mg, 0.19 mmol), 3-amino-5-bromopyrazole (47 mg, 0.29 mmol, 1.5 equiv.), **L-1** (6 mg, 0.02 mmol, 10 mol%), CuI (2 mg, 0.01 mmol, 5 mol%), and K_2CO_3 (52 mg, 0.38 mmol, 2 equiv.); **30m** (44 mg, 0.127 mmol, 67%); ^{1}H NMR (DMSO-d_6, 500 MHz) δ 8.65 (d, J = 7.8 Hz, 1H), 8.01 (s, 1H), 7.41 (s, 1H), 6.48 (d, J = 7.8 Hz, 1H), 5.79 (bs, 1H), 5.76 (s, 1H), 3.51 ("bs", 4H), 1.98 ("bs", 4H); ^{13}C NMR (DMSO-d_6, 125 MHz) δ 153.8, 145.4, 139.9, 139.6, 136.6, 107.6, 99.0, 74.0, 47.4, 25.8, 24.8; HRMS Calcd. for $C_{13}H_{15}N_7Br$ [M + H]: 348.0572; Found: 348.0566 (Δ = 1.7 ppm).

N-(2-(2-(2-(2-Azidoethoxy)ethoxy)ethoxy)ethyl)-5-(pyrrolidin-1-yl)pyrazolo[1,5-a]pyrimidin-3-amine (30n)

3-Bromo-5-(pyrrolidin-1-yl)pyrazolo[1,5-*a*]pyrimidine (50 mg, 0.19 mmol), 2-(2-(2-(2-azidoethoxy)ethoxy)ethoxy)ethan-1-amine (63 mg, 0.29 mmol, 1.5 equiv.), **L-1** (6 mg, 0.02 mmol, 10 mol%), CuI (2 mg, 0.01 mmol, 5 mol%), and K_2CO_3 (52 mg, 0.38 mmol, 2 equiv.); **30n** (51 mg, 0.126 mmol, 66%); ^{1}H NMR ($CDCl_3$, 500 MHz) δ 8.02 (d, J = 7.7 Hz, 1H), 7.60 (s, 1H), 6.04 (d, J = 7.7 Hz, 1H), 3.70–3.66 (m, 12H), 3.54 ("bs", 4H), 3.37 (t, J = 5.0 Hz, 2H), 3.33 (t, J = 5.3 Hz, 2H), 2.03–2.00 (m, 4H); ^{13}C NMR ($CDCl_3$, 125 MHz) δ 152.1, 137.5, 135.6, 134.7, 117.6, 97.8, 70.30, 70.29, 70.23, 70.17, 70.13, 69.7, 50.4, 47.3, 47.0, 25.4; HRMS Calcd. for $C_{18}H_{29}N_8O_3$ [M + H]: 405.2363; Found: 405.2382 (Δ = 4.7 ppm).

5-(N-(Isopropyl)amino)-3-(2-N-(thiophen-2-yl)ethylamino)pyrazolo[1,5-a]pyrimidine (30o)

3-Bromo-5-(N-(isopropyl)amino)pyrazolo[1,5-*a*]pyrimidine (50 mg, 0.20 mmol), 2-(thiophen-2-yl)ethan-1-amine (38 mg, 0.30 mmol, 1.5 equiv.), **L-1** (6 mg, 0.02 mmol, 10 mol%), CuI (2 mg, 0.01 mmol, 5 mol%), and K_2CO_3 (52 mg, 0.38 mmol, 2 equiv.); **30o** (54 mg, 0.180 mmol, 90%); ^{1}H NMR (DMSO-d_6, 500 MHz) δ 8.22 (d, J = 7.6 Hz, 1H), 7.47 (s, 1H), 7.31 (dd, J = 5.1, 1.2 Hz, 1H), 7.02 (d, J = 7.3 Hz, 1H), 6.94 (dd, J = 5.1, 3.4 Hz, 1H), 6.90 (dd, J = 3.3 Hz, Jz = 1.0 Hz, 1H), 6.05 (d, J = 7.6 Hz, 1H), 4.12–4.06 (m, 1H), 3.82 (s, 1H), 3.32 (t, J = 7.1 Hz, 2H), 3.02 (t, J = 7.2 Hz, 2H), 1.17 (d, J = 6.5 Hz, 6H); ^{13}C NMR (DMSO-d_6, 125 MHz) δ 153.1, 142.8, 137.8, 135.0, 134.8, 127.4, 125.4, 124.1, 117.3, 100.3, 49.6, 42.0, 30.5, 22.7; HRMS Calcd. for $C_{15}H_{20}N_5S$ [M + H]: 302.1439; Found: 302.1428 (Δ = 3.6 ppm).

3-(4-(tert-Butyl)anilin-1-yl)-5-(N-(isopropyl)amino)pyrazolo[1,5-a]pyrimidine (30p)

3-Bromo-5-(N-(isopropyl)amino)pyrazolo[1,5-*a*]pyrimidine (50 mg, 0.20 mmol), 4-*tert*-butylaniline (45 mg, 0.30 mmol, 1.5 equiv.), **L-1** (6 mg, 0.02 mmol, 10 mol%), CuI (2 mg, 0.01 mmol, 5 mol%), and K_2CO_3 (52 mg, 0.38 mmol, 2 equiv.); **30p** (56 mg, 0.173 mmol, 87%); ^{1}H NMR (DMSO-d_6, 500 MHz) δ 8.35 (d, J = 7.6 Hz, 1H), 7.71 (s, 1H), 7.22 (d, J = 7.6 Hz, 1H), 7.06 (d, J = 8.7 Hz, 2H), 6.84 (s, 1H), 6.56 (d, J = 8.7 Hz, 2H), 6.16 (d, J = 7.6 Hz, 1H), 4.06–4.00 (m, 1H), 1.20 (s, 9H), 1.12 (d, J = 6.5 Hz, 6H); ^{13}C NMR (DMSO-d_6, 125 MHz) δ 154.5, 146.5, 143.1, 141.6, 138.9, 135.5, 125.7, 112.9, 108.9, 100.6, 41.9, 33.9, 31.9, 22.7; HRMS Calcd. for $C_{19}H_{26}N_5$ [M + H]: 324.2188; Found: 324.2174 (Δ = 4.3 ppm).

5-(N-(Isopropyl)amino)-3-(morpholin-1-yl)pyrazolo[1,5-a]pyrimidine (30q)

3-Bromo-5-(N-(isopropyl)amino)pyrazolo[1,5-*a*]pyrimidine (50 mg, 0.20 mmol), morpholine (26 mg, 0.30 mmol, 1.5 equiv.), **L-1** (6 mg, 0.02 mmol, 10 mol%), CuI (2 mg, 0.01 mmol, 5 mol%), and K_2CO_3 (52 mg, 0.38 mmol, 2 equiv.); **30q** (47 mg, 0.180 mmol, 90%); ^{1}H NMR (DMSO-d_6, 500 MHz) δ 8.28 (d, J = 7.6 Hz, 1H), 7.53 (s, 1H), 7.14 (d, J = 7.0 Hz, 1H), 6.11 (d, J = 7.6 Hz, 1H), 4.06–4.00 (m, 1H), 3.72 (t, J = 4.6 Hz, 4H), 3.08 ("bs", 4H), 1.17 (d, J = 6.5 Hz, 6H); ^{13}C NMR (DMSO-d_6, 125 MHz) δ 152.0, 137.5, 134.3, 133.0, 119.8, 99.3, 65.5, 50.6, 41.1, 21.5; HRMS Calcd. for $C_{13}H_{20}N_5O$ [M + H]: 262.1668; Found: 262.1656 (Δ = 4.6 ppm).

3-(N-(Isopropyl)amino)-5-(4-(methoxy)anilin-1-yl)pyrazolo[1,5-a]pyrimidine (30r)

3-Bromo-5-(*4-(methoxy)anilin-1-yl*)-pyrazolo[1,5-*a*]pyrimidine (50 mg, 0.16 mmol), isopropylamine (14 mg, 0.24 mmol, 1.5 equiv.), **L-1** (4.4 mg, 0.016 mmol, 10 mol%), CuI (1.5 mg, 0.008 mmol, 5 mol%), and K_2CO_3 (44 mg, 0.32 mmol, 2 equiv.); **30r** (43 mg, 0.145 mmol, 91%); ^{1}H NMR (DMSO-d_6, 500 MHz) δ 9.31 (s, 1H), 8.40 (d, *J* = 7.5 Hz, 1H), 7.78 (d, *J* = 8.9 Hz, 2H), 7.55 (s, 1H), 6.91 (d, *J* = 8.9 Hz, 2H), 6.29 (d, *J* = 7.6 Hz, 1H), 3.74 (s, 3H), 3.65–3.60 (bs, 1H), 3.53 (sept, *J* = 6.3 Hz, 1H), 1.12 (d, *J* = 6.2 Hz, 6H); ^{13}C NMR (DMSO-d_6, 125 MHz) δ 154.6, 150.7, 137.2, 136.3, 135.7, 134.4, 120.5, 117.8, 114.3, 100.7, 55.6, 47.9, 23.7; HRMS Calcd. for $C_{16}H_{20}N_5O$ [M + H]: 298.1668; Found: 298.1668 (Δ = 0 ppm).

5-(4-(Methoxy)anilin-1-yl)-3-(morpholin-1-yl)pyrazolo[1,5-a]pyrimidine (30s)

3-Bromo-5-(*4-(methoxy)anilin-1-yl*)-pyrazolo[1,5-*a*]pyrimidine (50 mg, 0.16 mmol), morpholine (21 mg, 0.24 mmol, 1.5 equiv.), **L-1** (4.4 mg, 0.016 mmol, 10 mol%), CuI (1.5 mg, 0.008 mmol, 5 mol%), and K_2CO_3 (44 mg, 0.32 mmol, 2 equiv.); **30s** (47 mg, 0.145 mmol, 91%); ^{1}H NMR (DMSO-d_6, 500 MHz) δ 9.41 (s, 1H), 8.47 (d, *J* = 7.6 Hz, 1H), 7.69 (d, *J* = 9.0 Hz, 2H), 7.65 (s, 1H), 6.92 (d, *J* = 9.1 Hz, 2H), 6.35 (d, *J* = 7.6 Hz, 1H), 3.78 (t, *J* = 4.6 Hz, 4H), 3.74 (s, 3H), 3.78 (t, *J* = 4.6 Hz, 4H); ^{13}C NMR (DMSO-d_6, 125 MHz) δ 154.9, 150.7, 137.3, 136.0, 134.3, 134.0, 121.9, 120.7, 114.4, 101.0, 66.6, 55.7, 51.7; HRMS Calcd. for $C_{17}H_{20}N_5O_2$ [M + H]: 326.1617; Found: 326.1611 (Δ = 1.8 ppm).

5-(4-(Methoxy)anilin-1-yl)-3-(anilin-1-yl)pyrazolo[1,5-a]pyrimidine (30t)

3-Bromo-5-(*4-(methoxy)anilin-1-yl*)-pyrazolo[1,5-*a*] pyrimidine (50 mg, 0.16 mmol), aniline (22 mg, 0.24 mmol, 1.5 equiv.), **L-1** (4.4 mg, 0.016 mmol, 10 mol%), CuI (1.5 mg, 0.008 mmol, 5 mol%), and K_2CO_3 (44 mg, 0.32 mmol, 2 equiv.); **30t** (43 mg, 0.13 mmol, 81%); ^{1}H NMR (DMSO-d_6, 500 MHz) δ 9.47 (s, 1H), 8.54 (d, *J* = 7.6 Hz, 1H), 7.86 (s, 1H), 7.64 (d, *J* = 9.0 Hz, 2H), 7.21 (s, 1H), 7.09 (t, *J* = 7.6 Hz, 2H), 6.75 (d, *J* = 9.0 Hz, 2H), 6.70 (d, *J* = 7.9 Hz, 2H), 6.61 (t, *J* = 7.3 Hz, 1H), 6.41 (d, *J* = 7.6 Hz, 1H), 3.69 (s, 3H); ^{13}C NMR (DMSO-d_6, 125 MHz) δ 154.8, 152.1, 148.5, 141.6, 141.5, 136.2, 133.9, 129.2, 120.9, 116.9, 114.2, 113.6, 109.9, 101.2, 55.6; HRMS Calcd. for $C_{19}H_{18}N_5O$ [M + H]: 332.1511; Found: 332.1524 (Δ = 3.9 ppm).

5-(4-(Methoxy)anilin-1-yl)-3-(N-(pyrimidin-2-yl)amino)pyrazolo[1,5-a]pyrimidine (30u)

3-Bromo-5-(*4-(methoxy)anilin-1-yl*)pyrazolo[1,5-*a*]pyrimidine (50 mg, 0.16 mmol), 2-aminopyrimidine (23 mg, 0.24 mmol, 1.5 equiv.), **L-1** (4.4 mg, 0.016 mmol, 10 mol%), CuI (1.5 mg, 0.008 mmol, 5 mol%), and K_2CO_3 (44 mg, 0.32 mmol, 2 equiv.); **30u** (42 mg, 0.13 mmol, 81%); ^{1}H NMR (DMSO-d_6, 500 MHz) δ 9.48 (s, 1H), 8.63 (s, 1H), 8.54 (d, *J* = 7.6 Hz, 1H), 8.35 (d, *J* = 4.7 Hz, 2H), 8.04 (s, 1H), 7.73 (d, *J* = 8.9 Hz, 2H), 6.83 (d, *J* = 9.0 Hz, 2H), 6.71 (t, *J* = 4.7 Hz, 1H), 6.41 (d, *J* = 7.6 Hz, 1H), 3.72 (s, 3H); ^{13}C NMR (DMSO-d_6, 125 MHz) δ 166.9, 163.2, 159.6, 156.9, 145.7, 145.4, 140.8, 138.7, 125.7, 119.0, 116.3, 112.4, 106.0; HRMS Calcd. for $C_{17}H_{16}N_7O$ [M + H]: 334.1416; Found: 334.1421 (Δ = 1.5 ppm).

3-(N-(Isopropyl)amino)-5-(4-(isopropyl)anilin-1-yl)pyrazolo[1,5-a]pyrimidine (30v)

3-Bromo-5-(*4-(isopropyl)anilin-1-yl*)pyrazolo[1,5-*a*]pyrimidine (50 mg, 0.15 mmol), isopropylamine (14 mg, 0.23 mmol, 1.5 equiv.), **L-1** (4 mg, 0.015 mmol, 10 mol%), CuI (1.4 mg, 0.007 mmol, 5 mol%), and K_2CO_3 (41 mg, 0.30 mmol, 2 equiv.); **30v** (42 mg, 0.136 mmol, 90%); ^{1}H NMR (DMSO-d_6, 500 MHz) δ 8.10 (d, *J* = 7.6 Hz, 1H), 7.61 (s, 1H), 7.34 (d, *J* = 8.4 Hz, 2H), 7.19 (s, 1H), 7.16 (d, *J* = 8.4 Hz, 2H), 6.58 (bs, 1H), 6.13 (d, *J* = 7.6 Hz, 1H), 3.38 (sept, *J* = 6.3 Hz, 1H), 2.84 (sept, *J* = 6.9 Hz, 1H), 1.20 (d, *J* = 6.9 Hz, 6H), 1.15 (d, *J* = 6.3 Hz, 6H); ^{13}C NMR (DMSO-d_6, 125 MHz) δ 150.8, 142.2, 138.8, 137.2, 136.5, 135.8, 126.9, 119.0, 117.9, 100.9, 48.0, 33.3, 24.5, 23.6; HRMS Calcd. for $C_{18}H_{24}N_5$ [M + H]: 310.2032; Found: 310.2027 (Δ = 1.6 ppm).

5-(4-(Isopropyl)anilin-1-yl)-3-(morpholin-1-yl)pyrazolo[1,5-a]pyrimidine (30w)

3-Bromo-5-(*4-(isopropyl)anilin-1-yl*)pyrazolo[1,5-*a*]pyrimidine (50 mg, 0.15 mmol), morpholine (20 mg, 0.23 mmol, 1.5 equiv.), **L-1** (4 mg, 0.015 mmol, 10 mol%), CuI (1.4 mg, 0.007 mmol, 5 mol%), and K_2CO_3 (41 mg, 0.30 mmol, 2 equiv.); **30w** (44 mg, 0.13 mmol, 87%); ^{1}H NMR (DMSO-d_6, 500 MHz) δ 9.52 (s, 1H), 8.46 (d, *J* = 7.6 Hz, 1H), 7.68 (d, *J* = 8.4 Hz, 2H), 7.66 (s, 1H), 7.18 (d, *J* = 8.5 Hz, 2H), 6.39 (d, *J* = 7.6 Hz, 1H), 3.77 (t, *J* = 4.6 Hz, 4H), 3.13 (t, *J* = 4.5 Hz, 4H), 2.84 (sept, *J* = 6.9 Hz, 1H), 1.18 (d, *J* = 6.9 Hz, 6H); ^{13}C NMR (DMSO-d_6, 125 MHz) δ 150.7, 142.5, 138.5, 137.4, 136.0, 134.4, 126.9, 119.2, 101.2, 66.5, 51.6, 33.3, 24.5; HRMS Calcd. for $C_{19}H_{24}N_5O$ [M + H]: 338.1981; Found: 338.1984 (Δ = 0.9 ppm).

5-(4-(Isopropyl)anilin-1-yl)-3-(anilin-1-yl)pyrazolo[1,5-a]pyrimidine (30x)

3-Bromo-5-(4-(isopropyl)anilin-1-yl)pyrazolo[1,5-*a*]pyrimidine (50 mg, 0.15 mmol), aniline (21 mg, 0.23 mmol, 1.5 equiv.), **L-1** (4 mg, 0.015 mmol, 10 mol%), CuI (1.4 mg, 0.007 mmol, 5 mol%), and K_2CO_3 (41 mg, 0.30 mmol, 2 equiv.); **30x** (41 mg, 0.12 mmol, 80%); ^{1}H NMR (DMSO-d_6, 500 MHz) δ 9.56 (s, 1H), 8.54 (d, *J* = 7.5 Hz, 1H), 7.86 (s, 1H), 7.61 (d, *J* = 8.2 Hz, 2H), 7.20 (s, 1H), 7.10 (t, *J* = 7.2 Hz, 2H), 7.02 (d, *J* = 7.9 Hz, 2H), 6.70 (d, *J* = 8.1 Hz, 2H), 6.61 (t, *J* = 7.2 Hz, 1H), 6.44 (d, *J* = 7.5 Hz, 1H), 2.78 (sept, *J* = 6.8 Hz, 1H), 1.14 (d, *J* = 6.8 Hz, 6H); ^{13}C NMR (DMSO-d_6, 125 MHz) δ 152.1, 148.3, 142.5, 141.5, 141.4, 138.3, 136.2, 129.2, 126.7, 119.5, 117.0, 113.6, 110.1, 101.3, 33.2, 24.4; HRMS Calcd. for $C_{21}H_{22}N_5$ [M + H]: 344.1875; Found: 344.1865 (Δ = 2.9 ppm).

5-(4-(Isopropyl)anilin-1-yl)-3-(N-(pyrimidin-2-yl)amino)pyrazolo[1,5-a]pyrimidine (30y)

3-Bromo-5-(4-(isopropyl)anilin-1-yl)pyrazolo[1,5-*a*]pyrimidine (50 mg, 0.15 mmol), 2-aminopyrimidine (22 mg, 0.23 mmol, 1.5 equiv.), **L-1** (4 mg, 0.015 mmol, 10 mol%), CuI (1.4 mg, 0.007 mmol, 5 mol%), and K_2CO_3 (41 mg, 0.30 mmol, 2 equiv.); **30y** (36 mg, 0.104 mmol, 69%); ^{1}H NMR (DMSO-d_6, 500 MHz) δ 9.54 (s, 1H), 8.64 (s, 1H), 8.56 (d, *J* = 7.5 Hz, 1H), 8.36 (d, *J* = 4.7 Hz, 2H), 8.05 (s, 1H), 7.71 (d, *J* = 8.5 Hz, 2H), 7.11 (d, *J* = 8.5 Hz, 2H), 6.71 (t, *J* = 4.7 Hz, 1H), 6.44 (d, *J* = 7.6 Hz, 1H), 2.83 (sept, *J* = 6.9 Hz, 1H), 1.18 (d, *J* = 6.9 Hz, 6H); ^{13}C NMR (DMSO-d_6, 125 MHz) δ 162.1, 158.5, 152.1, 142.4, 140.9, 140.6, 138.4, 136.1, 126.8, 119.5, 111.6, 107.8, 101.3, 33.3, 24.5; HRMS Calcd. for $C_{19}H_{20}N_7$ [M + H]: 346.1780; Found: 346.1784 (Δ = 1.2 ppm).

5-(4-(Chloro)anilin-1-yl)-3-(N-isopropylamino)pyrazolo[1,5-a]pyrimidine (30z)

3-Bromo-5-(4-(chloro)anilin-1-yl)pyrazolo[1,5-*a*]pyrimidine (50 mg, 0.16 mmol), isopropylamine (14 mg, 0.24 mmol, 1.5 equiv.), **L-1** (4.4 mg, 0.016 mmol, 10 mol%), CuI (1.5 mg, 0.008 mmol, 5 mol%), and K_2CO_3 (44 mg, 0.32 mmol, 2 equiv.); **30z** (42 mg, 0.14 mmol, 87%); ^{1}H NMR (DMSO-d_6, 500 MHz) δ 9.60 (s, 1H), 8.47 (d, *J* = 7.5 Hz, 1H), 7.91 (d, *J* = 7.0 Hz, 2H), 7.59 (s, 1H), 7.36 (d, *J* = 8.9 Hz, 2H), 6.33 (d, *J* = 7.6 Hz, 1H), 3.70 (bs, 1H), 3.54 (m, 1H), 1.13 (d, *J* = 6.3 Hz, 6H); ^{13}C NMR (DMSO-d_6, 125 MHz) δ 150.3, 140.1, 136.5, 136.23, 136.18, 128.9, 125.3, 120.4, 118.5, 100.7, 47.8, 23.7; HRMS Calcd. for $C_{15}H_{17}ClN_5$ [M + H]: 302.1172; Found: 302.1179 (Δ = 2.3 ppm).

5-(4-(Chloro)anilin-1-yl)-3-(morpholin-1-yl)pyrazolo[1,5-a]pyrimidine (30a′)

3-Bromo-5-(4-(isopropyl)anilin-1-yl)pyrazolo[1,5-*a*]pyrimidine (50 mg, 0.16 mmol), morpholine (21 mg, 0.24 mmol, 1.5 equiv.), **L-1** (4.4 mg, 0.016 mmol, 10 mol%), CuI (1.5 mg, 0.008 mmol, 5 mol%), and K_2CO_3 (44 mg, 0.32 mmol, 2 equiv.); **30a′** (48 mg, 0.146 mmol, 91%); ^{1}H NMR (DMSO-d_6, 500 MHz) δ 9.75 (s, 1H), 8.51 (d, *J* = 7.6 Hz, 1H), 7.80 (d, *J* = 8.9 Hz, 2H), 7.69 (s, 1H), 7.37 (d, *J* = 8.7 Hz, 2H), 6.40 (d, *J* = 7.6 Hz, 1H), 3.77 (t, *J* = 4.5 Hz, 4H), 3.12 (t, *J* = 4.4 Hz, 4H); ^{13}C NMR (DMSO-d_6, 125 MHz) δ 150.4, 139.7, 136.9, 136.4, 134.4, 129.0, 125.7, 122.4, 120.6, 101.2, 66.5, 51.6; HRMS Calcd. for $C_{16}H_{17}ClN_5O$ [M + H]: 130.1122; Found: 330.1118 (Δ = 1.2 ppm).

5-(4-(Chloro)anilin-1-yl)-3-(anilin-1-yl)pyrazolo[1,5-a]pyrimidine (30b′)

3-Bromo-5-(4-(isopropyl)anilin-1-yl)pyrazolo[1,5-*a*] (50 mg, 0.16 mmol), aniline (22 mg, 0.24 mmol, 1.5 equiv.), **L-1** (4.4 mg, 0.016 mmol, 10 mol%), CuI (1.5 mg, 0.008 mmol, 5 mol%), and K_2CO_3 (44 mg, 0.32 mmol, 2 equiv.); **30b′** (39 mg, 0.12 mmol, 75%); ^{1}H NMR (DMSO-d_6, 500 MHz) δ 9.77 (s, 1H), 8.63 (d, *J* = 7.5 Hz, 1H), 7.92 (s, 1H), 7.75 (d, *J* = 7.0 Hz, 2H), 7.29 (s, 1H), 7.16 (d, *J* = 7.0 Hz, 2H), 7.11 (dd, *J* = 8.3, 7.4 Hz, 2H), 6.72 (d, *J* = 7.7 Hz, 2H), 6.63 (t, *J* = 7.3 Hz, 1H), 6.46 (d, *J* = 7.6 Hz, 1H); ^{13}C NMR (DMSO-d_6, 125 MHz) δ 151.6, 148.1, 141.5, 140.8, 139.6, 136.7, 129.2, 128.8, 125.6, 120.7, 117.1, 113.8, 110.6, 101.3; HRMS Calcd. for $C_{18}H_{15}ClN_5$ [M + H]: 336.1016; Found: 336.1009 (Δ = 2.1 ppm).

5-(4-(Chloro)anilin-1-yl)-3-(N-(pyrimidin-2-yl)amino)pyrazolo[1,5-a]pyrimidine (30c′)

3-Bromo-5-(4-(isopropyl)anilin-1-yl)pyrazolo[1,5-*a*]pyrimidine (50 mg, 0.16 mmol), 2-aminopyrimidine (23 mg, 0.24 mmol, 1.5 equiv.), **L-1** (4.4 mg, 0.016 mmol, 10 mol%), CuI (1.5 mg, 0.008 mmol, 5 mol%), and K_2CO_3 (44 mg, 0.32 mmol, 2 equiv.); **30c′** (40 mg, 0.12 mmol, 75%); ^{1}H NMR (DMSO-d_6, 500 MHz) δ 9.77 (s, 1H), 8.76 (s, 1H), 8.62 (d, *J* = 7.5 Hz, 1H), 8.37 (d, *J* = 4.7 Hz, 2H), 8.12 (s, 1H), 7.86 (d, *J* = 8.9 Hz, 2H), 7.26 (d, *J* = 8.9 Hz, 2H), 6.73 (t, *J* = 4.7 Hz, 1H), 6.46 (d, *J* = 7.6 Hz, 1H); ^{13}C NMR (DMSO-d_6, 125 MHz) δ 161.9, 158.5, 151.7, 140.8, 139.9, 139.7, 136.4, 128.8, 125.6, 120.8, 111.6, 108.4, 101.4; HRMS Calcd. for $C_{16}H_{13}ClN_7$ [M + H]: 338.0921; Found: 338.0912 (Δ = 2.7 ppm).

5-(Pyrrolidin-1-yl)pyrazolo[1,5-a]pyrimidine (31)

Similar to the method for preparing **29a–e**. 5-chloropyrazolo[1,5-*a*]pyrimidine (1.0 g, 6.5 mmol), pyrrolidine (925 mg, 13 mmol, 2.0 equiv), CsF (988 mg, 6.5 mmol, 1.0 equiv), and $BnNEt_3Cl$ (148 mg, 0.65 mmol,10 mol%), in DMSO (30 mL) was stirred in a sealed pressure vessel (350 mL) and heated at 100 °C for 8 hrs. After cooling to ambient temperature, H_2O (300 mL) was added. The mixture was diluted using ethyl acetate, and the organic layer was separated. The aqueous layer was partitioned with EtOAc (2 × 300 mL), and the combined organic layers were washed with saturated sodium bicarbonate solution (2 × 300 mL) and then water (2 × 300 mL). The organic layer was dried over anhydrous magnesium sulfate, filtered, and then evaporated under reduced pressure. The crude product was purified using flash chromatography (EtOAc/Hexanes) to give pure **31** (1.13 g, 6.0 mmol, 92%); ^{1}H NMR (DMSO-d_6, 500 MHz) δ 8.59 (dd, *J* = 7.7, 0.5 Hz, 1H), 7.81 (d, *J* = 2.1 Hz, 1H), 6.36 (d, *J* = 7.7 Hz, 1H), 5.98 (d, *J* = 1.5 Hz, 1H), 3.48 (t, *J* = 6.5 Hz, 4H), 1.95 ("bs", 4H); ^{13}C (DMSO, 125 MHz) δ 154.0, 148.8, 144.5, 135.6, 98.3, 91.1, 47.1, 25.4; HRMS Calcd. for $C_{10}H_{13}N_4$ [M + H]: 189.1140; Found: 189.1132 (Δ = 4.2 ppm).

4.5. 3-Halo-5-(Pyrrolidin-1-yl)Pyrazolo[1,5-a]Pyrimidines 32

General Procedure: To a stirred solution of 5-(pyrrolidin-1-yl)pyrazolo[1,5-*a*]pyrimidine **(31)** (100 mg) in CH_3CN:CH_2Cl_2 (1:3, 2 mL) chilled to 0 °C was added N-halosuccinimide (1.0 equiv in 0.5 mL CH_3CN), over 30 min. The mixture was stirred while slowly warming to ambient temperature (approx. 2 h). The solvents were removed under reduced pressure, and the crude residue was purified by flash chromatography (20% EtOAc/CH_2Cl_2).

3-Chloro-5-(pyrrolidin-1-yl)pyrazolo[1,5-a]pyrimidine (32a)

5-(Pyrrolidin-1-yl)pyrazolo[1,5-*a*]pyrimidine (100 mg, 0.53 mmol) in 2 mL (CH_3CN:CH_2Cl_2, 1:3), N-chlorosuccinimide (70 mg, 0.53 mmol) in 0.5 mL CH_3CN; **32a** (110 mg, 0.49 mmol, 93%); ^{1}H NMR (DMSO-d_6, 500 MHz) δ 8.59 (d, *J* = 7.8 Hz, 1H), 7.93 (s, 1H), 6.44 (d, *J* = 7.7 Hz, 1H), 3.52 ("bs", 4H), 1.96 ("bs", 4H); ^{13}C NMR (DMSO-d_6, 125 MHz) δ 154.3, 144.2, 142.2, 136.2, 99.2, 92.1, 47.2, 25.7, 24.9; HRMS Calcd. for $C_{10}H_{12}ClN_4$ [M + H]: 223.0750; Found: 223.0761 (Δ = 4.9 ppm).

3-Iodo-5-(pyrrolidin-1-yl)pyrazolo[1,5-a]pyrimidine (32b)

5-(Pyrrolidin-1-yl)pyrazolo[1,5-*a*]pyrimidine (100 mg, 0.53 mmol) in 2 mL (CH_3CN:CH_2Cl_2, 1:3); N-iodosuccinimide (119 mg, 0.53 mmol) in 0.5 mL CH_3CN; **32b** (160 mg, 0.51 mmol; 96%); ^{1}H NMR (DMSO-d_6, 500 MHz) δ 8.60 (d, J = 7.8 Hz, 1H), 7.87 (s, 1H), 6.40 (d, J = 7.7 Hz, 1H), 3.52 ("bs", 4H), 1.96 ("bs", 4H); ^{13}C NMR (DMSO-d_6, 125 MHz) δ 154.7, 148.2, 148.0, 136.2, 99.1, 47.2, 43.3, 25.7, 24.9; HRMS Calcd. for $C_{10}H_{12}IN_4$ [M + H]: 315.0107; Found: 315.0096 (Δ = 3.5 ppm).

5. Conclusions

We have developed an efficient method for rapid generation of diverse arrays of 3,5-bis-aminated pyrazolo[1,5-*a*]pyrimidine derivatives. The reaction proceeds in only two steps from commercially available 3-bromo-5-chloropyrazolo[1,5-*a*]pyrimidine and features efficient copper catalyzed C-3 amination of C-5 aminated precursors. 3,5-bis-aminated products consisting of both 3,5-bis arylamine, 3,5-bis alkylamine, or 3,5-bis alkyl/aryl amine combinations were efficiently obtained in only two steps utilizing our method. This represents the most comprehensive substrate scope for 3,5-bis-aminated pyrazolo[1,5-*a*]pyrimidines ever reported. Advantages of this method include rapid reaction time (1 h for C-3 amination), high yield, avoidance of toxic reagents such as $NaBH_3CN$ commonly used for reductive amination, and utilization of inexpensive CuI catalyst (5 mol%) in place of more expensive and/or less efficient air and moisture-sensitive palladium catalysts previously utilized for C-3 amination (Figure 2). With these advantages now in hand, access to broad, diversity-rich libraries of 3,5-bis-aminated pyrazolo[1,5-*a*]pyrimidines for screening novel or enhanced biological, medicinal, or fluorometric properties is now readily available.

Supplementary Materials: The following supporting information can be downloaded at: https://www.mdpi.com/article/10.3390/molecules30030458/s1, Experimental procedures for preparing **L-1–L-4**; Scheme S1: Synthesis of ligands **L-1–L-4**; spectral data for compounds **29a–e**, **30a–c′**, **31**, **32a–b**; and representative conversion-optimization data. References [30,32–34] are cited the supplementary materials.

Author Contributions: Conceptualization, T.H.I. and M.A.P.; methodology, T.H.I., B.A.T. and L.G; compound characterization, T.H.I., B.A.T. and L.O.G.; writing—original draft preparation, M.A.P.; writing—review and editing, T.H.I. and M.A.P.; supervision, T.H.I. and M.A.P.; project administration, M.A.P.; funding acquisition, M.A.P. All authors have read and agreed to the published version of the manuscript.

Funding: Generous support from the Simmons Center for Cancer Research and the College of Computational, Mathematical, and Physical Sciences at Brigham Young University is gratefully acknowledged.

Institutional Review Board Statement: Not applicable.

Informed Consent Statement: Not applicable.

Data Availability Statement: The original contributions presented in this study are included in the article/Supplementary Material. Further inquiries can be directed to the corresponding author.

Acknowledgments: We gratefully acknowledge Matthew Eli Stecher for technical support in preparing bulk quantities of starting materials.

Conflicts of Interest: The authors declare no conflict of interest.

References

1. Shen, J.; Deng, X.; Sun, R.; Tavallaie, M.S.; Wang, J.; Cai, Q.; Lam, C.; Lei, S.; Fu, L.; Jiang, F. Structural optimization of pyrazolo [1,5-*a*]pyrimidine derivatives as potent and highly selective DPP-4 inhibitors. *Eur. J. Med. Chem.* **2020**, *208*, 112850. [CrossRef]

2. Arias-Gomez, A.; Godoy, A.; Portilla, J. Functional pyrazolo [1,5-a]pyrimidines: Current approaches in synthetic transformations and uses as an antitumor scaffold. *Molecules* **2021**, *26*, 2708. [CrossRef] [PubMed]
3. Jismy, B.; Guillaumet, G.; Akssira, M.; Tikad, A.; Abarbri, M. Efficient microwave-assisted Suzuki–Miyaura cross-coupling reaction of 3-bromo pyrazolo [1,5-*a*]pyrimidin-5(4H)-one: Towards a new access to 3,5-diarylated 7-(trifluoromethyl)pyrazolo [1,5-*a*]pyrimidine derivatives. *RSC Adv.* **2021**, *11*, 1287–1302. [CrossRef] [PubMed]
4. Yamaguchi-Sasaki, T.; Tokura, S.; Ogata, Y.; Kawaguchi, T.; Sugaya, Y.; Takahashi, R.; Iwakiri, K.; Abe-Kumasaka, T.; Yoshida, I.; Arikawa, K.; et al. Discovery of a potent dual inhibitor of wild-type and mutant respiratory syncytial virus fusion proteins. *ACS Med. Chem. Lett.* **2020**, *11*, 1145–1151. [CrossRef] [PubMed]
5. Fouda, A.M.; Abbas, H.-A.; Ahmed, E.H.; Shati, A.A.; Alfaifi, M.Y.; Elbehairi, S.E.I. Synthesis, in vitro antimicrobial and cytotoxic activities of some new Pyrazolo [1,5-*a*]pyrimidine derivatives. *Molecules* **2019**, *24*, 1080. [CrossRef]
6. Hassan, A.S.; Masoud, D.M.; Sroor, F.M.; Askar, A.A. Synthesis and biological evaluation of pyrazolo [1,5-*a*]pyrimidine-3-carboxamide as antimicrobial agents. *Med. Chem. Res.* **2017**, *26*, 2909–2919. [CrossRef]
7. Al-Adiwish, W.M.; Tahir, M.I.M.; Siti-Noor-Adnalizawati, A.; Hashim, S.F.; Ibrahim, N.; Yaacob, W.A. Synthesis, antibacterial activity and cytotoxicity of new fused pyrazolo [1,5-*a*]pyrimidine and pyrazolo [5,1-*c*][1,2,4]triazine derivatives from new 5-aminopyrazoles. *Eur. J. Med. Chem.* **2013**, *64*, 464–476. [CrossRef]
8. Novinson, T.; Bhooshan, B.; Okabe, T.; Revankar, G.R.; Robins, R.K.; Senga, K.; Wilson, H.R. Novel heterocyclic nitrofurfural hydrazones. In vivo antitrypanosomal activity. *J. Med. Chem.* **1976**, *19*, 512–516. [CrossRef] [PubMed]
9. Tsai, P.C.; Wang, I.J. Synthesis and solvatochromic properties of some disazo dyes derived from pyrazolo [1,5-*a*]pyrimidine derivatives. *Dye. Pigm.* **2005**, *64*, 259–264. [CrossRef]
10. Tigreros, A.; Aranzazu, S.-L.; Bravo, N.F.; Zapata-Rivera, J.; Portilla, J. Pyrazolo [1,5-*a*]pyrimidines-based fluorophores: A comprehensive theoretical-experimental study. *RSC Adv.* **2020**, *10*, 39542–39552. [CrossRef]
11. Tigreros, A.; Rosero, H.A.; Castillo, J.-C.; Portilla, J. Integrated pyrazolo [1,5-*a*]pyrimidine-hemicyanine system as a colorimetric and fluorometric chemosensor for cyanide recognition in water. *Talanta* **2019**, *196*, 395–401. [CrossRef] [PubMed]
12. Ding, R.; He, Y.; Xu, J.; Liu, H.; Wang, X.; Feng, M.; Qi, C.; Zhang, J.; Peng, C. Preparation and bioevaluation of ^{99m}Tc nitrido radiopharmaceuticals with pyrazolo [1,5-*a*]pyrimidine as tumor imaging agents. *Med. Chem. Res.* **2011**, *21*, 523–530. [CrossRef]
13. Zhang, Y.; Liu, Y.; Zhou, Y.; Zhang, Q.; Han, T.; Tang, C.; Fan, W. Pyrazolo [1,5-*a*]pyrimidine based trk inhibitors: Design, synthesis, biological activity evaluation. *Bioorg. Med. Chem. Lett.* **2021**, *31*, 127712. [CrossRef] [PubMed]
14. Kosugi, T.; Mitchell, D.R.; Fujino, A.; Imai, M.; Kambe, M.; Kobayashi, S.; Makino, H.; Matsueda, Y.; Oue, Y.; Komatsu, K.; et al. Mitogen-activated protein kinase-activated protein kinase 2 (mapkap-k2) as an antiinflammatory target: Discovery and in vivo activity of selective pyrazolo [1,5-*a*]pyrimidine inhibitors using a focused library and structure-based optimization approach. *J. Med. Chem.* **2012**, *55*, 6700–6715. [CrossRef]
15. Loew, G.; Toll, L.; Lawson, J.; Uyeno, E.; Kaegi, H. Pyrazolo [1,5-*a*]pyrimidines: Receptor binding and anxiolytic behavioral studies. *Pharmacol. Biochem. Behav.* **1984**, *20*, 343–348. [CrossRef] [PubMed]
16. Selleri, S.; Bruni, F.; Costagli, C.; Costanzo, A.; Guerrini, G.; Ciciani, G.; Gratteri, P.; Besnard, F.; Costa, B.; Montali, M.; et al. A novel selective GABAAα1 receptor agonist displaying sedative and anxiolytic-like properties in rodents. *J. Med. Chem.* **2005**, *48*, 6756–6760. [CrossRef]
17. Farago, A.F.; Demetri, G.D. Larotrectinib, a selective tropomyosin receptor kinase inhibitor for adult and pediatric tropomyosin receptor kinase fusion cancers. *Future Oncol.* **2020**, *16*, 417–425. [CrossRef] [PubMed]
18. Sava, J. FDA Grants Breakthrough Therapy Designation to Repotrectinib for ROS1+ Metastatic NSCLC. Available online: https://www.targetedonc.com/view/fda-grants-breakthrough-therapy-designation-to-repotrectinib-for-ros1-metastatic-nsclc (accessed on 20 January 2024).
19. A Study to Test the Safety of the Investigational Drug Selitrectinib in Children and Adults That May Treat Cancer. Available online: https://clinicaltrials.gov/ct2/show/NCT03215511 (accessed on 20 January 2024).
20. Novotna, K.; Thomas, A.G.; Stepanek, O.; Murphy, B.; Hin, N.; Skacel, J.; Mueller, L.; Tenora, L.; Pal, A.; Alt, J.; et al. Neutral sphingomyelinase 2 inhibitors based on the pyrazolo [1,5-*a*]pyrimidin-3-amine scaffold. *Eur. J. Med. Chem.* **2023**, *259*, 115674. [CrossRef] [PubMed]
21. Hu, Y.; Wu, D.; Peng, W.; Li, X.; Hu, F.; Huang, B.; Zhu, J.; Wu, Y. Heterocyclic compound, application thereof and pharmaceutical composition comprising same. *PCT Int. Appl.* **2019**, *2019*, 158107.
22. Pal, K.; Ciblat, S.; Albert, V.; Bruneau-Latour, N.; Boudreault, J. Preparation of 5-(2-(2,5-difluorophenyl)pyrrolidin-1-yl)-3-(1H-pyrazol-1-yl)pyrazolo [1,5-*a*]pyrimidine derivatives and related compounds as Trk kinase inhibitors for treating cancer. *PCT Int. Appl.* **2019**, *2019*, 118584.
23. Ellermann, M.; Valot, G.; Cancho Grande, Y.; Hassfeld, J.; Kinzel, T.; Koebberling, J.; Beyer, K.; Roehrig, S.; Sperzel, M.; Stampfuss, J.; et al. Piperidinylpyrazolopyrimidinones as plasminogen inhibitors and their preparation. *PCT Int. Appl.* **2016**, *2016*, 071216.
24. Magavi, S.S.; Parks, D.J.; Tait, B.D.; Cho, J.; Agrawal, R.; Shaw, P.R. Preparation of substituted pyrazolopyridines as platelet-derived growth factor receptor (PDGFR) alpha inhibitors and uses thereof. *PCT Int. Appl.* **2023**, *2023*, 081923.

25. Henning, N.J.; Nomura, D.K.; Boike, L.; Marquess, D.; Keitz, P. Preparation of bifunctional compounds as protein-stabilizing deubiquitinase-targeting chimeras useful in treatment of cystic fibrosis and other diseases. *PCT Int. Appl.* **2024**, *2024*, 097355.
26. Iorkula, T.H.; Tolman, B.A.; Singleton, J.D.; Peterson, M.A. An efficient microwave assisted copper catalyzed C-3 amination of 3-bromopyrazolo [1,5-*a*]pyrimidine. *Tetrahedron Lett.* **2023**, *118*, 154393. [CrossRef]
27. Reynolds, M.; Eary, C.T.; Spencer, S.; Juengst, D.; Hache, B.; Jiang, Y.; Haas, J.; Andrews, S.W. Preparation of (S)-N-(5-((R)-2-(2,5-Difluorophenyl)pyrrolidin-1-yl)pyrazolo [1,5-a]pyrimidin-3-yl)-3-hydroxypyrrolidine-1-carboxamide. WO, 201241, 2017. U.S. Patent 11,214,571, 4 January 2022.
28. Flick, A.C.; Leverett, C.A.; Ding, H.X.; McInturff, E.; Fink, S.J.; Helal, C.J.; DeForest, J.C.; Morse, P.D.; Mahapatra, S.; O'Donnell, C.J. Synthetic approaches to new drugs approved during 2018. *J. Med. Chem.* **2020**, *63*, 10652–10704. [CrossRef] [PubMed]
29. Chi, L.-L.; Hao, L.-L.; Cai, Z.-Q.; Kong, D.-L.; Wang, Y.-N.; Qin, W.-T.; Gao, Y.; Qu, Z.-Z. Design, synthesis, and biological evaluation of novel pyrazolo [1,5-*a*]pyrimidine and 1,3-Benzodiazine derivatives as potent antitumor agents. *Russ. J. Gen. Chem.* **2022**, *92*, 2698–2707. [CrossRef]
30. Hong, P.; Zhu, X.; Lai, X.; Gong, Z.; Huang, M.; Wan, Y. Room-temperature CuI-catalyzed *N*-arylation of cyclopropylamine. *J. Org. Chem.* **2024**, *89*, 57–67. [CrossRef]
31. Hong, P.; Zhu, X.; Chen, F.; Huang, M.; Wan, Y. $CuSO_4$/*N*-(9*H*-carbazol-9-yl)picolinamide-catalyzed C-O coupling of (hetero)aryl chlorides with phenols on water. *Org Lett.* **2024**, *26*, 7202–7206. [CrossRef]
32. FrØyen, P. Formation of acyl bromides from carboxylic acids and N-bromosuccinimide; some reactions of bromocyanotriphenylphosphorane. *Phosphorus Sulfur Silicon Relat. Elem.* **1995**, *102*, 253–259. [CrossRef]
33. Wangngae, S.; Duangkamol, C.; Pattarawarapana, M.; Phakhodee, W. Significance of reagent addition sequence in the amidation of carboxylic acids mediated by PPh_3 and I_2. *RSC Adv.* **2015**, *5*, 25789–25793. [CrossRef]
34. Cui, H.; Wang, L.; Bai, G.; Li, D.; Lin, P. Synthesis of N-Amino-Carbazole. *Yingyong Huagong* **2006**, *35*, 295–297. [CrossRef]
35. Iorkula, T.H.; Tolman, B.A.; Burt, S.R.; Peterson, M.A. An Efficient synthesis of C-6 aminated 3-bromoimidazo [1,2-*b*]pyridazines. *Synth. Comm.* **2024**, *54*, 121–132. [CrossRef]

molecules

MDPI

Article

Concise Synthesis of Naphthalene-Based 14-Aza-12-Oxasteroids

Smriti Srivastava [1], Jun Luo [1], Daniel Whalen [1], Katherine N. Robertson [2] and Amitabh Jha [1,*]

[1] Department of Chemistry, Acadia University, Wolfville, NS B4P 2R6, Canada
[2] Department of Chemistry, Saint Mary's University, Halifax, NS B3H C3C, Canada
* Correspondence: amitabh.jha@acadiau.ca; Tel.: +1-902-585-1515

Abstract: A concise, transition metal-free four-step synthetic pathway has been developed for the synthesis of tetracyclic heterosteroidal compounds, 14-aza-12-oxasteroids, starting from readily available 2-naphthol analogues. After conversion of 2-naphthols to 2-naphthylamines by the Bucherer reaction, subsequent selective C-acetylation was achieved via the Sugasawa reaction and reduction of the acetyl group using borohydride, which resulted into the corresponding amino-alcohols. The naphthalene-based amino-alcohols underwent double dehydrations and double intramolecular cyclization with oxo-acids leading to one-pot formation of a C-N bond, a C-O bond and an amide bond in tandem, to generate two additional rings completing the steroidal framework. A series of 14-aza-12-oxasteroids were synthesized using our developed synthetic strategy in moderate yields, and the structure of one of the final products, 12a-Methyl-11-phenyl-11,12a-dihydro-1*H*-naphtho[2,1-d]pyrrolo[2,1-b][1,3]oxazin-3(2*H*)-one, was further confirmed by single crystal X-ray crystallography.

Keywords: heterosteroids; 14-aza-12-oxosteroid; Sugasawa reaction; Dean–Stark condensation; double dehydration

1. Introduction

Steroids serve as an important class of natural products by virtue of their ability to penetrate cells and bind to membranes and nuclear receptors [1,2]. Steroidal hormones are classified as glucocorticoids, mineralocorticoids, estrogen, progestogens, and androgens because of their binding to a particular receptor to express their biological response [3,4]. Estrogens (Figure 1), an aromatic ring containing steroidal hormone, bind with estrogen receptors (ER) and trigger them in promoting a variety of physiological responses such as bone maturation, neuroprotective effects, reproductive functions, modulation of blood lipid profile, and breast cell proliferation [5,6]. Progesterone is a pro-gestational steroid hormone secreted by the female reproductive system. It is associated with the menstrual cycle, pregnancy, and development of an embryo (Figure 1) [7]. Medicinal chemists are intrigued with the biological activities of various modified steroids for the drug design and development process. While retaining the steroidal framework, the structural modifications allow for modulation of the biological activities for the resulting molecules. The naturally occurring steroidal frameworks have been modified in various ways with the aim of identifying active compounds with improved efficacy and reduced toxicity [8,9]. Heterosteroids are one such modification of the natural steroids, where one or more carbons are replaced by heteroatom(s). The incorporation of heteroatoms leads to distinct chemical and biological properties, enabling a wide range of potential therapeutic applications of these compounds. The electron-rich nitrogen atoms in N-containing heterocycles are capable of easily accepting or donating protons. Due to their ability to form diverse weak interactions, they can also readily interact with enzymes and receptors in biological targets [10].

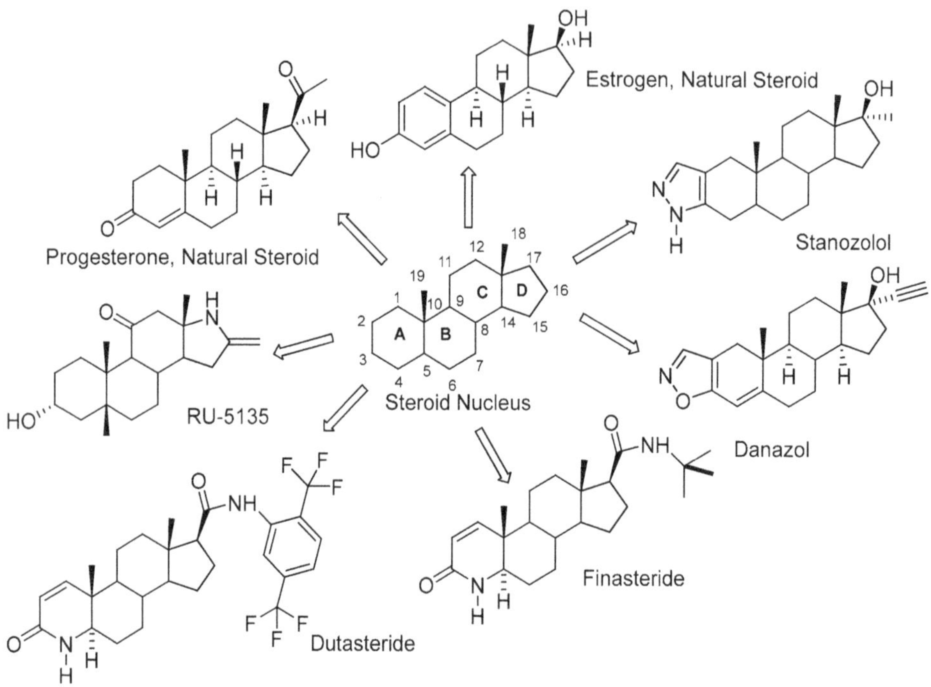

Figure 1. Representative natural steroids and clinically used heterosteroids.

The medicinal chemistry of heterosteroids has been extensively studied [11–15]. A few of the heterosteroids have already culminated into successful drugs, such as stanozolol (hereditary angioedema) [16], danazol (antiestrogen) [17], RU-5135 (antagonist of GABA in the CNS) [18], finasteride (antihyperplastics) [19], and dutasteride (benign prostatic hyperplasia) [20] (Figure 1).

Although synthesis of these structurally complex molecules in a short route from readily available starting materials is a challenge in the field of organic chemistry, several aza- and oxa-steroids have been synthesized by various research groups in the past by following different synthetic protocols. Oumzil et al. [21] reported total synthesis of an 11-oxasteroid having a pyridine ring as the A ring, using an intramolecular Diels–Alder reaction. Singh and Panda [22] described an efficient approach for the synthesis of 14-azasteroids using L-proline via an intramolecular S_N2' cyclization reaction as a key step for the construction of the nitrogen-containing ring C. Bernath et al. [23] reported the stereospecific synthesis of 8-aza-12-oxasteroids. In continuation of our efforts in developing complex heterosteroids [24,25], we herein report a concise synthesis of 14-aza-12-oxosteroids in four simple steps. We have previously reported the synthesis of tricyclic 7-oxa-2-azatricyclo[7.4.0.02,6]trideca-1(9),10,12-trien-3-ones and their homologs, by the reaction of keto acids with methyl chloroformate and variously substituted o-aminobenzyl alcohols using triethylamine as a base in toluene at room temperature [26]; a similar type of synthesis with a different approach is reported here.

2. Results

We envisioned a synthetic strategy to obtain our designed 14-aza-12-oxasteroids with general structure **A**, the retrosynthetic analysis of which is presented in Scheme 1. Double dehydration and cyclization between **B** and the oxo-acids **C** would yield the target 14-aza-12-oxasteroids **A**. Naphthalene-based amino-alcohols **B** can be prepared by the reduction of 1-acyl-2-aminonaphthalenes **D**, which in turn can be synthesized via the Sugasawa reaction [27] of **E** with various aryl and alkyl nitriles **F**. 2-Aminonaphthalenes **E** can be obtained from commercially available 2-naphthols **G** via the Bucherer reaction [28].

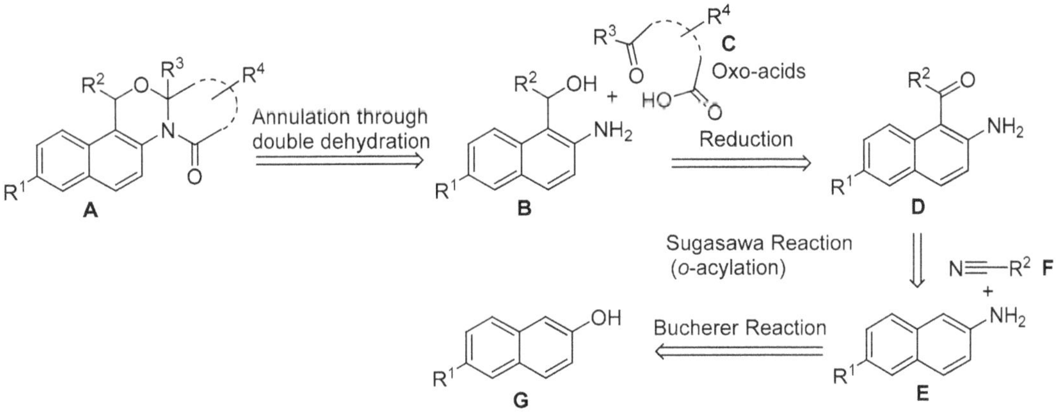

Scheme 1. Retrosynthetic analysis for the synthesis of 14-aza-12-oxasteroids **A**.

Our synthetic scheme started from commercially available 2-naphthol and 6-bromo-2-naphthol (Scheme 2), which were converted into 2-napthylamine (**1a**, 78% yield, Caution: Carcinogenic!) and 6-bromo-2-napthylamine (**1e**, 85% yield), respectively, by using the Bucherer reaction [24,28]. The Friedel–Craft acylation is a common technique to attach acetyl groups on aromatic rings [29]. However, in an aniline system, coordination of the amino group with the Lewis acid results in a deactivated aromatic system. Sugasawa [27] took a unique approach to achieve *ortho*-acylation of anilines by using nitriles and two Lewis acids for the reaction. We successfully achieved *ortho*-acylation of 2-napthylamines (**1a** and **1e**) by using various nitriles (benzyl nitrile, benzonitrile, acetonitrile, and butyronitrile) by the Sugasawa protocol employing the BCl_3 and $AlCl_3$ system to obtain our desired amino-ketone compounds **2a–h** (Scheme 2). The reactions were performed under inert atmosphere since moisture in the air would react with the Lewis acids to form the respective hydroxides. During the work-up of this reaction, dilute acid was used to convert imine intermediates to the desired ketones, although in the case of compounds **2b** and **2f**, extensive amounts of conc. acids and significantly longer heating times were required. Requirement of the forcing conditions for imine hydrolysis was probably due to poor solubility or greater stability of these imines having conjugation with naphthalene and benzene rings. The ortho-acylated compounds **2a–h** were obtained in 50–95% yields after purification by column chromatography. Furthermore, these acylated 2-aminonaphthalenes **2a–h** were reduced to the corresponding amino-alcohol analogues **3a–h**, by borohydride reduction [30,31]. The crude compounds **3a–h** were obtained in 93–98% yields; these compounds were used for the next reaction without further purification. A greener approach not involving any reagents, with recoverable solvent and lesser reaction time, was explored to induce the double dehydration reaction between keto acids and the amino-alcohols **3a–h** utilizing a Dean–Stark apparatus as compared to our previous report for the synthesis of similar tricyclic compounds [26]. It was gratifying that the attempt to achieve the double dehydration and subsequent formation of three bonds in tandem between naphthalene-based amino-alcohols **3a–h** and various keto acids (levulinic acid/2-carboxybenzaldehyde/4-acetylbutyric acid/(2-oxopropyl sulfanyl)-acetic acid) was successful. Formation of new C-N, C-O, and amide bonds along with two cyclization reactions lead to the synthesis of our desired 14-aza-12-oxasteroid compounds **4a–l**, in moderate to decent yields (30–78%, Scheme 2). The structures of all the synthesized compounds **1a**, **1e**, **2a–h**, **3a–h**, and **4a–l** were unambiguously established based on their spectral (^{1}H-, ^{13}C NMR, and HRMS) data analysis. The structures of known compounds **1a** and **1e** were further confirmed on the basis of comparison of their physical and spectral data with those reported in the literature [24,28].

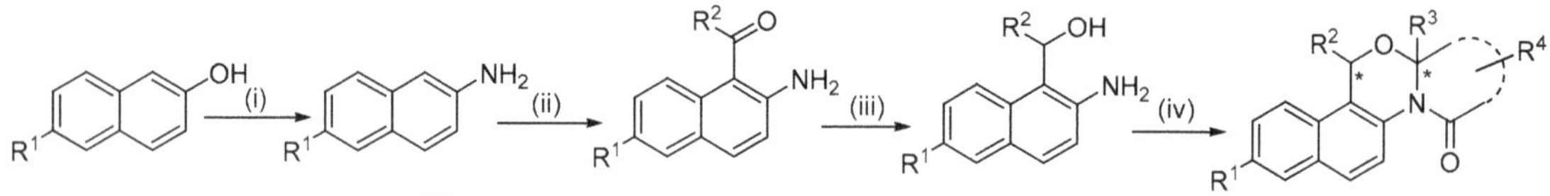

R^1: H or Br; **1a** (78%), **1e** (85%); **2a–h** (50–95%); **3a–h** (93–98%); **4a–l** (30–78%)

Suffix	R^1	R^2	R^3	R^4
a	H	Bn	Me	CH_2CH_2
b	H	Ph	Me	CH_2CH_2
c	H	Me	Me	CH_2CH_2
d	H	*n*-Pr	Me	CH_2CH_2
e	Br	Bn	Me	CH_2CH_2
f	Br	Ph	Me	CH_2CH_2
g	Br	Me	Me	CH_2CH_2
h	Br	*n*-Pr	Me	CH_2CH_2
i	H	Me	H	*o*-C_6H_4
j	Br	Me	H	*o*-C_6H_4
k	H	Me	Me	$CH_2CH_2CH_2$
l	H	Me	Me	CH_2SCH_2

Scheme 2. Reagents and conditions: (i) $(NH_4)_2SO_3$ (2.0 equiv.), 28% aq. NH_3, 100 psi, 150 °C, 5–6 h; (ii) (a) R^2CN (6.5 equiv.), BCl_3 (1.2 equiv.), $AlCl_3$ (1.2 equiv.), N_2, toluene, 2–4 h; (b) 1.0 M HCl (1.2 equiv. except for **2b** and **2f** where 12 M, 500.0 equiv. used), 30 min–4 h; (iii) $NaBH_4$ (6.0 equiv., 12.0 equiv. for **2b** and **2f**, respectively), MeOH:THF (1:1), N_2, 1 h; (iv) keto acids [1.5 equiv., levulinic acid/2-carboxybenzaldehyde/4-acetylbutyric acid/(2-oxopropylsulfanyl)-acetic acid], toluene, reflux, Dean–Stark trap, 2 h. "*" represents chiral centers. The yields reported are unoptimized.

3. Discussion

The most plausible mechanism for the cyclization and dehydration reaction is shown in Scheme 3, taking **4b** as a representative example. The amino-alcohol **3b** contains two nucleophiles with varying nucleophilicity. The amine group is more nucleophilic than the alcohol group; therefore, it attacks the more electrophilic ketone carbonyl on levulinic acid first, eventually leading to imine formation via dehydration (**I** and **II**). Subsequent nucleophilic attack from the alcoholic hydroxy group onto the imine carbon results in intramolecular cyclization, forming the amino-acetal C-ring (**III**). After proton transfer, the final attack from the newly formed more nucleophilic secondary amino group, on the carboxylic acid under azeotropic distillation conditions, forces the formation of the lactam ring in **4b**.

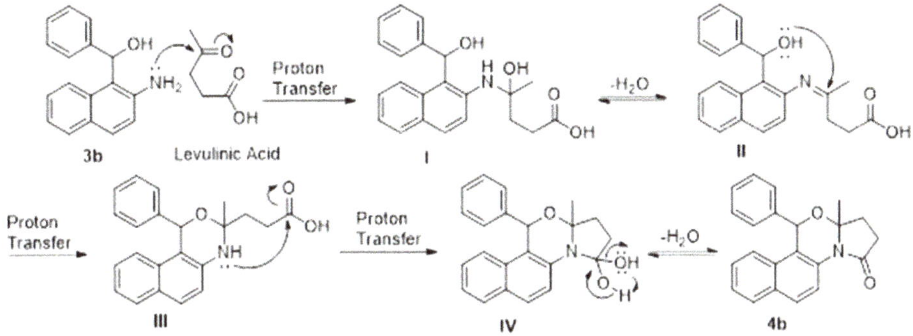

Scheme 3. A plausible mechanism for synthesis of 14-aza-12-oxasteroids considering synthesis of **4b** as the representative example. The intermediates are labelled as **I–IV**.

The heterosteroidal products **4a–l** obtained following this synthetic pathway were expected to give a mixture of up to four stereoisomeric products: the (*R*,*R*) and (*S*,*S*) enantiomeric pair and the (*R*,*S*) and (*S*,*R*) enantiomeric pair, as we did not incorporate any means for enantioselection. In most cases (except in **4i** and **4j**), there was no evidence for the formation of mixtures of diastereomers in the NMR data. The final compounds **4a–h** and **4k–l** showed only one set of peaks in their NMR spectra (see Supporting Information), which indicates that these molecules were synthesized as one set of enantiomeric pairs.

The diastereoselectivity observed in the formation of the target compounds indicated that among the diastereomeric pairs, the thermodynamically more stable pair of enantiomers were exclusively obtained except in the cases of **4i** and **4j** where diastereoselectivities of 6:4 and 8:2, respectively, were obtained. Inspection of the 3D structure of the target molecules reveals that the *RS*/*SR* enantiomeric pair (e.g., see structure of *SR*-**4b** in Figure 2) would be more thermodynamically stable as the two bulky substituents on the ring C at position C4 (Me/Ph) and C6 (the part of pyrrolidine ring) are *cis* to each other [32]. Such thermodynamic stability is expected to diminish in the case of 4i and 4j where the angular Me at position C4 is replaced by H. This explains the formation of all four stereoisomers (albeit diastereoselectively) in the cases of **4i** and **4j**.

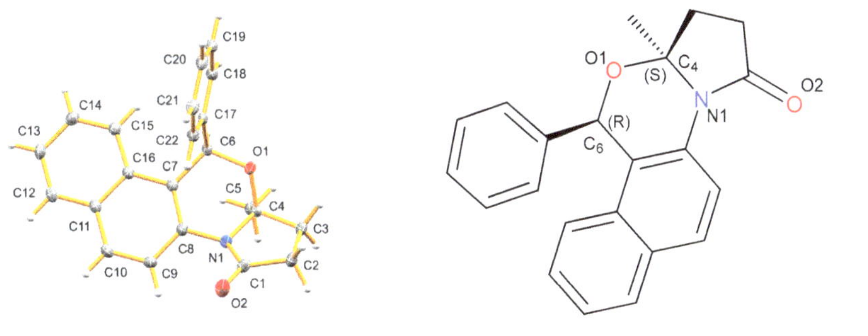

Figure 2. (**a**) ORTEP diagram of compound **4b** drawn in 50% thermal probability ellipsoids as a single (4*S*,6*R*) enantiomer; (**b**) chemical structure of **4b** with similar numbering and atom colouring patterns as in the X-ray crystal structure.

To address the actual stereochemistry of the compounds **4a–l**, X-ray crystallographic studies on a representative compound (**4b**) were undertaken. The analysis of the X-ray crystallographic data of compound **4b** (Figure 2) showed it to crystallize in the noncentrosymmetric, polar space group $Pna2_1$, with two molecules in the independent unit. The crystal is not a racemic mixture, containing only one of the possible enantiomeric pairs. The *R* (C6 and C28), *S* (C4 and C28) enantiomer (see SI) may have crystallized preferentially, or a crystal of this type may have been randomly selected for analysis from a mixture of crystal forms.

While there are two chiral carbon centers in each molecule, the nitrogen atom is not pyramidal and therefore not chiral. In each molecule, it is best described as being sp^2 hybridized, with bond angles close to 120°, distorted slightly by the geometry restrictions caused by being part of two different ring systems. Neither of the rings is planar but the nitrogen atom does not lie any further out of the ring planes than the other atoms in any of the rings. The sum of the bond angles around the nitrogen center in each molecule is 354°, which is again characteristic of trigonal planar hybridization; pyramidal nitrogen centers have total angle sums of closer to 340° [33]. The Cx=O and Cx-N bond lengths (1.218(2) and 1.376(2) Å in both molecules) suggest donation of electron density from the former into the latter bond, possible from the sp^2 hybridization at the nitrogen centers. These bond lengths are similar to those found in lactams [34], with the Cx=O bonds slightly shorter and the latter Cx-N bonds slightly longer, suggesting less donation in the molecules of **4b**. Finally, the crystal structure of compound **4b** shows that the methyl group (bonded to C4) and the phenyl group (bonded to C6) are in a trans geometry indicating that the most thermodynamically stable product has been formed. The crystallographic data of compound **4b** were deposited in the Cambridge Crystallographic Data Centre with CCDC no. 2006680. The ORTEP diagram of one of the independent molecules in compound **4b** is shown in Figure 2.

4. Materials and Methods

General: All the chemicals and solvents used for these syntheses were obtained from commercial sources and were used without modification, except for toluene and methanol. Toluene was dried with sodium metal overnight and distilled before reactions. Methanol was dried with molecular sieves overnight before use. (2-Oxopropylsulfanyl)-acetic acid was synthesized by following a literature procedure [35] starting from mercaptoacetic acid. TLCs were performed on pre-coated Merck silica gel 60F254 plates with the spots detected under UV light. The silica gel used in column chromatography was 230–400 mesh. All column chromatographies were performed using DCM: methanol gradient as the eluent and the TLCs were analyzed using UV chamber by comparing the Rf values of the starting material and the product. NMR spectra were recorded using a Bruker AC-300 Avance spectrometer (Bruker Corporation, Billerica, MA, USA) at 300 MHz for ^{1}H NMR and 75 MHz for ^{13}C NMR. The ^{1}H and ^{13}C NMR spectra were recorded in either $CDCl_3$ or DMSO-d_6 solvents. ^{1}H NMR spectra were reported relative to $CHCl_3$ (δ: 7.26) or DMSO (δ: 2.50). ^{13}C NMR spectra were reported relative to $CDCl_3$ (δ: 77.16) or DMSO-d_6 (δ: 39.52). HRMS (ESI) spectra were recorded using an Orbitrap Q-exactive analyzer (Thermo Fisher Scientific, San Jose, CA, USA; Software: Xcalibur, version 3.1). In X-ray crystallography, the crystal chosen was attached to the tip of a MicroLoop with Paratone-N oil. Measurements were made on a Bruker D8 VENTURE diffractometer equipped with a PHOTON III CMOS detector using monochromated Cu Kα radiation (λ = 1.54178 Å) from an Incoatec micro-focus sealed tube at 100 K (Bruker AXS Inc., Madison, WI, USA).

General procedure for the synthesis of 1-acyl-2-aminonaphthalenes (**2a–h**):

In a N_2-flushed 250 mL RB flask, 2-naphthylamines (2.00 g, **1a**: 13.96 mmol; **1e**: 9.01 mmol, 1.0 equiv.) and $AlCl_3$ (1.2 equiv.) were added. The flask was flushed again with N_2. Toluene was added to the reaction mixture followed by a BCl_3 heptane solution (1.0 M, 1.2 equiv.). The R^2CN (acetonitrile/butyronitrile/benzonitrile/benzyl nitrile, 6.5 equiv.)

was then added to the reaction mixture. The mixture was refluxed and was monitored with TLC (2% MeOH in DCM). Compounds **2a–b**, **2c–d**, **2e–f**, and **2g–h** required 3.5 h, 2.0 h, 4.0 h, and 2.5 h, respectively, for reaction completion. After cooling to rt, aq. HCl (1.0 M, 1.2 equiv., **2b** and **2f**: 12.0 M, ~500 equiv.) was added to the resultant mixture forming a yellow-orange mixture, which was heated at 80 °C for 30 min (4 h for **2b** and **2f**). During the work-up, the reaction mixture was cooled to rt, and the yellow-orange mixture was poured into ice water (ca. 50 mL). This was followed by the addition of aq. NaOH (5.0 M) until pH > 13. More ice and NaOH pellets were used as needed with **2b** and **2f**, to bring the pH > 13. The mixture was extracted with EtOAc (3 × 50 mL). The organic layers were added together and washed with brine (50 mL) and dried with anh. Na_2SO_4. The solvent was removed in vacuo to form the crude product. The crude product was purified by column chromatography (using DCM as eluent and slowly increasing the polarity with MeOH) to yield **2a–h**.

1-(2-Aminonaphthalen-1-yl)-2-phenylethanone (**2a**): Dark orange solid; 50% yield (1.81 g); m.p. 72–73 °C; ^{1}**H NMR** ($CDCl_3$, 300 MHz): δ_H 7.85 (d, $^3J_{H\text{-}H}$ = 9.0 Hz, 1H, Ar*H*), 7.75 (d, $^3J_{H\text{-}H}$ = 9.0 Hz, 1H, Ar*H*), 7.70 (d, $^3J_{H\text{-}H}$ = 9.0 Hz, 1H, Ar*H*), 7.51 (t, $^3J_{H\text{-}H}$ = 7.5 Hz, 1H, Ar*H*), 7.39–7.28 (m, 6H, Ar*H*), 6.83 (d, $^3J_{H\text{-}H}$ = 9.0 Hz, 1H, Ar*H*), 5.48 (brs, 2H, NH_2), 4.36 (s, 2H, CH_2Ph); ^{13}**C NMR** ($CDCl_3$, 75 MHz): δ_C 203.6 (CO), 146.0, 135.6, 133.4, 129.4, 128.7, 128.4, 127.4, 126.6, 124.1, 122.6, 122.6, 119.2, 115.4 (ArCs), 50.0 (CH_2Ph); **HRMS (ESI$^+$)** MeOH/$CHCl_3$, *m/z* (RI): 262.1227 ([M + H]$^+$, 100%). [M + H]$^+$ calcd. for $C_{18}H_{15}NO$, 262.1226.

(2-Aminonaphthalen-1-yl)(phenyl)methanone (**2b**): Yellow crystals; 60% yield (2.06 g); m.p. 166–167 °C; ^{1}**H NMR** (DMSO-d$_6$, 300 MHz): δ_H 7.78 (d, $^3J_{H\text{-}H}$ = 9.0 Hz, 1H, Ar*H*), 7.71 (d, $^3J_{H\text{-}H}$ = 6.0 Hz, 1H, Ar*H*), 7.66 (d, $^3J_{H\text{-}H}$ = 6.0 Hz, 2H, Ar*H*), 7.59 (t, $^3J_{H\text{-}H}$ = 6.0 Hz, 1H, Ar*H*), 7.46 (t, $^3J_{H\text{-}H}$ = 7.5 Hz, 2H, Ar*H*), 7.17–7.06 (m, 4H, Ar*H*), 5.93 (brs, 2H, NH_2); ^{13}**C NMR** (DMSO-d$_6$, 75 MHz): δ_C 197.8 (CO), 145.7, 138.6, 132.8, 132.2, 131.4, 129.0, 128.5, 128.0, 126.3, 125.9, 123.3, 121.2, 118.9, 112.4 (ArCs); **HRMS (ESI$^+$)** MeOH/$CHCl_3$, *m/z* (RI): 248.1070 ([M + H]$^+$, 100%). [M + H] calcd. for $C_{17}H_{13}NO$, 248.1070.

1-Acetyl-2-aminonaphthalene (**2c**): Orange solid; 63% yield (1.62 g); m.p. 107–108 °C; ^{1}**H NMR** ($CDCl_3$, 300 MHz): δ_H 7.82 (d, $^3J_{H\text{-}H}$ = 9.0 Hz, 1H, Ar*H*), 7.68 (d, $^3J_{H\text{-}H}$ = 6.0 Hz, 1H, Ar*H*), 7.65 (d, $^3J_{H\text{-}H}$ = 6.0 Hz, 1H, Ar*H*), 7.46 (t, $^3J_{H\text{-}H}$ = 9.0 Hz, 1H, Ar*H*), 7.27 (t, $^3J_{H\text{-}H}$ = 9.0 Hz, 1H, Ar*H*), 6.85 (d, $^3J_{H\text{-}H}$ = 9.0 Hz, 1H, Ar*H*), 5.80 (brs, 2H, NH_2), 2.70 (s, 3H, CH_3); ^{13}**C NMR** ($CDCl_3$, 75 MHz): δ_C 202.9 (CO), 146.4, 133.8, 132.6, 128.7, 127.7, 127.4, 124.4, 122.6, 119.5, 115.2 (ArCs), 32.3 (CH_3); **HRMS (ESI$^+$)** MeOH/$CHCl_3$, *m/z* (RI): 186.2337 ([M + H]$^+$, 100%). [M + H] calcd. for $C_{12}H_{11}NO$, 186.2335.

1-(2-Aminonaphalen-1-yl)butan-1-one (**2d**): Orange oil; 50% yield (1.33 g); ^{1}**H NMR** ($CDCl_3$, 300 MHz): δ_H 7.74–7.68 (m, 2H, Ar*H*), 7.66 (d, $^3J_{H\text{-}H}$ = 9.0 Hz, 1H, Ar*H*), 7.46 (t, $^3J_{H\text{-}H}$ = 9.0 Hz, 1H, Ar*H*), 7.28 (t, $^3J_{H\text{-}H}$ = 7.5 Hz, 1H, Ar*H*), 6.87 (d, $^3J_{H\text{-}H}$ = 9.0 Hz, 1H, Ar*H*), 5.17 (brs, 2H, NH_2), 2.99 (t, $^3J_{H\text{-}H}$ = 7.5 Hz, 2H, C$H_2$$CH_2CH_3$), 1.86 (sx, $^3J_{H\text{-}H}$ = 7.8 Hz, 2H, CH_2C$H_2$$CH_3$), 0.97 (t, $^3J_{H\text{-}H}$ = 6.0 Hz, 3H, CH_2CH_2CH_3); ^{13}**C NMR** ($CDCl_3$, 75 MHz): δ_C 207.7 (CO), 144.8, 132.8, 131.2, 128.6, 127.6, 127.2, 124.0, 122.5, 119.6, 116.4 (ArCs), 46.2, 19.3, 13.8 (*n*-Pr); **HRMS (ESI$^+$)** MeOH/$CHCl_3$, *m/z* (RI): 214.1231 ([M + H]$^+$, 100%). [M + H]$^+$ calcd. for $C_{14}H_{15}NO$, 214.1226.

1-(2-Amino-6-bromonaphthalen-1-yl)-2-phenylethanone (**2e**): Dark purple solid; 75% yield (2.29 g); m.p. 101–102 °C; ^{1}**H NMR** ($CDCl_3$, 300 MHz): δ_H 7.84 (s, 1H, Ar*H*), 7.65 (d, $^3J_{H\text{-}H}$ = 9.0 Hz, 1H, Ar*H*), 7.55 (m, 2H, Ar*H*), 7.34–7.26 (m, 5H, Ar*H*), 6.83 (d, $^3J_{H\text{-}H}$ = 12.0 Hz, 1H, Ar*H*), 5.46 (brs, 2H, NH_2), 4.26 (s, 2H, CH_2Ph); ^{13}**C NMR** ($CDCl_3$, 75 MHz): δ_C 203.2 (CO), 146.0, 135.3, 132.3, 130.6, 130.4, 129.3, 128.7, 128.5, 126.8, 125.7, 120.3, 115.9 (ArCs),

50.1 (CH_2Ph); **HRMS (ESI$^+$)** MeOH/$CHCl_3$, *m/z* (RI): 340.0335 ([M + H]$^+$, 98%). [M + H]$^+$ calcd. for $C_{18}H_{14}BrNO$, 340.0332.

(2-Amino-6-bromonaphthalen-1-yl)(phenyl)methanone (**2f**): Yellow crystals; 65% yield (1.90 g); m.p. 120–121 °C; **^{1}H NMR** (DMSO-d_6, 300 MHz): δ_H 7.97 (d, $^4J_{H\text{-}H}$ = 1.5 Hz, 1H, Ar*H*), 7.77 (d, $^3J_{H\text{-}H}$ = 12.0 Hz, 1H, Ar*H*), 7.64 (d, $^3J_{H\text{-}H}$ = 6.0 Hz, 2H, Ar*H*), 7.60 (d, $^3J_{H\text{-}H}$ = 6.0 Hz, 1H, Ar*H*), 7.46 (t, $^3J_{H\text{-}H}$ = 7.5 Hz, 2H, Ar*H*), 7.27 (d, $^3J_{H\text{-}H}$ = 6.0 Hz, 1H, Ar*H*), 7.16 (d, $^3J_{H\text{-}H}$ = 9.0 Hz, 1H, Ar*H*), 7.00 (d, $^3J_{H\text{-}H}$ = 9.0 Hz, 1H, Ar*H*), 6.03 (s, 2H, NH_2); **^{13}C NMR** (DMSO-d_6, 75 MHz): δ_C 197.6 (CO), 146.4, 138.6, 133.3, 131.1, 130.9, 129.9, 129.3, 128.9, 127.4, 125.6, 120.4, 113.9, 112.3 (Ar*C*s); **HRMS (ESI$^+$)** MeOH/$CHCl_3$, *m/z* (RI): 326.0184 ([M + H]$^+$, 100%). [M + H]$^+$ calcd. for $C_{17}H_{12}BrNO$, 326.0175.

1-Acetyl-2-amino-6-bromonaphthalene (**2g**): Orange solid; 95% yield (2.25 g); m.p. 135–137 °C. **^{1}H NMR** ($CDCl_3$, 300 MHz): δ_H 7.80 (s, 1H, Ar*H*), 7.68 (d, $^3J_{H\text{-}H}$ = 9.0 Hz, 1H, Ar*H*), 7.55 (d, $^3J_{H\text{-}H}$ = 9.0 Hz, 2H, Ar*H*), 6.84 (d, $^3J_{H\text{-}H}$ = 9.0 Hz, 1H, Ar*H*), 5.94 (brs, 2H, NH_2), 2.66 (s, 3H, CH_3); **^{13}C NMR** ($CDCl_3$, 75 MHz): δ_C 202.3 (CO), 147.0, 132.7, 131.2, 130.6, 130.4, 130.0, 128.8, 128.2, 125.9, 120.6 (Ar*C*s), 32.2 (CH_3); **HRMS (ESI$^+$)** MeOH/$CHCl_3$, *m/z* (RI): 265.1291 ([M + H]$^+$, 100%). [M + H]$^+$ calcd. for $C_{12}H_{11}BrNO$, 265.1295.

1-(2-Amino-6-bromonaphalen-1-yl)butan-1-one (**2h**): Brown solid; 60% yield (1.57 g); m.p. 42–43 °C; **^{1}H NMR** ($CDCl_3$, 300 MHz): δ_H 7.81 (s, 1H, Ar*H*), 7.59–7.47 (m, 3H, Ar*H*), 6.86 (d, $^3J_{H\text{-}H}$ = 9.0 Hz, 1H, Ar*H*), 5.02 (brs, 2H, NH_2), 2.92 (t, $^3J_{H\text{-}H}$ = 7.5 Hz, 2H, $CH_2CH_2CH_3$), 1.79 (dd, $^3J_{H\text{-}H}$ = 15.0 Hz, 9.0 Hz, 2H, $CH_2CH_2CH_3$), 0.95 (t, $^3J_{H\text{-}H}$ = 7.5 Hz, 3H, $CH_2CH_2CH_3$); **^{13}C NMR** ($CDCl_3$, 75 MHz): δ_C 207.1 (CO), 145.1, 131.8, 130.7, 130.5, 130.2, 128.7, 125.6, 120.3, 116.1, 115.7 (Ar*C*s), 46.2, 19.3, 13.8 (*n*-Pr); **HRMS (ESI$^+$)** MeOH/$CHCl_3$, *m/z* (RI): 292.0337 ([M + H]$^+$, 100%). [M + H]$^+$ calcd. for $C_{14}H_{14}BrNO$, 292.0332.

General procedure for the synthesis of 1-(1-hydroxyalkyl)-2-aminonaphthalenes (**3a–h**): In a 50 mL RB flask, **2a–h** (500 mg, 1.0 equiv.) was dissolved in a 1:1 mixture of methanol and THF (10 mL). The reaction flask was flushed with N_2 gas and cooled to 0 °C in an ice bath. While stirring, $NaBH_4$ powder (6.0 equiv., except **2b** and **2f**: 12.0 equiv.) was added in small batches. The reaction was stirred in an RB flask with a N_2 balloon for 5 min. in the ice bath, and further stirred at rt until no starting material was observed on the TLC (ca. 1 h). The solvent was removed in vacuo leaving a solid. Distilled water (ca. 5 mL) was added, and the products were extracted with EtOAc (2 × 25 mL) and washed with distilled water (2 × 10 mL). The organic layer was dried with anh. Na_2SO_4 and the solvent was removed in vacuo yielding the products in pure form.

1-(2-Amimonaphthalen-1-yl)-2-phenylethanol (**3a**): Orange solid; 93% yield (467 mg); m.p. 129–131 °C; **^{1}H NMR** ($CDCl_3$, 300 MHz): δ_H 7.82 (d, $^3J_{H\text{-}H}$ = 6.0 Hz, 1H, Ar*H*), 7.73 (d, $^3J_{H\text{-}H}$ = 6.0 Hz, 1H, Ar*H*), 7.61 (d, $^3J_{H\text{-}H}$ = 9.0 Hz, 1H, Ar*H*), 7.44 (t, $^3J_{H\text{-}H}$ = 7.5 Hz, 1H, Ar*H*), 7.39–7.25 (m, 6H, Ar*H*), 6.89 (d, $^3J_{H\text{-}H}$ = 9.0, 1H, Ar*H*), 5.85 (dd, $^3J_{H\text{-}H}$ = 9.0 Hz, 3.0 Hz, 1H, Ar*H*), 3.81 (brs, 3H, NH_2 and O*H*), 3.39 (dd, $^2J_{H\text{-}H}$ = 15.0 Hz, $^3J_{H\text{-}H}$ = 9.0 Hz, 1H, CH_2Ph), 3.12 (dd, $^2J_{H\text{-}H}$ = 12.0 Hz, $^3J_{H\text{-}H}$ = 3.0 Hz, 1H, CH_2Ph); **^{13}C NMR** ($CDCl_3$, 75 MHz): δ_C 143.1, 138.6, 131.8, 129.5, 129.1, 128.8, 128.5, 128.1, 126.6, 126.6, 121.8, 120.6, 120.3, 115.9 (Ar*C*s), 72.0 (CHOH), 40.7 (CH_2Ph); **HRMS (ESI$^+$)** MeOH/$CHCl_3$, *m/z* (RI): 264.1383 ([M + H]$^+$, 26%). [M + H]$^+$ calcd. for $C_{18}H_{17}NO$, 264.1383.

(2-Aminonaphthalen-1-yl)(phenyl)methanol (**3b**): Pale orange solid; 94% yield (474 mg); m.p. 131–133 °C; **^{1}H NMR** (DMSO-d_6, 300 MHz): δ_H 7.97 (d, $^3J_{H\text{-}H}$ = 9.0 Hz, 1H, Ar*H*), 7.67 (d, $^3J_{H\text{-}H}$ = 9.0 Hz, 1H, Ar*H*), 7.59 (d, $^3J_{H\text{-}H}$ = 9.0 Hz, 1H, Ar*H*), 7.38 (d, $^3J_{H\text{-}H}$ = 6.0 Hz, 2H, Ar*H*), 7.31–7.23 (m, 3H, Ar*H*), 7.19–7.09 (m, 2H, Ar*H*), 7.02 (d, $^3J_{H\text{-}H}$ = 9.0 Hz, 1H, Ar*H*), 6.62 (s, 1H, Ar*H*), 6.16 (s, 1H, O*H*), 5.61 (brs, 2H, NH_2); **^{13}C NMR** (DMSO-d_6, 75 MHz): δ_C 144.7, 144.4, 132.8, 128.4, 128.3, 127.8, 127.0, 126.3, 126.0, 125.9, 122.1, 120.6, 119.9, 116.0

(Ar*Cs*), 68.5 (CHOH); **HRMS (ESI$^+$)** MeOH/$CHCl_3$, *m/z* (RI): 250.1214 ([M + H]$^+$, 36%). [M + H]$^+$ calcd. for $C_{17}H_{15}NO$, 250.1226.

1-(1-Hydroxyethyl)-2-aminonaphthalene (**3c**): Colorless solid; 98% yield (499 mg); m.p. 89.9–91.2 °C; **^{1}H NMR** ($CDCl_3$, 300 MHz): δ_H 7.78 (d, $^3J_{H\text{-}H}$ = 9.0 Hz, 1H, Ar*H*), 7.70 (d, $^3J_{H\text{-}H}$ = 9.0 Hz, 1H, Ar*H*), 7.58 (d, $^3J_{H\text{-}H}$ = 9.0 Hz, 1H, Ar*H*), 7.41 (t, $^3J_{H\text{-}H}$ = 6.0 Hz, 1H, Ar*H*), 7.23 (t, $^3J_{H\text{-}H}$ = 6.0 Hz, 1H, Ar*H*), 6.87 (d, $^3J_{H\text{-}H}$ = 9.0 Hz, 1H, Ar*H*), 5.91 (q, $^3J_{H\text{-}H}$ = 7.5 Hz, 1H, Ar*H*), 4.82 (brs, 1H, O*H*), 3.33 (brs, 2H, NH_2), 1.65 (d, $^3J_{H\text{-}H}$ = 7.5 Hz, 3H, CH_3); **^{13}C NMR** ($CDCl_3$, 75 MHz): δ_C 142.8, 131.6, 128.8, 128.7, 127.9, 126.6, 121.8, 120.6, 120.2, 117.4 (Ar*Cs*), 67.0 (*C*HOH), 20.2 (CH_3); **HRMS (ESI$^+$)** MeOH/$CHCl_3$, *m/z* (RI): 188.1068 ([M + H]$^+$, 100%). [M + H]$^+$ calcd. for $C_{12}H_{13}NO$, 188.1070.

1-(2-Aminonaphthalen-1-yl)butan-1-ol (**3d**): Orange solid; 95% yield (480 mg); m.p. 86–88 °C; **^{1}H NMR** (DMSO-d_6, 300 MHz): δ_H 7.87 (d, $^3J_{H\text{-}H}$ = 6.0 Hz, 1H, Ar*H*), 7.64 (d, $^3J_{H\text{-}H}$ = 9.0 Hz, 1H, Ar*H*), 7.52 (d, $^3J_{H\text{-}H}$ = 9.0 Hz, 1H, Ar*H*), 7.32 (t, $^3J_{H\text{-}H}$ = 7.5 Hz, 1H, Ar*H*), 7.10 (t, $^3J_{H\text{-}H}$ = 7.5 Hz, 1H, Ar*H*), 6.97 (d, $^3J_{H\text{-}H}$ = 9.0, 1H, Ar*H*), 5.64 (brs, 2H, C*H*OH, O*H*), 5.53 (brs, 2H, NH_2), 2.01–1.89 (m, 1H, $CH_2CH_2CH_3$), 1.73–1.62 (m, 1H, $CH_2CH_2CH_3$), 1.57–1.46 (m, 1H, $CH_2CH_2CH_3$), 1.34–1.23 (m, 1H, $CH_2CH_2CH_3$), 0.89 (t, $^3J_{H\text{-}H}$ = 7.5 Hz, 3H, $CH_2CH_2CH_3$); **^{13}C NMR** (DMSO-d_6, 75 MHz): δ_C 144.5, 132.2, 128.4, 127.7, 126.9, 121.2, 120.5, 120.0, 116.5, 68.1 (CHOH), 37.8, 19.1, 14.1 ($CH_2CH_2CH_3$ and CH_3); **HRMS (ESI$^+$)** MeOH/$CHCl_3$, *m/z* (RI): 216.1375 ([M + H]$^+$, 100%). [M + H]$^+$ calcd. for $C_{14}H_{17}NO$, 216.1383.

1-(2-Amimo-6-bromonaphthalen-1-yl)-2-phenylethanol (**3e**): Purple solid; 93% yield (468 mg); m.p. 144–146 °C; **^{1}H NMR** (DMSO-d_6, 300 MHz): δ_H 7.86 (s, 1H, Ar*H*), 7.54 (d, *J* = 9.0 Hz, 1H, Ar*H*), 7.35 (d, $^3J_{H\text{-}H}$ = 9.0 Hz, 1H, Ar*H*), 7.25–7.21 (m, 4H, Ar*H*), 7.18–7.12 (m, 2H, Ar*H*), 7.06 (d, $^3J_{H\text{-}H}$ = 9.0, 1H, Ar*H*), 5.80 (brs, 2H, NH_2), 5.61 (brs, 1H, O*H*), δ 5.69 (s, 1H, CHOH), 3.20 (dd, $^2J_{H\text{-}H}$ = 15.0 Hz, $^3J_{H\text{-}H}$ = 9.0 Hz, 1H, CH_2Ph), 2.96 (dd, $^3J_{H\text{-}H}$ = 12.0 Hz, $^3J_{H\text{-}H}$ = 6.0 Hz, 1H, CH_2Ph); **^{13}C NMR** (DMSO-d_6, 75 MHz): δ_C 145.1, 140.5, 139.3, 130.7, 129.9, 129.5, 128.4, 128.2, 127.9, 127.3, 125.8, 121.1, 115.9, 113.0 (ArCs), 69.6 (CHOH), 40.4 (CH_2Ph); **HRMS (ESI$^+$)** MeOH/$CHCl_3$, *m/z* (RI): 342.0495 ([M + H]$^+$, 100%). [M + H]$^+$ calcd. for $C_{18}H_{16}BrNO$, 342.0488.

(2-Amino-6-bromonaphthalen-1-yl)(phenyl)methanol (**3f**): Very pale orange solid; 94% yield (474 mg); m.p. 171–172 °C; **^{1}H NMR** (DMSO-d_6, 300 MHz): δ_H 7.94–7.89 (m, 2H, Ar*H*), 7.58 (d, $^3J_{H\text{-}H}$ = 9.0 Hz, 1H, Ar*H*), 7.34 (d, $^3J_{H\text{-}H}$ = 9.0 Hz, 3H, Ar*H*), 7.25 (t, $^3J_{H\text{-}H}$ = 7.5 Hz, 2H, Ar*H*), 7.16 (t, $^3J_{H\text{-}H}$ = 6.0 Hz, 1H, Ar*H*), 7.06 (d, $^3J_{H\text{-}H}$ = 9.0 Hz, 1H, Ar*H*), 6.54 (s, 1H, Ar*H*), 6.17 (s, 1H), 5.72 (brs, 2H, NH_2); **^{13}C NMR** (DMSO-d_6, 75 MHz): δ_C 145.2, 144.2, 131.4, 129.8, 128.4, 127.8, 127.7, 126.3, 125.9, 124.9, 120.9, 116.1, 113.1 (Ar*Cs*), 68.2 (CHOH); **HRMS (ESI$^+$)** MeOH/$CHCl_3$, *m/z* (RI): 328.0319 ([M + H]$^+$, 76%). [M + H]$^+$ calcd. for $C_{17}H_{14}BrNO$, 328.0332.

1-(1-Hydroxyethyl)-2-amino-6-bromonaphthalene (**3g**): Colorless solid; 98% yield (495 mg); m.p. 120–123 °C; **^{1}H NMR** (DMSO-d_6, 300 MHz): δ_H 7.90 (s, 1H, Ar*H*), 7.87 (d, $^3J_{H\text{-}H}$ = 3.0 Hz, 1H, Ar*H*), 7.50 (d, $^3J_{H\text{-}H}$ = 9.0 Hz, 1H, Ar*H*), 7.39 (d, $^3J_{H\text{-}H}$ = 3.0 Hz, 1H, Ar*H*), 6.99 (d, $^3J_{H\text{-}H}$ = 9.0 Hz, 1H, Ar*H*), 5.74 (brs, 2H, NH_2), 5.59 (q, $^3J_{H\text{-}H}$ = 6.0 Hz, 1H, C*H*OH), 5.52 (brs, 1H, O*H*), 1.41 (d, $^3J_{H\text{-}H}$ = 6.0 Hz, 3H, CH_3); **^{13}C NMR** (DMSO-d_6, 75 MHz): δ_C 144.6, 130.3, 129.9, 128.4, 128.2, 126.9, 123.7, 121.0, 117.1, 112.9 (Ar*Cs*), 64.4 (CHOH), 21.1 (CH_3); **HRMS (ESI$^+$)** MeOH/$CHCl_3$, *m/z* (RI): 266.0173 ([M + H]$^+$, 100%). [M + H]$^+$ calcd. for $C_{12}H_{12}BrNO$, 266.0175.

1-(2-Amino-6-bromonaphthalen-1-yl)butan-1-ol (**3h**): Brown solid; 98% yield (493 mg); m.p. 138–140 °C; **^{1}H NMR** (DMSO-d_6, 300 MHz): δ_H 7.87 (d, $^4J_{H\text{-}H}$ = 1.5 Hz, 2H, Ar*H*), 7.52 (d, $^3J_{H\text{-}H}$ = 9.0 Hz, 1H, Ar*H*), 7.40 (d, $^3J_{H\text{-}H}$ = 9.0 Hz, 1H, Ar*H*), 7.01 (d, $^3J_{H\text{-}H}$ = 9.0, 1H, Ar*H*),

5.80 (brs, 2H, NH_2), 5.61 (brs, 1HO*H*), 5.42 (t, $^3J_{H\text{-}H}$ = 6.0 Hz, 1H, C*H*OH), 1.96–1.84 (m, 1H, $CH_2CH_2CH_3$), 1.69–1.56 (m, 1H, $CH_2CH_2CH_3$), 1.54–1.42 (m, 1H, $CH_2CH_2CH_3$), 1.30–1.18 (m, 1H, $CH_2CH_2CH_3$), 0.87 (t, $^3J_{H\text{-}H}$ = 7.5 Hz, 3H, $CH_2CH_2CH_3$); **^{13}C NMR** (DMSO-d_6, 75 MHz): δ_C 144.7, 130.8, 129.9, 128.4, 127.0, 121.1, 116.9, 113.0 (ArC*s*), 68.0 (CHOH), 36.9, 19.1, 14.1 ($CH_2CH_2CH_3$); **HRMS (ESI$^+$)** MeOH/$CHCl_3$, *m/z* (RI) (RI): 294.0490 ([M + H]$^+$, 3%), 276.0390 ([M + H-H_2O]$^+$, 100%). [M + H]$^+$ calcd. for $C_{14}H_{16}BrNO$, 294.0488.

General procedure for the synthesis of 14-aza-12-oxasteroid analogues (**4a–l**):

In a 50 mL RB flask with a stir bar, **3a–h** (200 mg, 1 equiv.) was dissolved in toluene (ca. 5 mL), and then the appropriate keto acid [levulinic acid/2-carboxybenzaldehyde/4-acetylbutyric acid/(2-oxopropyl sulfanyl)-acetic acid, 1.5 equiv.] was added. The reaction mixture was refluxed for 2 h using a Dean–Stark apparatus, with dry toluene in the vertical column. At this point, TLC (2% MeOH in DCM) revealed complete consumption of the starting materials and formation of the product. The solvent was removed from the reaction mixture in vacuo and the crude product was purified using column chromatography (using DCM as eluent and slowly increasing the polarity with MeOH) to yield the respective 14-aza-12-oxasteroid analogues.

rac-(11R,12aS)-11-benzyl-12a-methyl-11,12a-dihydro-1H-naphtho[2,1-d]pyrrolo[2,1-b][1,3]oxazin-3(2H)-one (**4a**): Orange solid; 40% yield (104 mg); R_f 0.55 (2% MeOH in DCM); m.p. 124–126 °C; **^{1}H NMR** ($CDCl_3$, 300 MHz): δ_H 8.16 (d, $^3J_{H\text{-}H}$ = 9.0 Hz, 1H, Ar*H*), 7.93 (t, $^3J_{H\text{-}H}$ = 7.5 Hz, 2H, Ar*H*), 7.84 (d, $^3J_{H\text{-}H}$ = 9.0 Hz, 1H, Ar*H*), 7.62 (t, $^3J_{H\text{-}H}$ = 7.5 Hz, 1H, Ar*H*), 7.52 (t, $^3J_{H\text{-}H}$ = 7.5 Hz, 1H, Ar*H*), 7.27–7.22 (m, 3H, Ar*H*), 7.07 (d, $^4J_{H\text{-}H}$ = 3.0, 2H, Ar*H*), 5.85 (d, $^3J_{H\text{-}H}$ = 3.0 Hz, 1H, $PhCH_2CH$), 3.43 (d, $^2J_{H\text{-}H}$ = 12.0 Hz, 1H, $PhCH_2CH$), 3.06 (dd, $^2J_{H\text{-}H}$ = 15.0 Hz, $^3J_{H\text{-}H}$ = 6.0 Hz, 1H, $PhCH_2CH$), 2.58 (t, $^3J_{H\text{-}H}$ = 7.5 Hz, 2H, CH_2CH_2CO), 2.37–2.20 (m, 2H, CH_2CH_2CO), 1.42 (s, 3H, CH_3); **^{13}C NMR** (DMSO, 75 MHz): δ_C 172.1 (CO), 137.7, 131.4, 129.5, 129.1, 128.4, 127.8, 126.5, 126.5, 124.9, 122.6, 121.8, 120.7 (ArC*s*), 89.3 (C12), 72.1 (C11), 42.8 ($PhCH_2$), 32.6, 29.8, 22.4 (CH_2CH_2 and CH_3); **HRMS (ESI$^+$)** MeOH/$CHCl_3$, *m/z* (RI): 344.1644 ([M + H]$^+$, 100%). [M + H]$^+$ calcd. for $C_{23}H_{21}NO_2$, 344.1645.

rac-(11R,12aS)-12a-methyl-11-phenyl-11,12a-dihydro-1H-naphtho[2,1-d]pyrrolo[2,1-b][1,3]oxazin-3(2H)-one (**4b**): Yellow-orange powder; 45% yield (119 mg); R_f 0.46 (2% MeOH in DCM); m.p. 189–191 °C; **^{1}H NMR** ($CDCl_3$, 300 MHz): δ_H 8.46 (d, $^3J_{H\text{-}H}$ = 9.0 Hz, 1H, Ar*H*), 7.90 (d, $^3J_{H\text{-}H}$ = 9.0 Hz, 1H, Ar*H*), 7.83 (d, $^3J_{H\text{-}H}$ = 9.0 Hz, 1H, Ar*H*), 7.48 (d, $^3J_{H\text{-}H}$ = 9.0 Hz, 1H, Ar*H*), 7.36 (t, $^3J_{H\text{-}H}$ = 6.0 Hz, 1H, Ar*H*), 7.32–7.26 (m, 6H, Ar*H*), 6.45 (s, 1H, PhC*H*), 2.71–2.66 (m, 2H, CH_2CH_2CO), 2.29 (t, $^3J_{H\text{-}H}$ = 9.0 Hz, 2H, CH_2CH_2CO), 1.64 (s, 3H, CH_3); **^{13}C NMR** (DMSO-d_6, 75 MHz): δ_C 172.6 (CO), 141.6, 132.3, 131.5, 129.7, 129.5, 129.2, 129.0, 129.0, 128.8, 126.6, 125.2, 124.6, 120.7, 120.6 (ArC*s*), 90.6 (C12), 75.8 (C11), 33.1, 30.6, 22.0 (CH_2CH_2 and CH_3); **HRMS (ESI$^+$)** MeOH/$CHCl_3$, *m/z* (RI): 330.1489 ([M + H]$^+$, 100%). [M + H]$^+$ calcd. for $C_{22}H_{19}NO_2$, 330.1489.

rac-(11R,12aS)-11,12a-dimethyl-11,12a-dihydro-1H-naphtho[2,1-d]pyrrolo[2,1-b][1,3]oxazin-3(2H)-one (**4c**): Viscous orange liquid; 34% yield (96 mg); R_f 0.39 (2% MeOH in DCM); **^{1}H NMR** ($CDCl_3$, 300 MHz): δ_H 8.35 (d, $^4J_{H\text{-}H}$ = 1.5 Hz, 1H, Ar*H*), 7.87–7.78 (m, 3H, Ar*H*), 7.53–7.43 (m, 2H, Ar*H*), 5.66 (q, $^3J_{H\text{-}H}$ = 6.0 Hz, 1H, CH_3CH), 2.67 (t, $^3J_{H\text{-}H}$ = 9.0 Hz, 2H, CH_2CH_2CO), 2.30 (t, $^3J_{H\text{-}H}$ = 9.0 Hz, 2H, CH_2CH_2CO), 1.65 (d, $^3J_{H\text{-}H}$ = 6.0 Hz, 3H, CH_3), 1.47 (s, 3H, CH_3); **^{13}C NMR** ($CDCl_3$, 75 MHz): δ_C 172.1 (CO), 131.2, 130.2, 129.2, 128.9, 128.2, 126.2, 124.8, 123.3, 123.1, 120.4 (ArC*s*), 89.3 (C12), 68.0 (C11), 32.8, 30.1, 23.1, 21.5 (CH_2CH_2 and CH_3); **HRMS (ESI$^+$)** MeOH/$CHCl_3$, *m/z* (RI): 268.1332 ([M + H]$^+$, 100%). [M + H]$^+$ calcd. for $C_{17}H_{17}NO_2$, 268.1331.

rac-(11R,12aS)-12a-methyl-11-propyl-11,12a-dihydro-1H-naphtho[2,1-d]pyrrolo[2,1-b][1,3]oxazin-3(2H)-one (**4d**): Orange solid; yield 45% (123 mg); R_f 0.40 (2% MeOH in DCM); m.p. 123–

125 °C; ^{1}**H NMR** ($CDCl_3$, 300 MHz): δ_H 8.27 (d, $^{3}J_{H\text{-}H}$ = 9.0 Hz, 1H, Ar*H*), 7.87–7.76 (m, 3H, Ar*H*), 7.52 (t, $^{3}J_{H\text{-}H}$ = 9.0 Hz, 1H, Ar*H*), 7.45 (t, $^{3}J_{H\text{-}H}$ = 9.0 Hz, 1H, Ar*H*), 5.58 (dd, $^{3}J_{H\text{-}H}$ = 6.0 Hz, $^{4}J_{H\text{-}H}$ = 1.5 Hz, 1H, CH_3CH), 2.67 (t, $^{3}J_{H\text{-}H}$ = 7.5 Hz, 2H, CH_2CH_2CO), 2.30 (t, $^{3}J_{H\text{-}H}$ = 7.5 Hz, 2H, CH_2CH_2CO), 2.12–2.01 (m, 1H, $CH_2CH_2CH_3$), 1.87–1.75 (m, 1H, $CH_2CH_2CH_3$), 1.55–1.44 (m, 1H, $CH_2CH_2CH_3$), 1.44 (s, 3H, CH_3), 1.32–1.21 (m, 1H, $CH_2CH_2CH_3$), 0.87 (t, $^{3}J_{H\text{-}H}$ = 6.0 Hz, 3H, $CH_2CH_2CH_3$); ^{13}**C NMR** (DMSO-d_6, 75 MHz): δ_C 172.3 (CO), 131.2, 131.0, 129.2, 128.9, 128.1, 126.2, 124.8, 122.9, 122.6, 120.5 (Ar*C*s), 89.3 (*C*12), 71.3 (*C*11), 38.8, 32.8, 30.1, 21.9, 18.0, 13.9 (CH_2CH_2, CH_3, *n*-Pr); **HRMS (ESI$^+$)** MeOH/$CHCl_3$, *m/z* (RI): 296.1631 ([M + H]$^+$, 100%). [M + H]$^+$ calcd. for $C_{19}H_{21}NO_2$, 296.1645.

rac-(11R,12aS)-11-benzyl-8-bromo-12a-methyl-11,12a-dihydro-1H-naphtho[2,1-d]pyrrolo[2,1-b][1,3] oxazin-3(2H)-one (**4e**): Purple solid; 33% yield (82 mg); R_f 0.49 (2% MeOH in DCM); m.p. 106–108 °C; ^{1}**H NMR** ($CDCl_3$, 300 MHz): δ_H 8.16 (d, $^{3}J_{H\text{-}H}$ = 9.0 Hz, 1H, Ar*H*), 8.05 (d, $^{4}J_{H\text{-}H}$ = 1.5 Hz, 1H), 7.79 (d, $^{3}J_{H\text{-}H}$ = 9.0 Hz, 1H, Ar*H*), 7.72 (d, $^{3}J_{H\text{-}H}$ = 9.0 Hz, 1H, Ar*H*), 7.66 (d, $^{3}J_{H\text{-}H}$ = 9.0 Hz, 1H, Ar*H*), 7.23–7.19 (m, 3H, Ar*H*), 7.00–6.99 (m, 2H, Ar*H*), 5.79 (dd, $^{3}J_{H\text{-}H}$ = 6.0 Hz, $^{4}J_{H\text{-}H}$ = 3.0 Hz, 1H, $PhCH_2CH$), 3.36 (dd, $^{2}J_{H\text{-}H}$ = 15.0 Hz, $^{3}J_{H\text{-}H}$ = 3.0 Hz, 1H, $PhCH_2CH$), 3.04 (dd, $^{2}J_{H\text{-}H}$ = 15.0 Hz, $^{3}J_{H\text{-}H}$ = 6.0 Hz, 1H, $PhCH_2CH$), 2.56 (t, $^{3}J_{H\text{-}H}$ = 7.5 Hz, 2H, CH_2CH_2CO), 2.18–2.34 (m, 2H, CH_2CH_2CO), 1.40 (s, 3H, CH_3); ^{13}**C NMR** (DMSO-d_6, 75 MHz): δ_C 172.0 (CO), 137.2, 132.3, 131.7, 131.0, 129.8, 129.4, 127.8, 127.5, 126.6, 124.4, 122.0, 121.8, 118.8 (Ar*C*s), 71.9 (*C*12), 89.3 (*C*11), 42.8 ($PhCH_2$), 32.6, 29.8, 22.3 (CH_2CH_2, CH_3); **HRMS (ESI$^+$)** MeOH/$CHCl_3$, *m/z* (RI): 421.0740 ([M + H]$^+$, 100%). [M + H]$^+$ calcd. for $C_{23}H_{20}BrNO_2$, 421.0677.

rac-(11R,12aS)-8-bromo-12a-methyl-11-phenyl-11,12a-dihydro-1H-naphtho[2,1-d]pyrrolo[2,1-b][1,3] oxazin-3(2H)-one (**4f**): Purple-brown solid; 30% yield (75 mg); R_f 0.46 (2% MeOH in DCM); m.p. 203–205 °C; ^{1}**H NMR** ($CDCl_3$, 300 MHz): δ_H 8.49 (d, $^{3}J_{H\text{-}H}$ = 9.0 Hz, 1H, Ar*H*), 7.97 (s, 1H, Ar*H*), 7.79 (d, $^{3}J_{H\text{-}H}$ = 9.0 Hz, 1H, Ar*H*), 7.31–7.27 (m, 5H, Ar*H*), 7.23–7.21 (m, 2H, Ar*H*), 6.39 (s, 1H, Ph*CH*), 2.70–2.65 (m, 2H, CH_2CH_2CO), 2.29 (t, $^{3}J_{H\text{-}H}$ = 6.0 Hz, 2H, CH_2CH_2CO), 1.64 (s, 3H, CH_3); ^{13}**C NMR** (DMSO-d_6, 75 MHz): δ_C 172.2 (CO), 140.9, 132.3, 132.2, 130.6, 129.5, 128.9, 128.5, 128.1, 127.8, 125.9, 121.4, 120.3, 118.8 (Ar*C*s), 90.2 (*C*12), 75.2 (*C*11), 32.7, 30.1, 21.6 (CH_2CH_2, CH_3); **HRMS (ESI$^+$)** MeOH/$CHCl_3$, *m/z* (RI): 408.0584 ([M + H]$^+$, 100%). [M + H]$^+$ calcd. for $C_{22}H_{18}BrNO_2$, 408.0594.

rac-(11R,12aS)-8-bromo-11,12a-dimethyl-11,12a-dihydro-1H-naphtho[2,1-d]pyrrolo[2,1-b][1,3]-oxazin- 3(2H)-one (**4g**): Viscous orange liquid; 35% yield (92 mg); R_f 0.41 (2% MeOH in DCM); ^{1}**H NMR** ($CDCl_3$, 300 MHz): δ_H 8.35 (d, $^{4}J_{H\text{-}H}$ = 1.5 Hz, 1H, Ar*H*), 7.98 (s, 1H, Ar*H*), 7.69–7.39 (m, 3H, Ar*H*), 5.58 (q, $^{3}J_{H\text{-}H}$ = 6.0 Hz, 1H, CH_3CH), 2.65 (t, $^{3}J_{H\text{-}H}$ = 7.5 Hz, 2H, CH_2CH_2CO), 2.27 (t, $^{3}J_{H\text{-}H}$ = 7.5 Hz, 2H, CH_2CH_2CO), 1.59 (d, $^{3}J_{H\text{-}H}$ = 6.0 Hz, 3H, CH_3), 1.43 (s, 3H, CH_3); ^{13}**C NMR** ($CDCl_3$, 75 MHz): δ_C 172.2 (CO), 132.3, 130.7, 129.4, 127.7, 127.2, 124.8, 123.5, 121.4, 119.0, 118.7 (Ar*C*s), 89.3 (*C*12), 67.7 (*C*11), 32.7, 30.0, 23.1, 21.4 (CH_2CH_2, CH_3); **HRMS (ESI$^+$)** MeOH/$CHCl_3$, *m/z* (RI): 346.0433 ([M + H]$^+$, 100%). [M + H]$^+$ calcd. for $C_{17}H_{16}BrNO_2$, 346.0437.

rac-(11R,12aS)-8-Bromo-12a-methyl-11-propyl-11,12a-dihydro-1H-naphtho[2,1-d]pyrrolo[2,1-b][1,3]- oxazin-3(2H)-one (**4h**): Purple oil; 40% yield (102 mg); R_f 0.62 (2% MeOH in DCM); ^{1}**H NMR** ($CDCl_3$, 300 MHz): δ_H 8.29 (d, $^{3}J_{H\text{-}H}$ = 9.0 Hz, 1H, Ar*H*), 7.99 (d, $^{4}J_{H\text{-}H}$ = 1.5 Hz, 1H, Ar*H*), 7.70–7.62 (m, 2H, Ar*H*), 7.56 (d, $^{3}J_{H\text{-}H}$ = 9.0 Hz, 1H, Ar*H*), 5.53 (dd, $^{3}J_{H\text{-}H}$ = 6.0 Hz, 3.0 Hz, 1H, CH_3CH), 2.66 (t, $^{3}J_{H\text{-}H}$ = 7.5 Hz, 2H, CH_2CH_2CO), 2.29 (t, $^{3}J_{H\text{-}H}$ = 7.5 Hz, 2H, CH_2CH_2CO), 2.06–1.95 (m, 1H, $CH_2CH_2CH_3$), 1.82–1.70 (m, 1H, $CH_2CH_2CH_3$), 1.51–1.39 (m, 1H, $CH_2CH_2CH_3$), 1.43 (s, 3H, CH_3), 1.23–1.15 (m, 1H, $CH_2CH_2CH_3$), 0.85 (t, $^{3}J_{H\text{-}H}$ = 7.5 Hz, 3H, $CH_2CH_2CH_3$); ^{13}**C NMR** (DMSO-d_6, 75 MHz): δ_C 172.3 (CO), 132.4, 131.3, 130.8, 129.4, 127.6, 127.2, 124.6, 122.7, 121.5, 118.7 (Ar*C*s), 89.3 (*C*12), 71.1 (*C*11), 38.8, 32.8, 30.0, 21.8,

17.9, 13.8 (CH_2CH_2, CH_3, *n*-Pr); **HRMS (ESI$^+$)** MeOH/$CHCl_3$, *m/z* (RI): 374.0739 ($[M + H]^+$, 100%). $[M + H]^+$ calcd. for $C_{19}H_{20}BrNO_2$, 374.0750.

7-Methyl-7H-naphtho[2′,1′:4,5][1,3]oxazino[2,3-a]isoindol-13(8aH)-one (**4i**): Viscous orange liquid; 78% yield (249 mg). R_f 0.79 (2% MeOH in DCM); mixture of two diastereomers (6:4). Data for the major diastereomer are presented here; **^{1}H NMR** ($CDCl_3$, 300 MHz): δ_H 8.56 (d, $^4J_{H\text{-}H}$ = 1.5 Hz, 1H, Ar*H*), 7.94–7.43 (m, 9H, Ar*H*), 5.95 (q, $^3J_{H\text{-}H}$ = 9.0 Hz, 1H, CH_3CH), 5.79 (s, 1H, NC*H*O), 1.72 (d, $^3J_{H\text{-}H}$ = 9.0 Hz, 3H, CH_3); **^{13}C NMR** ($CDCl_3$, 75 MHz): δ_C 164.8 (CO), 140.2, 132.7, 130.2, 128.9, 128.5, 126.6, 126.4, 124.7, 124.6, 124.0, 123.5, 123.2, 123.1, 122.4, 120.8, 119.0 (ArCs), 83.1, 72.4, 23.3 (CH_3); **HRMS (ESI$^+$)** MeOH/$CHCl_3$, *m/z* (RI): 302.1173 ($[M + H]^+$, 100%). $[M + H]^+$ calcd. for $C_{20}H_{15}NO_2$, 302.1176.

4-Bromo-7-methyl-7H-naphtho[2′,1′:4,5][1,3]oxazino[2,3-a]isoindol-13(8aH)-one (**4j**): Colorless solid; 40% yield (112 mg); R_f 0.85 (2% MeOH in DCM); m.p. 230–233 °C. Mixture of two diastereomers (8:2). Data for the major diastereomer are presented here; **^{1}H NMR** ($CDCl_3$, 300 MHz): δ_H 8.58 (d, $^3J_{H\text{-}H}$ = 9.0 Hz, 1H, Ar*H*), 8.03 (s, 1H, Ar*H*), 7.95 (d, $^3J_{H\text{-}H}$ = 9.0 Hz, 1H, Ar*H*), 8.03–7.59 (m, 6H, Ar*H*), 5.94 (q, $^3J_{H\text{-}H}$ = 6.0 Hz, 1H, CH_3CH), 5.87 (s, 1H, NC*H*O), 1.70 (d, $^3J_{H\text{-}H}$ = 6.0 Hz, 3H, CH_3); **^{13}C NMR** ($CDCl_3$, 75 MHz): δ_C 164.9 (CO), 140.2, 132.9, 132.2, 131.0, 130.5, 130.0, 129.7, 128.2, 127.7, 124.9, 124.2, 123.6, 122.5, 120.3, 118.6 (ArCs), 83.2 (C-12), 72.3 (C-11), 23.4 (CH_3); **HRMS (ESI$^+$)** MeOH/$CHCl_3$, *m/z* (RI): 380.0284 ($[M + H]^+$, 96%). $[M + H]^+$ calcd. for $C_{20}H_{14}BrNO_2$, 380.0281.

rac-(12R,13aS)-12,13a-dimethyl-1,2,3,13a-tetrahydronaphtho[2,1-d]pyrido[2,1-b][1,3]oxazin-4(12H)-one (**4k**): Viscous orange liquid; 30% yield (91 mg); R_f 0.61 (2% MeOH in DCM); **^{1}H NMR** ($CDCl_3$, 300 MHz): δ_H 7.87–7.70 (m, 4H, Ar*H*), 7.55–7.46 (m, 2H, Ar*H*), 5.65 (q, $^3J_{H\text{-}H}$ = 6.0 Hz, 1H, CH_3CH), 2.63 (t, $^3J_{H\text{-}H}$ = 9.0 Hz, 2H, $CH_2CH_2CH_2$), 2.13 (t, $^3J_{H\text{-}H}$ = 9.0 Hz, 2H, $CH_2CH_2CH_2$), 1.63 (d, $^3J_{H\text{-}H}$ = 6.0 Hz, 3H, CH_3), 1.38 (s, 3H, CH_3), 0.83 (t, $^3J_{H\text{-}H}$ = 9.0 Hz, 2H, $CH_2CH_2CH_2$); **^{13}C NMR** ($CDCl_3$, 75 MHz): δ_C 170.1 (CO), 132.8, 131.6, 128.9, 128.8, 127.6, 127.1, 126.2, 125.2, 122.5 (ArCs), 86.1 (C12), 67.3 (C11), 37.0, 34.1, 25.0, 23.5, 17.0 ($CH_2CH_2CH_2$, CH_3); **HRMS (ESI$^+$)** MeOH/$CHCl_3$, *m/z* (RI): 282.3619 ($[M + H]^+$, 100%). $[M + H]^+$ calcd. for $C_{18}H_{19}NO_2$: 282.3625.

rac-(12R,13aS)-12,13a-dimethyl-12,13a-dihydro-1H-naphtho[2,1-d][1,4]thiazino[3,4-b][1,3]oxazin-4(3H)-one (**4l**): Viscous orange liquid; 30% yield (92 mg); R_f 0.62 (2% MeOH in DCM); **^{1}H NMR** ($CDCl_3$, 300 MHz): δ_H 7.87–7.66 (m, 4H, Ar*H*), 7.56–7.47 (m, 2H, Ar*H*), 5.65 (q, $^3J_{H\text{-}H}$ = 7.5 Hz, 1H, CH_3CH), 3.60 (m, 2H, CH_2SCH_2), 3.06 (s, 2H, CH_2SCH_2), 1.68 (d, $^3J_{H\text{-}H}$ = 7.5 Hz, 3H, CH_3), 1.54 (s, 3H, CH_3); **^{13}C NMR** ($CDCl_3$, 75 MHz): δ_C 165.6 (CO), 132.1, 131.6, 128.8, 128.7, 127.2, 126.3, 125.5, 124.6, 122.5 (ArCs), 87.7 (C12), 67.5 (C11), 39.0, 33.8, 23.4, 23.3 (CH_2SCH_2), 2xCH_3); **HRMS (ESI$^+$)** MeOH/$CHCl_3$, *m/z* (RI): 300.1047 ($[M + H]^+$, 100%). $[M + H]^+$ calcd. for $C_{17}H_{17}NO_2S$: 300.1053.

5. Conclusions

We have devised a concise synthesis of 14-aza-12-oxasteroids, **4a–l** in four simple steps, starting from inexpensive and readily available 2-naphthol analogues. The Bucherer reaction was employed to convert naphthol analogues to their corresponding naphthylamines, which in turn were ortho-acylated using the Sugasawa reaction. Furthermore, the 1-acylated-2-aminonaphthylenes were then reduced to the corresponding amino-alcohols. Double dehydrations and double intramolecular cyclizations of the synthesized amino-alcohols with four different oxo-acids resulted in a one-pot formation of a C-N bond, a C-O bond, and an amide bond in tandem. A series of twelve 14-aza-12-oxasteroid analogues were synthesized in moderate yields. The structure, stereochemistry, and connectivity of the bonds were further confirmed by an X-ray crystal structure determination of one of the analogues, 12a-methyl-11-phenyl-11,12a-dihydro-1*H*-naphtho[2,1-d]pyrrolo[2,1-b][1,3]oxazin-3(2*H*)-one (**4b**).

Supplementary Materials: The following supporting information can be downloaded at: https://www.mdpi.com/article/10.3390/molecules30020415/s1, ^{1}H and ^{13}C NMR spectra of all compounds reported in this article. Figure S1: Structure of compound **4b** showing the two molecules of the independent unit in the correct relative orientation. Figure S2: Separate diagrams of the two molecules in compound **4b** showing the full heavy atom labelling. Table S1: Crystal data and structure refinement for Compound **4b**. Refs [36–41] are cited in Supplementary Materials.

Author Contributions: Conceptualization, A.J.; methodology, J.L. and D.W.; X-ray crystallography, K.N.R.; writing—original draft preparation, S.S., J.L., and D.W.; writing—review and editing, A.J.; supervision, A.J. and S.S.; All authors have read and agreed to the published version of the manuscript.

Funding: This research was funded by the Natural Sciences and Engineering Research Council (NSERC) of Canada (salary support to JL and DW) and Beatrice Hunter Cancer Research Institute (BHCRI), Halifax (salary support to SS).

Institutional Review Board Statement: Not applicable.

Informed Consent Statement: Not applicable.

Data Availability Statement: All spectroscopic and X-ray data obtained during this research are included in the Supplementary Materials (see above).

Acknowledgments: Paraza Pharma, Inc. is thanked for their support in recording HRMS data of reported products.

Conflicts of Interest: The authors declare no conflicts of interest.

References

1. Ibrahim-Ouali, M. Recent advances in oxasteroids chemistry. *Steroids* **2007**, *72*, 475–508. [CrossRef] [PubMed]
2. Cole, T.J.; Short, K.L.; Hopper, S.B. The science of Steroid. *Semin. Fetal Neonatal Med.* **2019**, *24*, 170–175. [CrossRef] [PubMed]
3. White, R.; Parker, M.G. Molecular mechanisms of steroid hormone action. *Endocr.-Relat. Cancer* **1998**, *5*, 1–14. [CrossRef]
4. Frye, C.A. Steroids, reproductive endocrine function, and affect: A review. *Minerva Ginecol.* **2009**, *61*, 541–562. [PubMed]
5. Wall, E.H.; Hewitt, S.C.; Case, L.K.; Lin, C.-Y.; Korach, K.S.; Teuscher, C. The role of genetics in estrogen responses: A critical piece of an intricate puzzle. *FASEB J.* **2014**, *12*, 5042–5054. [CrossRef]
6. Chen, L.; Xu, T.; Lou, J.; Zhang, T.; Wu, S.; Xie, R.; Xu, J. The beneficial roles and mechanism of estrogens in immune health and infection disease. *Steroids* **2024**, *207*, 109426. [CrossRef]
7. Taraborrelli, S. Physiology, production and action of progesterone. *Acta Obstet. Gynecol. Scand.* **2015**, *94*, 8–16. [CrossRef] [PubMed]
8. Bansal, R.; Acharya, P.C. Man-made cytotoxic steroids: Exemplary agents for cancer therapy. *Chem. Rev.* **2014**, *114*, 6986–7005. [CrossRef] [PubMed]
9. Guon, Z. The modification of natural products for medical use. *Acta Pharm. Sin. B* **2017**, *7*, 119–136. [CrossRef]
10. Kerru, N.; Gummidi, L.; Maddila, S.; Gangu, K.K.; Jonnalagadda, S.B. A review on recent advances in nirogen containing molecules and their biological applications. *Molecules* **2020**, *25*, 1909. [CrossRef]
11. Singh, H.; Jindal, D.P.; Yadav, M.R.; Kumar, M. Heterosteroids and drug research. In *Progress in Medicinal Chemistry*; Ellis, G.P., West, G.B., Eds.; Elsevier: Amsterdam, The Netherlands, 1991; pp. 233–300. [CrossRef]
12. Ibrahim-Ouali, M.; Rocheblave, L. Recent advances in azasteroids chemistry. *Steroids* **2008**, *73*, 375–407. [CrossRef] [PubMed]
13. Ibrahim-Ouali, M.; Santelli, M. Recent advances in thiasteroids chemistry. *Steroids* **2006**, *71*, 1025–1044. [CrossRef] [PubMed]
14. Burbiel, J.; Bracher, F. Azasteroids as antifungals. *Steroids* **2003**, *68*, 587–594. [CrossRef]
15. Dembitsky, V.M. Steroidsbearing heteroatom as potential drugs for medicine. *Biomediciens* **2023**, *11*, 2698–2785. [CrossRef]
16. Helfman, T.; Falanga, D. Miami and Florida, Stanozolol as a novel therapeutic agent in dermatology. *J. Am. Acad. Dermatol.* **1995**, *33*, 254–258. [CrossRef] [PubMed]
17. Cottreau, C.M.; Ness, R.B.; Modugno, F.; Allen, G.O.; Goodman, M.T. Endometriosis and its treatment with danazol or lupron in relation to ovarian cancer. *Clin. Cancer Res.* **2003**, *9*, 5142–5144. [PubMed]
18. Curtis, D.R.; Malik, R. Glycine antagonism by RU 5135. *Eur. J. Pharmacol.* **1985**, *110*, 383–384. [CrossRef] [PubMed]

19. Edwards, J.; Moore, R.A. Finasteride in the treatment of clinical benign prostatic hyperplasia: A systematic review of randomised trials. *BMC Urol.* **2002**, *2*, 14–30. [CrossRef] [PubMed]
20. Pohlman, G.D.; Pohlman, E.A.; Crawford, E.D. Dutasteride: A review of its use in the management of prostate disorders. *Clin. Med. Insights Ther.* **2011**, *3*, 172–177. [CrossRef]
21. Oumzil, K.; Ibrahim-Ouali, M.; Santelli, M. First total synthesis of (±)-3-aza-11-oxa-1,3,5(10)-trieno steroids. *Steroids* **2006**, *71*, 886–894. [CrossRef] [PubMed]
22. Singh, R.; Panda, G. L-Proline derived nitrogenous steroidal systems: An asymmetric approach to 14-azasteroids. *RSC Adv.* **2013**, *3*, 19533–19544. [CrossRef]
23. Bernath, G.; Fueloep, F.; Argay, G.; Kalman, A.; Sohar, P. Stereochemical studies. Saturated heterocycles. A simple stereospecific synthesis of oxazasteroids. *Tetrahedron Lett.* **1981**, *22*, 3797–3800. [CrossRef]
24. Abdelkhalik, A.M.; Paul, N.K.; Jha, A. Concise synthesis of 12a-methul-11-aryl-1, 2-dihydrobenzo[f]pyrrolo [1,2-a]quinolin-3(12aH)-ones as racemic 14-azaestrongen analogues. *Steroids* **2015**, *98*, 107–113. [CrossRef]
25. Jha, A.; Chou, T.-Y.; Jaroudi, Z.A.L.; Ellis, B.D.; Cameron, T.S. Aza-Diels–Alder reaction between N-aryl-1-oxo-1H-isoindolium ions and tert-enamides: Steric effects on reaction outcome. *Beilstein J. Org. Chem.* **2014**, *10*, 848–857. [CrossRef] [PubMed]
26. Jha, A.; Naidu, A.B.; Abdelkhalik, A.M. Transition metal-free one-pot cascade synthesis of 7-oxa-2-azatricyclo [7.4.0.0^{2,6}] trideca-1(9), 10, 12-trien-3-ones from biomass-derived levulinic acid under mild conditions. *Org. Biomol. Chem.* **2013**, *11*, 7559–7565. [CrossRef]
27. Sugasawa, T.; Toyoda, T.; Adachi, M.; Sasakura, K. Aminohaloborane in organic synthesis. 1. Specific ortho substitution reaction of anilines. *J. Am. Chem. Soc.* **1978**, *100*, 4842–4852. [CrossRef]
28. Canete, A.; Melendrez, M.X.; Saitz, C.; Zanocco, A.L. Synthesis of amino naphthalene derivatives using the Bucherer reaction under microwave irradiation. *Synth. Commun.* **2001**, *31*, 2143–2148. [CrossRef]
29. Sartori, G.; Maggi, R. Use of solid catalysts in Friedel−Crafts acylation reactions. *Chem. Rev.* **2006**, *106*, 1077–1104. [CrossRef] [PubMed]
30. González-Morales, A.; Díaz-Coutiño, D.; Fernánez-Zertuche, M.; García-Barradas, O.; Ordóñez, M. Preparation of dimethyl (R)- and (S)-2-(2-aminophenyl)-2-hydroxyethylphosphonate from anthranilic acid. *Tetrahedron Asymmetry* **2004**, *15*, 457–463. [CrossRef]
31. Zhao, Y.; Huang, B.; Yang, C.; Chen, Q.; Xia, W. Sunlight-driven forging of amide/ester bonds from three independent components: An approach to carbamates. *Org. Lett.* **2016**, *18*, 5572–5575. [CrossRef] [PubMed]
32. Orloff, H.D. The stereoisomerism of cyclohexane derivatives. *Chem. Rev.* **1954**, *54*, 347–447. [CrossRef]
33. Imoto, H.; Fujii, R.; Naka, K. Solid-state emissive diaminomaleimide dimers and a polymer—Syntheses, structures and optical properties. *Eur. J. Org. Chem.* **2019**, *2019*, 3086–3092. [CrossRef]
34. Allen, F.H.; Kennard, O.; Watson, D.G.; Brammer, L.; Orpen, A.G.; Taylor, R. Tables of bond lengths determined by X-Ray and neutron diffraction. Part I. Bond lengths in organic compounds. *J. Chem. Soc. Perkin Trans.* **1987**, *2*, S1–S19. [CrossRef]
35. Brink, M. "Long range"—Kopplungen in einigen acetonylderivaten. *Tetrahedron Lett.* **1971**, *29*, 2753–2756. [CrossRef]
36. *APEX 3*; Bruker AXS Inc.: Madison, WI, USA, 2018.
37. *SAINT*; Bruker AXS Inc.: Madison, WI, USA, 2016.
38. *SADABS*; Bruker AXS Inc.: Madison, WI, USA, 2016.
39. Sheldrick, G.M. SHELXT—Integrated space-group and crystal structure determination. *Acta Cryst.* **2015**, *A71*, 3–8. [CrossRef] [PubMed]
40. Sheldrick, G.M. Crystal Structure Refinement with SHELXL. *Acta Cryst.* **2015**, *C71*, 3–8. [CrossRef]
41. Spek, A.L. Structure validation in chemical crystallography. *Acta Cryst.* **2009**, *D65*, 148–155. [CrossRef]

molecules

MDPI

Article

Synthesis and Optical Properties of *N*-Arylnaphtho- and Anthra[2,3-*d*]oxazol-2-amines

Yuki Murata , Masato Kawakubo, Ayumi Maruyama, Mio Matsumura and Shuji Yasuike *

School of Pharmaceutical Sciences, Aichi Gakuin University, 1-100 Kusumoto-cho, Chikusa-ku, Nagoya 464-8650, Japan; y-murata@dpc.agu.ac.jp (Y.M.); ag213a02@dpc.agu.ac.jp (M.K.); yaohuaxue870@gmail.com (A.M.); m-matsu@dpc.agu.ac.jp (M.M.)

* Correspondence: s-yasuik@dpc.agu.ac.jp; Tel.: +81-52-757-6774

Abstract: Oxazole, a versatile and significant heteroarene, serves as a bridge between synthetic organic chemistry and applications in the medicinal, pharmaceutical, and industrial fields. Polycyclic aromatic compounds with amino groups substituted at the 2-position of an oxazole, such as 2-aminonaphthoxazoles, are expected to be functional probes, but their synthetic methods are extremely limited. Herein, we describe electrochemical reactions of 3-amino-2-naphthol or 3-amino-2-anthracenol and isothiocyanates in DMSO, using a graphite electrode as an anode and a platinum electrode as a cathode in the presence of potassium iodide (KI), which afford *N*-arylnaphtho- and *N*-arylanthra[2,3-*d*]oxazol-2-amines via cyclodesulfurization. This reaction is the first example of synthesis of 2-aminoxazole-based polycyclic compounds using an electrochemical reaction. An examination of the spectroscopic properties of polycyclic oxazoles revealed that the λ_{abs} value of the tetracyclic oxazoles was redshifted relative to that of the tricyclic oxazoles. Moreover, synthesized naphthalene/anthracene-fused tricyclic and tetracyclic oxazoles exhibited extended π-conjugated skeletons and fluoresced in the 340–430 nm region in chloroform.

Keywords: electrochemical reaction; polycyclic benzene-fused oxazole; cyclodesulfurization; thiourea; optical property

1. Introduction

Oxazoles are important building blocks for the synthesis of functional organic materials and biologically active compounds [1–5]. Among them, *N*-substituted benzo[*d*]oxazol-2-amines (2-aminobenzoxazoles) have garnered considerable attention due to their potential applications in biological and pharmaceutical therapies. For example, suvorexant was recently approved as a medication for treating insomnia [6,7]. Furthermore, 2-aminobenzoxazole derivatives have demonstrated significant inhibitory effects on 5-lipoxygenase and p90 ribosomal S6 kinase, as well as strong α-glucosidase inhibition activity [8–10]. As a result, numerous synthetic methods for producing 2-aminobenzoxazoles have been developed [11–13]. Moreover, 2-aminonaphthoxazole, which has one more fused benzene ring, has also been reported to be a potential probe for detecting γ-hydroxybutyric acid in soft drinks and alcoholic beverages based on color and fluorescence changes [14]. However, only a few methods are known for the synthesis of its parent 2-aminonaphthoxazole [14–18]. For instance, in 2018, synthesis of *N*-phenylnaphtho[2,3-*d*]oxazol-2-amine via triphenylbismuth dichloride-promoted cyclodesulfurization of thioureas derived from 3-amino-2-naphthol and phenyl isothiocyanate was reported [15]. Since then, Cu_2O/tetrabutylammonium bromide [16] and elemental sulfur/K_2CO_3 systems [17] have been used as reagents for the oxidative cyclodesulfurization reactions, and artificial hemoglobin-containing cobalt porphyrin [18] has been reported to serve as a catalyst in these reactions. However, these studies were more focused on the synthesis of benzoxazoles than on the synthesis of polycyclic *N*-arylnaphtho- and anthra[2,3-*d*]oxazol-2-amines. Moreover, the spectroscopic properties of the synthesized compounds were not investigated despite their potential as functional materials. Electrochemical synthesis attracted attention as an environmentally friendly method [19–21]. Bicyclic 2-aminobenzoxazoles have been synthesized through tandem electrochemical reactions from 2-aminophenols and isothiocyanates via thiourea [22]. Inspired by these works, this study aimed to synthesize *N*-aryloxazol-2-amines fused with naphthalene and anthracene rings, namely *N*-arylnaphtho- and *N*-arylanthra[2,3-*d*]oxazol-2-amines, via electrochemical cyclodesulfurization from 3-amino-2-naphthol or 3-amino-2-anthracenol and isothiocyanates. The optical properties of the synthesized compounds were also investigated.

2. Results and Discussion

In the electrolytic synthesis of 2-aminobenzoxazoles from 2-aminophenols and isothiocyanates, as reported by Wacharasindhu et al., two reagents, NaI and NaCl, were used as electrolytes and/or reaction mediators with a mixed ethanol–water solvent [22]. Since polycyclic aromatic heterocycles are poorly soluble in water, we attempted to carry out the reaction under simpler electrolysis conditions while considering the solubility of the products. The electrocyclization of thioamide (**3a**) generated from 3-amino-2-naphthol (**1**) and phenyl isothiocyanate (**2a**) in the presence of various electrolytes was carried out to obtain suitable reaction conditions (Table 1). To clarify the effect of the electrolyte, MeCN was used as the solvent with a current of 20 mA using a graphite electrode as the anode [C(+)] and a platinum electrode as the cathode [Pt(−)] (entries 1–5). KI was found to be the best electrolyte for the reaction in terms of the yield of the desired oxazole (**4**). Solvent screening experiments suggested that the reaction proceeded efficiently in DMSO (87%), MeCN (83%), DMF (79%), and DMA (69%) (entries 4 and 6–8), whereas the reaction did not proceed in EtOH or THF (entries 9 and 10). Next, several combinations of the graphite and platinum electrodes were examined. As evident in Table 1, slightly higher yields were obtained when graphite and platinum were used as the anode and cathode, respectively (entries 4 and 11–13). The yield decreased when the KI concentration was reduced (entry 14). When the applied current was reduced from 20 to 5 mA, the reaction time increased (entry 15). Thus, the best result was obtained when **3a**, generated from **1** and **2a**, reacted in DMSO in the presence of a KI (0.5 M) electrolyte at a current density of 20 mA using a graphite [C(+)] anode and a platinum [Pt(−)] cathode (entry 8).

The generality of this protocol was evaluated by examining the reactions of 3-amino-2-naphthol (**1**) or 3-amino-2-anthracenol (**5**) with aryl isothiocyanate (**2**); the results are summarized in Table 2. The reaction of **1** with aryl isothiocyanate (**2**) gave the corresponding tricyclic naphthoxazoles (**6–8**) in high yields. When 3-amino-2-anthracenols (**5**) were used as a substrate instead of **1**, the desired tetracyclic anthroxazoles (**9–12**) were obtained, albeit in low yields. These lower yields can be attributed to the low solubility of the thiourea intermediates (**3**) derived from **5** and **2**.

The mechanism of this electrochemical synthesis is unclear at present. However, a plausible mechanism has been proposed (Scheme 1). Iodine (I_2) is generated from the iodide (I^-) of an electrolyte such as nBuN$_4$I, KI, and NaI at the positive electrode of the electrolytic system [23–25]. The thiolic form (**A**), a tautomer of thiourea derived from aminophenol and isothiocyanate, reacts with iodine to produce **B**. The electrochemical reaction of 1,3-diphenylthiourea (**13**) under standard conditions affords carbodiimide (**14**) in a 62% yield (Scheme 2a), and the reductive elimination of **B** generates carbodiimide (**C**), sulfur, and hydroiodic acid (Route A). Finally, **C** undergoes intramolecular cyclization to give polycyclic oxazole. Another possible route (Route B) could be the intramolecular cyclization of **B** to give oxazole through dehydrosulfurization via **D**. Reactions of thiourea (**3a**) using catalytic or equivalent amounts of iodine, without electrolysis, gave the corresponding product (**4**) in a 43% yield (Scheme 2b). These results suggest that the addition of iodine alone is less effective and that electrolytic reactions proceed more efficiently using KI, although the reaction mechanism remains unclear.

Table 1. Screening of reaction conditions [a].

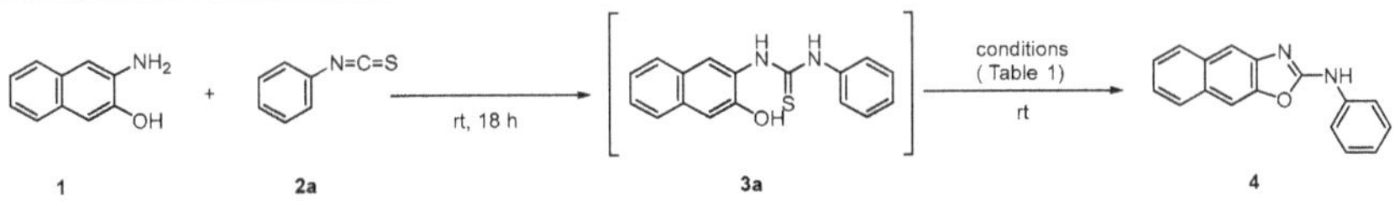

Entry	Electrolyte	Electrode	Current (mA)	Solvent	Time (h)	Yield (%) [b]
1	NaI	C(+)/Pt(−)	20	MeCN	3	64
2	NH_4I	C(+)/Pt(−)	20	MeCN	3	77
3	nBu$_4$NI	C(+)/Pt(−)	20	MeCN	4	76
4	KI	C(+)/Pt(−)	20	MeCN	3	83
5	KBr	C(+)/Pt(−)	20	MeCN	24	---
6	KI	C(+)/Pt(−)	20	DMF	3	79
7	KI	C(+)/Pt(−)	20	DMA	3	69
8	KI	C(+)/Pt(−)	20	DMSO	2	87
9	KI	C(+)/Pt(−)	20	EtOH	24	---
10	KI	C(+)/Pt(−)	20	THF	24	---
11	KI	C(+)/C(-)	20	DMSO	2	72
12	KI	Pt(+)/Pt(−)	20	DMSO	3	81
13	KI	Pt(+)/C(-)	20	DMSO	4	81
14 [c]	KI	C(+)/Pt(−)	20	DMSO	3	60
15	KI	C(+)/Pt(−)	5	DMSO	24	80

[a] **1a** (0.4 mmol), **2a** (0.4 mmol), electrolyte (0.5 mmol, 0.1 M), and solvent (5 mL). [b] Yield of isolated product. [c] Electrolyte (0.024 M).

Table 2. Synthesis of polycyclic *N*-aryloxazol-2-amines [a,b].

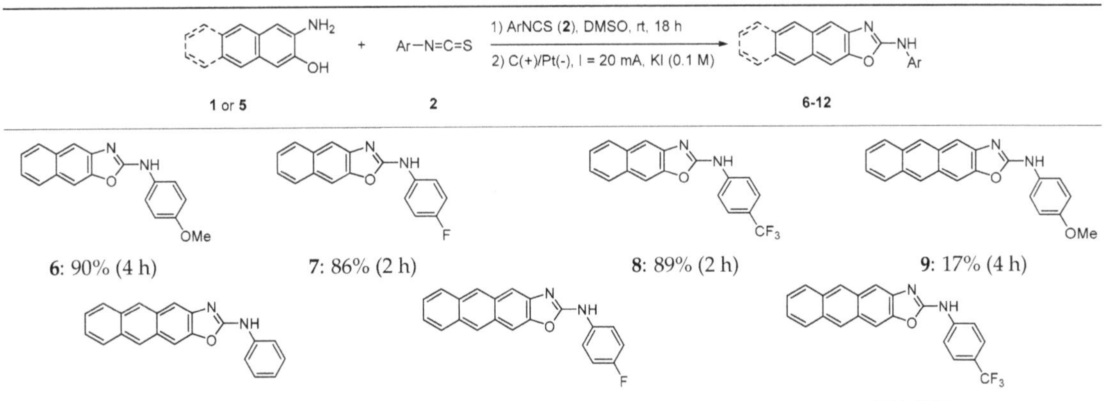

6: 90% (4 h)
7: 86% (2 h)
8: 89% (2 h)
9: 17% (4 h)
10: 36% (2 h)
11: 47% (2 h)
12: 53% (2 h)

[a] All reactions were carried out using **1** or **5** (0.4 mmol), **2** (0.4 mmol), KI (0.5 mmol, 0.1 M), and DMSO (5 mL). [b] Yield of isolated product.

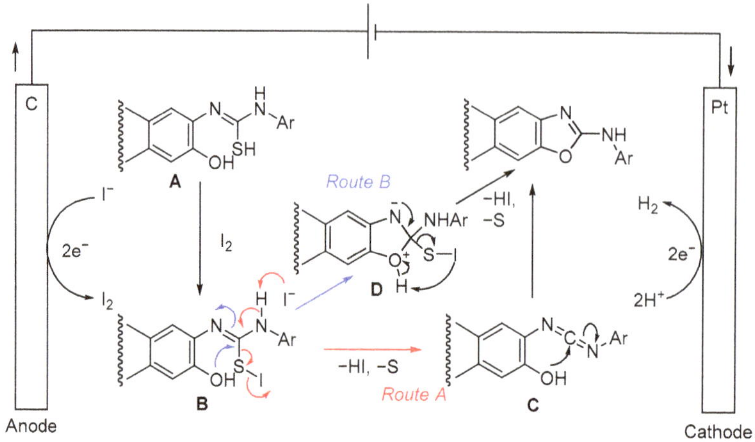

Scheme 1. Proposed mechanism.

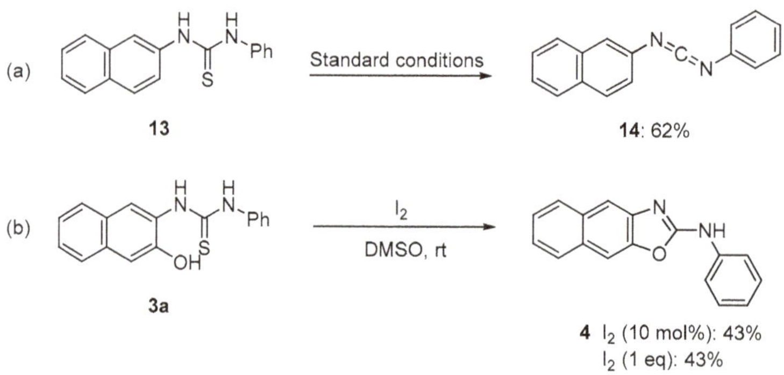

Scheme 2. Control reactions. (**a**) Desulfurization of thiourea. (**b**) Experiments using catalytic or equivalent amounts of iodine.

The optical properties of tricyclic and tetracyclic oxazole derivatives **4** and **6–12** were evaluated based on their UV–Vis absorption and fluorescence spectra in chloroform (Figures 1 and 2 and Table 3). Tricyclic oxazoles **4** and **6–8** exhibited an absorption band with a maximum (λ_{abs}) at around 330 nm. The absorption maxima were largely unaffected by the substituents at the 4-position of the *N*-phenyl ring. Tetracyclic oxazoles **9–12** displayed three bands (λ_{abs}) ranging from 350 to 400 nm, which were largely unaffected by the substituents at the 4-position of the *N*-phenyl ring. It is evident that the λ_{abs} value of the tetracyclic oxazoles was redshifted by ~70 nm relative to that of the tricyclic oxazoles, as presented in Table 3. Compounds **4** and **6–8** exhibited fluorescence maxima (λ_{em}) at 339–357 nm. The 4-methoxyphenyl derivative (**6**) exhibited negligible fluorescence, while the 4-trifluoromethyl derivative (**8**) showed stronger fluorescence (Φ_F = 32%) than compounds **4** and **7**. Furthermore, the λ_{em} of **4** and **6–8** was redshifted from 339 to 357 nm with an increase in the electron-donating ability of the substituents at the 4-position on the *N*-phenyl ring. Tetracyclic oxazoles **9–12** exhibited two fluorescence emission bands at around 410 and 430 nm, with Φ_F values of 14–27%. The λ_{em} values of tetracyclic oxazoles **9–12** were not affected by the electronic property of the substituent on the *N*-phenyl ring.

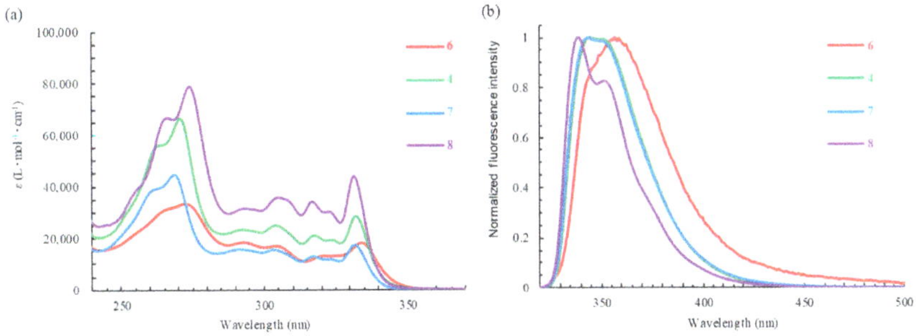

Figure 1. The (**a**) absorption and (**b**) fluorescence spectra of *N*-arylnaphtho[2,3-*d*]oxazol-2-amines **4** and **6–8** in $CHCl_3$. The excitation wavelength was 310 nm.

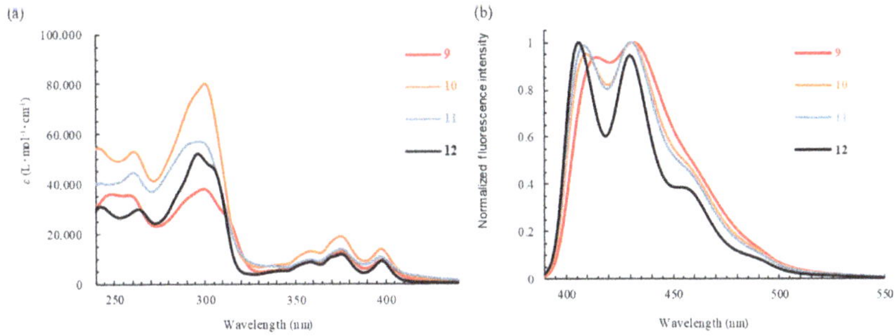

Figure 2. The (**a**) absorption and (**b**) fluorescence spectra of *N*-arylanthra[2,3-*d*]oxazol-2-amines **9–12** in $CHCl_3$. The excitation wavelength was 380 nm.

Table 3. Absorption (C = 1.30×10^{-5}–2.98×10^{-5} M) and fluorescence (C = 1.51×10^{-6}–4.73×10^{-6} M) data for **4** and **6–12** in $CHCl_3$ at room temperature.

Compound	Ar	λ_{abs} (nm)	λ_{em} (nm) [a]	Φ_F (%) [a,b]
6	4-$MeOC_6H_4$	334	357	<1
4	C_6H_5	332	345	15
7	4-FC_6H_4	332	344	11
8	4-$CF_3C_6H_4$	332	339	32
9	4-$MeOC_6H_4$	359, 376, 398	414, 431	27
10	C_6H_5	359, 376, 398	409, 431	18
11	4-FC_6H_4	359, 376, 398	408, 430	14
12	4-$CF_3C_6H_4$	359, 376, 398	406, 430	20

[a] Excitation at **310** (**4** and **6–8**) or **380** (**9–12**) nm. [b] The relative quantum yields were estimated using a solution of anthracene in ethanol (Φ_F = 0.27) for **4** and **6–8** and using 9,10-diphenyl anthracene in cyclohexane (Φ_F = 0.90) for **9–12** [26].

3. Materials and Methods

3.1. General Information

Unless otherwise stated, all reagents and solvents were purchased from commercial suppliers and used without further purification. ^1H NMR (DMSO: δ 2.48 ppm as an internal standard), ^{13}C NMR (DMSO-d_6: δ 39.5 ppm as an internal standard), and ^{19}F NMR (trifluoromethylbenzene: δ −64.0 ppm as an external standard) spectra were recorded on a JEOL ECZ-400S (400 MHz, 100 MHz, and 376 MHz) spectrometer (JEOL Ltd., Tokyo, Japan) in DMSO-d_6. Melting points were measured on a Yanagimoto micro melting point hot stage apparatus (Yanaco Technical Science Co., Ltd., Tokyo, Japan) and were not corrected. GC-MS (EI) spectra were recorded on an Agilent 5977E Diff-SST MSD-230V spectrometer

(Agilent Technologies Japan, Ltd., Tokyo, Japan). HRMS (ESI) were measured on an Agilent 6230 TOF mass spectrometer (Agilent Technologies Japan, Ltd., Tokyo, Japan). IR spectra were recorded on an FTIR-8400S system from Shimadzu (SHIMADZU Corp., Kyoto, Japan) and are reported using the frequency of absorption (cm^{-1}). UV–Vis spectra were recorded at room temperature on a HITACHI U-2800A spectrophotometer (Hitachi High-Tech Corporation, Tokyo, Japan, C = 1.30×10^{-5}–2.98×10^{-5} M in $CHCl_3$), and fluorescence spectra were recorded on a JASCO FP-8300 luminescence spectrometer (JASCO Corporation, Tokyo, Japan, C = 1.51×10^{-6}–4.73×10^{-6} M in $CHCl_3$). Only selected IR absorbencies are reported. All chromatographic separations were accomplished with Silica Gel 60N (Kanto Chemical Co., Inc., Tokyo, Japan). Thin-layer chromatography (TLC) was performed with Macherey-Nagel Sil G25 UV254 pre-coated TLC plates. 3-Amino-2-naphthol (**1**) and isothiocyanates (**2**) were purchased from TCI Fine Chemicals, Japan. Aminophenol (**5**) [27] was prepared according to the reported procedure. All electrochemical reactions were carried out in an IKA ElectraSyn 2.0 using ElectraSyn 2.0 undivided cells (a 5 mL vial) equipped with standard IKA ElectraSyn 2.0 electrodes (IKA Japan K.K., Osaka, Japan).

3.2. Synthesis of N-Arylnaphtho- and Anthra[2,3-d]oxazol-2-amines

In a 5.0 mL ElectraSyn vial, a mixture of aminophenol (**1** or **5**: 0.4 mmol) and arylisothiocyanate (**2**: 0.4 mmol) in DMSO (5.0 mL) was stirred well for 18 h. After complete conversion of the reaction, KI (83.0 mg, 0.5 mmol, and 0.1 M) was added to the reaction mixture. The vial was sealed with an ElectraSyn vial cap fitted with graphite SK-50 as an anode and platinum foil as a cathode. Constant-current electrolysis was performed at 20 mA for 2–4 h. The reaction mixture was diluted with H_2O (20 mL) and AcOEt (20 mL), and the aqueous phase was extracted with AcOEt (3×30 mL). The combined organic phase was washed with brine (20 mL) and dried over $MgSO_4$. Evaporation of the solvent provided the crude product. The crude product was then purified by column chromatography on silica gel (copies of ^{1}H and ^{13}C NMR see Supplementary Materials).

3.3. Characterization Data

3.3.1. *N*-Phenylnaphtho[2,3-*d*]oxazol-2-amine (**4**) [15]

Yellow prisms (90.4 mg, 87%), m.p.: 198–200 °C (from CH_2Cl_2-Hexane), $R_f = 0.45$ (Hexane/AcOEt = 8:2). ^{1}H NMR (400 MHz, DMSO-d_6): δ 10.9 (brs, 1H, NH), 7.93–7.90 (m, 3H, Ar-H), 7.84 (s, 1H, Ar-H), 7.79 (d, J = 8.0 Hz, 2H, Ar-H), 7.42–7.37 (m, 4H, Ar-H), and 7.05 (t, J = 7.6 Hz, 1H, Ar-H) ppm. ^{13}C NMR (100 MHz, DMSO-d_6): δ 159.3 (C), 146.9 (C), 142.9 (C), 138.4 (C), 131.3 (C), 129.6 (C), 129.0 (CH), 127.6 (CH), 127.5 (CH), 124.4 (CH), 124.0 (CH), 122.6 (CH), 118.0 (CH), 112.3 (CH), and 104.5 (CH) ppm. HRMS: m/z $[M + H]^+$ calculated for $C_{17}H_{13}N_2O$: 261.1022. Found: 261.1019. IR (KBr): ν 3019 (NH) cm^{-1}.

3.3.2. *N*-(4-Methoxyphenyl)naphtho[2,3-*d*]oxazol-2-amine (**6**) [16]

Colorless needles (104 mg, 90%), m.p.: 190–191 °C (from CH_2Cl_2), $R_f = 0.40$ (Hexane/AcOEt = 8:2). ^{1}H NMR (400 MHz, DMSO-d_6): δ 10.6 (brs, 1H, NH), 7.92–7.88 (m, 3H, Ar-H), 7.78 (s, 1H, Ar-H), 7.68 (td, J = 9.2, 2.9 Hz, 2H, Ar-H), 7.41–7.35 (m, 2H, Ar-H), 6.96 (td, J = 9.2, 2.9 Hz, 2H, Ar-H), and 3.73 (s, 3H, OCH_3) ppm. ^{13}C NMR (100 MHz, DMSO-d_6): δ 160.2 (C), 155.5 (C), 147.7 (C), 143.7 (C), 132.1 (C), 131.9 (C), 130.0 (C), 128.2 (CH), 128.0 (CH), 124.9 (CH), 124.5 (CH), 120.2 (CH), 114.9 (CH), 112.5 (CH), 104.9 (CH), and 55.9 (CH_3) ppm. HRMS: m/z $[M + H]^+$ calculated for $C_{18}H_{15}N_2O_2$: 291.1128. Found: 291.1118. IR (KBr): ν 3018 (NH) cm^{-1}.

3.3.3. *N*-(4-Fluorophenyl)naphtho[2,3-*d*]oxazol-2-amine (**7**)

Colorless prisms (95.6 mg, 86%), m.p.: 250–252 °C (from CH_2Cl_2-Hexane), $R_f = 0.4$ (Hexane/AcOEt = 9:1). ^{1}H NMR (400 MHz, DMSO-d_6): δ 10.9 (brs, 1H, NH), 7.93–7.89 (m,

3H, Ar-H), 7.82–7.78 (m, 3H, Ar-H), 7.42–7.36 (m, 2H, Ar-H), and 7.23 (tt, J = 9.2, 3.6 Hz, 2H, Ar-H) ppm. ^{13}C NMR (100 MHz, DMSO-d_6): δ 159.9 (C), 158.3 (d, $^1J_{C,F}$ = 237 Hz, C), 147.6 (C), 143.4 (C), 135.4 (C), 131.9 (C), 130.1 (C), 128.2 (CH), 128.0 (CH), 125.0 (CH), 124.6 (CH), 120.2 (d, $^3J_{C,F}$ = 7.7 Hz, CH), 116.2 (d, $^2J_{C,F}$ = 22.0 Hz, CH), 112.9 (CH), and 105.1 (CH) ppm. ^{19}F NMR (376 MHz, DMSO-d_6): δ −120.4 ppm. HRMS: m/z [M + H]$^+$ calculated for $C_{17}H_{12}FN_2O$: 279.0928. Found: 279.0923. IR (KBr): ν 3057 (NH) cm^{-1}.

3.3.4. *N*-[4-(Trifluoromethyl)phenyl]naphtho[2,3-*d*]oxazol-2-amine (**8**)

Colorless prisms (116 mg, 89%), m.p.: 201–202 °C (from CH_2Cl_2-Hexane), R_f = 0.43 (Hexane/AcOEt = 8:2). ^{1}H NMR (400 MHz, DMSO-d_6): δ 11.3 (brs, 1H, NH), 8.00–7.90 (m, 6H, Ar-H), 7.76 (d, J = 9.2 Hz, 2H, Ar-H), and 7.45–7.39 (m, 2H, Ar-H) ppm. ^{13}C NMR (100 MHz, DMSO-d_6): δ 159.3 (C), 147.4 (C), 142.9 (C), 142.6 (C), 131.8 (C), 130.4 (C), 128.3 (CH), 128.2 (CH), 127.0 (q, $^3J_{C,F}$ = 3.8 Hz, CH), 125.4 (q, $^1J_{C,F}$ = 270 Hz, C), 125.1 (CH), 124.9 (CH), 123.0 (q, $^2J_{C,F}$ = 31.6 Hz, C), 118.4 (CH), 113.5 (CH), and 105.4 (CH) ppm. ^{19}F NMR (376 MHz, DMSO-d_6): δ −60.0 ppm. HRMS: m/z [M + H]$^+$ calculated for $C_{18}H_{12}F_3N_2O$: 329.0896. Found: 329.0888. IR (KBr): ν 3018 (NH) cm^{-1}.

3.3.5. *N*-(4-Methoxyphenyl)anthra[2,3-*d*]oxazol-2-amine (**9**)

Yellow powder (23.1 mg, 17%), m.p. > 300 °C (from THF-Hexane), R_f = 0.40 (Hexane/AcOEt = 7:3). ^{1}H NMR (400 MHz, DMSO-d_6): δ 10.8 (brs, 1H, NH), 8.52 (d, J = 7.9 Hz, 2H, Ar-H), 7.99 (t, J = 4.6 Hz, 3H, Ar-H), 7.88 (s, 1H, Ar-H), 7.69 (d, J = 8.6 Hz, 2H, Ar-H), 7.43–7.39 (m, 2H, Ar-H), 6.98 (d, J = 9.1 Hz, 2H, Ar-H), and 3.74 (s, 3H, OCH_3) ppm. ^{13}C NMR (100 MHz, DMSO-d_6): δ 160.5 (C), 155.7 (C), 148.3 (C), 144.6 (C), 137.9 (C), 131.8 (C), 131.7 (C), 130.9 (CH), 130.5 (C), 129.2 (C), 128.14 (CH), 128.05 (CH), 126.1 (CH), 125.5 (CH), 125.3 (CH), 120.4 (CH), 114.9 (CH), 111.2 (CH), 103.8 (CH), and 55.9 (CH_3) ppm. HRMS: m/z [M + H]$^+$ calculated for $C_{22}H_{17}N_2O_2$: 341.1285. Found: 341.1275. IR (KBr): ν 2955 (NH) cm^{-1}.

3.3.6. *N*-Phenylanthra[2,3-*d*]oxazol-2-amine (**10**) [15]

Yellow powder (44.7 mg, 36%), m.p.: 278–281 °C (from AcOEt-Hexane), R_f = 0.37 (Hexane/AcOEt = 8:2). ^{1}H NMR (400 MHz, DMSO-d_6): δ 10.9 (brs, 1H, NH), 8.56 (d, J = 4.4 Hz, 2H, Ar-H), 8.03–8.00 (m, 3H, Ar-H), 7.95 (s, 1H, Ar-H), 7.81 (d, J = 7.2 Hz, 2H, Ar-H), 7.45–7.38 (m, 4H, Ar-H), and 7.08 (t, J = 7.6 Hz, 1H, Ar-H) ppm. ^{13}C NMR (100 MHz, DMSO-d_6): δ 159.7 (C), 147.5 (C), 143.7 (C), 138.2 (C), 130.3 (C), 130.2 (C), 129.9 (C), 129.0 (CH), 128.7 (CH), 127.6 (CH), 127.5 (CH), 125.5 (CH), 125.1 (CH), 124.9 (CH), 124.7 (CH), 122.8 (CH), 118.2 (CH), 111.0 (CH), and 103.3 (CH) ppm. HRMS: m/z [M + H]$^+$ calculated for $C_{21}H_{15}N_2O$: 311.1179. Found: 311.1168. IR (KBr): ν 2997 (NH) cm^{-1}.

3.3.7. *N*-(4-Fluorophenyl)anthra[2,3-*d*]oxazol-2-amine (**11**)

Yellow prisms (61.7 mg, 47%), m.p.: 285–286 °C (from AcOEt-Hexane), R_f = 0.25 (Hexane/AcOEt = 8:2). ^{1}H NMR (400 MHz, DMSO-d_6): δ 11.0 (brs, 1H, NH), 8.55 (d, J = 4.8 Hz, 2H, Ar-H), 8.03–8.00 (m, 3H, Ar-H), 7.94 (s, 1H, Ar-H), 7.84–7.81 (m, 2H, Ar-H), 7.46–7.41 (m, 2H, Ar-H), and 7.26 (t, J = 8.0 Hz, Ar-H) ppm. ^{13}C NMR (100 MHz, DMSO-d_6): δ 160.3 (C), 158.4 (d, $^1J_{C,F}$ = 237 Hz, C), 148.1 (C), 144.2 (C), 135.2 (C), 130.9 (C), 130.8 (C), 130.5 (C), 129.2 (C), 128.13 (CH), 128.11 (CH), 126.1 (C), 125.7 (C), 125.5 (CH), 125.3 (CH), 120.3 (d, $^3J_{C,F}$ = 7.7 Hz, CH), 116.3 (d, $^2J_{C,F}$ = 23.0 Hz, CH), 111.6 (CH), and 103.9 (CH) ppm. ^{19}F NMR (376 MHz, DMSO-d_6): δ -120.1 ppm. HRMS: m/z [M$^+$] calculated for $C_{21}H_{14}FN_2O$: 329.1085. Found: 329.1076. IR (KBr): ν 2928 (NH) cm^{-1}.

3.3.8. *N*-[4-(Trifluoromethyl)phenyl]anthra[2,3-*d*]oxazol-2-amine (**12**)

Yellow powder (80.1 mg, 53%), m.p. > 300 °C (from THF-Hexane), R_f = 0.35 (Hexane/AcOEt = 8:2). ^{1}H NMR (400 MHz, DMSO-d_6): δ 11.4 (brs, 1H, NH), 8.57 (s, 2H, Ar-H), 8.08 (s, 1H, Ar-H), 8.03–8.00 (m, 5H, Ar-H), 7.77 (d, J = 8.2 Hz, 2H, Ar-H), and 7.44–7.42 (m, 2H, Ar-H) ppm. ^{13}C NMR (100 MHz, DMSO-d_6): δ 159.8 (C), 147.9 (C), 143.8 (C), 142.4 (C), 130.9 (C), 130.73 (C), 130.69 (C), 129.4 (C), 128.1 (q, $^3J_{C,F}$ = 4.8 Hz, CH), 127.0 (q, $^4J_{C,F}$ = 3.8 Hz, CH), 126.2 (CH), 125.9 (CH), 125.7 (CH), 125.5 (CH), 124.9 (q, $^1J_{C,F}$ = 273 Hz, C), 123.3 (q, $^2J_{C,F}$ = 31.3 Hz, C), 118.7 × 2 (CH), 112.4 (CH), and 104.3 (CH) ppm. ^{19}F NMR (376 MHz, DMSO-d_6): δ −60.0 ppm. HRMS: m/z [M + H]$^+$ calculated for $C_{22}H_{14}F_3N_2O$: 379.1053. Found: 379.1047. IR (KBr): ν 3059 (NH) cm^{-1}.

4. Conclusions

In conclusion, we have demonstrated a simple method for the synthesis of polycyclic benzene-fused oxazoles using electrochemical cyclodesulfurization reactions. These are environmentally friendly reactions that use commercially available and inexpensive KI as an electrolytic mediator. Oxazoles **4** and **6–12** exhibited fluorescence in the 340–430 nm region in chloroform. Detailed mechanistic investigations of these electrochemical reactions, structural characterization, and elucidation of the physical properties of the polycyclic oxazoles are currently underway in our laboratory.

Supplementary Materials: The following supporting information can be downloaded at https://www.mdpi.com/article/10.3390/molecules30020319/s1.

Author Contributions: Conceptualization, Y.M. and S.Y.; methodology, Y.M. and S.Y.; validation, A.M. and Y.M.; formal analysis, M.K. and M.M.; investigation, A.M. and Y.M.; writing—original draft preparation, Y.M. and S.Y.; writing—review and editing, Y.M., M.M. and S.Y.; supervision, S.Y.; project administration, S.Y.; funding acquisition, M.K., Y.M., M.M. and S.Y.; All authors have read and agreed to the published version of the manuscript.

Funding: This research was supported by a research grant from the Institute of Pharmaceutical Life Sciences, Aichi Gakuin University, and a Nagai Memorial Research Scholarship from the Pharmaceutical Society of Japan (M.K.).

Institutional Review Board Statement: Not applicable.

Informed Consent Statement: Not applicable.

Data Availability Statement: Data are available from the authors upon reasonable request.

Conflicts of Interest: The authors declare no conflicts of interest.

References

1. Yan, X.; Wen, J.; Zhou, L.; Fan, L.; Wang, X.; Xu, Z. Current scenario of 1,3-oxazole derivatives for anticancer activity. *Curr. Top. Med. Chem.* **2020**, *20*, 1916–1937. [CrossRef] [PubMed]
2. Demmer, C.S.; Bunch, L. Benzoxazoles and oxazolopyridines in medicinal chemistry studies. *Eur. J. Med. Chem.* **2015**, *97*, 778–785. [CrossRef]
3. Kaur, A.; Wakode, S.; Pathak, D.P. Benzoxazole: The molecule of diverse pharmacological importance. *Int. J. Pharm. Pharm. Sci.* **2015**, *7*, 16–23.
4. Singh, S.; Veeraswamy, G.; Bhattarai, D.; Goo, J.; Lee, K.; Choi, Y. Recent advances in the development of pharmacologically active compounds that contain a benzoxazole scaffold. *Asian J. Org. Chem.* **2015**, *4*, 1338–1361. [CrossRef]
5. Oliveira, E.; Santos, H.M. An overview on sensing materials depending on the electromagnetic spectra region applied. *Dye. Pigment.* **2016**, *135*, 3–25. [CrossRef]
6. Flick, A.C.; Ding, H.X.; Leverett, C.A.; Kyne, R.E., Jr.; Liu, K.K.C.; Fink, S.J.; O'Donnell, C.J. Synthetic approaches to the 2014 new drugs. *Bioorg. Med. Chem.* **2016**, *24*, 1937–1980. [CrossRef] [PubMed]
7. Dubey, A.K.; Handu, S.S.; Mediratta, P.K. Suvorexant: The first orexin receptor antagonist to treat insomnia. *J. Pharmacol. Pharmacother.* **2015**, *6*, 118–121. [CrossRef]

8. Song, H.; Oh, S.R.; Lee, H.K.; Han, G.; Kim, J.H.; Chang, H.W.; Doh, K.E.; Rhee, H.K.; Choo, H.Y.P. Synthesis and evaluation of benzoxazole derivatives as 5-lipoxygenase inhibitors. *Bioorg. Med. Chem.* **2010**, *18*, 7580–7585. [CrossRef] [PubMed]
9. Costales, A.; Mathur, M.; Ramurthy, S.; Lan, J.; Subramanian, S.; Jain, R.; Atallah, G.; Setti, L.; Lindvall, M.; Appleton, B.A.; et al. 2-Amino-7-substituted benzoxazole analogs as potent RSK2 inhibitors. *Bioorg. Med. Chem. Lett.* **2014**, *24*, 1592–1596. [CrossRef]
10. Wang, G.; Peng, Z.; Wang, J.; Li, J.; Li, X. Synthesis, biological evaluation and molecular docking study of *N*-arylbenzo[*d*]oxazol-2-amines as potential α-glucosidase inhibitors. *Bioorg. Med. Chem.* **2016**, *24*, 5374–5379. [CrossRef]
11. Kadagathur, M.; Shaikh, A.S.; Jadhav, G.S.; Sigalapalli, D.K.; Shankaraiah, N.; Tangellamudi, N.D. Cyclodesulfurization: An enabling protocol for synthesis of various heterocycles. *ChemstrySelect* **2021**, *6*, 2621–2640. [CrossRef]
12. Gu, Y.; Li, Y.D.; Ge, Y.; Huang, J.L.; Xu, H.J.; Hu, Y. Hypervalent iodine mediated synthesis of 2-aminobenzazoles and 2-aminobenzothiazoles. *Asian J. Org. Chem.* **2024**, *13*, e202400076. [CrossRef]
13. Kant, K.; Patel, C.K.; Banerjee, S.; Naik, P.; Padhi, A.; Sharma, V.; Singh, V.; Almeer, R.; Keremane, K.S.; Atta, A.K.; et al. HFIP-mediated cyclodesulfurization approach for the synthesis of 2-aminobenzoxazole and 2-aminobenzothiazole derivatives. *Asian J. Org. Chem.* **2024**, *13*, e202400223. [CrossRef]
14. Rodríguez-Nuévalos, S.; Costero, A.M.; Arroyo, P.; Sáez, J.A.; Parra, M.; Sancenón, F.; Martínez-Máñez, R. Protection against chemical submission: Naked-eye detection of γ-hydroxybutyric acid (GHB) in soft drinks and alcoholic beverages. *Chem. Commun.* **2020**, *56*, 12600–12603. [CrossRef] [PubMed]
15. Murata, Y.; Matsumoto, N.; Miyata, M.; Kitamura, Y.; Kakusawa, N.; Matsumura, M.; Yasuike, S. One-pot reaction for the synthesis of *N*-substituted 2-aminobenzoxazoles using triphenylbismuth dichloride as cyclodesulfurization reagent. *J. Organomet. Chem.* **2018**, *859*, 18–23. [CrossRef]
16. Zhang, J.; Chen, L.; Dong, Y.; Yang, J.; Wu, Y. A Cu_2O/TBAB-promoted approach to synthesize heteroaromatic 2-amines *via* one-pot cyclization of aryl isothiocyanates with *ortho*-substituted amines in water. *Org. Biomol. Chem.* **2020**, *18*, 7425–7430. [CrossRef] [PubMed]
17. Tran, D.T.; Huynh, T.N.; Nguyen, P.C.; Phan, N.T.S.; Nguyen, T.T. Synthesis of 2-aminobenzoxazoles from elemental sulfur mediated cyclization of 2-aminophenols and aryl isothiocyanates. *Tetrahedron Lett.* **2023**, *122*, 154510. [CrossRef]
18. Xu, Y.; Li, F.; Zhao, N.; Su, J.; Wang, C.; Wang, C.; Li, Z.; Wang, L. Environment-friendly and efficient synthesis of 2-aminobenzoxazoles and 2-aminobenzothiazoles catalyzed by *Vitreoscilla* hemoglobin incorporating a cobalt porphyrin cofactor. *Green Chem.* **2021**, *23*, 8047–8052. [CrossRef]
19. Listratova, A.V.; Sbei, N.; Voskressensky, L.G. Catalytic electrosynthesis of *N,O*-heterocycles—Recent advances. *Eur. J. Org. Chem.* **2020**, *2020*, 2012–2027. [CrossRef]
20. Sbei, N.; Listratova, A.V.; Titov, A.A.; Voskressensky, L.G. Recent advances in electrochemistry for the synthesis of *N*-heterocycles. *Synthesis* **2019**, *51*, 2455–2473. [CrossRef]
21. Jiang, Y.; Xu, K.; Zeng, C. Use of electrochemistry in the synthesis of heterocyclic structures. *Chem. Rev.* **2018**, *118*, 4485–4540. [CrossRef]
22. Huynh, T.N.T.; Tankam, T.; Koguchi, S.; Rerkrachaneekorn, T.; Sukwattanasinitt, M.; Wacharasindhu, S. Electrochemical NaI/NaCl-mediated one-pot synthesis of 2-aminobenzoxazoles in aqueous media *via* tandem addition–cyclization. *Green Chem.* **2021**, *23*, 5189–5194. [CrossRef]
23. Liu, K.; Song, C.; Lei, A. Recent advances in iodine mediated electrochemical oxidative cross-coupling. *Org. Biomol. Chem.* **2018**, *16*, 2375–2387. [CrossRef]
24. Bentlety, C.L.; Bond, A.M.; Hollenkamp, A.F.; Mahon, P.J.; Zhang, J. Electrochemistry of iodide, iodine, and iodine monochloride in chloride containing nonhaloaluminate ionic liquids. *Anal. Chem.* **2016**, *88*, 1915–1921. [CrossRef] [PubMed]
25. Zhang, X.; Wang, C.; Jiang, H.; Sun, L. Convenient synthesis of selenyl-indoles *via* iodide ion-catalyzed electrochemical C–H selenation. *Chem. Commun.* **2018**, *54*, 8781–8784. [CrossRef] [PubMed]
26. Eaton, D.F. Reference materials for fluorescence measurement. *Pure Appl. Chem.* **1988**, *60*, 1107–1114. [CrossRef]
27. Partes, C.; Yildirim, C.; Schuster, S.; Kind, M.; Bats, J.W.; Zharnikov, M.; Terfort, A. Self-assembled monolayers of pseudo-C_{2v}-Symmetric, low-band-gap areneoxazolethiolates on gold surfaces. *Langmuir* **2016**, *32*, 11474–11484. [CrossRef] [PubMed]

molecules

MDPI

Article

Synthesis of Tumor Selective Indole and 8-Hydroxyquinoline Skeleton Containing Di-, or Triarylmethanes with Improved Cytotoxic Activity

Dóra Hegedűs [1], Nikoletta Szemerédi [2], Krisztina Petrinca [1], Róbert Berkecz [3,4], Gabriella Spengler [2] and István Szatmári [1,5,*]

1 Institute of Pharmaceutical Chemistry, University of Szeged, Eötvös u. 6, H-6720 Szeged, Hungary
2 Department of Medical Microbiology, Albert Szent-Györgyi Health Center and Albert Szent-Györgyi Medical School, University of Szeged, Semmelweis utca 6, H-6725 Szeged, Hungary; spengler.gabriella@med.u-szeged.hu (G.S.)
3 Institute of Pharmaceutical Analysis, University of Szeged, Somogyi u. 4, H-6720 Szeged, Hungary; berkecz.robert@szte.hu
4 Department of Forensic Medicine, Albert Szent-Györgyi Health Center, Kossuth Lajos sgt. 40, H-6724 Szeged, Hungary
5 HUN-REN–SZTE Stereochemistry Research Group, University of Szeged, Eötvös u. 6, H-6720 Szeged, Hungary
* Correspondence: szatmari.istvan@szte.hu

Abstract: The reaction between glycine-type aminonaphthol derivatives substituted with 2- or 1-naphthol and indole or 7-azaindole has been tested. Starting from 2-naphthol as a precursor, the reaction led to the formation of ring-closed products, while in the case of a 1-naphthol-type precursor, the desired biaryl ester was isolated. The synthesis of a bifunctional precursor starting from 5-chloro-8-hydroxyquinoline, morpholine, and ethyl glyoxylate via modified Mannich reaction is reported. The formed Mannich base **10** was subjected to give bioconjugates with indole and 7-azaindole. The effect of the aldehyde component and the amine part of the Mannich base on the synthetic pathway was also investigated. In favor of having a preliminary overview of the structure-activity relationships, the derivatives have been tested on cancer and normal cell lines. In the case of bioconjugate **16**, as the most powerful scaffold in the series bearing indole and a 5-chloro-8-hydroxyquinoline skeleton, a potent toxic activity against the resistant Colo320 colon adenocarcinoma cell line was observed. Furthermore, this derivative was selective towards cancer cell lines showing no toxicity on non-tumor fibroblast cells.

Keywords: modified Mannich reaction; bioconjugation; 8-hydroxyquinoline skeleton; anticancer activity; indole skeleton

1. Introduction

The Mannich reaction is a meaningful method in organic synthesis [1,2]. The modified Mannich reaction (*m*MR) is considered to be a particular variation of this C–C bond-forming reaction. 8-Hydroxyquinoline (8-HQ) as a unique electron-rich aromatic compound can be interpreted as a potential substrate of the modified Mannich reaction, and correlations between modification of the substitution pattern of the scaffold and biological activities were examined [3–6]. The indole skeleton proves to be a potent biological moiety through antitumor activity considering its effects on the proliferation, migration, invasion of different cancer cell lines [7], aryl hydrocarbon receptors [8], and ABCB1 efflux pump modifying [9] properties. It also exhibits immunoregulatory, anti-inflammatory, and antioxidative effects (lncRNAs and miRNAs expression) [10]. Additionally, symmetrical and un-symmetrical diindolylmethane derivatives were synthesized and tested at GPR84 (G protein-coupled receptor). It was found that the activity can be fine-tuned by substitution of indole skeleton [11].

The synthesis of indole- and naphthol-containing triarylmethane derivatives has been previously described. These compounds were synthesized by using different synthetic methods: (i) through the reaction of naphthol and the Mannich base of indole or (ii) applying the reaction of indole and the Mannich base of naphthol [12,13]. Recently, kynurenic acid derivatives were successfully applied as electron-rich aromatic compounds in this latter reaction [14].

As the result of our previous research, some C-3-coupled indole and azaindole derivatives showed toxicity on sensitive (Colo205) and ABCB1 efflux pump expressing (Colo320) colon adenocarcinoma cell lines and a normal, non-cancerous fibroblast cell line MRC-5 [15]. In addition, we previously studied the synthesis of new bifunctional precursors through the reaction of 2- or 1-naphthol, morpholine, and ethyl glyoxylate by using a modified Mannich-type synthetic method and their transformations via [4 + 2] cycloaddition with the use of different cyclic imines as dienophiles. Regarding the biological results, in the case of some compounds, inhibition of the efflux pump system in susceptible and methicillin-resistant *Staphylococcus aureus* strains was observed [16].

Based on the previous results, our first aim was the synthesis of a new glycine-type precursor bearing the 8-hydroxyquinoline skeleton. Our further aim was to investigate the reaction between some selected 8-hydroxyquinoline Mannich bases and indole or 7-azaindole forming new diarylmethane derivatives. Finally, their activity on Colo205 and Colo320 cell lines as well as on the normal MRC-5 cell line was also tested.

2. Results and Discussions

2.1. Synthesis

Regarding the synthetic and biological results in the field of aminonaphthol and naphthoxazino derivatives [16] as well as compounds bearing the indole moiety [11–15], we focused our efforts on reacting ethyl 2-(2-hydroxynaphthalen-1-yl)-2-morpholinoacetate with indole and 7-azaindole (Scheme 1). Having in hand 2-naphthol-substituted glycine precursor **1** reported previously and indole (**2**), the reaction was performed at 100 °C in 1,4-dioxane under microwave conditions. *p*-Toluenesulfonic acid (*p*-TSA) is considered to be an acid catalyst, and is applied frequently in the modified Mannich reaction [17], therefore, we decided to investigate its effect on the reaction. After a reaction time of 1.5 h, the formation of **3** through an ester intermediate was observed. The synthesis of 1-(1*H*-indol-3-yl)naphtho[2,1-*b*]furan-2(1*H*)-one (**3**) was already described by Jadhav et al. utilizing a different synthetic pathway, starting from ethyl2-(4-fluorophenylamino)-2-(1*H*-indol-3-yl) acetate and 2-naphthol using scandium triflate in 1,2 dichloroethane via arylation-cyclization [18]. Applying different reaction conditions (modification of the solvent, temperature, and additive) led to the same product. In our next experiment, precursor **1** was reacted with 7-azaindole (**4**) in the presence of 10 mol% *p*-TSA at 100 °C under MW irradiation in 1,4-dioxane. Based on ^{1}H NMR analysis of the crude reaction mixture and thin-layer chromatography (TLC), we concluded that the synthesis did not result in the desired derivative. The reaction was then repeated in toluene as the solvent, and the formation of a multi-spot reaction mixture was monitored by TLC. Unfortunately, the desired product could not be isolated even by using column chromatography. Next, the reaction was carried out in solvent-free conditions under microwave irradiation at 100 °C. A detailed NMR spectroscopic analysis of the product indicated the formation of a lactone structure (compound **5**) via intramolecular loss of ethanol. It is worth mentioning, that the formation and structure of this lactone as a side product was reported in our previous study [16].

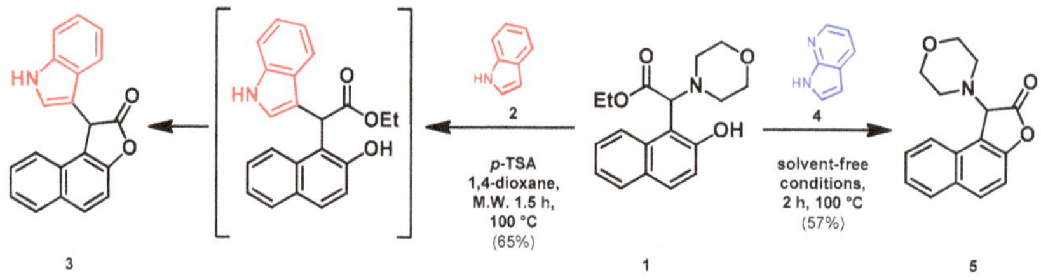

Scheme 1. Reactions of 2-(2-hydroxynaphthalen-1-yl)-2-morpholinoacetate with indole and 7-azaindole.

In view of these preliminary findings, we focused on testing the reaction from 1-naphthol-substituted precursor **6** (Scheme 2). Accordingly, **6** and indole (**2**) were reacted in the presence of 10 mol% *p*-TSA at 100 °C for 60 min under MW irradiation in 1,4-dioxane. The progress of the synthesis was monitored by TLC showing the formation of a new spot. On the basis of ^{1}H NMR analysis of the crude reaction mixture, the formation of **7** as a single product was assumed. On the basis of previous observations, the reaction was carried out under other conditions as well, applying toluene as the solvent in the presence of 10 mol% *p*-TSA. However, formation of the desired product did not take place. A similar result was found by running the reaction under solvent-free conditions. In favor of examining the possibility of extending the reaction scope, 7-azaindole (**4**) was reacted with precursor **6**. However, even by applying different solvents (toluene, 1,4-dioxane), using solvent-free conditions, and testing different temperatures (80 °C, 100 °C, 120 °C, 150 °C), the target compound did not form.

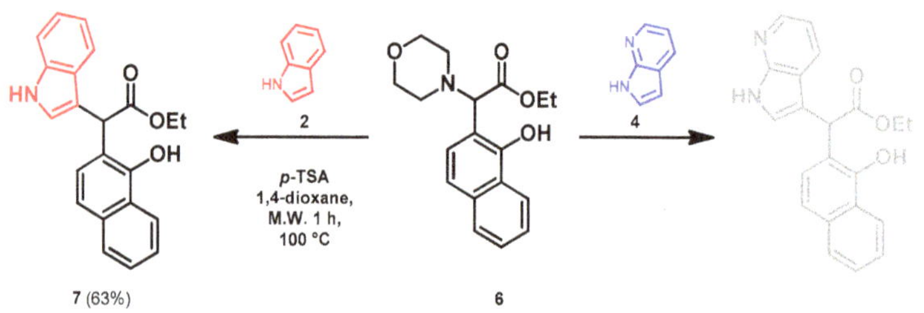

Scheme 2. Preparation of target compound **7** from 1-naphthol-substituted precursor **6**.

Based on literature data, 8-hydroxyquinoline is a biologically active moiety [5] interpreted as a potential 1-naphthol analogue. Chen et al. observed the most potent MMP-2/9 inhibitor effect, anti-proliferation activities against several cancer cell lines, and anti-migration/invasion activities on A549 cells in the case of Mannich bases of 5-chloro-8-hydroxyquinoline (5-Cl-8-HQ) [19]. Regarding earlier biological results, we proposed to examine the behavior of 5-Cl-8-HQ in the Mannich reaction. Accordingly, derivative **8** was reacted with morpholine (**9**) in the presence of ethyl glyoxylate as an aldehyde component (Scheme 3). According to TLC and NMR investigation of the crude reaction mixture, the formation of a single product and the presence of initial compounds were observed. New compound **10** was isolated and purified by crystallization and recrystallization from *i*Pr_2O.

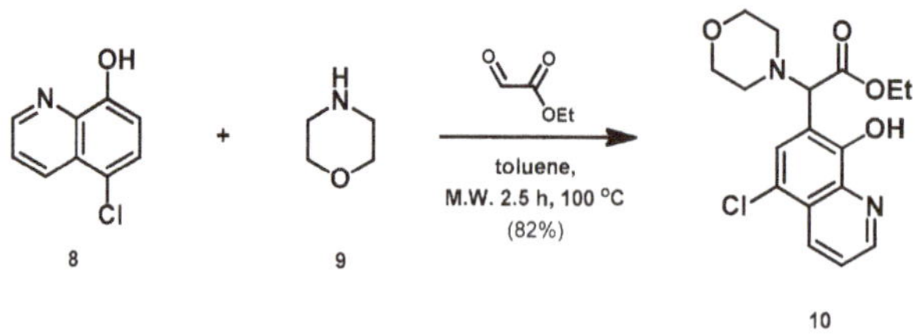

Scheme 3. Synthesis of ethyl 2-(5-chloro-8-hydroxyquinolin-7-yl)-2-morpholinoacetate (**10**).

After introducing the newly synthesized precursor bearing the 8-hydroxyquinoline moiety, the preparation of **11** starting from **10** and indole (**2**) in the presence of 10 mol% *p*-TSA was accomplished (Scheme 4). In this case, **11** was isolated with a yield of 70% in 4 h at 120 °C. To study the scope and limitations of the reaction, the synthesis of **12** was planned through the reaction of the precursor **10** with 7-azaindole (**4**). The reaction was tested first in toluene at 120 °C by monitoring the conversion of the starting compounds by TLC analysis. Since the yield was not satisfactory, the reaction was repeated at 150 °C (Scheme 4). By using a higher temperature and a longer reaction time (3 h), the desired product was isolated by crystallization from the reaction mixture.

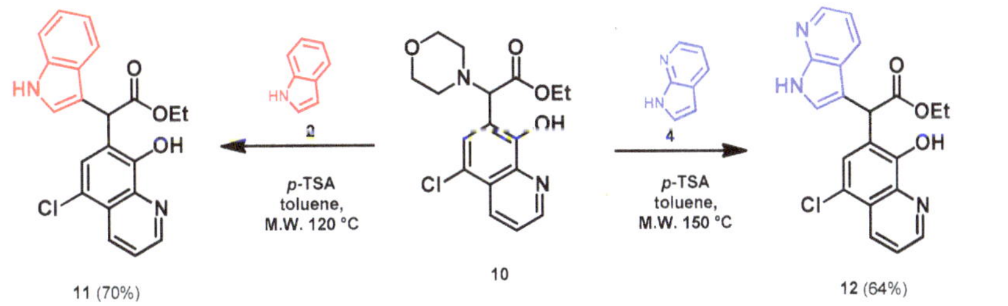

Scheme 4. The transformation of compound **10** bearing the 5-chloro-8-hydroxyquinoline moiety with indole and 7-azaindole.

In order to study the effect of the aldehyde component on the synthetic pathway, 5-Cl-8-HQ and morpholine were fixed and the aldehyde moiety was varied among benzaldehyde and paraformaldehyde. The re-synthesis of precursor **13** was achieved based on the previously published synthetic method [20]. When precursor **13** was reacted with indole (**2**) at 150 °C for 4 h, the desired triarylmethane **14** was isolated in a yield of 67% after purification by column chromatography (Scheme 5). In contrast, the reaction between **13** and 7-azaindole (**4**) did not show any transformation upon testing various reaction conditions (solvent, reaction time, reaction temperature) (Scheme 5).

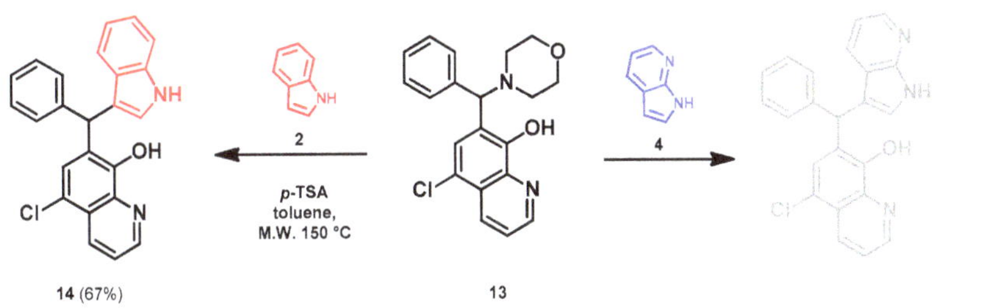

Scheme 5. The reaction of precursor **13** with indole and 7-azaindole.

In the next step, we focused our efforts on investigating the transformations of precursors that were originally prepared from paraformaldehyde as the aldehyde component. For this reason, the re-synthesis of Mannich base **15** was accomplished by applying a synthetic method published previously [20]. Starting from precursor **15**, indole (**2**), and 7-azaindole (**4**), the formation of target compounds **16** and **17** were isolated in low yield with relatively long reaction times at high temperatures (Scheme 6).

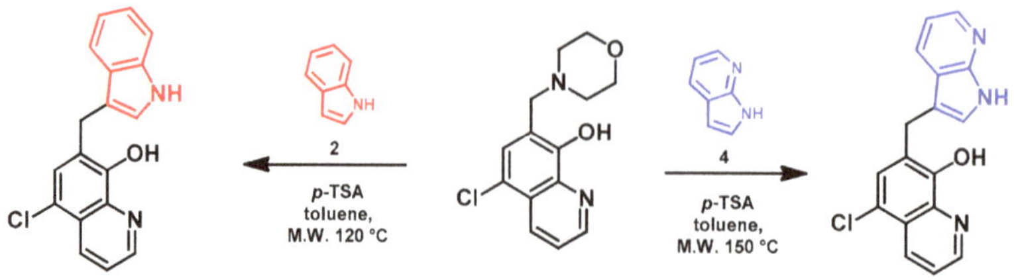

16 (23%) 15 17 (35%)

Scheme 6. The transformation of **15** containing paraformaldehyde as the aldehyde component.

Since derivative **16** is postulated to be formed in the reaction of indole with *ortho*-quinone methide evolving from Mannich base **15**, we carried out a systematic investigation of the synthesis of **16**. According to previous observations, the amine moiety of the Mannich base can be interpreted as a leaving group. Consequently, two precursors have been selected. These were compound **15** bearing the morpholine skeleton and Mannich base **18** having the L-proline motif. This latter precursor was re-synthesized by using our synthetic pathway published earlier [21], starting from 5-chloro-8-hydroxyquinoline, L-proline, and aqueous formaldehyde. The parallel arylation of precursors **15** and **18** with indole was also investigated (Scheme 7).

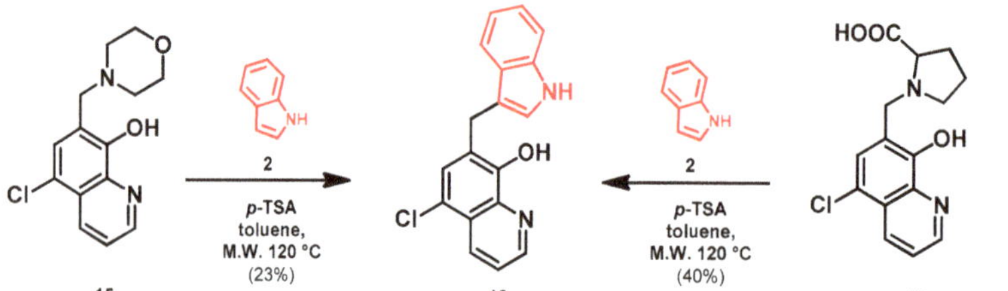

15 16 18

Scheme 7. Synthesis of compound **16** starting from different precursors.

This latter reaction offered a possibility to compare the effect of the leaving groups on conversion (Figure 1). The reaction was monitored by TLC and NMR analysis of the crude reaction mixtures every 30 min. These systematic investigations allowed us to conclude that starting from precursor **18** the desired diaryl derivative was formed in higher conversion. This may be explained by the better leaving group character of L-proline compared with that of morpholine.

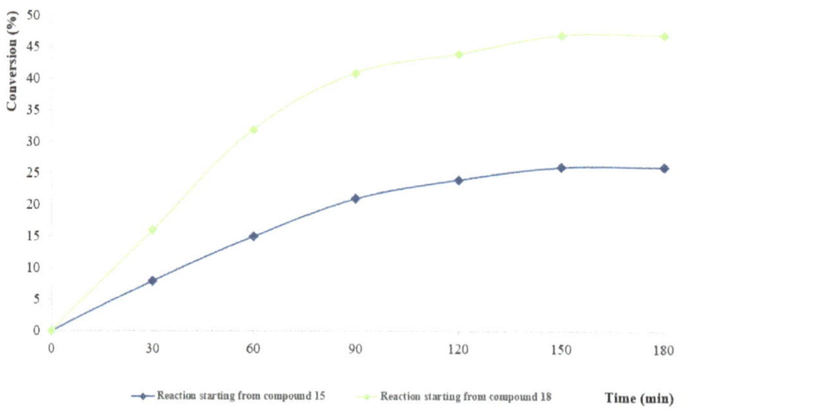

Figure 1. Systematic investigation of compounds **15** and **18** to determine the effect of the leaving groups.

2.2. Biological Evaluations

Cytotoxicity Assay

In this study, the potential ABCB1 cancer efflux pump-inhibiting activity of the derivatives was investigated. For this reason, the doxorubicin-sensitive Colo205 and the doxorubicin-resistant, ABCB1-expressing Colo320 colon adenocarcinoma cell lines were selected for biological studies. Based on the obtained results, the compounds showed a potent toxic activity against the sensitive Colo205 and resistant Colo320 cell lines (IC_{50} values were between 1.72 μM and 53.64 μM) (Table 1). Overall, it was observed that the tested compounds were more effective on the Colo320 cell line, with the exception of compound **14**. It is worth noting that **7** and **3** showed no effect on Colo205 cells, and **3** had no effect on Colo320 either. Only derivatives **11**, **12**, and **14** exerted a mild cytotoxic effect on the normal MRC-5 cell line compared to the cancer cell lines. This means that the compounds can be considered selective, as they were not toxic to normal fibroblast cells (Table 1). Doxorubicin was applied as a positive control, and DMSO was the solvent control. In our research, the selectivity of the compounds towards cancer cells (with the activity compared to normal cells) was investigated. The potential collateral sensitivity (CS) of the derivatives was also tested by applying sensitive Colo 205 colon adenocarcinoma cells and ABCB1- and LRP-expressing resistant Colo 320 cells [22]. The expression of the MDR transporter ABCB1 or P-glycoprotein is associated with a poor overall response to chemotherapy and unfavorable prognosis. Collateral sensitivity, which refers to the ability of certain compounds to selectively kill MDR cells over their parental counterparts, presents a potential alternative approach to overcoming MDR. This approach may lead to the development of highly selective and potent agents capable of effectively targeting MDR cells and resensitizing tumors to treatment [23].

Table 1. Cytotoxic effect of the compounds on sensitive (Colo205) and resistant (Colo320) colon adenocarcinoma, and MRC-5 normal embryonal fibroblast cell lines based on the IC_{50} values. Doxorubicin was used as a positive control. AV: average of 3 or 2 parallel experiments, SD: standard deviation. DMSO was applied as a solvent control that had no toxic effect at the highest concentration ap-plied in the assay (2% (v/v) concentration).

		Colo205 (μM)				Colo320 (μM)				MRC-5 (μM)		
					AV ± SD				AV ± SD			AV ± SD
3	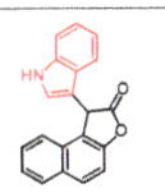	-	-	-	>100 ± -	-	-	-	>100 ± -	-	-	>100 ± -
7	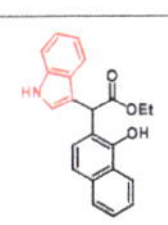	-	-	-	>100 ± -	53.82	54.92	52.18	53.64 ± 1.38	-	-	>100 ± -
10	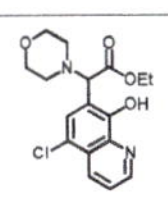	2.07	2.06	2.02	2.05 ± 0.02	1.565	1.813	1.79	1.72 ± 0.14	-	-	>100 ± -
11	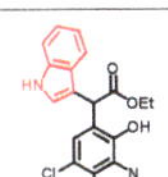	5.51	4.71	4.42	4.88 ± 0.56	4.636	4.832	4.957	4.81 ± 0.16	54.46	55.08	54.77 ± 0.44
12	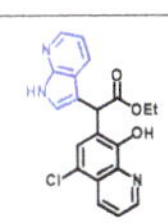	8.05	7.42	8.84	8.11 ± 0.71	5.157	4.923	5.109	5.06 ± 0.12	46.9	45.56	46.23 ± 0.95

Table 1. *Cont.*

		Colo205 (μM)				Colo320 (μM)				MRC-5 (μM)		
					AV ± SD				AV ± SD			AV ± SD
14	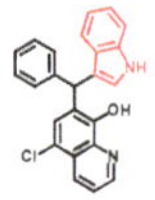	2.49	2.49	2.63	2.53 ± 0.08	2.552	2.84	2.781	2.72 ± 0.15	30.08	29.28	29.68 ± 0.57
16	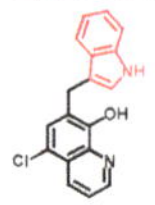	13.56	13.37	12.26	13.06 ± 0.70	4.967	4.74	4.91	4.87 ± 0.12	-	-	>100 ± -
17	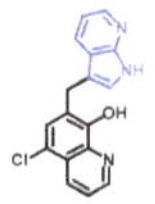	36.49	39.42	40.40	38.77 ± 2.03	22.98	21.89	22.54	22.47 ± 0.55	-	-	>100 ± -
	DOX	0.84	0.89	0.81	0.85 ± 0.04	3.117	3.12	2.961	3.07 ± 0.09	-	-	>8.62 ± -
	DMSO	-	-	-	>2% ± -	-	-	-	>2% ± -	-	-	>2% ± -

Relative resistance (RR) values were calculated as the ratio between the IC_{50} of a compound against the resistant cells and the IC_{50} against the corresponding sensitive cells (Table 2). Compounds with RR < 1 show selectivity against resistant cells, whereas RR $\leq$ 0.5 means that the compounds are highly selective towards resistant cells. All derivatives, with the exception of **14**, were selective towards resistant tumor cells. Furthermore, **16** exerted high potency to eliminate the MDR Colo320 cells (RR = 0.37), and this derivative showed no toxicity on normal MRC-5 cells being the most powerful scaffold in the series.

Table 2. Selectivity indexes (SI) and relative resistance (RR) of the compounds. NA: not applicable (if the precise value cannot be determined because of the lack of a precise IC_{50}).

		MRC-5/Colo205	MRC-5/Colo320	Colo320/Colo205
		SI		RR
3	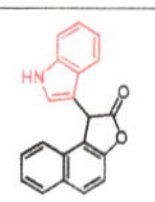	-	-	NA
7	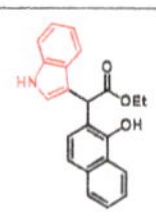	-	NA	NA
10	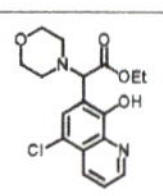	>6	>6	0.83
11	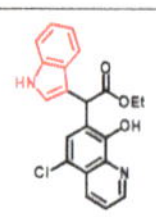	11.2	11.4	0.98

Table 2. *Cont.*

		MRC-5/Colo205	MRC-5/Colo320	Colo320/Colo205
		SI		RR
12	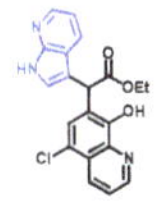	5.7	9.1	0.62
14	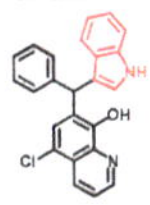	11.7	11	1.07
16	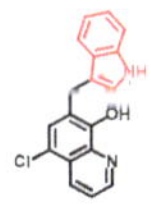	>6	>6	0.37
17	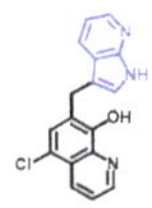	NA	NA	0.57

Considering the correlation between structure and biological activity, it can be concluded in the case of powerful scaffolds that the biaryl structural element and 5-chloro-8-hydroxyquinoline skeleton could be identified as a significant moiety (compound **16**). For improved cytotoxic activity, the presence of a morpholine-bearing cationic centre and 5-chloro-8-hydroxyquinoline is beneficial (compound **10**).

For comparison, in the case of C-3-indole- and azaindole-coupled cyclic amine derivatives, it was found that the simultaneous presence of the indole skeleton and thieno[3,2-*c*]pyridine or *β*-carboline is favourable for cytotoxic activity. The IC_{50} values were determined as 21.81 μM and 24.71 μM on the Colo205 cells, and 12.94 μM and 13.55 μM on the Colo320 cells [15]. Our present results support that the 5-chloro-8-hydroxyquinoline skeleton connected to the indole via methylene carbon is preferable in terms of cytotoxicity (Table 1, compound **16**; IC_{50}: 13.06 μM on Colo205 cells and 4.87 μM on Colo320 cells) compared to the previously studied cyclic amines coupled to indole derivatives.

Furthermore, 8-hydroxyquinoline-D-proline and homo-proline hybrids and their half-sandwich Ru(η^6-*p*-cymene) and Rh(η^5-C_5Me_5) complexes were synthesized, and their in vitro cytotoxic activity on two human colon adenocarcinoma cell lines (Colo205 and Colo320) and on one non-tumoral human lung fibroblast cell line (MRC-5) were investigated by Pivarcsik et al. Regarding the biological activity, D-amino acid hybrids exhibited significant cytotoxicity (IC_{50} = 12–21 μM) against the tested cancer cell lines [24]. In parallel with our research, with Mannich base **10** having the morpholine instead of L-proline motif, it was observed that ethyl 2-(5-chloro-8-hydroxyquinolin-7-yl)-2-morpholinoacetate (**10**) showed a potent toxic activity against the doxorubicin-sensitive Colo205 (IC_{50} = 2.05 μM) and -resistant Colo320 (IC_{50} = 1.72 μM) cell lines, and no toxicity on normal MRC-5 cells. These results allowed us to conclude that morpholine as an amine component can improve the cytotoxic activity. Moreover, cytotoxic activity of substituted 2-(4-hydroxyquinolin-2-yl) acetates was evaluated using doxorubicin-sensitive and -resistant colon adenocarcinoma cell lines (Colo205 and Colo320, respectively) and normal human embryonic MRC-5 fibroblasts [25]. Interesting to note that for those compounds only the Knoevenagel products (para-substituted benzylidene derivatives) showed improved activity which supports that the presence of an 8-hydroxyquinoline skeleton is necessary for cytotoxic activity.

3. Materials and Methods

3.1. Biological Assays

3.1.1. Cell Lines and Their Maintenance

The doxorubicin-sensitive Colo205 (ATCC-CCL-222) and the doxorubicin-resistant ABCB1- and LRP-expressing Colo320/MDR-LRP (ATCC-CCL-220.1) human colon adenocarcinoma cell lines were obtained from LGC Promochem in Teddington, UK. These cell lines were cultivated in RPMI 1640 medium supplemented with 10% heat-inactivated foetal bovine serum (FBS), 2 mM L-glutamine, 1 mM Na-pyruvate, 10 mM HEPES, along with nystatin, and gentamicin. The normal MRC-5 (ATCC CCL-171) human embryonic lung fibroblast cell line was purchased from Sigma-Aldrich (Merck KGaA, Darmstadt, Germany). The MRC-5 cells were cultured in EMEM medium, supplemented with 1% non-essential amino acid (NEAA) mixture, a selection of vitamins, 10% heat-inactivated FBS, 2 mM L-glutamine, 1 mM Na pyruvate, nystatin, and gentamicin. The cell lines were incubated in humidified atmosphere (5% CO_2, 95% air) at 37 °C. The cells were detached with a Trypsin-Versene (EDTA) solution for 5 min at 37 °C.

3.1.2. MTT Assay

The effects of increasing concentrations of the compounds on cell growth were tested in 96-well flat-bottomed microtiter plates. Namely, 1×10^4 of human colonic adenocarcinoma cells in 100 µL of the medium (RPMI 1640) were added to each well, except for the medium control wells. The adherent human embryonic lung fibroblast cell line was seeded in the EMEM medium for 24 h before the assay. The two-fold serial dilutions of the compounds were made in a separate plate (100–0.19 µM) and then transferred to the plates containing the cells. The starting concentration of the solvent DMSO was 2% v/v, and the highest concentration of the compounds in the plate was 100 µM. Plates were incubated at 37 °C for 24 h. At the end of the incubation, 20 µL of MTT (thiazolyl blue tetrazolium bromide) solution (from a 5 mg/mL stock solution) was added to each well. After incubation at 37 °C for 4 h, 100 µL of sodium dodecyl sulfate (SDS) solution (10% SDS in 0.01 M HCl) was added to each well, and the plates were further incubated at 37 °C overnight. Cell growth was determined by measuring the optical density (OD) at 540 nm (ref. 630 nm) with a Multiscan EX ELISA reader (Thermo Labsystems, Cheshire, WA, USA). Inhibition of cell growth was expressed as IC_{50} values, defined as the inhibitory dose that reduces the growth of the cells exposed to the tested compounds by 50%. IC_{50} values and the SD of triplicate and duplicate (in the case of MRC-5) experiments were calculated by using GraphPad Prism software version 5.00 for Windows, with a non-linear regression curve fit (GraphPad Software, San Diego, CA, USA; www.graphpad.com, accessed on 12 July 2021). Doxorubicin (from a 2 mg/mL stock solution, Teva Pharmaceuticals) was used as a positive control. The solvent (DMSO) did not have any effect on the cell growth in the tested concentrations. The relative resistance (RR) was calculated as the ratio of the IC_{50} value in the resistant cancer cells and the IC_{50} value in the sensitive cancer cell lines. The selectivity indexes (SI) were calculated as the ratio of the IC_{50} value in the non-tumor cells and the IC_{50} value in the cancer cell lines. The activity of the compounds towards cancer cells is considered to be strongly selective if the selectivity index (SI) value is higher than 6, moderately selective if $3 < SI < 6$, slightly selective if $1 < SI < 3$, and non-selective if the SI is lower than 1 [22].

3.2. Preparation Protocols for the Synthesis of the New Derivatives

Melting points were determined using a Hinotek X-4 melting point apparatus. Merck Kieselgel $60F_{254}$ plates were applied for TLC (Merck KGaA, Darmstadt, Germany). Microwave reactions were carried out with a CEM Discover SP microwave reactor.

^1H and ^{13}C-NMR spectra were recorded in DMSO-d_6 or $CDCl_3$ solutions in 5 mm tubes at room temperature (RT) with a Bruker DRX-500 spectrometer (Bruker Biospin, Karlsruhe, Baden Wurttemberg, Germany) at 500 (^1H) and 125 (^{13}C) MHz, with the deuterium signal of the solvent as the lock and TMS as the internal standard (^1H, ^{13}C). All spectra (^1H,

^{13}C, and NOESY) were acquired and processed with the standard BRUKER software (TopSpin 3.6.2.).

The HRMS flow injection analysis was performed with Thermo Scientific™ Orbitrap Exploris™ 240 hybrid quadrupole-Orbitrap (Thermo Fisher Scientific, Waltham, MA, USA) mass spectrometer coupled to a Waters Acquity I-Class UPLC™ (Waters, Manchester, UK). The mass spectrometer was operated in negative (**7**) and positive (**10, 11, 12, 14, 16, 17**) ionization mode.

The FTIR analysis was performed with Thermoscientific Nicolet Summit FTIR Spectrometer.

Ethyl 2-(1-hydroxynaphthalen-2-yl)-2-(1H-indol-3-yl)acetate (**7**)

In a 35 mL pressurized reaction vial, the mixture of ethyl 2-(1-hydroxynaphthalen-2-yl)-2-morpholinoacetate (**6**; 50.0 mg, 0.16 mmol) and indole (**2**; 17.5 mg, 0.16 mmol) in the presence of 10 mol% (2.8 mg, 0.016 mmol) *p*-TSA in 1,4-dioxane was heated at 100 °C for 1 h under MW irradiation. Following the removal of the solvent, the residue was purified by column chromatography (*n*-hexane:EtOAc, 4:1); R_f value = 0.32 (*n*-hexane:EtOAc, 4:1); 34.9 mg, (63%); oil; ^{1}H-NMR (CDCl$_3$): 1.35 (t, 3H, *J* = 7.15 Hz), 4.24–4.40 (m, 2H), 5.36 (s, 1H), 7.09 (t, 1H, *J* = 7.51 Hz), 7.19–7.21 (m, 1H), 7.34 (t, 2H, *J* = 8.86 Hz), 7.39 (d, 1H, *J* = 8.51 Hz), 7.44–7.52 (m, 3H), 7.74–7.80 (m, 1H), 8.08 (brs, 1H), 8.31–8.36 (m, 1H), 8.99 (s, 1H).^{13}C NMR (CDCl$_3$): 14.1; 29.7; 48.7; 111.1; 111.4; 115.9; 118.7; 119.9; 120.1; 122.5; 122.7; 123.6; 125.3; 126.3; 126.3; 126.5; 127.2; 128.9; 134.4; 136.5; 151.6; 176.0 (Figures S1 and S2). HRMS calculated for $[M-H]^-$ = 344.1292 *m*/*z*, $[2M-H]^-$ = 689.2657 *m*/*z* found 344.1287 *m*/*z* (Δm_i = −1.3 ppm) and 689.2659 *m*/*z* (Δm_i = 0.3 ppm). The ethoxy loss followed by ring-closing yielded proposed fragment ions with 298.0875 (Δmi = 0.3 ppm) *m*/*z* during in-source fragmentation (Figure S3). FTIR in Figure S4.

Ethyl 2-(5-chloro-8-hydroxyquinolin-7-yl)-2-morpholinoacetate (**10**)

A mixture of 5-chloro-8-hydroxyquinoline (**8**; 1.0 g, 5.56 mmol), ethyl glyoxylate (1.2 g, 5.93 mmol; 50% in toluene) and morpholine (**9**; 0.5 g, 5.54 mmol) in toluene (15 mL) was placed into a 35-mL pressurized reaction vial and heated at 100 °C for 2.5 h under MW irradiation. The solvent was removed in vacuo. The desired product was isolated by crystallization with EtOAc (10 mL), recrystallized from *i*Pr$_2$O (5 mL); R_f value = 0.43 (CH$_2$Cl$_2$: EtOAc, 4:1); 1.6 g, (82%); white crystals; m.p. 124–126 °C; ^{1}H-NMR (CDCl$_3$): 1.23 (t, 3H, *J* = 7.12 Hz), 2.55–2.68 (m, 4H), 3.74–3.81 (m, 4H), 4.12–4.26 (m, 2H), 4.60 (s, 1H), 7.54–7.59 (m, 1H), 7.68 (s, 1H), 8.50 (d, 1H, *J* = 8.51 Hz), 8.87 (d, 1H, *J* = 4.36 Hz). ^{13}C NMR (CDCl$_3$): 14.1; 51.3; 61.4; 66.8; 68.2; 116.5; 120.8; 122.8; 126.3; 127.1; 133.2; 139.0; 149.0; 150.8; 170.1 (Figures S5 and S6). HRMS calculated for $[M+H]^+$ =351.1106 *m*/*z*, $[2M+Na]^+$ =723.1959 *m*/*z* found 351.1091 *m*/*z* (Δmi = −4.3 ppm) and 723.1927 *m*/*z* (Δmi = −4.4 ppm). The loss of the morpholino group can explain the observed 264.0411 *m*/*z* (Δmi = −4.2 ppm) peak during in-source fragmentation (Figure S7). FTIR in Figure S8.

Ethyl 2-(5-chloro-8-hydroxyquinolin-7-yl)-2-(1H-indol-3-yl)acetate (**11**)

Bifunctional precursor **10** (50.0 mg, 0.14 mmol) and indole (**2**; 17.0 mg, 0.14 mmol) in the presence of 10 mol% (2.4 mg, 0.014 mmol) *p*-TSA were dissolved in toluene (10 mL) in a 35 mL pressurized reaction vial and heated at 120 °C for 4 h under MW irradiation. Following column chromatography purification (EtOAc:*n*-hexane, 1:1), the eluent was removed in vacuo; R_f value = 0.63 (EtOAc:*n*-hexane, 1:1); 37.7 mg, (70%); oil; ^{1}H-NMR (CDCl$_3$): 1.27 (t, 3H, *J* = 7.16 Hz), 4.21–4.29 (m, 2H), 5.92 (s, 1H), 7.07 (t, 1H, *J* = 7.46 Hz), 7.19 (t, 1H, *J* = 7.65 Hz), 7.36–7.40 (m, 2H), 7.50–7.53 (m, 1H), 7.58–7.61 (m, 2H), 8.15 (brs, 1H), 8.45 (d, 1H, *J* = 8.49 Hz), 8.82 (d, 1H, *J* = 4.33 Hz). ^{13}C NMR (CDCl$_3$): 14.2; 41.4; 61.3; 111.2; 112.9; 119.1; 119.9; 120.4; 121.0; 122.3; 122.5; 122.9; 125.5; 126.7; 128.0; 133.3; 136.2; 138.5; 148.2; 148.4; 172.5 (Figures S9 and S10). HRMS calculated for $[M+H]^+$ = 381.1000 *m*/*z*, $[2M+Na]^+$ = 783.1748 *m*/*z* found 381.0986 *m*/*z* (Δmi = −3.7 ppm) and 783.1714 *m*/*z* (Δmi = −4.3 ppm) (Figure S11). FTIR in Figure S12.

Ethyl 2-(5-chloro-8-hydroxyquinolin-7-yl)-2-(7-azaindole-3-yl)acetate (**12**)

In a 35 mL pressurized reaction vial a mixture of product **10** (50.0 mg, 0.14 mmol) and 7-azaindole (**4**; 17.0 mg, 0.14 mmol) in the presence of 10 mol% (2.4 mg, 0.014 mmol) *p*-TSA in toluene was heated at 150 °C for 3 h under MW irradiation. The solvent was removed under reduced pressure. The desired product was isolated by crystallisation with Et_2O (10 mL); R_f value = 0.52 (*n*-hexane:EtOAc, 1:4); 34.8 mg, (64%); beige crystals; m.p. 228–230 °C; ^{1}H-NMR (DMSO): 1.16 (t, 3H, *J* = 7.11 Hz), 4.13–4.20 (q, 2H, *J* = 7.11 Hz), 5.73 (s, 1H), 7.03 (dd, 1H, J_1 = 4.64 Hz, J_2 = 7.91 Hz), 7.40 (s, 1H), 7.45–7.48 (m, 1H), 7.70–7.75 (m, 1H), 7.84 (d, 1H, *J* = 7.83 Hz), 8.22 (d, 1H, *J* = 4.79 Hz), 8.46 (d, 1H, *J* = 8.54 Hz), 8.98 (d, 1H, *J* = 4.25 Hz), 11.69 (brs, 1H). ^{13}C NMR (DMSO): 14.5; 42.4; 61.3; 110.3; 115.9; 118.7; 119.0; 122.7; 123.5; 125.0; 125.3, 127.3; 127.9; 133.0, 139.1; 143.5; 149.0; 149.7; 150.2, 172.1 (Figures S13 and S14). HRMS calculated for $[M + H]^+$ m/z = 382.0953, found m/z = 382.0939 (Δm_i = −3.7 ppm) (Figure S15). FTIR in Figure S16.

7-((1H-indol-3-yl)(phenyl)methyl)-5-chloroquinolin-8-ol (**14**)

Glycine type precursor **13** (50.0 mg, 0.14 mmol) and indole (**2**; 16.5 mg, 0.14 mmol) in the presence of 10 mol% (2.4 mg, 0.014 mmol) *p*-TSA in toluene were heated in a 35 mL pressurized reaction vial at 150 °C for 4 h under MW irradiation. Following the removal of the solvent, the residue was purified by column chromatography (*n*-hexane:EtOAc, 3:1); R_f value = 0.44 (*n*-hexane:EtOAc, 3:1); 36.2 mg, (67%); oil; ^{1}H-NMR ($CDCl_3$): 6.34 (s, 1H), 6.70 (s, 1H), 6.99 (t, 1H, *J* = 7.50 Hz), 7.17 (t, 1H, *J* = 7.33 Hz), 7.21–7.25 (m, 1H), 7.27–7.34 (m, 5H), 7.35–7.40 (m, 2H), 7.48–7.53 (m, 1H), 7.99 (brs, 1H), 8.46 (d, 1H, *J* = 8.42 Hz), 8.80 (d, 1H, *J* = 4.13 Hz). ^{13}C NMR ($CDCl_3$): 41.3; 111.1; 118.6; 119.5; 119.8; 120.0; 122.0; 122.3; 123.9; 125.0; 125.9; 126.4; 127.0; 128.4; 128.5; 128.9; 133.3; 136.8; 138.7; 142.8; 148.1; 148.4 (Figures S17 and S18). HRMS calculated for $[M + H]^+$ m/z = 385.1102, found m/z = 385.1088 (Δm_i= −3.6 ppm). The in-source fragmentation resulted in 268.0514 m/z (−3.7 ppm) of fragment ion, and the exit of indole moiety might be responsible for its formation (Figure S19). FTIR in Figure S20.

General procedure for the synthesis of 7-((1H-indol-3-yl)methyl)-5-chloroquinolin-8-ol (**16**) from precursor **15** and **18**

Indole (**2**; 20.0 mg, 0.17 mmol) and Mannich base **15** or **18** (0.17 mmol) in the presence of 10 mol% (2.9 mg, 0.017 mmol) *p*-TSA in toluene were heated at 120 °C for 3 h under MW irradiation. Following column chromatography purification (EtOAc:*n*-hexane, 1:2 in the case of reaction starting from precursor **15**; EtOAc:*n*-hexane, 1:3 in the case of reaction starting from precursor **18**), the eluent was removed in vacuo; R_f value = 0.68 (EtOAc:*n*-hexane, 1:2); 12.1 mg, (23%) in the case of reaction starting from precursor **15**; R_f value = 0.51 EtOAc:*n*-hexane, 1:3); 21.0 mg (40%) in the case of reaction starting from precursor **18**; oil; ^{1}H-NMR ($CDCl_3$): 4.32 (s, 2H), 7.06–7.11 (m, 2H), 7.18 (t, 1H, *J* = 7.44 Hz), 7.36 (d, 1H, *J* = 8.21 Hz), 7.43 (s, 1H), 7.46–7.51 (m, 1H), 7.65 (d, 1H, *J* = 7.88 Hz), 7.99 (brs, 1H), 8.45 (d, 1H, *J* = 8.39 Hz), 8.81 (d, 1H, *J* = 4.28 Hz). ^{13}C NMR ($CDCl_3$): 24.8; 111.1; 114.5; 119.2; 119.5; 120.0; 121.7; 122.2; 122.5; 123.4; 124.8; 127.4; 129.3; 133.3; 136.4; 138.6; 148.1; 148.3 (Figures S21 and S22). HRMS calculated for $[M + H]^+$ 309.0789, found m/z = 309.0779 (Δm_i = −3.2 ppm) (Figure S23). FTIR in Figure S24.

7-((7-azaindole-3-yl)methyl)-5-chloroquinolin-8-ol (**17**)

A mixture of precursor **15** (50.0 mg, 0.18 mmol) and 7-azaindole (**4**; 21.0 mg, 0.18 mmol) in the presence of 10 mol% (3.1 mg, 0.018 mmol) *p*-TSA in toluene (15 mL) was placed into a 35-mL pressurized reaction vial and heated at 150 °C for 3 h under MW irradiation. Following column chromatography purification (EtOAc:*n*-hexane, 47:3), the eluent was removed in vacuo; oil; R_f value = 0.79 (EtOAc:*n*-hexane, 47:3); 19.5 mg, (35%); ^{1}H-NMR ($CDCl_3$): 5.68 (s, 2H), 6.48 (d, 1H, *J* = 3.33 Hz), 7.07–7.12 (m, 1H), 7.38 (d, 1H, *J* = 3.37 Hz), 7.45 (s, 1H), 7.49–7.54 (m, 1H), 7.93 (d, 1H, *J* = 7.71 Hz), 8.38 (d, 1H, *J* = 4.34 Hz), 8.45 (d, 1H, *J* = 8.20 Hz), 8.85 (s, 1H). ^{13}C NMR ($CDCl_3$): 42.5; 100.2; 115.9; 120.1; 120.7; 120.7; 122.5; 126.0; 128.1; 128.3; 129.2; 133.2; 139.2; 142.8; 147.4; 148.8; 149.4 (Figures S25 and S26).

HRMS calculated for $[M + H]^+$ m/z = 310.0742, found m/z = 310.0729 (Δm_i = −4.2 ppm) (Figure S27). FTIR in Figure S28.

4. Conclusions

Glycine derivatives bearing 2- and 1-naphthol were reacted with indole and 7-azaindole. Reactions were found to depend on the naphthol skeleton. Starting from the 2-naphthol-substituted precursor, lactam-type ring-closed products were isolated. The transformation of 1-naphthol, in turn, led to the formation of the desired biaryl ester **7**. Starting from 5-chloro-8-hydroxyquinoline, morpholine, and ethyl glyoxylate as the aldehyde component, a new bifunctional precursor was formed using a modified Mannich-type synthetic pathway. To further investigate the scope and limitations of the reaction, ethyl 2-(5-chloro-8-hydroxyquinolin-7-yl)-2-morpholinoacetate was reacted with indole and 7-azaindole resulting in the desired diarylmethanes **11** and **12**. The series of 8-hydroxyquinoline skeleton containing bioactive derivatives was extended by reacting 7-aminobenzyl-8-hydroxyquinoline and 7-aminomethyl-8-hydroxyquinoline Mannich bases with indole and 7-azaindole. The synthesis of 7-((1H-indol-3-yl)methyl)-5-chloroquinolin-8-ol from different precursors bearing the morpholine skeleton or L-proline as amines, was also achieved. These studies represent a systematic investigation of the effect of leaving groups on conversion. The reactions were performed either under microwave irradiation or solvent-free conditions. The effect of *p*-toluenesulfonic acid on the reactions was investigated. On the basis of biological evaluations, the synthesized compounds showed toxic activity against the Colo205 and Colo320 cell lines, being more toxic on the resistant cell line expressing the ABCB1 and LRP multidrug transporters. The possible interaction between the derivatives and the MDR transporters should be investigated by further functional and docking studies. Regarding the cytotoxic effect on the normal MRC-5 cell line, the compounds can be considered tumor-selective, exhibiting low or no toxicity on non-tumor fibroblasts. The reaction of 5-chloro-8-hydroxyquinoline, morpholine, and ethyl glyoxylate resulted in the potent toxic activity against the sensitive Colo205 and resistant Colo320 cell lines. The compound can be considered selective, as it was not toxic to normal fibroblast cells. Besides the presence of 5-chloro-8-hydroxyquinoline, the morpholine moiety furnishing a cationic centre have been found to be relevant in anticancer activity. 7-((1H-indol-3-yl)methyl)-5-chloroquinolin-8-ol was detected as the most powerful scaffold in the series. This derivative exerted high potency to eliminate the resistant Colo320 cells and showed no toxicity on normal MRC-5 cells. The biaryl structure bearing 5-chloro-8-hydroxyquinoline and an indole skeleton can be considered favorable in the aspect of toxic activity against cancer cell lines.

Supplementary Materials: The ^{1}H, ^{13}C NMR spectra, FTIR spectra and HRMS spectra as supporting information can be downloaded at: https://www.mdpi.com/article/10.3390/molecules29174176/s1.

Author Contributions: Conceptualization, I.S. and G.S.; investigation, D.H., N.S., K.P. and R.B.; visualization, D.H.; writing—original draft preparation, I.S., D.H., N.S. and G.S.; writing—review and editing, I.S. and G.S. All authors have read and agreed to the published version of the manuscript.

Funding: This research received no external funding.

Institutional Review Board Statement: Not applicable.

Informed Consent Statement: Not applicable.

Data Availability Statement: The study did not report data.

Acknowledgments: The authors thank the Hungarian Research Foundation (OTKA No. K-138871), the Ministry of Human Capacities, Hungary grant, TKP-2021-EGA-32. N.S. was supported by the ÚNKP-23-4-SZTE-347 New National Excellence Program of the Ministry for Innovation and Technology from the source of the National Research, Development and Innovation Fund. G.S. was supported by the János Bolyai Research Scholarship (BO/00158/22/5) of the Hungarian Academy of Sciences and by the ÚNKP-23-5-SZTE-677 New National Excellence Program of the Ministry for Culture and Innovation from the source of the National Research, Development and Innovation Fund. The FTIR analysis was performed by Éva Anna Enyedy.

Conflicts of Interest: The authors declare no conflict of interest.

References

1. Tramontini, M.; Angiolini, L. Further advances in the chemistry of mannich bases. *Tetrahedron* **1990**, *46*, 1791–1837. [CrossRef]
2. Paqett, L.A. (Ed.) *Encyclopedia of Reagents for Organic Synthesis*; Wiley: UK, 2001; Volume 4, p. 21982.
3. Kos, J.; Ku, C.F.; Kapustikova, I.; Oravec, M.; Zhang, H.; Jampilek, J. 8-Hydroxyquinoline-2-carboxanilides as antiviral agents against avian influenza virus. *ChemistrySelect* **2019**, *4*, 4582–4587. [CrossRef]
4. Krawczyk, M.; Pastuch-Gawolek, G.; Mrozek-Wilczkiewicz, A.; Kuczak, M.; Skonieczna, M.; Musiol, R. Synthesis of 8-hydroxyquinoline glycoconjugates and preliminary assay of their B1,4-GalT inhibitory and anti-cancer properties. *Bioorg. Chem.* **2019**, *84*, 326–338. [CrossRef] [PubMed]
5. Pape, V.F.S.; Palkó, R.; Tóth, S.; Szabó, M.J.; Sessler, J.; Dormán, G.; Enyedy, É.A.; Soós, T.; Szatmári, I.; Szakács, G. Structure–activity relationships of 8-hydroxyquinoline-derived Mannich bases with tertiary amines targeting multi-drug-resistant cancer. *J. Med. Chem.* **2022**, *65*, 7729–7745. [CrossRef]
6. Xu, H.; Chen, W.; Zhan, P.; Liu, X. 8-Hydroxyquinoline: A privileged structure with a broad-ranging pharmacological poten-tial. *MedChemComm* **2015**, *6*, 61–74. [CrossRef]
7. Baez-Gonzalez, A.S.; Carrazco-Carrillo, J.A.; Figueroa-Gonzalez, G.; Quintas-Granados, L.I.; Padilla-Benavides, T.; Reyes-Hernandez, O.D. Functional effect of indole-3 carbinol in the viability and invasive properties of cultured cancer cells. *Biochem. Biophys. Rep.* **2023**, *35*, 101492. [CrossRef] [PubMed]
8. Amare, D.E.; Bovee, T.F.H.; Mulder, P.P.J.; Hamers, A.; Hoogenboom, R.L.A.P. Acid condensation products of in-dole-3-carbinol and their in-vitro (anti)estrogenic, (anti)androgenic and aryl hydrocarbon receptor activities. *Arab. J. Chem.* **2020**, *13*, 7199–7211. [CrossRef]
9. Puri, S.; Stefan, K.; Khan, S.L.; Pahnke, J.; Stefan, S.M.; Juvale, K. Indole derivatives as new structural class of potent and antiproliferative inhibitors of Monocarboxylate Transporter 1 (MCT1; SLC16A1). *J. Med. Chem.* **2023**, *66*, 657–676. [CrossRef]
10. El-Boghdady, N.A.; El-Hakk, S.A.; Abd-Elmawla, M.A. The lncRNAs UCA1 and CRNDE target miR-145/TLR4/NF-κB/TNF-α axis in acetic acid-induced ulcerative colitis model: The beneficial role of 3,3-diindolylmethane. *Int. Immunopharmacol.* **2023**, *121*, 110541. [CrossRef]
11. Pillaiyar, T.; Köse, M.; Sylvester, K.; Weighardt, H.; Thimm, D.; Borges, G.; Förster, I.; Kügelgen, I.; Müller, C.E. Diindol-ylmethane derivatives: Potent agonists of the immunostimulatory Orphan G Protein-Coupled Receptor GPR84. *J. Med. Chem.* **2017**, *60*, 3636–3655. [CrossRef]
12. Deb, M.L.; Pegu, C.D.; Deka, B.; Dutta, P.; Kotmale, A.S.; Baruah, P.K. Brønsted-Acid-Mediated Divergent Reactions of Betti Bases with Indoles: An Approach to Chromeno[2,3-b]indoles through Intramolecular Dehydrogenative C2-Alkoxylation of Indole. *Eur. J. Org. Chem.* **2016**, *2016*, 3441–3448. [CrossRef]
13. Pegu, C.D.; Nasrin, S.B.; Deb, M.L.; Das, D.J.; Saikia, K.K.; Baruah, P.K. CAN-Catalyzed microwave promoted reaction of indole with betti bases under solvent-free condition and evaluation of antibacterial activity of the products. *Synth. Commun.* **2017**, *21*, 2007–2014. [CrossRef]
14. Lőrinczi, B.; Simon, P.; Szatmári, I. Synthesis of indole-coupled KYNA derivatives via C–N bond cleavage of Mannich bases. *Int. J. Mol. Sci.* **2022**, *23*, 7152. [CrossRef]
15. Hegedűs, D.; Szemerédi, N.; Gábor, M.; Sas, J.; Belasri, K.; Szatmári, I.; Spengler, G. Cyclic amines coupled to indole deriva-tives with improved efflux pump inhibiting activity in bacteria and cancer cells. *Anticancer Res.* **2024**, *44*, 1149–1160. [CrossRef]
16. Hegedűs, D.; Szemerédi, N.; Spengler, G.; Szatmári, I. Application of partially aromatic ortho-quionone-methides for the synthesis of novel naphthoxazines with improved antibacterial activity. *Eur. J. Med. Chem.* **2022**, *237*, 114391. [CrossRef] [PubMed]
17. Csütörtöki, R.; Szatmári, I.; Mándi, A.; Kurtán, T.; Fülöp, F. Synthesis of hydroxynaphthyl-substituted-amino acid derivatives via a modified Mannich reaction. *Synlett* **2011**, *13*, 1940–1946. [CrossRef]
18. Jadhav, S.D.; Singh, A. Synthesis of unsymmetrical α,α-diarylacetates. *J. Org. Chem.* **2016**, *81*, 522–531. [CrossRef] [PubMed]
19. Chen, C.; Yang, X.; Fang, H.; Hou, X. Design, Synthesis and preliminary bioactivity evaluations of 8-hydroxyquinoline de-rivatives as Matrix Metalloproteinase (MMP) inhibitors. *Eur. J. Med. Chem.* **2019**, *181*, 111563. [CrossRef] [PubMed]
20. Liu, J.O.; Shim, J.S.; Chong, C.R.; Bhat, S. Preparation of Quinoline Derivatives for Use as Human Methionine Aminopeptidase, sirT1 and Angiogenesis Inhibitors. Patent WO2010042163, 15 April 2010.
21. Mészáros, J.P.; Poljarević, J.M.; Szatmári, I.; Csuvik, O.; Fülöp, F.; Szoboszlai, N.; Spengler, G.; Enyedy, É.A. An 8-hydroxyquinoline-proline hybrid with multidrug resistance reversal activity and the solution chemistry of its half-sandwich organometallic Ru and Rh complexes. *Dalton Trans.* **2020**, *49*, 7977–7992. [CrossRef]
22. Sancha, S.A.R.; Szemerédi, N.; Spengler, G.; Ferreira, M.-J.U. Lycorine carbamate derivatives for reversing P-glycoprotein-mediated multidrug resistance in human Colon adenocarcinoma cells. *Int. J. Mol. Sci.* **2023**, *24*, 2061. [CrossRef]
23. Pluchino, K.M.; Hall, M.D.; Goldsborough, A.S.; Callaghan, R.; Gottesman, M.M. Collateral sensitivity as a strategy against cancer multidrug resistance. *Drug Resist. Updates* **2012**, *15*, 98–105. [CrossRef] [PubMed]

24. Pivarcsik, T.; Dömötör, O.; Mészáros, J.P.; May, N.V.; Spengler, G.; Csuvik, O.; Szatmári, I.; Enyedy, É.A. 8-Hydroxyquinoline Amino Acid Hybrids and Their Half-Sandwich Rh and Ru Complexes: Synthesis, Anticancer Activities, Solution Chemistry and Interaction with Biomolecules. *Int. J. Mol. Sci.* **2021**, *22*, 11281. [CrossRef] [PubMed]
25. Csuvik, O.; Szemerédi, N.; Spengler, G.; Szatmári, I. Synthesis of 4-Hydroxyquinolines as Potential Cytotoxic Agents. *Int. J. Mol. Sci.* **2022**, *23*, 9688. [CrossRef] [PubMed]

back of this strategy is the preparation of the starting tetranitroethene via the decomposition of unstable hexanitroethane [40,41] (Scheme 5).

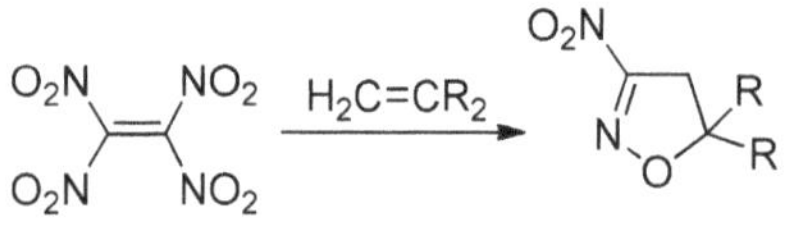

R = H, Alk, OR, Ar

Scheme 5. Conversion of tetranitroethene into 3-nitroisoxazoline derivatives.

All of the aforementioned strategies are not universally applicable and cannot be treated as general procedures for the preparation of a wide range of 3-nitro-2-isoxazilines. The [3+2] cycloaddition reaction involving formonitrile N-oxide **1**, which can be obtained via the thermal decomposition of dinitrofuroxan [42] (Scheme 6), can be considered a potential universal protocol for the preparation of these compounds.

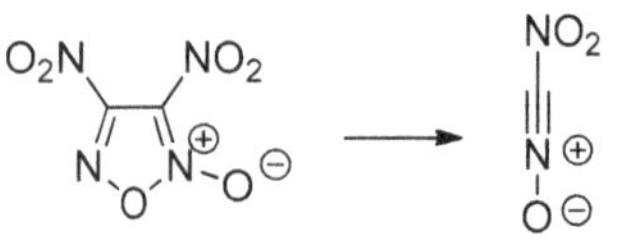

1

Scheme 6. Decomposition of dinitrofuroxane leading to formonitrile N-oxide **1**.

Unfortunately, these reactions have not yet been the subject of deeper systematic studies. In the literature, one may only find short communications regarding the 32CA reaction of selected COOMe-substituted alkenes [43]. This study does not, however, conduct a systematic exploration of the main problem. It should be noted that at this point the synthesis and properties of other, simply substituted nitrile N-oxides are understood substantially better (F [44,45], Cl [46–48], Br [49–51], and CN [52–54]).

Some similar reactions involving nitrile N-oxide **1** were tested under catalytic conditions with the presence of an RTIL (Room-Temperature Ionic Liquid). It should, however, be noted that the latest discoveries in the field of RTIL-catalyzed cycloaddition reactions indicate that its selectivity and molecular mechanism is completely different from analogous thermal reactions [55–57]. So, the main problem requires a deeper and systematic exploration. Due to the aforementioned issues, in the present work, we carried out a comprehensive DFT computational study regarding the 32CA processes involving formonitrile N-oxide **1** and a series of electron-rich mono- and gem-substituted alkenes.

Due to the expected electrophilic nature of the nitro-substituted formonitrile N-oxide **1** determined by the presence of a strong EWG (*Electron Withdrawing Group*) character of the nitro group, we selected nucleophilic-activated alkenes, such as isobutene **2a**, methylenecyclopentane **2b**, ethyl-vinyl ether **2c**, and N,N-dimethyl-vinyl amine **2d**, as model alkenes for carrying out the main cycloadditions. These alkenes were recently tested by us as components of different types of cycloadditions involving electrophilic agents [58–60] (Scheme 7).

Independently of the energetic aspects and their influence on the regioselectivity, the main processes also require a detailed mechanistic evaluation. According to the actual state of knowledge, different types of molecular mechanisms can occur during the [3+2] cycloaddition reactions: (a) polar mechanisms (one-step synchronous mechanism, one-step–two-stage asynchronous mechanism, and stepwise zwitterionic mechanism) [61–64], or (b) non-polar mechanisms (one-step synchronous mechanism, one-step–two-stage asynchronous mechanism, and stepwise biradical mechanism) [65,66].

Thus, in this work, we aimed to (i) analyze the of the global and power local interactions in the light of the Conceptual Density Functional Theory (CDFT) for predicting the nature of cycloaddition and regioselectivity; (ii) consider thermodynamic factors; and (iii) perform a full exploration of the reaction profiles for understanding the molecular mechanism as well as kinetic aspects of the main processes. The aforementioned analyses were performed according to DFT quantum chemical calculations.

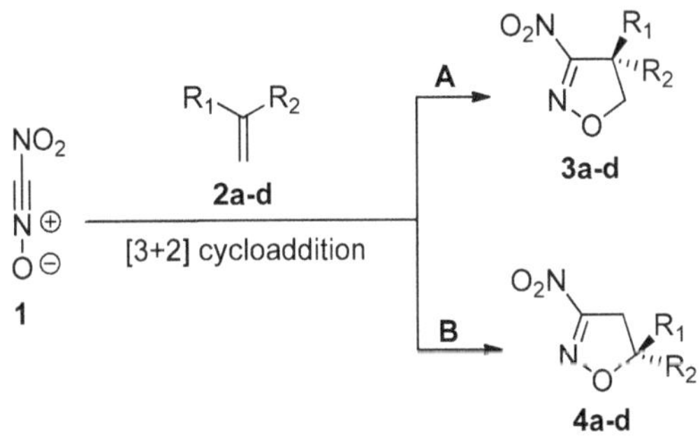

(**a**) R_1 = Me, R_2 = Me, (**b**) $R_1 = R_2$ = -$(CH_2)_4$-, (**c**) R_1 = H, R_2 = OEt, (**d**) R_1 = H, R_2 = NMe_2

Scheme 7. Theoretically possible paths of the cycloaddition of nitro-substituted formonitrile N-oxide **1** with electron-rich alkenes **2a–d**.

2. Results and Discussion

2.1. Power of the Global and Local Interactions in the Context of the CDFT

Analysis of the electronic properties of organic molecules is an important and universal tool for predicting their reactivity [67–69]. This approach was recently and successfully applied to interpret the reactivity of different types of molecular segments and the course of many bimolecular organic processes [70–74].

Global and local electronic properties of the main compounds were estimated according to equations recommended in the literature (see Computational Details) and presented in Table 1. The analysis of the values of electronic chemical potential clearly shows that, in the case of all considered reaction pairs, the electron density transfers from the alkene molecule to nitrile N-oxide. Thus, the considered processes—according to the *Domingo* convention [75]—should be classified as Forward Electron Density Flux (FEDF) processes. The global electrophilicity of nitrile N-oxide **1** is equal to 3.68 eV. Therefore, it should be classified as a strong electrophile [76]. In contrast, the global nucleophilicity index of **1** is less than 1 eV. Thus, the potential nucleophilic properties of this compound is evidently marginal. On the other hand, all four alkenes, **2a–d**, are characterized by a global electrophilicity index equal to or less than 0.6 eV. Hence, their electrophilic properties are marginal. In contrast, all these compounds exhibit greater nucleophilic properties as their respective global nucleophilicities are equal to or more than 2.5 eV. The most nucleophilic agent within the considered group is N,N-dimethylvinylamine **2d**, whereas the weakest nucleophilic agent is isobutene **2a** (Scheme 8).

The analysis of the electronic properties allow for the prediction of the potential regioselectivity of bimolecular processes. This is possible because these types of reactions occur under the control of local interactions between the most electrophilic reaction center on the first molecule and the most nucleophilic center at on second one [72,77]. It was found that the more electrophilic center within the CNO moiety of N-oxide **1** is localized on the oxygen atom (ω_O = 0.34 eV). On the other hand, the more nucleophilic center at the >C=C< moiety of the alkene is always localized on the terminal carbon atom (N_1 = 1.68–2.23 eV). The interaction of these centers stimulates the energetical preference for the formation of 3-nitro-4,4-substituted 2-isoxazolines **3a–d**.

Table 1. Global (electronic chemical potentials μ, chemical hardness η, global electrophilicities ω, and global nucleophilicities N) and local (local Parr spin density functions P^{+}_{k} and P^{-}_{k}, local nucleophilicities N_k, and local electrophilicities ω_k) electronic properties of nitro-substituted formonitrile N-oxide **1** and electron-rich alkenes **2a–d**.

	Global Properties				Local Properties							
					 **1**				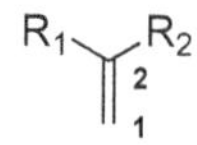 **2a-d**			
	μ [eV]	η [eV]	ω [eV]	N [eV]	P^{-}_{2}	P^{-}_{1}	N_2 [eV]	N_1 [eV]	P^{+}_{C}	P^{+}_{O}	ω_C [eV]	ω_O [eV]
1	−5.98	4.85	3.68	0.72					0.048	0.091	0.18	0.34
2a	−2.83	7.37	0.55	2.60	0.295	0.646	0.77	1.68				
2b	−2.87	6.95	0.59	2.77	0.266	0.636	0.74	1.76				
2c	−2.39	6.97	0.41	3.25	0.057	0.577	0.18	1.87				
2d	−1.87	6.50	0.27	3.99	0.110	0.557	0.44	2.23				

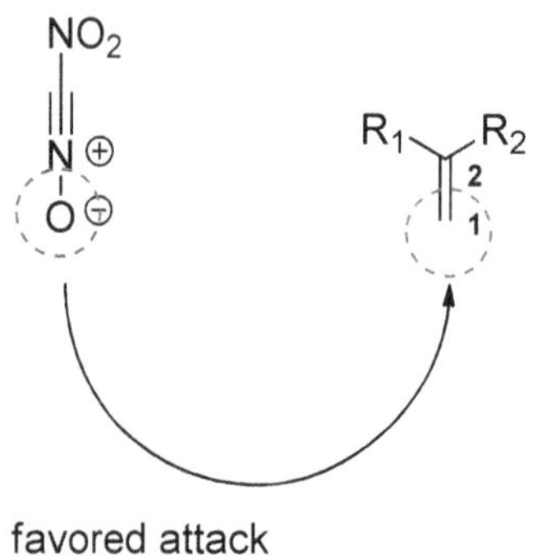

Scheme 8. Most efficient local electronic interactions of nitro-substituted formonitrile N-oxide **1** with electron-rich alkenes **2a–d**.

2.2. Thermodynamic Considerations

Within the framework of the thermodynamic analysis of the main cycloadditions, we treated all competing reactions as independent processes. To perform this analysis, the values of the thermodynamic functions of the substrates and products were required. Unfortunately, for the compounds that are the subject of our study, these data are not available in the chemical literature. Therefore, we used data from DFT quantum chemical calculations to determine them. Based on estimated enthalpies and entropies of formation of substrates and products, we calculated the enthalpy and entropy values for each cycloaddition reaction, followed by the *Gibbs* free energies of the reaction. The results of thermodynamic calculations for the considered reaction paths are presented in Table 2.

The data obtained via DFT calculations show (Table 2) that, independently of the alkene structures, in the toluene solution, the Gibbs free energies of the reaction are strongly negative for both considered cycloaddition paths. During this period, the reaction channels leading to less sterically crowded 5,5-disubstituted 2-isoxazolines (path **B**) are always preferred. The replacement of toluene with the more polar solvent (nitromethane) did not change this tendency. Thus, the thermodynamic preference for the formation of 3-nitro-5,5-di(R)-2-isoxazolines can be considered as a general rule in the context of the main reactions.

Table 2. Key thermodynamic and kinetic parameters for 32CA of nitro-substituted formonitrile N-oxide **1** with electron-rich alkenes **2a–d** according to the wb97xd/6-311+G(d) (PCM) calculations (ΔH and ΔG are in kcal/mol; ΔS are in cal/molK).

Reaction	Transition	Toluene			Nitromethane		
		ΔH	ΔG	ΔS	ΔH	ΔG	ΔS
1+2a	**1+2a→MCA**	−2.3	5.9	−27.4	−1.8	6.3	−27.4
	1+2a→TSA	7.7	19.6	−39.8	8.5	20.4	−39.9
	1+2a→3a	−63.5	−47.9	−52.2	−64.7	−49.2	−51.9
	1+2a→MCB	−2.5	6.6	−30.4	−3.4	5.7	−30.7
	1+2a→TSB	3.6	13.6	−33.3	4.1	13.8	−32.7
	1+2a→4a	−70.0	−54.6	−51.7	−71.4	−55.7	−52.6
1+2b	**1+2b→MCA**	−4.4	4.3	−29.1	−3.3	4.9	−27.3
	1+2b→TSA	8.4	20.5	−40.9	9.1	21.6	41.8
	1+2b→3b	−64.3	−49.6	−49.3	−65.4	−50.7	−49.2
	1+2b→MCB	−3.9	4.7	−28.7	−2.9	5.4	−27.9
	1+2b→TSB	2.0	12.6	−35.6	2.3	13.0	−35.8
	1+2b→4b	−69.2	−56.6	−42.1	−70.4	−58.2	−41.0
1+2c	**1+2c→MCA**	−7.0	4.6	−38.9	−5.5	5.5	−37.0
	1+2c→TSA	5.7	18.2	−41.8	6.5	18.9	−41.4
	1+2c→3c	−59.9	−45.1	−49.4	−60.6	−45.7	−49.8
	1+2c→MCB	−5.0	3.7	−29.2	−1.9	3.8	−19.1
	1+2c→TSB	0.5	11.5	−37.1	1.4	12.0	−35.6
	1+2c→4c	−70.5	−55.9	−49.1	−71.0	−56.4	−48.9
1+2d	**1+2d→MCA**	−4.3	3.1	−24.8	−3.5	4.1	−25.3
	1+2d→TSA	5.1	15.9	−36.0	5.7	16.5	−36.2
	1+2d→3d	−58.5	−44.8	−46.2	−58.9	−44.8	−47.2
	1+2d→MCB	−4.6	3.1	−25.8	−3.7	4.2	−26.5
	1+2d→TSB	−4.0	6.6	−35.7	−3.6	6.8	−35.2
	1+2d→4d	−67.4	−53.5	−46.5	−68.3	−54.3	−46.8

2.3. Exploration of Reaction Profiles

Firstly, we analyzed the energy reaction profiles for the 32CA of nitro-substituted formonitrile N-oxide **1** and isobutene **2a**. It was found that the nature of the reaction trajectories are qualitatively similar for both regioisomeric cycloaddition channels considered. In particular, in the framework of both considered profiles, two critical points (pre-reaction molecular complex, MC, and the transition state, TS) between the valleys of the starting molecules and products were detected. The quantitative description of the aforementioned transformations is, however, substantially different.

In the initial phase of the reaction progress, respective pre-reaction molecular complexes are formed in the case of both 32CA pathways (**MCA** and **MCB** for paths **A** and **B**, respectively). The reduction in enthalpy of the molecular system by about 2.3–2.5 kcal/mol is a result of this transformation (Table 2, Figure 1). At the same time, however, the entropy of the molecular system is substantially reduced. As a result, the Gibbs free energies of formation for both MCs are positive. This excludes the possibility of the existence of **MCA** and **MCB** complexes as thermodynamically stable intermediates. Within the MC, substructures of addends adopt orientations that stimulate maximum stable coulombic interactions. Any new sigma bond, however, are not created at this stage. The key interatomic distances are beyond the range typical for C-C and O-C bonds formed in the transition state (Table 3, Figure 1). Additionally, the electron density transfer between substructures is not observed. Thus, the detected structures cannot be considered as charge–transfer complexes [78,79]. Similar intermediates were recently detected in many 32CAs involving different types of TACs [80,81].

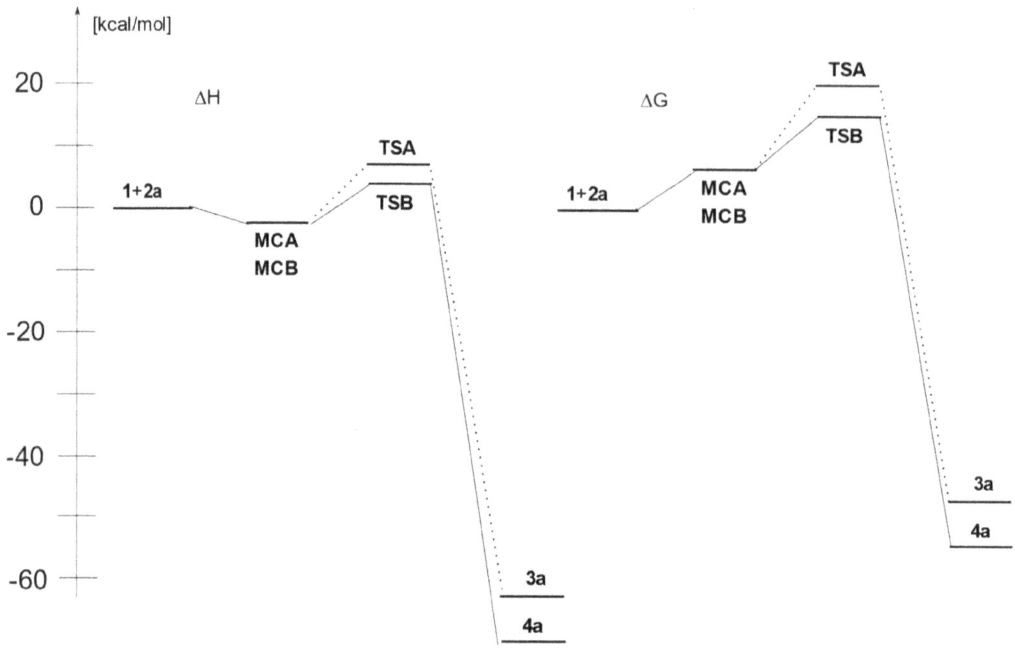

Figure 1. Enthalpy and *Gibbs* free energy profiles the [3+2] cycloaddition reaction between the nitro-substituted formonitrile N-oxide **1** and isobutene **2a** in the toluene solution according to the wb97xd/6-311+G(d) (PCM) calculations.

Table 3. Key parameters of critical structures of the 32CA of nitro-substituted formonitrile N-oxide **1** with electron-rich alkenes **2a–d** in the toluene solution according to the wb97xd/6-311+G(d) (PCM) calculations (ΔH and ΔG are in kcal/mol; ΔS are in cal/molK).

Reaction	Structure	Interatomic Distances [Å]					GEDT [e]
		O1–N2	N2–C3	C3–C4	C4–C5	C5–O1	
1+2a	**1**	1.181	1.162				
	2a				1.332		
	MCA	1.179	1.163	3.299	1.332	3.344	0.00
	TSA	1.195	1.215	2.360	1.364	2.434	0.16
	3a	1.347	1.271	1.510	1.546	1.455	
	MCB	1.184	1.157	3.216	1.333	3.416	0.00
	TSB	1.196	1.199	2.292	1.357	2.865	0.13
	4a	1.343	1.271	1.487	1.542	1.489	
1+2b	**2b**				1.329		
	MCA	1.179	1.167	3.596	1.330	3.565	0.00
	TSA	1.195	1.213	2.342	1.361	2.461	0.20
	3b	1.356	1.272	1.503	1.534	1.450	
	MCB	1.179	1.167	3.519	1.330	3.556	0.00
	TSB	1.199	1.198	2.268	1.354	3.037	0.14
	4b	1.344	1.271	1.488	1.532	1.489	

Table 3. *Cont.*

Reaction	Structure	Interatomic Distances [Å]					GEDT [e]
		O1–N2	N2–C3	C3–C4	C4–C5	C5–O1	
1+2c	**2c**				1.329		
	MCA	1.179	1.168	3.854	1.331	5.153	0.00
	TSA	1.201	1.217	2.264	1.365	2.484	0.12
	3c	1.349	1.272	1.510	1.531	1.453	
	MCB	1.179	1.168	3.739	1.328	3.475	0.00
	TSB	1.200	1.192	2.357	1.351	3.045	0.13
	4c	1.360	1.269	1.488	1.524	1.467	
1+2d	**2d**				1.341		
	MCA	1.187	1.164	3.271	1.347	4.557	0.00
	TSA	1.204	1.227	2.680	1.376	2.227	0.22
	3d	1.351	1.272	1.520	1.533	1.453	
	MCB	1.192	1.164	2.890	1.349	4.070	0.00
	TSB	1.206	1.186	2.370	1.365	3.643	0.13
	4d	1.346	1.272	1.487	1.533	1.515	

The gradual reduction in interatomic distances between the nitrile N-oxide **1** and isobutene **2a** substructures along the reaction coordinate leads directly to the transition state (**TSA** and **TSB** for paths **A** and **B,** respectively). This stage is associated with an increase in enthalpy of the reaction system by 7.7 kcal/mol in the case of path **A** and 3.6 kcal/mol in the case of path **B** (Table 2, Figure 1). Including entropic factors in the energetic considerations yield Gibbs free energies of activation equal to 19.9 kcal/mol and 13.6 kcal/mol, respectively, for paths **A** and **B**. Thus, in the considered system of competing reactions, the formation process of isoxazoline **3a** should be considered as forbidden from a kinetic point of view. Therefore, the predicted regioselectivity of the cycloaddition is fully different from that suggested by earlier analysis of local electronic interactions. The regiodirection of the considered reaction is predominantly determined by steric repulsion involving the germinal methyl groups derived from the isobutene molecule. In contrast, analogous 32CAs involving electron-rich nitrile N-oxides and electrophilic alkenes are fully determined by the nature of global and local electronic interactions [23,24,70].

Within both optimized TSs, two new sigma-bonds are formed. The synchronicity of this process is, however, substantially different in the case of **TSA** than in the case of **TSB** (Table 2, Figure 2). In particular, in **TSA**, key interatomic distances C3-C4 and C5-O1 are equal to 2.360Å and 2.434Å, respectively. However, within **TSB,** respective interatomic distances are equal to 2.292Å and 2.865Å, respectively. The asynchronicity is, however, not sufficient to enforce a stepwise, zwitterionic mechanism [82–85]. The formation of new bonds is accompanied by the redistribution of electron density. In particular, the electron density is transferred gradually from the alkene substructure to the nitrile oxide substructure (GEDT = 0.16–0.16 e). This confirms that the currently analyzed reaction should be classified as a Forward Electron Density Flux (FEDF) process. Molecular electrostatic potential maps for **TS4** and **TS5** structures are presented in Figure 3. This analysis evidently confirms the polar nature of localized TSs. Further conversion of the TS structures leads directly to the following respective products—**3a** in the case of pathway **A** and **4a** in the case of pathway **B**.

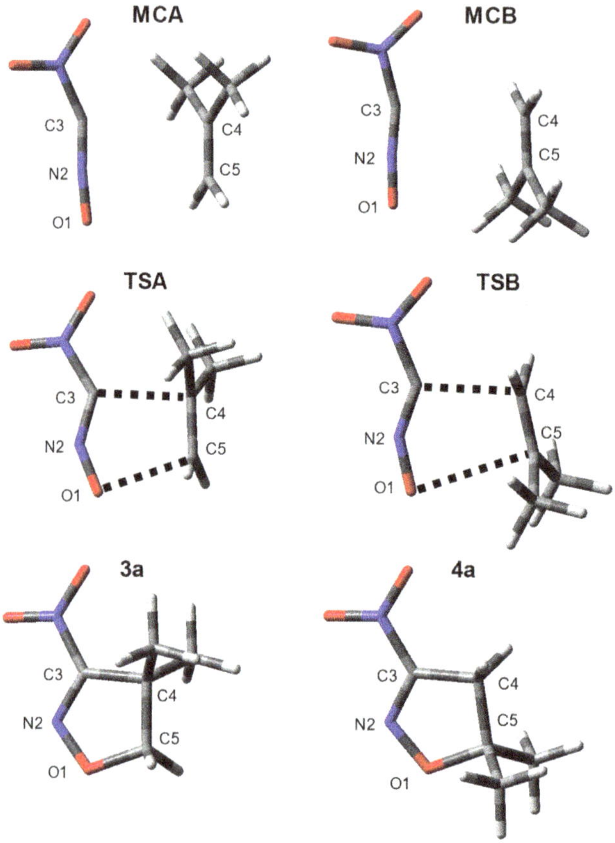

Figure 2. Views of critical structures of the [3+2] cycloaddition reaction between the nitro-substituted formonitrile N-oxide **1** and isobutene **2a** in the toluene solution according to the wb97xd/6-311+G(d) (PCM) calculations.

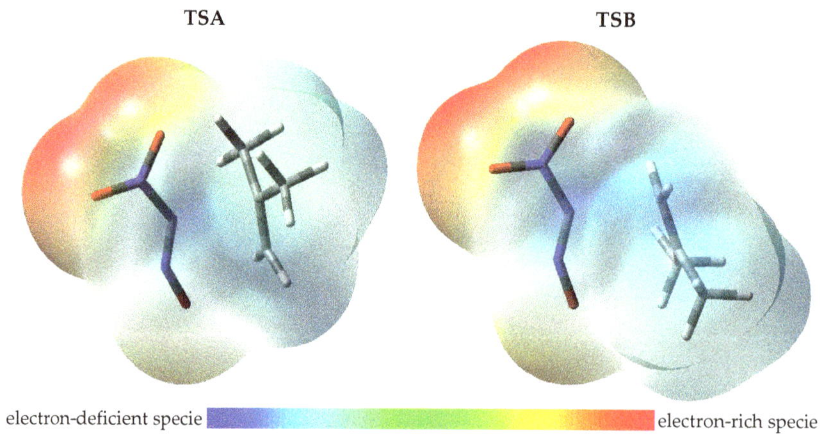

Figure 3. Molecular electrostatic potential maps for **TS4** and **TS5** structures of the [3+2] cycloaddition reaction between the nitro-substituted formonitrile N-oxide **1** and isobutene **2a**. Negative charges are colored in red, while negligible charges are colored in green.

Replacement of isobutene **3a** in the 32CA environment with the more nucleophilic alkenes, such as **3b**, **3c** and **3d**, does not change the nature of the energy profiles for all considered cycloaddition channels. The description of respective profiles as well as optimized critical points are, however, changed. In particular, the energy barrier on the kinetically favored cycloaddition pathway **B** is clearly reduced with the increase in the global nucleophilicity of the alkene. At the same time, the polarity and asynchronicity of respective transition structures are more evident. However, we cannot detect theoretically possible zwitterionic intermediates. A similar effect is also determined due to the influence of the polarity solvent on the reaction progress (Tables 2–4). Thus, the described scheme and mechanism should be treated as a general rule for a wide range 32CAs involving nitro-substituted formonitrile N-oxide **1**.

Table 4. Key parameters of critical structures of the 32CA of nitro-substituted formonitrile N-oxide **1** with electron-rich alkenes **2a–d** in the nitromethane solution according to the wb97xd/6-311+G(d) (PCM) calculations (ΔH and ΔG are in kcal/mol; ΔS are in cal/molK).

Reaction	Structure	Interatomic Distances [Å]					GEDT [e]
		O1–N2	N2–C3	C3–C4	C4–C5	C5–O1	
1+2a	**1**	1.183	1.158				
	2a				1.333		
	MCA	1.180	1.161	3.308	1.333	3.334	0.00
	TSA	1.195	1.214	2.369	1.364	2.458	0.18
	3a	1.346	1.271	1.509	1.547	1.459	
	MCB	1.181	1.164	3.416	1.333	4.015	0.00
	TSB	1.198	1.197	2.283	1.357	2.995	0.15
	4a	1.342	1.272	1.485	1.543	1.495	
1+2b	**2b**				1.330		
	MCA	1.181	1.164	3.586	1.331	3.602	0.00
	TSA	1.196	1.212	2.352	1.360	2.484	0.21
	3b	1.354	1.272	1.502	1.535	1.455	
	MCB	1.181	1.164	3.529	1.331	3.711	0.00
	TSB	1.201	1.196	2.271	1.353	3.166	0.16
	4b	1.341	1.272	1.486	1.534	1.495	
1+2c	**2c**				1.330		
	MCA	1.180	1.166	3.813	1.332	4.933	0.00
	TSA	1.200	1.216	2.283	1.364	2.513	0.13
	3c	1.346	1.273	1.510	1.531	1.457	
	MCB	1.182	1.161	3.417	1.330	3.381	0.00
	TSB	1.200	1.190	2.361	1.351	3.146	0.15
	4c	1.357	1.270	1.487	1.525	1.475	
1+2d	**2d**				1.343		
	MCA	1.186	1.153	3.290	1.348	4.536	0.00
	TSA	1.202	1.226	2.756	1.375	2.257	0.25
	3a	1.349	1.273	1.521	1.532	1.457	
	MCB	1.191	1.164	2.892	1.350	4.082	0.00
	TSB	1.205	1.179	2.443	1.363	3.737	0.23
	4d	1.343	1.273	1.486	1.533	1.526	

3. Computational Details

The DFT calculations were performed using the wb97xd functional and 6-311+G(d) basis sets and the software implemented within the Gaussian package [86]. The PlGrid infrastructure ("Ares" cluster) in the computing center "Cyfronet" was applied. A similar computational level has already been successfully used to explore the mechanistic aspects of different types of cycloaddition processes, including the [3+2] cycloaddition reactions involving allenyl-type TACs [87–89] as well as different-type transformations of hydrocarbons [90]. All localized stationary points have been characterized using vibrational analysis. It was found that the starting molecules, intermediates, and products had positive eigenvalues in their Hessian matrices, whereas all transition states (**TS**) showed only one negative eigenvalue in their Hessian matrices. Intrinsic reaction coordinate (IRC) calculations have been performed for all localized transition states. The presence of the solvent in the reaction environment (toluene and nitromethane) has been included using the IEFPCM algorithm [91]. The Global Electron Density Transfer (GEDT) [92] was calculated according to Formula (1):

$$\mathrm{GEDT} = -\Sigma qA \quad (1)$$

where qA is the net charge, and the sum is taken including all the atoms of nitroalkene.

Global electronic properties of the reactants were estimated according to the equations recommended by *Parr* and *Domingo* [64,66,67,93–96]. According to *Domingo*'s recommendation, for this purpose, the B3LYP/6-31G(d) level of theory was used. In particular, the electronic chemical potentials (μ) and chemical hardness (η) were evaluated in terms of one-electron energies of FMO (E_{HOMO} and E_{LUMO}) using Equations (2) and (3):

$$\mu \approx (E_{HOMO} + E_{LUMO})/2 \quad (2)$$

$$\eta \approx E_{LUMO} - E_{HOMO} \quad (3)$$

Next, the values of μ and η were used for the calculation of global electrophilicity (ω) according to Formula (4):

$$\omega = \mu^2/2\eta \quad (4)$$

Subsequently, global nucleophilicity (N) [66] could be expressed in terms of Equation (5):

$$N = E_{HOMO} - E_{HOMO\,(tetracyanoethene)} \quad (5)$$

The local electrophilicity (ω_k) [67] condensed to atom *k* was calculated by projecting the index ω onto any reaction center *k* in the molecule by using the *Parr* function $P^+{}_k$ (6):

$$\omega_k = P^+{}_k \cdot \omega \quad (6)$$

The local nucleophilicity (N_k) [67] condensed to atom *k* was calculated using global nucleophilicity N and *Parr* function $P^-{}_k$ according to Formula (7):

$$N_k = P^-{}_k \cdot N \quad (7)$$

4. Conclusions

Our comprehensive DFT study regarding [3+2] cycloaddition involving nitro-substituted formonitrile N-oxide **1** and electron-rich alkenes shed light on the reaction regioselectivity and molecular mechanism. In particular, the wb97xd/6-311+G(d) (PCM) calculations show that independently of the alkene structure as well as of the polarity of the solvent, the thermodynamic factors allow for the formation of stable cycloadducts within considered cycloadditions. Next, the analysis of the local electronic properties suggest the preference for the formation of 4-R-isoxazolines. In contrast, kinetic factors obtained via the full exploration of energy profiles lead to a preference for the formation of 5-substituted 2-isoxazolines. This clearly demonstrates the influence of steric factors on the course of the reaction. The respective transition states leading to the aforementioned products are always asynchronous, but all attempts to locate the hypothetical zwitterionic intermediates along the cycloaddition paths were unsuccessful. In summary, it should be underlined that the scheme and mechanism described in this study should be treated as a general occurrence for a wide range of 32CA involving nitro-substituted formonitrile N-oxide.

Author Contributions: Methodology, R.J.; Formal analysis, E.D. and A.W.; Investigation, R.J.; Data curation, E.D. and A.W.; Writing—original draft, R.J.; Supervision, R.J. All authors have read and agreed to the published version of the manuscript.

Funding: This research received no external funding.

Institutional Review Board Statement: Not applicable.

Informed Consent Statement: Not applicable.

Data Availability Statement: The raw data supporting the conclusions of this article will be made available by the authors on request.

Acknowledgments: We gratefully acknowledge Polish high-performance computing infrastructure PLGrid (HPC Centers: ACK Cyfronet AGH) for providing computer facilities and support within computational grant no. PLG/2023/016326.

Conflicts of Interest: The authors declare no conflict of interest.

References

1. Machado, M.A.; Campos, D.R.; Lopes, N.L.; Bastos, I.P.B.; Alves, M.S.R.; Correia, T.R.; Scott, F.B.; Fernandes, J.I. Efficacy of afoxolaner in the flea control in experimentally infested cats. *Rev. Bras. Parasitol. Vet.* **2019**, *28*, 760–763. [CrossRef] [PubMed]
2. Paarlberg, T.; Karadzovska, D.; Helbig, R. Efficacy of lotilaner (Credelio™) against experimentally induced infestations of the adult cat flea, *Ctenocephalides felis*, and flea eggs following oral administration to cats. *Parasites Vectors* **2021**, *14*, 139. [CrossRef] [PubMed]
3. Six, R.H.; Geurden, T.; Carter, L.; Everett, W.R.; McLoughlin, A.; Mahabir, S.P.; Myers, M.R.; Slootmans, N. Evaluation of the speed of kill of sarolaner (Simparica™) against induced infestations of three species of ticks (*Amblyomma maculatum*, *Ixodes scapularis*, *Ixodes ricinus*) on dogs. *Vet. Parasitol.* **2016**, *222*, 37–42. [CrossRef] [PubMed]
4. Fujinami, M.; Takahashi, Y.; Tanetani, Y.; Ito, M.; Nasu, M. Development of a rice herbicide, fenoxasulfone. *J. Pestic. Sci.* **2019**, *44*, 282–289. [CrossRef]
5. Kleinman, E.F. Isoxazoline Compoundsas Antnflammatoryagents. U.S. Patent 5,716,967 A, 10 February 1998.
6. Al-Abed, Y. Isoxazoline Compounds Having MIF Antagonist Activity. U.S. Patent 7,662,843 B2, 16 February 2004.
7. Sajadikhah, S.S.; Didehban, K. Synthesis of substituted isoxazolidines (microreview). *Chem. Heterocycl. Compd.* **2023**, *59*, 640–642. [CrossRef]
8. Nikol'skiy, V.V.; Starosotnikov, A.M. Synthesis of isoxazolo[4,5-b]pyridine derivatives. *Chem. Heterocycl. Comp.* **2023**, *59*, 240–242. [CrossRef]
9. Xu, J.; Liu, J.; Mei, H.; Soloshonok, V.A.; Han, J. A three-component cycloaddition of alkyl trifluorodiazoethane for the synthesis of trifluoromethylated isoxazolines. *Chem. Heterocycl. Comp.* **2023**, *59*, 465–471. [CrossRef]
10. Kras, J.; Sadowski, M.; Zawadzińska, K.; Nagatsky, R.; Woliński, P.; Kula, K.; Łapczuk, A. Thermal [3+2] cycloaddition reactions as most universal way for the effective preparation of five-membered nitrogen containing heterocycles. *Sci. Rad.* **2023**, *2*, 247–267. [CrossRef]
11. Łapczuk, A. The [3+2] cycloaddition reaction as an attractive way for the preparation of nicotine analogs. *Chem. Heterocycl. Comp.* **2023**, *59*, 109–111. [CrossRef]
12. Dresler, E. The Participation of Oleic Acid and its Esters in [3+2] Cycloaddition Reactions: A Mini-Review. *Sci. Rad.* **2024**, *3*, 53–61. [CrossRef]
13. Digamber, R.; Mukund, S. Recent Advances in Nitrile Oxide Cycloadditions. Synthesis of Isoxazolines. *Curr. Org. Synth.* **2011**, *8*, 616–627.
14. Ono, N. Cycloaddition Chemistry of Nitro Compounds. In *The Nitro Group in Organic Synthesis*; Feuer, H., Ono, N., Eds.; Wiley-VCH: Weinheim, Germany, 2001.
15. Jasiński, R. Synthesis of 1,2-oxazine N-oxides via non-catalyzed hetero-Diels–Alder reactions of nitroalkenes (microreview). *Chem. Heterocycl. Compd.* **2024**, *60*, 121–123.
16. Hao, F.; Nishiwaki, N. Recent Progress in Nitro-Promoted Direct Functionalization of Pyridones and Quinolones. *Molecules* **2020**, *25*, 673. [CrossRef]
17. Rocard, L.; Goujon, A.; Hudhomme, P. Nitro-Perylenediimide: An Emerging Building Block for the Synthesis of Functional Organic Materials. *Molecules* **2020**, *25*, 1402. [CrossRef] [PubMed]
18. Sadowski, M.; Kula, K. Nitro-functionalized analogues of 1,3-Butadiene: An overview of characteristic, synthesis, chemical transformations and biological activity. *Curr. Chem. Lett.* **2024**, *13*, 15–30. [CrossRef]
19. Kras, J.; Mikulska, M.; Allnajar, R.; Kula, K. Unsuccessful synthesis of individual 1,1-dinitroethene. *Sci. Rad.* **2024**, *3*, 9–13. [CrossRef]
20. Boguszewska-Czubara, A.; Kula, K.; Wnorowski, A.; Biernasiuk, A.; Popiołek, Ł.; Miodowski, D.; Demchuk, O.M.; Jasiński, R. Novel functionalized β-nitrostyrenes: Promising candidates for new antibacterial drugs. *Saudi Pharm. J.* **2019**, *27*, 593–601. [CrossRef] [PubMed]

21. Boguszewska-Czubara, A.; Lapczuk-Krygier, A.; Rykala, K.; Biernasiuk, A.; Wnorowski, A.; Popiolek, L.; Maziarka, A.; Hordyjewska, A.; Jasiński, R. Novel synthesis scheme and in vitro antimicrobial evaluation of a panel of (E)-2-aryl-1-cyano-1-nitroethenes. *J. Enzyme Inhib. Med. Chem.* **2016**, *31*, 900–907. [CrossRef]
22. Zawadzińska, K.; Gostyński, B. Nitrosubstituted analogs of isoxazolines and isoxazolidines: A surprising estimation of their biological activity via molecular docking. *Sci. Rad.* **2023**, *2*, 25–46. [CrossRef]
23. Kula, K.; Zawadzińska, K. Local nucleophile-electrophile interactions in [3+2] cycloaddition reactions between benzonitrile N-oxide and selected conjugated nitroalkenes in the light of MEDT computational study. *Curr. Chem. Lett.* **2021**, *10*, 9–16. [CrossRef]
24. Zawadzińska, K.; Kula, K. Application of β-Phosphorylated Nitroethenes in [3+2] Cycloaddition Reactions Involving Benzonitrile N-Oxide in the Light of a DFT Computational Study. *Organics* **2021**, *2*, 26–37. [CrossRef]
25. Jasiński, R.; Jasińska, E.; Dresler, E. A DFT computational study of the molecular mechanism of [3+2] cycloaddition reactions between nitroethene and benzonitrile N-oxides. *J. Mol. Model.* **2017**, *23*, 13. [CrossRef] [PubMed]
26. Zawadzińska, K.; Ríos-Gutiérrez, M.; Kula, K.; Woliński, P.; Mirosław, B.; Krawczyk, T.; Jasiński, R. The Participation of 3,3,3-Trichloro-1-nitroprop-1-ene in the [3+2] Cycloaddition Reaction with Selected Nitrile N-Oxides in the Light of the Experimental and MEDT Quantum Chemical Study. *Molecules* **2021**, *26*, 6774. [CrossRef] [PubMed]
27. Zawadzińska, K.; Gaurav, G.K.; Jasiński, R. Preparation of conjugated nitroalkenes: Short review. *Sci. Rad.* **2022**, *1*, 69–83. [CrossRef]
28. Halimehjani, A.Z.; Namboothiri, I.N.N.; Hooshmanda, S.E. Nitroalkenes in the synthesis of carbocyclic compounds. *RSC Adv.* **2014**, *4*, 31261–31299. [CrossRef]
29. Lam, P.Y.S.; Adams, J.J.; Clark, C.G.; Calhoun, W.J.; Luettgen, J.M.; Knabb, R.M.; Wexler, R.R. Discovery of 3-Amino-4-Chlorophenyl P1 as a novel and potent benzamidine mimic via solid-phase synthesis of an isoxazoline library. *Bioorg. Med. Chem. Lett.* **2003**, *13*, 1795–1799. [CrossRef] [PubMed]
30. Harding, S.L.; Marcuccio, S.M.; Savage, G.P. Aryl nitrile oxide cycloaddition reactions in the presence of pinacol boronic acid ester. *Beilstein J. Org. Chem.* **2012**, *8*, 606–612. [CrossRef] [PubMed]
31. Toczko, M.A.; Heathcock, C.H. Total Synthesis of (±)-Aspidospermidine. *J. Org. Chem.* **2000**, *65*, 2642–2645. [CrossRef] [PubMed]
32. Groundwater, P.W.; Nyerges, M.; Fejes, I.; Hibbs, D.E.; Bendell, D.; Anderson, R.J.; McKillop, A.; Sharif, T.; Zhang, W. Preparation and reactivity of some stable nitrile oxides and nitrones. *ARKIVOC* **2000**, *V*, 684–697. [CrossRef]
33. Shapiro, A.B.; Skripnichenko, L.N.; Pavlikov, V.V.; Rozantsev, E.G. Synthesis of nitroxyl radicals based on 4-ethynyl-4-hydroxy-2,2,6,6-tetramethylpiperidine. *Russ. Chem. Bull.* **1979**, *28*, 140–148. [CrossRef]
34. Shitov, O.P.; Baranin, S.V.; Tartakovsky, V.A. Esters of nitromethane nitronic acid. *Russ. Chem. Bull.* **2022**, *71*, 350–353. [CrossRef]
35. Kacka-Zych, A. Push-pull nitronates in the [3+2] cycloaddition with nitroethylene: Molecular Electron Density Theory study. *J. Mol. Graph. Model.* **2020**, *97*, 1075492. [CrossRef] [PubMed]
36. Wade, P.A.; D'Ambrosio, S.G.; Price, D.T. 2-Amino-2-deoxytetrose Derivatives. Preparation from 4,5-Dihydroisoxazoles via Reductive Cleavage. *J. Org. Chem.* **1995**, *60*, 6302–6308. [CrossRef]
37. Kloeckner, J.; Schmitz, J.; Holzgrabe, U. Convergent, short synthesis of the muscarinic superagonist iperoxo. *Tetrahedron Lett.* **2010**, *51*, 3470–3472. [CrossRef]
38. Wade, P.A. Synthesis of 3-substituted 2-isoxazolines and 5,6-dihydro-4H-1,2-oxazines. *J. Org. Chem.* **1978**, *43*, 2020–2023. [CrossRef]
39. Diamantini, G.; Duranti, E.; Tontini, A. Nitroisoxazoles by Manganese(IV) Oxide Oxidation of Nitro-4,5-dihydroisoxazoles. *Synthesis* **1993**, *1993*, 1104–1108. [CrossRef]
40. Baum, K.; Tzeng, D. Synthesis and reactions of tetranitroethylene. *J. Org. Chem.* **1985**, *50*, 2736–2739. [CrossRef]
41. Ivanova, O.A.; Ekaterina, M.; Budynina, E.M.; Averina, E.B.; Kuznetsova, T.S.; Zefirov, N.S. Application of a Thermal b-Elimination Reaction to N-Alkoxy-3,3-dinitroisoxazolidines: Synthesis of 3-Nitroisoxazolines. *Synthesis* **2006**, *2006*, 706–710. [CrossRef]
42. Ovchinnikov, I.V.; Makhova, N.N.; Khmel'nitskii, L.I. Generation of Nitro Formonitrile Oxide as an Intermediate for the Preparation of Dinitrofuroxan. *Mendeleev Commun.* **1993**, *3*, 210–211. [CrossRef]
43. Fershtat, L.L.; Khakimov, D.V.; Makhova, N.N. Dinitrofuroxan cycloreversion as a novel general approach for the synthesis of nitroazoles. *Russ. Chem. Bull.* **2015**, *64*, 415–422. [CrossRef]
44. Rodriguez-Betancourtt, V.-M.; Quezada-Navarro, V.-M.; Neff, M.; Rauhut, G. Anharmonic frequencies of [F,C,N,X] isomers (X = O,S) obtained from explicitly correlated coupled-cluster calculations. *Chem. Phys.* **2011**, *387*, 1–4. [CrossRef]
45. Jacobs, J.; Jülicher, B.; Schatte, G.; Willner, H.; Mack, H.G. CFNO-Isomere—Theoretische und experimentelle Studien. *Chem. Ber.* **1993**, *126*, 2167–2176. [CrossRef]
46. Zana, A.; Galbiati, A. Synthesis and Reactivity of 3-Halo-4,5-dihydroisoxazoles: An Overview. *ChemistrySelect* **2021**, *6*, 8249–8261. [CrossRef]
47. Lichau, H.; Gillies, C.W.; Gillies, J.Z.; Ross, S.C.; Winnewisser, B.P.; Winnewisser, M. On the Anharmonic XCN Bending Modes of the Quasilinear Molecules BrCNO and ClCNO. *J. Phys. Chem. A* **2001**, *105*, 10065–10079. [CrossRef]
48. Pasinszki, T.; Westwood, N.P.C. Substituted oximes and furoxans as precursors to unstable nitrile oxides. electronic and geometric structures by ultraviolet photoelectron spectroscopy, infrared spectroscopy and ab initio calculations. *J. Mol. Struct.* **1997**, *408–409*, 161–169. [CrossRef]

49. Yazdani, H.; Hooshmand, S.E.; Varma, R.S. Gold Nanoparticle-Catalyzed Multicomponent Reactions. *ACS Sustain. Chem. Eng.* **2021**, *9*, 16556–16569. [CrossRef]
50. Tamborini, L.; Mastronardi, F.; Presti, L.L.; Nielsen, B.; De Micheli, C.; Conti, P.; Pinto, A. Synthesis of l-Tricholomic Acid Analogues and Pharmacological Characterization at Ionotropic Glutamate Receptors. *ChemistrySelect* **2017**, *2*, 10295–10299. [CrossRef]
51. Singh, M.S.; Chowdhury, S.; Koley, S. Progress in 1,3-dipolar cycloadditions in the recent decade: An update to strategic development towards the arsenal of organic synthesis. *Tetrahedron* **2016**, *72*, 1603–1644. [CrossRef]
52. Cao, W.-L.; Tariq, Q.-u.-N.; Li, Z.-M.; Yang, J.-Q.; Zhang, J.-G. Recent advances on the nitrogen-rich 1,2,4-oxadiazole-azoles-based energetic materials. *Def. Technol.* **2022**, *18*, 344–367. [CrossRef]
53. Escapa, C.; Torres, T.; Neuparth, T.; Coimbra, R.N.; García, A.I.; Santos, M.M.; Otero, M. Zebrafish embryo bioassays for a comprehensive evaluation of microalgae efficiency in the removal of diclofenac from water. *Sci. Total Environ.* **2018**, *640–641*, 1024–1033. [CrossRef]
54. Robertson, E.G.; McNaughton, D. IR Spectroscopy of OP−X and Derivatives: Mistaken Identity on a Large Scale. *J. Phys. Chem. A* **2003**, *107*, 642–650. [CrossRef]
55. Jasiński, R. A stepwise, zwitterionic mechanism for the 1,3-dipolar cycloaddition between (Z)-C-4-methoxyphenyl-N-phenylnitrone and gem-chloronitroethene catalysed by 1-butyl-3-methylimidazolium ionic liquid cations. *Tetrahedron Lett.* **2015**, *56*, 532–535. [CrossRef]
56. Jasiński, R. First example of stepwise, zwitterionic mechanism for bicyclo[2.2.1]hept-5-ene (norbornene) formation process catalyzed by the 1-butyl-3-methylimidazolium cations. *Monatsh. Chem.* **2016**, *147*, 1207–1213. [CrossRef] [PubMed]
57. Kącka, A.; Jasiński, R. A dramatic change of kinetic conditions and molecular mechanism of decomposition processes of nitroalkyl carboxylates catalyzed by ethylammonium cations. *Comput. Theor. Chem.* **2017**, *1104*, 37–42. [CrossRef]
58. Woliński, P.; Kącka-Zych, A.; Mirosław, B.; Wielgus, E.; Olszewska, A.; Jasiński, R. Green, one-pot synthesis of 1,2-oxazine-type herbicides via non-catalyzed Hetero Diels-Alder reactions comprising (2E)-3-aryl-2-nitroprop-2-enenitriles. *J. Clean. Prod.* **2022**, *356*, 131878. [CrossRef]
59. Wen, C.; Dechsupa, N.; Yu, Z.; Zhang, X.; Liang, S.; Lei, X.; Xu, T.; Gao, X.; Hu, Q.; Innuan, P.; et al. Pentagalloyl Glucose: A Review of Anticancer Properties, Molecular Targets, Mechanisms of Action, Pharmacokinetics, and Safety Profile. *Molecules* **2023**, *28*, 4856. [CrossRef] [PubMed]
60. Jasiński, R.; Kubik, M.; Łapczuk-Krygier, A.; Kącka, A.; Dresler, E.; Boguszewska-Czubara, A. An experimental and theoretical study of the hetero Diels–Alder reactions between (E)-2-aryl-1-cyano-1-nitroethenes and ethyl vinyl ether: One-step or zwitterionic, two-step mechanism? *React. Kinet. Mech. Catal.* **2014**, *113*, 333–345. [CrossRef]
61. Dresler, E.; Wróblewska, A.; Jasiński, R. Understanding the Regioselectivity and the Molecular Mechanism of [3+2] Cycloaddition Reactions between Nitrous Oxide and Conjugated Nitroalkenes: A DFT Computational Study. *Molecules* **2022**, *27*, 8441. [CrossRef] [PubMed]
62. Zawadzińska, K.; Gadocha, Z.; Pabian, K.; Wróblewska, A.; Wielgus, E.; Jasiński, R. The First Examples of [3+2] Cycloadditions with the Participation of (E)-3,3,3-Tribromo-1-Nitroprop-1-Ene. *Materials* **2022**, *15*, 7584. [CrossRef]
63. Kula, K.; Łapczuk, A.; Sadowski, M.; Kras, J.; Zawadzińska, K.; Demchuk, O.M.; Gaurav, G.K.; Wróblewska, A.; Jasiński, R. On the Question of the Formation of Nitro-Functionalized 2,4-Pyrazole Analogs on the Basis of Nitrylimine Molecular Systems and 3,3,3-Trichloro-1-Nitroprop-1-Ene. *Molecules* **2022**, *27*, 8409. [CrossRef]
64. Dresler, E.; Woliński, P.; Wróblewska, A.; Jasiński, R. On the Question of Zwitterionic Intermediates in the [3+2] Cycloaddition Reactions between Aryl Azides and Ethyl Propiolate. *Molecules* **2023**, *28*, 8152. [CrossRef]
65. Domingo, L.R.; Ríos Gutiérrez, M.; Castellanos Soriano, J. Understanding the Origin of the Regioselectivity in Non-Polar [3+2] Cycloaddition Reactions through the Molecular Electron Density Theory. *Organics* **2020**, *1*, 19–35. [CrossRef]
66. Domingo, L.R.; Aurell, M.J.; Pérez, P. A DFT analysis of the participation of zwitterionic TACs in polar [3+2] cycloaddition reactions. *Tetrahedron* **2014**, *70*, 4519–4525. [CrossRef]
67. Domingo, L.R.; Ríos-Gutiérrez, M.; Pérez, P. Applications of the Conceptual Density Functional Theory Indices to Organic Chemistry Reactivity. *Molecules* **2016**, *21*, 748. [CrossRef] [PubMed]
68. Domingo, L.R. Molecular Electron Density Theory: A Modern View of Reactivity in Organic Chemistry. *Molecules* **2016**, *21*, 1319. [CrossRef] [PubMed]
69. Ríos-Gutiérrez, M.; Domingo, L.R. Unravelling the Mysteries of the [3+2] Cycloaddition Reactions. *Eur. J. Org. Chem.* **2019**, *2019*, 267–282. [CrossRef]
70. Dresler, E.; Wróblewska, A.; Jasiński, R. Understanding the Molecular Mechanism of Thermal and LA-Catalysed Diels–Alder Reactions between Cyclopentadiene and Isopropyl 3-Nitroprop-2-Enate. *Molecules* **2023**, *28*, 5289. [CrossRef] [PubMed]
71. Woliński, P.; Kącka-Zych, A.; Wróblewska, A.; Wielgus, E.; Dolot, R.; Jasiński, R. Fully Selective Synthesis of Spirocyclic-1,2-oxazine N-Oxides via Non-Catalysed Hetero Diels-Alder Reactions with the Participation of Cyanofunctionalysed Conjugated Nitroalkenes. *Molecules* **2023**, *28*, 4586. [CrossRef]
72. Sadowski, M.; Utnicka, J.; Wójtowicz, A.; Kula, K. The global and local Reactivity of C,N-diarylnitryle imines in [3+2] cycloaddition processes with trans-β-nitrostyrene according to Molecular Electron Density Theory: A computational study. *Curr. Chem. Lett.* **2023**, *12*, 421–430. [CrossRef]

73. Kącka-Zych, A. Understanding the uniqueness of the stepwise [4+1] cycloaddition reaction between conjugated nitroalkenes and electrophilic carbene systems with a molecular electron density theory perspective. *Int. J. Quantum. Chem.* **2021**, *121*, e26440. [CrossRef]
74. Aitouna, A.O.; Barhoumi, A.; Zeroual, A. A Mechanism Study and an Investigation of the Reason for the Stereoselectivity in the [4+2] Cycloaddition Reaction between Cyclopentadiene and Gem-substituted Ethylene Electrophiles. *Sci. Rad.* **2023**, *2*, 217–228. [CrossRef]
75. Domingo, L.R.; Ríos-Gutiérrez, M. A Useful Classification of Organic Reactions Based on the Flux of the Electron Density. *Sci. Rad.* **2023**, *2*, 1–24. [CrossRef]
76. Ríos-Gutiérrez, M.; Saz Sousa, A.; Domingo, L.R. Electrophilicity and nucleophilicity scales at different DFT computational levels. *J. Phys. Org. Chem.* **2023**, *36*, e4503. [CrossRef]
77. Ríos-Gutiérrez, M.; Domingo, L.R.; Ghodsi, F. Unveiling the Different Reactivity of Bent and Linear Three-Atom-Components Participating in [3+2] Cycloaddition Reactions. *Organics* **2021**, *2*, 274–286. [CrossRef]
78. Bent, H.A. Structural chemistry of donor-acceptor interactions. *Chem. Rev.* **1968**, *68*, 587–648. [CrossRef]
79. Foster, R. Electron donor-acceptor complexes. *J. Phys. Chem.* **1980**, *84*, 2135–2141. [CrossRef]
80. Ameur, S.; Barhoumi, A.; Ríos-Gutiérrez, M.; Aitouna, A.O.; El Alaoui El Abdallaoui, H.; Mazoir, N.; Belghiti, M.E.; Syed, A.; Zeroual, A.; Domingo, L.R. A MEDT study of the mechanism and selectivity of the hetero-Diels–Alder reaction between 3-benzoylpyrrolo[1,2-c][1,4]-benzoxazine-1,2,4-trione and vinyl acetate. *Chem. Heterocycl. Comp.* **2023**, *59*, 165–170. [CrossRef]
81. Krishna, C.; Seetharam, K.; Satyadev, T. Synthesis of β-amino alcohols by ring opening of epoxides with amines catalyzed by sulfated tin oxide under mild and solvent-free conditions. *Curr. Chem. Lett.* **2024**, *13*, 343–350. [CrossRef]
82. Jasiński, R.; Dresler, E. On the Question of Zwitterionic Intermediates in the [3+2] Cycloaddition Reactions: A Critical Review. *Organics* **2020**, *1*, 49–69. [CrossRef]
83. Siadati, S.A. Beyond the Alternatives that switch the Mechanism of the 1,3-Dipolar Cycloadditions from Concerted to Stepwise or Vice Versa: A Literature Review. *Prog. React. Kinet. Mech.* **2016**, *41*, 331–344. [CrossRef]
84. Siadati, S.A.; Rezazadeh, S. The extraordinary gravity of three atom 4π-components and 1,3-dienes to $_{C20\text{-}nXn}$ fullerenes; a new gate to the future of Nano technology. *Sci. Rad.* **2022**, *1*, 46–68. [CrossRef]
85. Mohtat, B.; Siadati, S.A.; Khalilzadeh, M.A.; Zareyee, D. The concern of emergence of multi-station reaction pathways that might make stepwise the mechanism of the 1,3-dipolar cycloadditions of azides and alkynes. *J. Mol. Struct.* **2018**, *1155*, 58–64. [CrossRef]
86. Frisch, M.J.; Trucks, G.W.; Schlegel, H.B.; Scuseria, G.E.; Robb, M.A.; Cheeseman, J.R.; Scalmani, G.; Barone, V.; Mennucci, B.; Petersson, G.A.; et al. *Gaussian 09*; Gaussian, Inc.: Wallingford, CT, USA, 2009.
87. Mondal, A.; Acharjee, N. Unveiling the exclusive stereo and site selectivity in [3+2] cycloaddition reactions of a tricyclic strained alkene with nitrile oxides from the molecular electron density theory perspective. *Chem. Heterocycl. Comp.* **2023**, *59*, 145–154. [CrossRef]
88. Abdoul-Hakim, M.; Idrissi, K.E.; Zeroual, A.; Garmes, H. Investigation of the solvent effect, regioselectivity, and the mechanism of the cycloaddition reaction between 2-chlorobenzimidazole and benzonitrile oxide. *Chem. Heterocycl. Comp.* **2023**, *59*, 155–164. [CrossRef]
89. Yousfi, Y.; Benchouk, W.; Mekelleche, S.M. Prediction of the regioselectivity of the ruthenium-catalyzed [3+2] cycloadditions of benzyl azide with internal alkynes using conceptual DFT indices of reactivity. *Chem. Heterocycl. Comp.* **2023**, *59*, 118–127. [CrossRef]
90. Wu, J.; Yu, D.; Liu, S.; Rong, C.; Zhong, A.; Chattaraj, P.K.; Shubin, L. Is It Possible to Determine Oxidation States for Atoms in Molecules Using Density-Based Quantities? An Information-Theoretic Approach and Conceptual Density Functional Theory Study. *J. Phys. Chem. A* **2019**, *123*, 6751–6760. [CrossRef] [PubMed]
91. Cossi, M.; Rega, N.; Scalmani, G.; Barone, V. Energies, structures, and electronic properties of molecules in solution with the C-PCM solvation model. *J. Comput. Chem.* **2003**, *24*, 669–681. [CrossRef] [PubMed]
92. Domingo, L.R. A new C–C bond formation model based on the quantum chemical topology of electron density. *RSC Adv.* **2014**, *4*, 32415–32428. [CrossRef]
93. Domingo, L.R.; Aurell, M.J.; Pérez, P.; Contreras, R. Quantitative characterization of the global electrophilicity power of common diene/dienophile pairs in Diels–Alder reactions. *Tetrahedron* **2002**, *58*, 4417–4423. [CrossRef]
94. Parr, R.G.; Szentpály, L.; Liu, S. Electrophilicity Index. *J. Am. Chem. Soc.* **1999**, *121*, 1922–1924. [CrossRef]
95. Domingo, L.R.; Chamorro, E.; Pérez, P. Understanding the Reactivity of Captodative Ethylenes in Polar Cycloaddition Reactions. A Theoretical Study. *J. Org. Chem.* **2008**, *73*, 4615–4624. [CrossRef] [PubMed]
96. Domingo, L.R.; Pérez, P.; Sáez, J.A. Understanding the local reactivity in polar organic reactions through electrophilic and nucleophilic Parr functions. *RSC Adv.* **2013**, *3*, 1486–1494. [CrossRef]

molecules

Article

Synthesis, Electronic, and Antibacterial Properties of 3,7-Di(hetero)aryl-substituted Phenothiazinyl *N*-Propyl Trimethylammonium Salts

Hilla Khelwati [1], Lasse van Geelen [2], Rainer Kalscheuer [2] and Thomas J. J. Müller [1,*]

[1] Institute of Organic Chemistry and Macromolecular Chemistry, Faculty of Mathematics and Natural Sciences, Heinrich Heine University Düsseldorf, Universitätsstrasse 1, D-40225 Düsseldorf, Germany; hilla.khelwati@hotmail.com

[2] Institute of Pharmaceutical Biology and Biotechnology, Faculty of Mathematics and Natural Sciences, Heinrich Heine University Düsseldorf, Universitätsstrasse 1, D-40225 Düsseldorf, Germany; lasse.geelen@hhu.de (L.v.G.); kalscheu@hhu.de (R.K.)

* Correspondence: thomasjj.mueller@hhu.de; Tel.: +49-211-81-12298

Abstract: In this study, a library of 3,7-di(hetero)aryl-substituted 10-(3-trimethylammoniumpropyl) 10*H*-phenothiazine salts is prepared. These title compounds and their precursors are reversible redox systems with tunable potentials. The Hammett correlation gives a very good correlation of the first oxidation potentials with σ_p parameters. Furthermore, the title compounds and their precursors are blue to green-blue emissive. Screening of the salts reveals for some derivatives a distinct inhibition of several pathogenic bacterial strains (*Mycobacterium tuberculosis*, *Staphylococcus aureus*, *Escherichia coli*, *Aconetobacter baumannii*, and *Klebsiella pneumoniae*) in the lower micromolar range.

Keywords: antibacterials; biological properties; cyclic voltammetry; fluorescence; heterocycles; Suzuki coupling

1. Introduction

Phenothiazine, i.e., 10*H*-dibenzo[b,e][1,4]thiazine, is a synthetic non-naturally occurring tricyclic heterocycle that has received recognition in many fields of molecular science, starting from dye chemistry in the late 19th century, via histochemical staining (as dye methylene blue), to pharmacology and biomedicine in the 20th century [1]. In the past decades, the extraordinary electronic structure of this electron-rich system has paved new alleys to functional materials in optoelectronics [2,3], dye-sensitized solar cells [4], and mesoporous silica hybrid materials [5]. And again, in the 21st century, the phenothiazine scaffold is receiving considerable attention for its multifaceted biological activity [6]. Phenothiazine derivatives exhibit promising antibacterial, antifungal, anticancer, antiviral, anti-inflammatory, antimalarial, antifilarial, trypanocidal, anticonvulsant, analgesic, immunosuppressive, and multidrug resistance reversal properties [7]. Increasingly, for phenothiazines, known as non-steroidal anti-inflammatory drugs or local anesthetics, antibacterial activity has been recognized [8]. In addition, among other privileged scaffolds, phenothiazines are considered to be photosensitizers in antimicrobial photodynamic therapy [9]. Derivatives bearing halogens or electron-withdrawing groups have been shown to have anticancer and antiprotozoal effects [10]. Diverse biological activities associated with phenothiazine hybrids make them particularly promising candidates for drug development [11].

For combatting multidrug resistance in treating infectious diseases and cancer, phenothiazines are also discussed as a cost-effective strategy against this resistance; however, simultaneously, their potential effects on bacterial resistance and dysbiosis have been considered [12]. In particular, in targeting Methicillin-Resistant *Staphylococcus aureus* (MRSA) with its multidrug resistance mechanisms, including efflux pumps and biofilm formation, phenothiazines, like chlorpromazine, which has received classic status in chemical neuroscience [13], can be considered to be efflux pump inhibitors [14].

In Gram-negative bacteria, e.g., *Escherichia coli* and *Pseudomonas aeruginosa*, the outermost membrane apparently forms a permeability barrier, causing their already intrinsic resistance to many active substances [15]. Molecules with a cationic or cationic–hydrophobic character can be considered for interaction with the phospholipid membrane of a bacterium.

Cationically substituted phenothiazines are (a) water-soluble and (b) amphiphilic due to the organic residue. In addition, the organic residue is (c) a chromophore (with absorption in near UV/Vis and emission in blue-green (fluorescence microscopy)) and (d) a redox system that is also capable of photo-induced electron transfer. Recently, we showed that phenothiazines with quaternary ammonium groups in their side chains are readily immobilized on both periodic mesoporous organosilica (PMO) and neat silica SBA-15 materials via ion exchange [16]. UV irradiation induced the formation of stable phenothiazine radical cations embedded in the hybrid materials, confirmed by cyclic voltammetry of the native materials and EPR spectroscopy of the oxidized specimen. Here, we generalize the modular approach to the synthesis of 3-trimethylammoniumpropyl sidechain-substituted 3,7-di(hetero)aryl-substituted phenothiazine to provide electronically diverse substitution in positions 3 and 7. Furthermore, we present and discuss the electronic properties investigated with cyclic voltammetry, optical absorption, and emission spectroscopy, and we disclose a screening of the antibacterial activity against seven selected pathogenic strains.

2. Results and Discussion

2.1. Synthesis

As in a previous work [16], the synthesis commences with the dibromination of phenothiazine (**1**) to give 3,7-dibromophenothiazine (**2**) in 88% yield (Scheme 1) [17]. The Michael addition of compound **2** in Triton B® furnishes 10-propionitrile phenothiazine **3** in 70% yield [16]. The indium-trichloride-mediated boranate reduction [18] gives 10-(3-amino propyl) 3,7-dibromophenothiazine (**4**) in 91% yield, that is, Boc-protected under standard conditions, to give the dibromo substrate **5** in 95% yield.

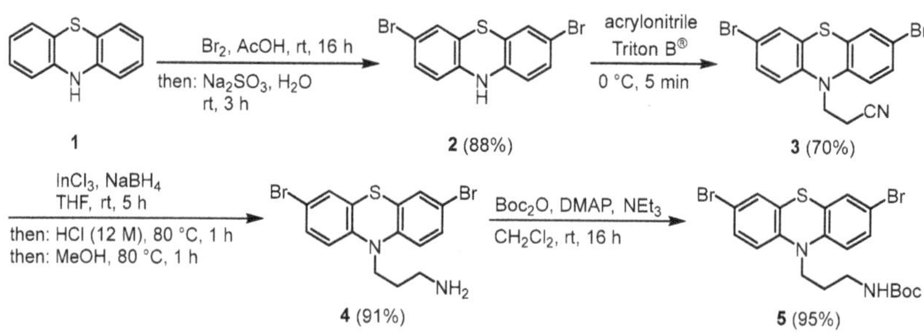

Scheme 1. Synthesis of *tert*-butyl (3-(3,7-dibromo-10*H*-phenothiazin-10-yl)propyl)carbamate (**5**).

The Boc-protected 10-(3-amino propyl) 3,7-dibromo phenothiazine **5** sets the stage for twofold Suzuki coupling with (hetero)aryl boronic acids or pinacol boronates **6** to give six 3,7-di(hetero)aryl-substituted phenothiazines **7** in good yield after isolation by flash chromatography (Scheme 2). The treatment of compounds **7** with an excess of trifluoroacetic acid leads to Boc deprotection; the ammonium salt is permethylated in the presence of an excess of methyl iodide (for the preparation of salt **8a**) or methyl triflate (for the preparation of salts **8b–f**), and potassium carbonate as a base gives rise to the isolation of six salts **8** in moderate to excellent yield. The structures of the salts (compound **8a** as a carbonate with two cations and compounds **8b–f** as triflates with one cation) is unambiguously assigned by the combination of ^{1}H and ^{13}C NMR spectroscopy, indicating mirror plane symmetry by the occurrence of a reduced set of signals, with IR spectroscopy identifying the counter anion by characteristic vibration bands (1454 cm^{-1} as a very dominant CO valence vibration for the carbonate ion in salt **8a**; ~1240 and ~1030 cm^{-1} as very dominant SO and CF valence vibrations for triflate ions in the salts **8b–f**), mass spectrometry unambiguously verifying the molecular peak of the cation, as well as HRMS identifying the molecular composition of the cations and/or combustion analysis for the molecular composition of the salts.

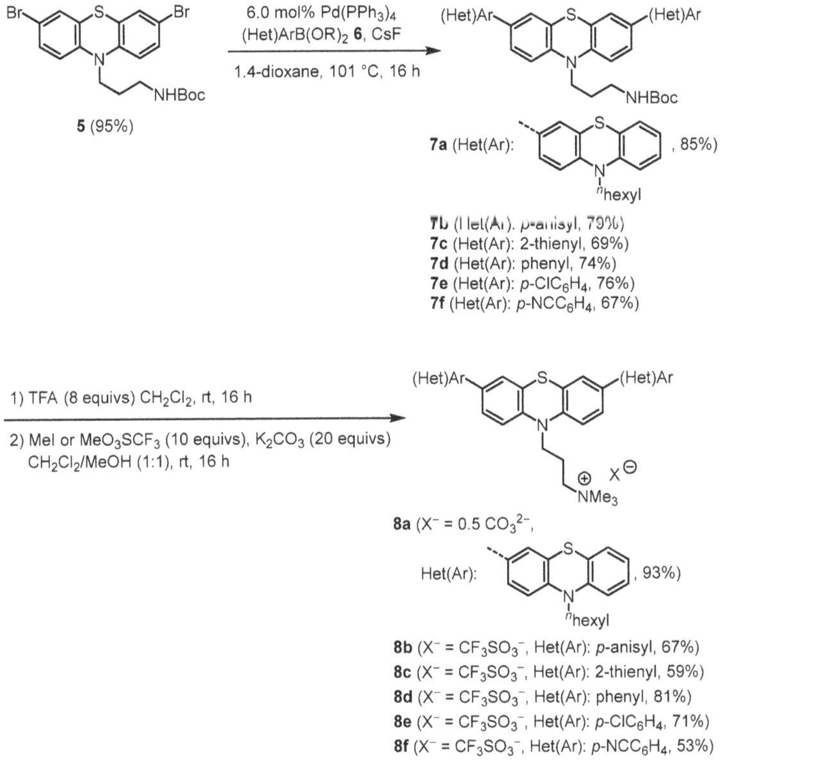

Scheme 2. Synthesis of Boc-protected 3,7-di(hetero)aryl-substituted 10-(3-aminopropyl) 10*H*-phenothiazines **7** and 3,7-di(hetero)aryl-substituted 10-(3-trimethylammoniumpropyl)10*H*-phenothiazine salts **8**.

2.2. *Electronic Properties of Compounds* **7** *and* **8**

As part of our ongoing interest in the structure–property relationships of 3-(hetero)aryl and 3,7-di(hetero)aryl-substituted phenothiazines [19,20], we became intrigued in studying the effect of Boc-protected 3-aminopropyl (compounds **7**) and 3-trimethylammoniumpropyl sidechains (compounds **8**) at the 10-position of 3,7-di(hetero)aryl-substituted phenothiazines on the electronic properties in the electronic ground and excited states. Therefore, cyclic voltammograms and absorption and emission spectra were recorded to determine the redox potentials (cyclic voltammograms) and UV/Vis absorption and fluorescence bands and thereof the Stokes shifts (Table 1).

The cyclic voltammetry of triads **7a** and **8a** discloses the expected behavior of three conjugatively coupled electrophores [21,22] with three reversible one-electron oxidations at potentials of 0.620, 0.740, and 0.850 V (**7a**) and 0.610, 0.720, and 0.830 V (**8a**), respectively (Table 1). As can be seen by the minute differences in the oxidation potentials, the effect of the remote 3-NH-Boc or 3-trimethylammonium substituent on the sidechain is only minor. Indeed, it can be considered that the first oxidation giving a radical cation, that is delocalized over all three phenothiazines, renders this potential to lie cathodically shifted.

For the consanguineous series **7b–f** and **8b–f**, the site of oxidation is clearly localized on the central phenothiazine unit, which certainly experiences the Coulombic interaction of the sidechain after oxidation. The first oxidation potentials $E_0^{0/+1}$ are found in a range from 0.640 to 0.830 V (compounds **7b–f**) and from 0.680 to 0.840 V (compounds **8b–f**), respectively. The second oxidations to the phenothiazine-centered dications are only found at potentials $E_0^{+1/+2}$ between 1.040 and 1.400 V for the more electron-rich substituted compounds **7b–d** and **8b–d**. For both series, a distinct anodic shift of $E_0^{0/+1}$ is discernable with altering the remote 3,7-substituents from electron-releasing to electron-withdrawing. For a closer inspection to estimate the transmission of the polar substituent effect of the remote substituent in the 3- and 7-position on the ease of oxidation, correlation analyses of the oxidation potentials $E_0^{0/+1}$ with Hammett parameters [23] σ_p, σ_R, σ_{R+}, σ_{p+}, and σ_{p-} were considered.

Table 1. Selected electronic data (oxidation potential, absorption maxima, emission maxima, Stokes shifts) of Boc-protected 3,7-di(hetero)aryl-substituted 10-(3-aminopropyl) 10*H*-phenothiazines **7** and 3,7-di(hetero)aryl-substituted 10-(3-trimethylammoniumpropyl)10*H*-phenothiazine salts **8**.

Compound	$E_0^{0/+1}$ [V] [a]	$E_0^{+1/+2}$ [V] [a]	$E_0^{+2/+3}$ [V] [a]	Absorption $\lambda_{max,abs}$ [nm] (ε [$M^{-1}cm^{-1}$]) [b]	Emission $\lambda_{max,em}$ [nm] [c]	Stokes Shift $\Delta\tilde{\nu}$ [cm^{-1}] [d]
7a	0.620	0.740	0.850	279 (111,000), 320 (45,000), **367** (37,000sh)	483, 499 (sh)	6500
7b	0.640	1.290	-	280 (108,000), **330** (26,000)	461	9300
7c	0.690	1.260	-	290 (42,000), **347** (14,000)	488	8300
7d	0.700	1.400	-	277 (17,000), **330** (41,000)	471	9500
7e	0.730	-	-	280 (35,000), **333** (9000)	488	9500
7f	0.830	-	-	292 (33,000), **373** (12,000)	520	7600
8a	0.610	0.720	0.830	276 (83,000), 322 (31,000), **368** (26,000)	478	6300
8b	0.680	1.040	-	278 (64,000), **324** (15,000)	455	8900
8c	0.720	1.260	-	278 (40,000), **326** (10,000)	466	9200
8d	0.730	1.270	-	275 (26,000), **323** (6000)	456	9000
8e	0.770	-	-	289 (53,000), **342** (18,000)	478	8300
8f	0.840	-	-	288 (54,000), **359** (7000)	500	7900

[a] Cyclic voltammograms were recorded in dichloromethane at T = 293 K. [b] Recorded at T = 293 K in dichloromethane at $c(\mathbf{7/8}) = 10^{-3}$ M. [c] Recorded in dichloromethane at T = 293 K at $c(\mathbf{7/8}) = 10^{-6}$ M. [d] $\Delta\tilde{\nu} = 1/\lambda_{max,abs} - 1/\lambda_{max,em}$ [cm^{-1}] (calculated using the bold absorption bands).

In the consanguineous series of compounds **7b–f**, indeed, the highest linear regression coefficients are obtained for parameters σ_p (R^2 = 0.9903), σ_{p+} (R^2 = 0.9631), and σ_{p-} (R^2 = 0.9491) (for details, see Supporting Information). The involvement of σ_p clearly indicates that the electronic substituent effects are transmitted via resonance and inductive pathways, as previously observed for other 3,7-disubstituted phenothiazines [19,24]. The same correlation can be established for the consanguineous series of the trimethylammonium salts **8b–f**, where the linear regression coefficients for σ_p (R^2 = 0.9855) and σ_{p+} (R^2 = 0.9839) are very similar and for σ_{p-} (R^2 = 0.9056) are less, indicating the similar transmission mechanism of remote polar substituents by resonance and inductive pathways for the stabilization of the formed radical dication.

The absorption spectra disclose the longest wavelength bands between 330 and 373 nm (compounds **7a–f**) and between 323 and 368 nm (compounds **8a–f**), with absorption coefficients in a range between 6000 and 41,000 $M^{-1}cm^{-1}$ (Table 1, Figure 1).

While the red shifts of the triads **7a** and **8a** amount to intense shoulders at 367 nm (**7a**) or a maximum at 368 nm (**8a**) attributed to the extended π-conjugation, the most redshifted bands appear for the *p*-cyano phenyl derivatives at 373 nm (**7f**) and 359 nm (**8f**). All compounds **7** and **8** luminesce intensively with a blue to green color in a range from 455 to 520 nm, and the Stokes shifts are quite broad, ranging from 6300 to 9500 cm^{-1}, which is quite typical for phenothiazine as a consequence of geometrical and electronic changes upon photonic excitation, namely the planarization of the phenothiazine core in the vibrationally relaxed excited state and considerable charge transfer from the central phenothiazine donor to the 3,7-substituents that act as acceptors in the excited state [25,26].

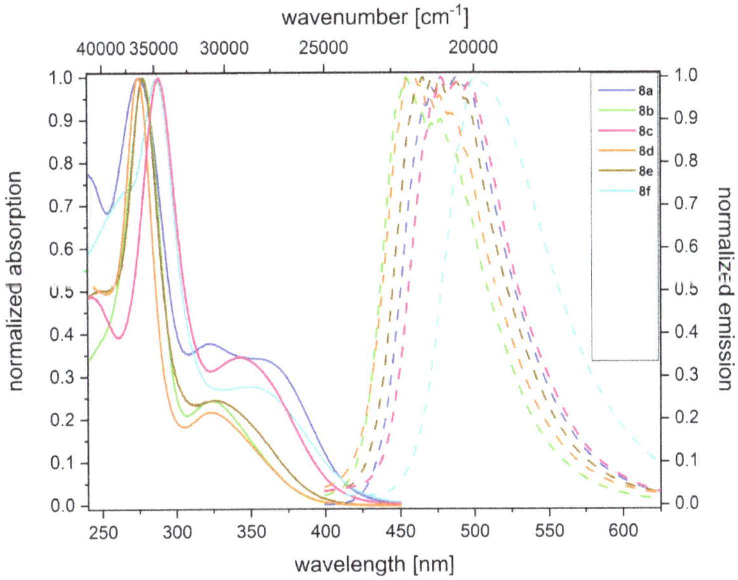

Figure 1. Normalized absorption (solid) and emission (dashed) spectra of 3,7-di(hetero)aryl-substituted 10-(3-trimethylammoniumpropyl)10*H*-phenothiazine salts **8** (recorded at T = 298 K for $c(\mathbf{8}) = 10^{-3}$ M (absorption) and $c(\mathbf{8}) = 10^{-6}$ M (emission) in dichloromethane).

Correlation analyses with the photophysical data $\lambda_{max,abs}$ (absorption), $\lambda_{max,em}$ (emission), and $\Delta\tilde{\nu}$ (Stokes shift) of the consanguineous series of compounds **7b–f** and **8b–f**, with Hammett parameters of σ_p, σ_R, σ_{R+}, σ_{p+}, and σ_{p-}, can be likewise probed for estimating the transmission of the polar substituent effect of the remote substituent in 3- and 7-position on excitation, radiative deexcitation, and structural changes in the vibrationally relaxed excited state [20]. While the data of the longest wavelength absorption bands $\lambda_{max,abs}$ only give mediocre (compounds **7b–f**) or poor (compounds **8b–f**) linear correlations with σ_{p-} and σ_p, respectively, the emission bands $\lambda_{max,em}$ correlate for both series with σ_p and σ_{p-} reasonably well (compounds **7b–f**, σ_{p-}: R^2 = 0.9564; compounds **8b–f**, σ_p: R^2 = 0.9129). This accounts for the transmission of the remote substituent effects in the planarized vibrationally relaxed excited state of phenothiazines [25,27]. Expectedly, the Stokes shifts indicating electronic or geometrical structural changes upon photonic excitation do not correlate in a meaningful sense due to the poor correlation of the absorption bands.

2.3. *Biological Properties of Compounds* **7** *and* **8**

We have tested the activity of the selected 3,7-di(hetero)aryl-substituted 10-(3-trimethylammoniumpropyl)10*H*-phenothiazine salts **8** and salt **9** [16] in a screening assay against common human pathogens, namely *Mycobacterium tuberculosis*, the Gram-positive bacterium *Staphylococcus aureus*, and four Gram-negative bacteria (*P. aeruginosa*, *E. coli*, *Actinobacter baumannii*, *Klebsiella pneumoniae*), to assess the antibacterial activity, i.e., determine the minimal inhibitory concentration (MIC) against the corresponding bacterial strains (Table 2).

Table 2. MIC of compounds **8a**, **8b**, **8d**, **8e**, **8f**, and **9** (stock solutions of the substances of approx. 10 mM in a volume of 1 mL DMSO (spectroscopy grade)).

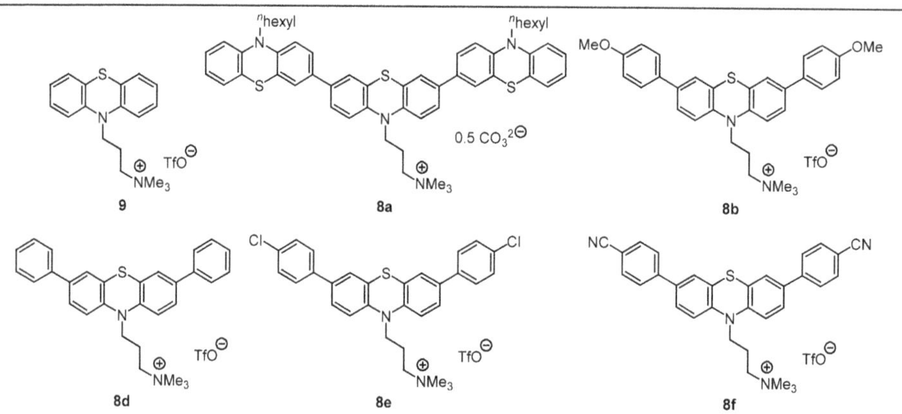

Compound	MW [g/mol]	Weighed Portion [mg]	Concentration [mM]	Antibacterial Activity MIC (μM)					
				M. tuberculosis	*S. aureus* ATCC 29213	*E. coli* ATCC 25922	*P. aeruginosa* ATCC 27853	*A. baumannii* ATCC 747	*K. pneumoniae* ATCC 13883
9	448.5	4.53	10.1	100	>100	>100	>100	>100	>100
8a	1784.6	9.89	5.5	>100	>100	>100	>100	>100	>100
8b	660.8	6.62	10.0	25	12.5	>100	>100	100	>100
8d	600.7	6.01	10.0	12.5	6.25	25	100	25	12.5
8e	669.6	6.69	10.0	12.5	12.5	>100	>100	50	>100
8f	650.7	6.50	10.0	25	12.5	>100	>100	>100	>100

The activity screening reveals that 3,7-diaryl-substituted 10-(3-trimethylammoniumpropyl) 10*H*-phenothiazine salts **8b–f** can effectively inhibit the growth of *M. tuberculosis* (MIC 12.5–25 μm) and *S. aureus* (MIC 6.25–12.5 μm), where compound **8d** is the most efficacious. In addition, this compound also inhibits the growth of *E. coli* (MIC 25 μm), *A. baumannii* (MIC 25 μm), and *K. pneumoniae* (MIC 12.5 μm). Apparently, *P. aeruginosa* is not inhibited by any of the substances under a threshold MIC of 100 μm. For *A. baumannii*, the aryl substituents obviously should be small, since bis(phenyl) derivative **8d** (MIC 25 μm) and bis(*p*-chlorophenyl) derivative **8e** (MIC 50 μm) at least show some inhibition. The steric demand predominantly applies for compound **8a**, which, despite its lipophilicity by the two *n*-hexyl phenothiazinyl substituents, does not show any activity of the tested pathogen strains. However, since the smallest compound in this series, compound **9**, also fails almost completely, a balanced decoration rendering lipophilicity and amphiphilicity as for 3,7-diaryl-substituted derivatives proves to give the most efficacious inhibition. This promises a good starting point for suitable cationically substituted phenothiazine derivatives.

3. Materials and Methods

3.1. *General Considerations*

Unless otherwise stated, all reactions were carried out in a nitrogen atmosphere using Schlenk, septum, and cannula techniques in heated single- or multi-neck flasks or Schlenk tubes. Nitrogen and argon were used as protective gases. Heated silicone baths were used at higher temperatures. At lower temperatures, cooling baths were used (acetone/dry ice, ice/water, or ice/salt mixtures). A high-vacuum pump from Büchi (Essen, Germany) was used for securation. Solvents were removed by distillation using vacuum pumps and rotary evaporators from Heidolph Instruments GmbH & Co. KG (Schwabach, Germany). The rotary evaporators were operated at a water bath temperature of 40 °C.

The chemicals used for this work, which were not synthesized in-house, were obtained commercially from the companies ABCR GmbH & Co. KG (Karlsruhe, Germany), Acros Organics (Geel, Belgium), Alfa-Aesar GmbH & Co. KG (Ward Hill, MA, USA), Carl Roth GmbH & Co. KG (Karlsruhe, Germany), Macherey-Nagel (Düren, Germany), Merck KGaA (Darmstadt, Germany), Sigma-Aldrich Chemie GmbH (St. Louis, MO, USA), or VWR (Radnor, PA, USA). Boronic acids and boronates **6** were synthesized according to the procedures outlined in the literature. The solvents used were purchased in pure form and removed as far as possible from water and other impurities according to standardized regulations or taken from a solvent drying system MB-SPS-800 from M. Braun Inertgas-Systeme GmbH (Garching, Germany).

For column chromatography, silica gel 60 M from Macherey-Nagel (grain size 0.04–0.063) or from Sigma-Aldrich (St. Louis, MO, USA, mesh 70–230, grain size 0.04–0.063, or mesh 230–400) was used as the stationary phase, and Celite® 545 was used as the adsorbent, and this was obtained from Carl Roth GmbH. Pure sea sand from AppliChem (Chicago, IL, USA) was also used. Column chromatography was carried out using flash technology at an overpressure of two bars of compressed air. Various solvents and solvent mixtures of distilled *n*-hexane, distilled *n*-hexane/acetone, distilled dichloromethane, and distilled dichloromethane/methanol were used as eluents.

The polarity of the eluent and the qualitative monitoring of the reaction progress were determined and carried out using thin-layer chromatography. Here, the R_f value of the product was adjusted so that it was approximately in the range of 0.3. For this purpose, silica-yellow-coated aluminum foil (60, F254) from Merck KGaA with a fluorescent indicator was used. Detection was carried out using UV light with wavelengths of 254 nm and 365 nm.

The ^{1}H and ^{13}C NMR spectra were recorded using the Avance III 300, DRX 500 and Avance III 600 spectrometer from Bruker (Karlsruhe, Germany). The solvents used were acetone-d_6 and DMSO-d_6, whereby all ^{1}H and ^{13}C spectra were locked to the signals of the non-deuterated solvents (acetone-d_6: ^{1}H NMR: δ 2.84, ^{13}C NMR: δ 206. 3 and δ 29.8; DMSO-d_6: ^{1}H NMR: δ 2.50, ^{13}C NMR: δ 39.51). The chemical shifts δ are given in ppm, and the coupling constants J are given in Hz. The individual signals were identified and assigned with the aid of the chemical shifts, the integrals, the multiplicity, and the coupling constants. When describing the multiplicity of the individual signals, the following common abbreviations were used: s (singlet), d (doublet), t (triplet), q (quartet), dd (doublet of doublet), dt (doublet of triplet), quint. (quintet), and m (multiplet). The carbon nuclei were assigned using ^{13}C NMR and 135-Dept spectra, with primary carbon nuclei designated as CH_3, secondary as CH_2, tertiary as CH, and quaternary as C_{quat}.

The electron ionization mass spectra were recorded using the TSQ 7000 triple quadrupole mass spectrometer from Finnigan MAT (Thermo Fisher Scientific, Waltham, MA, USA). The Ultraflex I spectrometer from Bruker DALTONICS was used to record the MALDI (TOF) spectra. High-resolution ESI measurements were carried out on a UHR-QTOF maXis 4G, also from Bruker Daltonics.

The infrared spectra were recorded on an IR-Affinity-1 using the attenuated total reflectance (ATR) technique from Shimadzu (Kyoto, Japan). The solids obtained were plotted and measured as such. The position of the absorption bands in the spectrum was indicated in wavenumbers $\widetilde{\nu}$ [cm^{-1}]. The intensities of the IR absorption bands obtained are indicated as s (strong), m (medium), and w (weak).

The elemental analyses were carried out using the Perkin Elmer Series II Analyzer 2400 from PerkinElmer (Waltham, MA, USA) or an Elementar vario MICRO CUBE at the Institute for Pharmaceutical and Medicinal Chemistry at Heinrich Heine University Düsseldorf, and the melting points were determined using the Melting Point B-540 apparatus from Büchi.

The UV/Vis spectra were recorded on a UV/VIS/NIR Lambda 19 spectrometer from Perkin Elmer. All compounds were measured at room temperature using high-purity solvents (HPLC or UVASOL). The extinction coefficients were determined using Lambert–Beer's law. For this purpose, five absorbance spectra of the respective compound were recorded at different concentrations. By plotting the absorbance against the concentration, the absorbance coefficient could be determined from the slope of the equalization line at the selected wavelength.

The emission spectra presented in this work were recorded on a calibrated Hitachi F-7000 fluorescence spectrometer (Tokyo, Japan). The fluorescence was always excited at the absorption maximum. All samples were measured at room temperature in dichloromethane (purity grade HPLC or UVASOL).

All experimentally recorded cyclic voltammograms within this work were measured in a small-volume glass cell (4 mL) with a three-electrode arrangement. The experiments were performed under argon atmosphere in dry and degassed dichloromethane at room temperature with feed rates of ν = 100, 250, 500, and 1000 $mV \cdot s^{-1}$. Tetra-*n*-butylammonium hexafluorophosphate (0.1 M in dichloromethane) was used as the electrolyte. A 2 mm platinum disk coated with glass was used as the working electrode. The counter electrode was a platinum wire, and the reference electrode consisted of a solid Ag/AgCl electrode filled with a sodium chloride solution (3.5 m). All potentials were referenced to the internal standard of decamethylferrocene/decamethylferrocenium ($[DMFc]/[DMFc]^+$, $E_0^{0/+1} = -0.095$ V). The absolute potential of ($[DMFc]/[DMFc]^+$ was determined against that of ferrocene ($[Fc]/[Fc]^+$, $E_0^{0/+1} = 0.450$ V) [28]. The redox-active substance was weighed into the measuring cell together with the conducting salt and degassed for five minutes by introducing argon while stirring. The model 263A galvanostat/potentiostat from EG&G Princeton Applied Research (Champaign, IL, USA) was used. The device was controlled by the PowerSuite Revision 2.12.1 program from Princeton Applied Research PerkinElmer Instruments.

3.2. Antibacterial Activity Testing

The cultivation of *M. tuberculosis* strain H37Rv was performed in PETG square bottles (Nalgene; Rochester, New York, NY, USA) in Middlebrock 7H9 liquid growth medium (BD; Franklin Lakes, NJ, USA) supplemented with 10% ADS (0.81% NaCl (*w*/*v*), 5% BSA (*w*/*v*), 2% dextrose (*w*/*v*)), 0.5% glycerol (*v*/*v*), and 0.05% tyloxapol (*v*/*v*). The MIC assay was performed in 96-well round-bottom polystyrene plates (Greiner; Kremsmünster, Austria) with a compound test range from 100 µM to 0.78 µM in a total volume of 100 µL. After static incubation for five days at 37 °C (5% CO_2, 80% humidity), growth was assessed using the resazurin reduction assay. For this, 10 µL of resazurin solution (100 µg/mL, Sigma Aldrich) was added into each well and incubated overnight. After the fixation of cells for 30 min at room temp following the addition of 10% (*v*/*v*) formalin, growth was quantified by measuring fluorescence using a microplate reader (TECAN; Maennedorf, Switzerland) (excitation: 540 nm, emission: 590 nm). After calculating growth relative to the DMSO solvent control (=100% growth) and uninoculated wells (subtraction of background fluorescence = 0% growth), the MIC value was determined as the actually tested lowest compound concentration, which resulted in 10% residual growth or less. Rifampicin was used as a positive control in all the experiments with *M. tuberculosis*.

The growth of the nosocomial bacteria *Staphylococcus aureus* (ATCC 29213), *Pseudomonas aeruginosa* (ATCC 27853), *E. coli* (ATCC 25922), *Acinetobacter baumannii* (ATCC BAA-747), and *Klebsiella pneumoniae* (ATCC 13883) was performed at 37 °C in Mueller–Hinton medium (BD; Franklin Lakes, NJ, USA). Moxifloxacin was used as the positive control. The antimicrobial activity against nosocomial bacteria was determined following the CLSI guidelines [29] in 96-well round-bottom polystyrene plates (Greiner; Kremsmünster, Austria) with a compound test ranging from 100 µm to 0.78 µm in a total volume of 100 µL. The MIC values were determined after overnight incubation as the actually tested lowest compound concentration that prevented visible growth by macroscopical evaluation.

3.3. *Synthesis of tert-Butyl (3-(3,7-Dibromo-10H-phenothiazin-10-yl)propyl)carbamate* (**5**) *[16]*

3.3.1. 3,7-Dibromo-10*H*-phenothiazine (**2**)

Phenothiazine (**1**) (25.0 g, 125 mmol) was placed in a dry two-necked round-bottom flask with a magnetic stir bar under nitrogen, and degassed acetic acid (600 mL) was added. Bromine (13.5 mL, 263 mmol) in acetic acid (200 mL) was added slowly and dropwise from the dropping funnel to the reaction mixture. The deep-colored solution was then stirred at room temperature for 16 h. A saturated aqueous solution of sodium sulfite (31.5 g, 250 mmol) was then added to the reaction mixture, which was then stirred at room temperature for 3 h. The initially violet suspension brightened to milky beige within this time. Then, the reaction mixture was poured on ice water (2 L). The resulting greenish precipitate was collected by suction and washed with a small amount of ice water. The precipitate was dried to weight constancy to give 3,7-dibromo-10*H*-phenothiazine (**2**) (50.7 g, 88%) as a greenish powder, Mp 186–188 °C. R_f (*n*-hexane/acetone 7:3): 0.42.

^{1}H NMR (300 MHz, DMSO-d$_6$): δ 6.58 (d, ^3J = 8.3 Hz, 2 H), 7.11 (d, ^4J = 2.3 Hz, 2 H), 7.14 (dd, ^4J = 2.3 Hz, ^3J = 8.3 Hz, 2 H), 8.84 (s, 1 H). ^{13}C NMR (151 MHz, DMSO-d$_6$): δ 112.7 (C$_{quat}$), 116.0 (C$_{quat}$), 118.2 (CH), 128.1 (CH), 130.3 (CH), 140.9 (C$_{quat}$). MS (EI) m/z (%): 359 (51, $[C_{12}H_7{}^{81}Br_2NS]^+$), 357 (100, $[C_{12}H_7{}^{79}Br^{81}BrNS]^+$), 355 (54, $[C_{12}H_7{}^{79}Br_2NS]^+$), 278 (78, $[C_{12}H_7{}^{81}BrNS]^+$), 276 (73, $[C_{12}H_7{}^{79}BrNS]^+$), 197 (81, $[C_{12}H_7NS]^+$). IR $\tilde{\nu}$ [cm^{-1}]: 3377 (vw), 3321 (w), 3069 (vw), 2849 (vw), 1566 (vw), 1441 (m), 1420 (m), 1383 (m), 1368 (w), 1288 (m), 1236 (w), 1136 (w), 1080 (m), 1057 (w), 1020 (w), 934 (w), 881 (m), 864 (w), 847 (w), 808 (s), 750 (m), 731 (m), 700 (w), 673 (m), 648 (w).

3.3.2. 3-(3,7-Dibromo-10*H*-phenothiazin-10-yl)propiononitrile (**3**)

3,7-Dibromo-10*H*-phenothiazine (**2**) (21.0 g, 59.0 mmol) was placed in a dry Schlenk flask with a magnetic stir bar under nitrogen and suspended in acrylonitrile (120 mL). The mixture was cooled to 0 °C (ice–water bath), and 40 wt.% of Triton B® in methanol (2.15 mL) was added slowly and dropwise under external cooling and vigorous stirring. After the addition was complete, the reaction mixture was stirred for 5 min. The progress of the reaction was monitored by thin-layer chromatography, and then the reaction mixture was poured into water (480 mL), and a beige solid precipitated. The precipitate was filtered off and dried under vacuum for 16 h. The crude product was adsorbed on Celite®, purified by column chromatography on silica gel (*n*-hexane/acetone 8:2), suspended in a small volume of acetone, filtered, and dried under vacuum to weight constancy to give 3-(3,7-dibromo-10*H*-phenothiazin-10-yl)propiononitrile (**3**) (16.9 g, 70%) as a colorless solid, Mp 146–148 °C. R_f (*n*-hexane/acetone 8:2): 0.26.

^{1}H NMR (600 MHz, DMSO-d$_6$): δ 2.90 (t, ^3J = 6.6 Hz, 2 H), 4.18 (t, ^3J = 6.6 Hz, 2 H), 7.04 (d, ^3J = 8.7 Hz, 2 H), 7.39 (dd, ^3J = 8.7, ^4J = 2.3 Hz, 2 H), 7.42 (d, ^4J = 2.3 Hz, 2 H). ^{13}C NMR (125 MHz, DMSO-d$_6$): δ 15.7 (CH$_2$), 42.5 (CH$_2$), 114.6 (C$_{quat}$), 117.7 (CH), 118.4 (C$_{quat}$), 126.1 (C$_{quat}$), 129.2 (CH), 130.3 (CH), 142.9 (C$_{quat}$). MS (EI) m/z (%): 412 (37, $[M^{81}Br_2]^+$), 410 (81, $[M^{81}Br^{79}Br]^+$), 408 (37, $[M^{79}Br]^+$), 372 (52, $[C_{13}H_8{}^{81}Br_2N^{32}S]^+$), 370 (100, $[C_{13}H_8{}^{81}Br^{79}BrN^{32}S]^+$), 368 (48, $[C_{13}H_8{}^{79}Br_2N^{32}S]^+$), 358 (44, $[(C_{12}H_6{}^{81}Br_2N^{32}S]^+$), 356 (89, $[(C_{12}H_6{}^{81}Br^{79}BrN^{32}S]^+$), 354 (42, $[(C_{12}H_6{}^{79}Br_2N^{32}S]^+$), 291 (59, $[C_{13}H_8{}^{81}BrN^{32}S]^+$), 289 (54, $[C_{13}H_8{}^{79}BrN^{32}S]^+$), 196 (65, $[C_{12}H_6N^{32}S]^+$). IR $\tilde{\nu}$ [cm^{-1}] = 3086 (vw), 2990 (vw), 2961 (vw), 2905 (vw), 2249 (vw), 1585 (vw), 1477 (w), 1452 (s), 1408 (w), 1391 (w), 1335 (m), 1317 (m), 1288 (w), 1248 (m), 1227 (w), 1194 (m), 1165 (w), 1155 (w), 1113 (w), 1096 (w), 1080 (w), 1070 (w), 1045 (w), 1028 (vw) 989 (vw), 901 (vw), 864 (m), 797 (s), 789 (m), 752 (m), 741 (w), 700 (vw), 650 (w), 606 (w).

3.3.3. 3-(3,7-Dibromo-10*H*-phenothiazin-10-yl)propylamine (**4**)

Indium trichloride (3.32 g, 15.0 mmol) and sodium borohydride (1.70 g, 45.0 mmol) were suspended in THF (50.0 mL) in a dry Schlenk tube with a magnetic stir bar and stirred at room temp for 1 h. After a careful and portion-wise addition of 3-(3,7-dibromo-10*H*-phenothiazin-10-yl)propylnitrile (**3**) (6.11 g, 15.0 mmol) to the grayish reaction mixture, the reaction mixture was stirred at room temp for 4 h. The reaction was stopped by the careful and dropwise addition of 3.00 m hydrochloric acid (50 mL), resulting in strong gas evolution. The reaction solution was then heated at 90 °C to reflux for 1 h. After cooling to room temp and the addition of methanol (25 mL) to the reaction mixture, heating to reflux at 90 °C was continued for 1 h. After cooling, the volatile organic components were removed under reduced pressure, and the precipitate was separated from the remaining aqueous phase by filtration. The crude product was adsorbed on Celite® and purified by column chromatography on silica gel (*n*-hexane/acetone/triethylamine 7:2:0.1). Then, the product was suspended in a small volume of methanol, filtered, and dried to weight constancy under vacuum to give 3-(3,7-dibromo-10*H*-phenothiazin-10-yl)propylamine (**4**) (6.09 g, 91%) as a pale beige powder, Mp 139–141 °C. R_f (*n*-hexane/acetone/triethylamine 1:1:0.01): 0.42.

^{1}H NMR (300 MHz, DMSO-d$_6$): δ 1.94 (m$_c$, 2 H). 2.92–2.78 (m, 2 H), 3.95 (t, ^3J = 6.9 Hz, 2 H), 7.02 (d, ^3J = 8.9 Hz, 2 H), 7.47–7.35 (m, 4 H), 7.93 (s, 2 H). ^{13}C NMR (151 MHz, DMSO-d$_6$): δ 24.3 (CH$_2$), 36.5 (CH$_2$), 43.9 (CH$_2$), 114.4 (C$_{quat}$), 117.9 (CH), 125.9 (C$_{quat}$), 129.2 (CH), 130.4 (CH), 143.6 (C$_{quat}$). MS (EI) m/z (%): 416 (36, [M^{81}Br$_2$]$^+$), 414 (70, [M^{81}Br^{79}Br]$^+$), 412 (34, [M^{79}Br$_2$]$^+$), 398 (20, [$C_{15}H_{12}{}^{81}Br_2N^{32}S$]$^+$), 396 (39, [$C_{15}H_{12}{}^{81}Br^{79}Br N^{32}S$]$^+$), 394 (19, [$C_{15}H_{12}{}^{79}Br_2N^{32}S$]$^+$), 372 (12, [$C_{13}H_8{}^{81}Br_2N^{32}S$]$^+$), 370 (24, [$C_{13}H_8{}^{81}Br^{79}BrN^{32}S$]$^+$), 368 (12, [$C_{13}H_8{}^{79}Br_2N^{32}S$]$^+$), 358 (48, [$C_{12}H_6{}^{81}Br_2N^{32}S$]$^+$), 356 (85, [$C_{12}H_6{}^{81}Br^{79}BrN^{32}S$]$^+$), 354 (42, [$C_{12}H_6{}^{79}Br_2N^{32}S$]$^+$), 291 (21, [$C_{13}H_8{}^{81}BrN^{32}S$]$^+$), 289 (20, [$C_{13}H_8{}^{79}BrN^{32}S$]$^+$), 277 (29, [$C_{12}H_6{}^{81}BrN^{32}S$]$^+$), 196 (73, [$C_{12}H_6N^{32}S$]$^+$). IR $\tilde{\nu}$ [cm^{-1}]: 3092 (vw), 3057 (vw), 2934 (vw), 1585 (w), 1479 (m), 1456 (s), 1387 (w), 1327 (w), 1294 (w), 1252 (m), 1229 (m), 1150 (w), 1109 (w), 1082 (w), 1057 (w), 868 (w), 806 (m), 750 (m), 650 (w).

3.3.4. *tert*-Butyl (3-(3,7-Dibromo-10*H*-phenothiazin-10-yl)propyl)carbamate (**5**)

In a dry Schlenk tube with a magnetic stir bar, 3-(3,7-dibromo-10*H*-phenothiazin-10-yl)propylamine (**4**) (3.97 g, 8.80 mmol), di-*tert*-butyl dicarbonate (1.92 g, 8. 80 mmol), and 4-dimethylaminopyridine (108 mg, 0.88 mmol) were placed under nitrogen, and dichloromethane (44 mL) and triethylamine (2.45 mL, 17.7 mmol) were added. The reaction mixture was stirred at room temp for 16 h. The reaction was monitored by thin-layer chromatography. Then, the solvent was removed under reduced pressure, and the crude product was adsorbed on Celite® and purified by column chromatography on silica gel (*n*-hexanes/acetone 9:1 to 7:3) to give *tert*-butyl (3-(3,7-dibromo-10*H*-phenothiazin-10-yl)propyl)carbamate (**5**) (4.30 g, 95%) as a beige-colored, glassy solid, Mp 70–72 °C. R_f (*n*-hexane/acetone/triethylamine 7:3:0.1): 0.39.

^{1}H NMR (300 MHz, DMSO-d$_6$): δ 1.35 (s, 9 H), 1.76 (m, 2 H), 3.01 (m, 2 H), 3.83 (t, ^3J = 6.5 Hz, 2 H), 6.86 (t, ^3J = 4.9 Hz, 1 H), 6.96 (d, ^3J = 7.7 Hz, 2 H), 7.38–7.33 (m, 4 H). ^{13}C NMR (75 MHz, DMSO-d$_6$): δ 26.4 (CH$_2$), 28.2 (CH$_3$), 37.6 (CH$_2$), 44.5 (CH$_2$), 77.5 (C$_{quat}$), 114.1 (C$_{quat}$), 117.6 (CH), 125.6 (C$_{quat}$), 129.1 (CH), 130.3 (CH), 143.8 (C$_{quat}$), 155.6 (C$_{quat}$). MS (EI) m/z (%): 516 (14, [M^{81}Br$_2$]$^+$), 514 (32, [M^{81}Br^{79}Br]$^+$), 512 (14, [M^{79}Br$_2$]$^+$), 460 (13, [$C_{16}H_{14}{}^{81}Br_2N_2O_2{}^{32}S$]$^+$), 458 (29, [$C_{16}H_{14}{}^{81}Br^{79}Br N_2O_2{}^{32}S$]$^+$), 456 (13, [$C_{16}H_{14}{}^{79}Br_2N_2O_2{}^{32}S$]$^+$), 357 (62, [$C_{12}H_7{}^{81}Br^{79}Br N^{32}S$]$^+$), 277 (30, [$C_{12}H_7{}^{79}BrN^{32}S$]$^+$), 196 (39, [$C_{12}H_6N^{32}S$]$^+$), 102 (100, [$C_5H_9O_2$]$^+$). IR $\tilde{\nu}$ [cm^{-1}]: 3395 (vw), 3379 (vw), 2976 (w), 2926 (w), 1694 (m), 1686 (m), 1516 (w), 1506 (m), 1481 (w), 1452 (vs), 1389 (m), 1364 (m), 1327 (w), 1267 (m), 1248 (s), 1202 (w), 1163 (s), 1109 (w), 1082 (w), 1040 (vw), 1003 (w), 959 (vw), 932 (vw), 868 (m), 802 (s), 779 (w), 750 (m), 652 (w). Anal. calcd. for $C_{20}H_{22}Br_2N_2O_2S$ [514.3]: C 46.71, H 4.31, N 5.45, S 6.23; Found: C 46.98, H 4.17, N 5.32, S 6.23.

3.4. *General Procedure (GP1) for the Suzuki Synthesis of Boc-Protected 3,7-Di(hetero)aryl-substituted 10-(3-aminopropyl) 10H-Phenothiazines* **7**

Phenothiazine dibromide **5** (1.00 equiv), boronic acid or boronic acid ester **6** (2.20–2.40 equivs), cesium fluoride (6.40 equivs), and tetrakis(triphenylphosphane)palladium (0) (6.00 mol%) were placed in a dry Schlenk tube with a magnetic stir bar under nitrogen atmosphere, and 1,4-dioxane was added (for experimental details, see Table 3). The reaction mixture was heated to reflux at 101 °C for 16 h to give a yellow solution or suspension. After cooling to room temp, dichloromethane (20 to 90 mL) was added, and the solution was washed with distilled water (2 × 20 to 100 mL) and with saturated brine (1 × 20 to 100 mL). The aqueous phases were again extracted with dichloromethane. The combined organic phases were dried (anhydrous magnesium sulfate), and the crude product was adsorbed on Celite® and purified by column chromatography on silica gel to give 3,7-di(hetero)aryl-substituted 10-(3-aminopropyl)-10*H*-phenothiazines **7**.

Table 3. Experimental details of the Suzuki synthesis of Boc-protected 3,7-di(hetero)aryl-substituted 10-(3-aminopropyl)10*H*-phenothiazines **7**.

Entry	Dibromide 5	Boronic Acid/Pinacol Boronate 6	CsF	$Pd(PPh_3)_4$	1,4-Dioxane	Compound 7 (Yield)
1	1.03 g (2.00 mmol)	1.81 g (4.40 mmol) of **6a**	1.94 g (12.0 mmol)	139 mg (0.12 mmol)	20 mL	1.56 g (85%) of **7a**
2	722 mg (1.40 mmol)	511 mg (3.36 mmol) of 4-methoxyphenyl boronic acid (**6b**)	1.36 g (8.97 mmol)	84 mg (0.07 mmol)	14 mL	630 mg (79%) of **7b**
3	514 mg (1.00 mmol)	462 mg (2.20 mmol) of pinacol 2-thienyl boronate (**6c**)	973 mg (6.40 mmol)	69 mg (0.06 mmol)	10 mL	510 mg (69%) of **7c**
4	771 mg (1.50 mmol)	402 mg (3.30 mmol) of phenyl boronic acid (**6d**)	1.45 g (9.60 mmol)	104 mg (0.09 mmol)	15 mL	560 mg (74%) of **7d**
5	771 mg (1.50 mmol)	787 mg (3.30 mmol) of 4-chlorophenyl boronic acid (**6e**)	1.46 g (9.60 mmol)	104 mg (0.09 mmol)	15 mL	660 mg (76%) of **7e**
6	722 mg (1.40 mmol)	1.36 g (8.97 mmol) of 4-cyanophenyl boronic acid (**6f**)	1.36 g (8.97 mmol)	84 mg (0.07 mmol)	14 mL	530 mg (67%) of **7f**

3.4.1. *tert*-Butyl (3-(10,10″-Dihexyl-10*H*,10′*H*,10″*H*-[3,3′:7′,3″-terphenothiazin]-10′-yl)propyl)carbamate (**7a**)

According to the GP1 and after flash chromatography on silica gel (*n*-hexane/acetone 8:2), compound **7a** (1.56 g, 85%) was obtained as yellow crystals, Mp 98–104 °C. R_f (*n*-hexane/acetone 8:2): 0.37.

^{1}H NMR (300 MHz, acetone-d$_6$): δ 0.85 (t, ^3J = 7.1 Hz, 6 H), 1.23–1.35 (m, 8 H), 1.38 (s, 9 H), 1.47 (m$_c$, 4 H), 1.81 (m$_c$, 4 H), 2.03 (m$_c$, 2 H), 3.26 (m$_c$, 2 H), 3.95 (t, ^3J = 7.0 Hz, 4 H), 4.05 (t, ^3J = 7.0 Hz, 2 H), 6.10 (t, ^3J = 5.8 Hz, 1 H), 6.90–6.98 (m, 2 H), 7.00–7.11 (m, 6 H), 7.13–7.23 (m, 4 H), 7.39 (m$_c$, 4 H), 7.77 (m$_c$, 4 H). ^{13}C NMR (75 MHz, acetone-d$_6$): δ 14.3 (CH_3), 23.3 (CH_2), 27.2 (CH_2), 27.6 (CH_2), 28.2 (CH_2), 28.6 (CH_3), 32.2 (CH_2), 39.0 (CH_2), 45.7 (CH_2), 47.8 (CH_2), 78.5 (C_{quat}), 116.6 (CH), 116.8 (CH), 116.8 (CH), 123.3 (CH), 125.1 (C_{quat}), 125.5 (CH), 125.6 (CH), 125.7 (C_{quat}), 126.0 (C_{quat}), 126.1 (CH), 126.2 (CH), 128.0 (CH), 128.3 (CH), 134.9 (C_{quat}), 135.1 (C_{quat}), 144.9 (C_{quat}), 145.3 (C_{quat}), 146.0 (C_{quat}), 156.8 (C_{quat}). MS (MALDI-TOF) m/z calcd. For $[C_{56}H_{62}N_4O_2S_3]^+$: 918.403; Found: 918.431. IR $\tilde{\nu}$ [cm^{-1}]: 2957 (m), 2926 (m), 2901 (w), 1709 (m), 1601 (w), 1576 (w), 1499 (m), 1456 (vs), 1414 (w), 1377 (m), 1364 (m), 1331 (m), 1240 (s), 1163 (m), 1140 (m), 1105 (w), 1065 (w), 1042 (w), 1011 (vw), 874 (m), 854 (vw), 806 (s), 779 (w), 745 (s), 706 (vw), 623 (w), 606 (m). Anal. Calcd. For $C_{56}H_{62}N_4O_2S_3$ [919.3]: C 73.16, H 6.80, N 6.09, S 10.46; Found: C 73.30, H 6.76, N 5.89, S 10.25.

3.4.2. *tert*-Butyl (3-(3,7-Bis(4-methoxyphenyl)-10*H*-phenothiazin-10-yl)propyl) carbamate (**7b**)

According to the GP1 and after flash chromatography on silica gel (*n*-hexane/acetone 8:2), compound **7b** (630 mg, 79%) was obtained as a colorless solid, Mp 151–156 °C. R_f (*n*-hexane/acetone 8:2): 0.31.

^{1}H NMR (300 MHz, DMSO-d$_6$): δ 1.36 (s, 9 H), 1.86 (m$_c$, 2 H), 3.08 (m$_c$, 2 H), 3.77 (s, 6 H), 3.91 (t, ^3J = 6.9 Hz, 2 H), 6.92 (t, ^3J = 5.4 Hz, 1 H), 6.98 (d, ^3J = 8.9 Hz, 4 H), 7.06 (d, ^3J = 8.4 Hz, 2 H), 7.39 (d, ^4J = 2.2 Hz, 2 H), 7.43 (dd, ^3J = 8.4 Hz, ^4J = 2.2 Hz, 2 H), 7.56 (d,

^3J = 8.9 Hz, 4 H). ^{13}C NMR (75 MHz, DMSO-d$_6$): δ 26.8 (CH_2), 28.3 (CH_3), 44.5 (CH_2), 55.2 (CH_3), 77.7 (C_{quat}), 114.4 (CH), 115.9 (CH), 123.8 (C_{quat}), 124.6 (CH), 125.3 (CH), 127.2 (CH), 131.4 (C_{quat}), 134.3 (C_{quat}), 143.2 (C_{quat}), 155.8 (C_{quat}), 158.7 (C_{quat}). MS (EI) *m/z* (%): 568 (40, $[M]^+$), 512 (42, $[C_{30}H_{27}N_2O_4S]^+$), 424 (9, $[C_{27}H_{22}NO_2S]^+$), 410 (100, $[C_{26}H_{20}NO_2S]^+$), 414 (15, $[C_{27}H_{16}N_3S]^+$), 400 (100, $[C_{26}H_{14}N_3S]^+$). IR $\tilde{\nu}$ [cm^{-1}]: 2970 (w), 2951 (w), 2930 (w), 2868 (vw), 2833 (w), 1697 (s), 1682 (m), 1607 (m), 1580 (w), 1518 (m), 1493 (m), 1460 (s), 1439 (m), 1430 (m), 1391 (m), 1364 (m), 1331 (m), 1281 (m), 1242 (vs), 1209 (m), 1179 (s), 1113 (m), 1086 (w), 1047 (m), 1024 (m), 1009 (m), 955 (w), 926 (w), 912 (w), 883 (w), 808 (s), 775 (m), 768 (m), 741 (w), 729 (w). 714 (w), 687 (w), 671 (m). Anal. calcd. for $C_{34}H_{30}N_4O_2S$ [568.7]: C: 71.80, H: 6.38, N: 4.93, S: 5.64, Found: C: 71.69, H: 6.49, N: 4.68, S: 5.48.

3.4.3. *tert*-Butyl (3-(3,7-Di(thien-2-yl)-10*H*-phenothiazin-10-yl)propyl)carbamate (**7c**) [16]

According to the GP1 and after flash chromatography on silica gel (*n*-hexane/acetone 8:2), compound **7c** (359 mg, 69%) was obtained as a yellow powder, Mp 138–141 °C. R_f (*n*-hexane/acetone 7:3): 0.34.

^{1}H NMR (600 MHz, DMSO-d$_6$): δ 1.36 (s, 9 H), 1.84 (m$_c$, 2 H), 3.07 (m$_c$, 2 H), 3.91 (t, ^3J = 7.0 Hz, 2 H), 6.92, (t, ^3J = 5.5 Hz, 1 H), 7.04 (d, ^3J = 9.1 Hz, 2 H), 7.04 (dd, ^3J = 9.1 Hz, 2 H), 7.46–7.42 (m, 6 H), 7.48 (dd, ^3J = 5.1 Hz, ^4J = 1.1 Hz, 2 H). ^{13}C NMR (151 MHz, DMSO-d$_6$): δ 26.6 (CH_2), 28.3 (CH_3), 37.7 (CH_2), 44.5 (CH_2), 77.5 (C_{quat}), 116.1 (CH), 123.1 (CH), 123.6 (C_{quat}), 123.7 (CH), 124.9 (CH), 125.0 (CH), 128.5 (CH), 128.5 (C_{quat}), 142.2 (C_{quat}), 143.4 (C_{quat}), 155.6 (C_{quat}). MS (EI) *m/z* (%): 520 (32, $[C_{28}H_{28}N_2O_2S_3]^+$), 464 (24, $[C_{24}H_{19}N_2O_2S_3]^+$), 446 (6, $[C_{24}H_{19}N_2OS_3]^+$), 420 (13, $[C_{23}H_{19}N_2S_3]^+$), 376 (8, $[C_{21}H_{14}NS_3]^+$), 362 (100, $[C_{20}H_{12}NS_3]^+$). IR $\tilde{\nu}$ [cm^{-1}]: 2974 (w), 2926 (w), 2864 (w), 1701 (m), 1680 (m), 1472 (vs), 1429 (m), 1400 (m), 1391 (m), 1364 (m), 1348 (w), 1333 (w), 1292 (w), 1261 (s), 1238 (s), 1207 (m), 1163 (s), 1109 (m), 1080 (w), 1042 (w), 1020 (w), 1007 (vw), 988 (w), 951 (vw), 951 (vw), 874 (m), 851 (m), 810 (s), 793 (m), 783 (w), 750 (m), 691 (vs), 650 (vw). Anal. Calcd. For $C_{28}H_{28}N_2O_2S_3$ [519.7]: C 64.58, H 5.42, N 5.38, S 18.47; Found: C 64.83, H 5.69, N 5.12, S 18.17.

3.4.4. *tert*-Butyl (3-(3,7-Diphenyl-10*H*-phenothiazin-10-yl)propyl)carbamate (**7d**)

According to the GP1 and after flash chromatography on silica gel (*n*-hexane/acetone 8:2), compound **7d** (560 mg, 74%) was obtained as yellow crystals, Mp 92–98 °C. R_f (*n*-hexane/acetone 8:2): 0.39.

^{1}H NMR (300 MHz, DMSO-d$_6$): δ 1.37 (s, 9 H), 1.88 (m$_c$, 2 H), 3.09 (m$_c$, 2 H), 3.95 (t, ^3J = 6.9 Hz, 2 H), 6.94 (t, ^3J = 5.3 Hz, 1 H), 7.11 (d, ^3J = 8.4 Hz, 2 H), 7.29–7.36 (m, 2 H). 7.39–7.48 (m, 6 H), 7.50 (dd, ^3J = 8.4 Hz, ^4J = 2.2 Hz, 2 H), 7.60–7.66 (m, 4 H). ^{13}C NMR (75 MHz, DMSO-d$_6$): δ 26.7 (CH_2), 28.3 (CH_3), 37.8 (CH_2), 44.5 (CH_2), 77.5 (C_{quat}), 116.0 (CH), 123.7 (C_{quat}), 125.0 (CH), 125.8 (CH), 126.1 (CH), 127.1 (CH), 128.9 (CH), 134.5 (C_{quat}), 138.9 (C_{quat}), 143.7 (C_{quat}), 155.7 (C_{quat}). MS (EI) *m/z* (%): 508 (36, $[M]^+$), 452 (32, $[C_{28}H_{23}N_2O_2S]$), 408 (4, $[C_{27}H_{23}N_2S]$), 364 (12, $[C_{25}H_{18}NS]$), 350 (100, $[C_{24}H_{16}NS]$). IR $\tilde{\nu}$ [cm^{-1}]: 2968 (w), 2901 (w), 2887 (w), 2870 (w), 1738 (w), 1678 (s), 1599 (w), 1585 (w), 1526 (m), 1462 (s), 1393 (m), 1364 (m), 1329 (w), 1275 (s), 1248 (s), 1202 (w), 1165 (s), 1157 (s), 1111 (m), 1078 (m), 1055 (m), 1042 (w), 1023 (m), 1003 (m), 986 (w), 939 (vw), 876 (m), 868 (m), 853 (w), 820 (m), 812 (m), 762 (vs), 737 (m), 700 (s), 689 (m), 613 (m). Anal. calcd. for $C_{32}H_{32}N_2O_2S$ [508.7]: C 75.56, H 6.34, N 5.51, S 5.92; Found: C 75.20, H 6.40, N 5.12, S 5.92.

3.4.5. *tert*-Butyl (3-(3,7-Bis(4-chlorophenyl)-10*H*-phenothiazin-10-yl)propyl)carbamate (**7e**)

According to the GP1 and after flash chromatography on silica gel (*n*-hexane/acetone 8:2), compound **7e** (660 mg, 76%) was obtained as yellow crystals, Mp 95–102 °C. R_f (*n*-hexane/acetone 8:2): 0.38.

^{1}H NMR (600 MHz, acetone-d$_6$): δ 1.38 (s, 9 H), 3.27 (m$_c$, 2 H), 4.07 (t, ^3J = 6.8 Hz, 2 H), 6.12 (t, ^3J = 5.3 Hz, 1 H), 7.13 (m$_c$, 2 H), 7.42–7.47 (m, 6 H), 7.49 (d, ^3J = 8.4 Hz, 2 H), 7.64 (d, ^3J = 7.8 Hz, 4 H). ^{13}C NMR (151 MHz, acetone-d$_6$): δ 28.1 (CH_2), 28.6 (CH_3), 39.0 (CH_2), 45.8

(CH_2), 78.6 (C_{quat}), 116.9 (CH), 125.7 (C_{quat}), 126.1 (CH), 126.8 (CH), 128.7 (CH), 129.7 (CH), 133.4 (C_{quat}), 134.9 (C_{quat}), 139.3 (C_{quat}), 145.4 (C_{quat}), 156.8 (C_{quat}). MS (EI) *m*/*z* (%): 576 (40, $[M]^+$), 520 (46, $[C_{28}H_{21}Cl_2N_2O_2S]$), 504 (6, $[C_{28}H_{21}Cl_2N_2OS]$), 476 (3, $[C_{27}H_{21}Cl_2N_2S]$), 432 (14, $[C_{25}H_{16}Cl_2NS]$), 419 (100, $[C_{24}H_{14}Cl_2NS]$). IR $\tilde{\nu}$ [cm^{-1}]: 3898 (vw), 3834 (vw), 3688 (vw), 3381 (vw), 3354 (vw), 3032 (vw), 2934 (vw), 2845 (vw), 2818 (vw), 2357 (vw), 1697 (m), 1684 (m), 1607 (w), 1585 (w), 1506 (m), 1460 (vs), 1418 (w), 1387 (m), 1364 (m), 1339 (w), 1271 (m), 1248 (s), 1200 (m), 1163 (s), 1092 (m), 1011 (m), 1003 (m), 959 (w), 910 (w), 885 (w), 876 (w), 808 (vs), 779 (m), 764 (m), 723 (w), 635 (w). Anal. Calcd. For $C_{32}H_{30}Cl_2N_2O_2S$ [577.6]: C 66.55, H 5.24, N 4.85, S 5.55; Found: C 66.83, H 5.51, N 4.56, S 5.54.

3.4.6. *tert*-Butyl (3-(3,7-Bis(4-cyanophenyl)-10*H*-phenothiazin-10-yl)propyl)carbamate (**7f**)

According to the GP1 and after flash chromatography on silica gel (*n*-hexane/acetone 7:3), compound **7f** (530 mg, 67%) was obtained as a yellow powder, Mp 183–185 °C. R_f (*n*-hexane/acetone 8:2): 0.22. ^{1}H NMR (300 MHz, acetone-d$_6$): δ 1.38 (s, 9 H), 2.07 (m, 2 H), 3.25–3.31 (m, 2 H), 4.11 (t, ^3J = 7.0 Hz, 2 H), 6.13 (t, ^3J = 5.3 Hz, 1 H), 7.19 (d, ^3J = 8.6 Hz, 2 H), 7.55 (d, ^4J = 2.1 Hz, 2 H), 7.60 (dd, ^3J = 8.6 Hz, ^4J = 2.1 Hz, 2 H), 7.82 (d, ^3J = 8.4 Hz, 4 H), 7.86 (d, ^3J = 8.4 Hz, 4 H). ^{13}C NMR (75 MHz, acetone-d$_6$): δ 28.1 (CH_2), 28.6 (CH_3), 38.9 (CH_2), 45.9 (CH_2), 78.6 (C_{quat}), 111.3 (C_{quat}), 117.1 (CH), 119.4 (C_{quat}), 125.7 (C_{quat}), 126.5 (CH), 127.4 (CH), 127.9 (CH), 133.5 (CH), 134.3 (C_{quat}), 144.8 (C_{quat}), 146.0 (C_{quat}), 156.8 (C_{quat}). MS (EI) *m*/*z* (%): 558 (1, $[M]^+$), 484 (50, $[C_{30}H_{22}N_4OS]^+$), 414 (15, $[C_{27}H_{16}N_3S]^+$), 400 (100, $[C_{26}H_{14}N_3S]^+$). IR $\tilde{\nu}$ [cm^{-1}]: 3393 (vw), 3364 (w), 2974 (w), 2932 (w), 2901 (vw), 2870 (vw), 2220 (s), 1676 (s), 1605 (m), 1585 (m), 1526 (m), 1518 (m), 1477 (vs), 1445 (w), 1418 (m), 1393 (s), 1364 (s), 1271 (s), 1244 (s), 1163 (s), 1125 (m), 1115 (m), 1038 (w), 1001 (m), 833 (s), 793 (vs), 648 (m). Anal. calcd. for $C_{34}H_{30}N_4O_2S$ [558.7]: C: 73.09, H: 5.41, N: 10.03, S: 5.74, Found: C: 72.86, H: 5.50, N: 9.78, S: 5.71.

3.5. *General Procedure (GP2) for the Synthesis of 3,7-Di(hetero)aryl-substituted 10-(3-Trimethylammoniumpropyl)10H-phenothiazine Salts* **8**

In a Schlenk flask with a magnetic stir bar, Boc-protected 3,7-di(hetero)aryl-substituted 10-(3-aminopropyl) 10*H*-phenothiazine **7** (1.00 equiv) was dissolved in dichloromethane (10.0 mL/mmol) under nitrogen atmosphere, and trifluoroacetic acid (8.00 equivs) was added, whereby the color immediately changed from yellow to green-brown (for experimental details, see Table 4). The reaction mixture was stirred at room temp for 16 h, after which the volatile components were removed under reduced pressure. The residue was dissolved in a 1:1 mixture of methanol and dichloromethane (10.0 mL/mmol), and potassium carbonate (20.0 equivs) was slowly added. Methyl iodide or methyl triflate (10.0 equivs) was added dropwise to the reaction mixture. The reaction solution was stirred at room temp for 16 h. The crude product was adsorbed on neutral alumina and purified by column chromatography on neutral alumina to give the trimethylammonium salts **8**.

3.5.1. Bis{3-(10,10″-dihexyl-10*H*,10′*H*,10″*H*-[3,3′:7′,3″-terphenothiazin]-10′-yl)-*N*,*N*,*N*-trimethylpropan-1-ammonium} Carbonate (**8a**)

According to the GP2 and after flash chromatography on neutral alumina (dichloro methane/methanol 100:0 to 100:1), compound **8a** (823 mg, 93%) was obtained as yellow crystals, Mp softening at 127 °C, melting between 148 and 152 °C. R_f(dichloromethane/methanol 100:1): 0.17.

^{1}H NMR (600 MHz, DMSO-d$_6$): δ 0.82 (t, ^3J = 6.9 Hz, 6 H), 1.21–1.29 (m, 8 H), 1.39 (m$_c$, 4 H), 1.69 (m$_c$, 4 H), 2.15 (m$_c$, 2 H), 3.04 (s, 9 H), 3.44 (m$_c$, 2 H), 3.88 (t, ^3J = 6.9 Hz, 4 H), 3.99 (t, ^3J = 6.9 Hz, 2 H), 6.92–6.97 (m, 2 H), 7.00–7.06 (m, 4 H), 7.13 (d, ^3J = 8.6 Hz, 2 H), 7.15 (dd, ^3J = 7.6 Hz, ^4J = 1.3 Hz, 2 H), 7.18–7.22 (m, 4 H), 7.42 (d, ^4J = 2.2 Hz, 2 H), 7.46–7.51 (m, 4 H). ^{13}C NMR (151 MHz, acetone-d$_6$): δ 14.3 (CH_3), 21.6 (CH_2), 23.3 (CH_2), 27.2 (CH_2), 27.5 (CH_2), 32.2 (CH_2), 44.6 (CH_2), 47.8 (CH_2), 53.8 (CH_3), 64.8 (CH_2), 116.6 (CH), 116.8 (CH), 117.3 (CH), 123.3 (CH), 125.1 (C_{quat}), 125.5 (CH), 125.6 (CH), 125.8 (C_{quat}),

126.0 (C_{quat}), 126.2 (CH), 126.5 (CH), 128.0 (CH), 128.4 (CH), 134.7 (C_{quat}), 135.3 (C_{quat}), 144.3 (C_{quat}), 145.3 (C_{quat}), 146.0 (C_{quat}). MS (MALDI-TOF) m/z calcd. for $[C_{54}H_{61}N_4S_3]^+$: 861.405; Found: 861.408. IR $\tilde{\nu}$ [cm^{-1}]: 2951 (w), 2924 (w), 2899 (w), 2866 (w), 2853 (w), 1601 (w), 1574 (w), 1454 (vs) 1416 (m), 1373 (m), 1331 (m), 1275 (w), 1238 (s), 1196 (m), 1134 (w), 1105 (m), 1078 (w), 1065 (w), 1040 (w), 1028 (w), 964 (w), 908 (w), 872 (w), 808 (m), 745 (s), 704 (w), 683 (w), 650 (w), 631 (w). Anal. calcd. for $(C_{54}H_{61}N_4S_3)_2CO_3$ [1784.6]: C 73.36, H 6.89, N 6.28, S 10.78; Found: C 73.23, H 6.83, N 5.91, S 10.41.

Table 4. Experimental details of the formation of 3,7-di(hetero)aryl-substituted 10-(3-trimethylammo niumpropyl)10*H*-phenothiazine salts **8**.

Entry	Compound 7	Dichloromethane	Trifluoro Acidic Acid	Potassium Carbonate	Methyl Iodide/Methyl Triflate	Compound 8 (Yield)
1	910 mg (0.99 mmol) of **7a**	10 mL	903 µL (7.92 mmol)	2.76 g (20.0 mmol)	623 µL (9.90 mmol) of methyl iodide	823 mg (93%) of **8a** [1]
2	329 mg (0.58 mmol) of **7b**	6 mL	354 µL (4.63 mmol)	1.60 g (11.6 mmol)	634 µL (5.79 mmol) of methyl triflate	256 mg (67%) of **8b** [2]
3	200 mg (0.39 mmol) of **7c**	4 mL	237 µL (3.08 mmol)	1.06 g (7.69 mmol)	420 µL (3.85 mmol) of methyl triflate	239 mg (59%) of **8c** [2]
4	250 mg (0.49 mmol) of **7d**	5 mL	302 µL (3.93 mmol)	1.36 g (9.82 mmol)	537 µL (4.91 mmol) of methyl triflate	239 mg (81%) of **8d** [2]
5	250 mg (0.43 mmol) of **7e**	4 mL	267 µL (3.49 mmol)	1.19 g (8.64 mmol)	473 µL (4.32 mmol) of methyl triflate	205 mg (71%) of **8e** [2]
6	200 mg (0.44 mmol) of **7f**	4 mL	267 µL (3.49 mmol)	1.36 g (9.82 mmol)	493 µL (4.36 mmol) of methyl triflate	150 mg (53%) of **8f** [2]

[1] Obtained as carbonate. [2] Obtained as a triflate.

3.5.2. 3-(3,7-Bis(4-methoxyphenyl)-10*H*-phenothiazin-10-yl)-*N,N,N*-trimethylpropan-1-ammoniumtriflate (**8b**)

According to the GP2 and after flash chromatography on neutral alumina (dichloro methane/methanol 100:1 to 100:2), compound **8b** (260 mg, 67%) was obtained as a pale beige powder, Mp 200–205 °C. R_f (*n*-hexane/acetone 4:6): 0.11.

^{1}H NMR (300 MHz, acetone-d$_6$): δ 2.50 (m$_c$, 2 H), 3.79 (m$_c$, 2 H), 3.34 (s, 9 H), 3.83 (s, 6 H), 4.21 (t, ^3J = 6.7 Hz, 2 H), 7.00 (d, ^3J = 8.8 Hz, 4 H), 7.17 (d, ^3J = 8.3 Hz, 2 H), 7.44–7.51 (m, 4 H), 7.56 (d, ^3J = 8.8 Hz, 4 H). ^{13}C NMR (75 MHz, acetone-d$_6$): δ 21.9 (CH_2), 44.8 (CH_2), 53.9 (CH_3), 53.9 (CH_3), 54.0 (CH_3), 55.7 (CH_3), 65.5 (CH_2), 115.3 (CH), 115.5 (C_{quat}), 117.2 (CH), 126.0 (CH), 126.6 (CH), 128.3 (CH), 131.6 (C_{quat}), 132.9 (C_{quat}), 136.7 (C_{quat}), 144.4 (C_{quat}), 160.4 (C_{quat}). MS (ESI) m/z calcd. for $[C_{32}H_{35}N_2O_2S]^+$: 511.2; Found: 511.5. IR $\tilde{\nu}$ [cm^{-1}]: 2930 (vw), 1609 (m), 1518 (w), 1491 (m), 1464 (s), 1443 (m), 1423 (w), 1393 (w), 1341 (w), 1238 (vs), 1225 (s), 1198 (m), 1177 (s), 1155 (s), 1111 (m), 1082 (m), 1045 (m), 1028 (vs), 964 (w), 937 (vw), 910 (m), 883 (w), 808 (s), 773 (w), 756 (m), 637 (vs). Anal. calcd. for $C_{33}H_{35}F_3N_2O_5S_2$ [660.8]: C 59.99, H 5.34, N 4.24, S 9.70; Found: C 60.24, H 5.39, N 4.05, S 9.47.

3.5.3. 3-(3,7-Di(thiophen-2-yl)-10*H*-phenothiazin-10-yl)-*N,N,N*-trimethylpropan-1-ammoniumtriflate (**8c**)

According to the GP2 and after flash chromatography on neutral alumina (dichloro methane/methanol 100:1 to 100:2), compound **8c** (140 mg, 59%) was obtained as a yellow powder, Mp 128–130 °C. R_f (methanol): 0.10.

^{1}H NMR (600 MHz, DMSO-d$_6$): δ 2.50 (m$_c$, 2 H), 3.80 (m$_c$, 2 H), 4.21 (t, ^3J = 6.9 Hz, 2 H), 7.04 (dd, ^3J = 3.6 Hz, ^4J = 1.1 Hz, 2 H), 7.11 (m$_c$, 2 H), 7.16 (d, ^3J = 8.3 Hz, 2 H), 7.43 (dd, ^3J = 5.1 Hz, ^4J = 0.9 Hz, 2 H), 7.52–7.49 (m, 4 H). ^{13}C NMR (125 MHz, methanol-d$_4$): δ 22.0 (CH_2), 45.0 (CH_2), 53.7 (CH_3), 65.8 (CH_2), 117.6 (CH), 123.9 (CH), 125.6 (CH), 125.6 (CH), 126.4 (CH), 127.5 (C_{quat}), 129.2 (CH), 131.5 (C_{quat}), 144.1 (C_{quat}), 145.1 (C_{quat}). IR $\tilde{\nu}$ [cm^{-1}]:

3034 (vw), 2963 (w), 1603 (vw), 1472 (s), 1427 (m), 1402 (m), 1346 (w), 1339 (w), 1258 (vs), 1223 (m), 1200 (w), 1153 (m), 1084 (s), 1045 (s), 1028 (vs), 1015 (s), 966 (w), 943 (vw), 910 (w), 872 (m), 851 (m), 799 (vs), 752 (m), 737 (w), 692 (s), 662 (vw), 637 (s), 610 (w). ESI-HRMS m/z calcd. for $[C_{26}H_{27}N_2S_3]^+$: 463.1331; Found: 463.1336.

3.5.4. 3-(3,7-Diphenyl-10*H*-phenothiazin-10-yl)-*N*,*N*,*N*-trimethylpropan-1-ammoniumtriflate (**8d**)

According to the GP2 and after flash chromatography on neutral alumina (dichloro methane/methanol 100:1 to 100:2), compound **8d** (240 mg, 81%) was obtained as yellow crystals, Mp softening at 110 °C, melting between 119 and 129 °C. R_f (*n*-hexane/acetone 4:6): 0.13.

^{1}H NMR (600 MHz, DMSO-d$_6$): δ 2.18 (m$_c$, 2 H), 3.05 (s, 9 H), 3.46 (m$_c$, 2 H), 4.02 (t, ^3J = 6.9 Hz, 2 H), 7.19 (d, ^3J = 8.5 Hz, 2 H), 7.33 (m$_c$, 2 H), 7.44 (m$_c$, 4 H), 7.52 (d, ^4J = 2.1 Hz, 2 H), 7.55 (dd, ^3J = 8.5 Hz, ^4J = 2.1 Hz, 2 H), 7.64 (d, ^3J = 7.5 Hz, 4 H). ^{13}C NMR (151 MHz, DMSO-d$_6$): δ 20.3 (CH_2), 43.7 (CH_2), 52.4 (CH_3), 63.2 (CH_2), 116.2 (CH), 124.1 (C$_{quat}$), 125.2 (CH), 126.0 (CH), 126.1 (CH), 127.3 (CH), 129.0 (CH), 134.9 (C$_{quat}$), 138.8 (C$_{quat}$), 143.4 (C$_{quat}$). MS (MALDI-TOF) m/z calcd. for $[C_{31}H_{31}F_3N_2O_3S_2]^+$: 451.220; Found: 451.325. IR $\tilde{\nu}$ [cm^{-1}]: 1599 (w), 1464 (s), 1393 (w), 1329 (w), 1252 (s), 1223 (m), 1196 (w), 1152 (m), 1109 (w), 1074 (w), 1028 (s), 966 (w), 935 (w), 914 (w), 883 (w), 826 (w), 760 (s), 737 (w), 696 (s), 637 (vs), 629 (m), 608 (m). Anal. calcd. for $C_{31}H_{31}F_3N_2O_3S_2$ [600.7]: C 61.98, H 5.20, N 4.66, S 10.67; Found: C 62.23, H 5.30, N 4.59, S 10.38.

3.5.5. 3-(3,7-Bis(4-chlorophenyl)-10*H*-phenothiazin-10-yl)-*N*,*N*,*N*-trimethylpropan-1-ammoniumtriflate (**8e**)

According to the GP2 and after flash chromatography on neutral alumina (dichloro methane/methanol 100:1 to 100:2), compound **8e** (210 mg, 71%) was obtained as a greenish powder, Mp softening at 151 °C, melting between 181 and 192 °C. R_f (*n*-hexane/acetone 4:6): 0.10.

^{1}H NMR (600 MHz, acetone-d$_6$): δ 2.49 (m$_c$, 2 H), 3.34 (s, 9 H), 3.82 (m$_c$, 2 H), 4.22 (t, ^3J = 6.9 Hz), 7.24 (d, ^3J = 8.5 Hz, 2 H), 7.45 (d, ^3J = 8.6 Hz, 4 H), 7.50 (d, ^4J = 2.2 Hz, 2 H), 7.55 (dd, ^3J = 8.5 Hz, ^4J = 2.2 Hz, 2 H), 7.65 (d, ^3J = 8.6 Hz, 4 H). ^{13}C NMR (151 MHz, acetone-d$_6$): δ 21.6 (CH_2), 44.7 (CH_2), 53.7 (CH_3), 53.7 (CH_3), 53.8 (CH_3), 65.0 (CH_2), 117.3 (CH), 126.1 (C$_{quat}$), 126.2 (CH), 127.1 (CH), 128.8 (CH), 129.8 (CH), 133.5 (C$_{quat}$), 135.3 (C$_{quat}$), 139.1 (C$_{quat}$), 145.0 (C$_{quat}$). MS (MALDI-TOF) m/z calcd. for $[C_{30}H_{29}{}^{35}Cl_2N_2S]^+$: 519.142; Found: 519.148. IR $\tilde{\nu}$ [cm^{-1}]: 3028 (vw), 2857 (vw), 1460 (s), 1414 (w), 1383 (w), 1341 (w), 1317 (vw), 1254 (s), 1225 (m), 1198 (w), 1157 (s), 1105 (w), 1092 (s), 1055 (w), 1028 (s), 1011 (m), 962 (w), 939 (vw), 910 (w), 891 (w), 810 (m), 791 (m), 768 (w), 756 (w), 745 (w), 723 (w), 691 (w), 637 (vs), 610 (w). Anal. calcd. for $C_{31}H_{29}Cl_2F_3N_2O_3S_2$ [669.6]: C 55.61, H 4.37, N 4.18, S 9.58; Found: C 55.53, H 4.55, N 3.98, S 9.28.

3.5.6. 3-(3,7-Bis(4-cyanophenyl)-10*H*-phenothiazin-10-yl)-*N*,*N*,*N*-trimethylpropan-1-ammoniumtriflate (**8f**)

According to the GP2 and after flash chromatography on neutral alumina (dichloro methane/methanol (100:1 to 100:2)), compound **8f** (150 mg, 53%) was obtained as a yellow powder, Mp softening at 145 °C, melting between 168 and 172 °C. R_f (acetone): 0.08.

^{1}H NMR (300 MHz, acetone-d$_6$): δ 2.52 (m$_c$, 2 H), 3.35 (s, 9 H), 3.82 (m$_c$, 2 H), 4.26 (t, ^3J = 6.9 Hz, 2 H), 7.29 (d, ^3J = 8.5 Hz, 2 H), 7.61 (d, ^4J = 2.2 Hz, 2 H), 7.66 (dd, ^3J = 8.5 Hz, ^4J = 2.2 Hz, 2 H), 7.85 (m$_c$, 8 H). ^{13}C NMR (151 MHz, acetone-d$_6$): δ 21.9 (CH_2), 44.9 (CH_2), 53.9 (CH_3), 54.0 (CH_3), 54.0 (CH_3), 65.3 (CH_2), 111.6 (C$_{quat}$), 117.5 (CH), 119.4 (C$_{quat}$), 126.4 (C$_{quat}$), 126.7 (CH), 127.7 (CH), 128.0 (CH), 133.6 (CH), 133.8 (C$_{quat}$), 134.9 (C$_{quat}$), 144.8 (C$_{quat}$), 145.8 (C$_{quat}$). MS (MALDI-TOF) m/z calcd. for $[C_{32}H_{29}N_4S]^+$: 501.211; Found: 501.222. IR $\tilde{\nu}$ [cm^{-1}]: 2930 (w), 2855 (m), 2224 (w), 1730 (w), 1605 (m), 1585 (w), 1468 (s), 1420 (m), 1364 (m), 1344 (w), 1248 (s), 1225 (m), 1198 (m), 1155 (s), 1109 (s), 1055 (m), 1030 (s), 962 (m), 914 (m), 881 (m), 843 (m), 810 (s), 743 (m), 689 (m), 637 (vs). Anal. calcd. for

$C_{33}H_{29}F_3N_4O_3S_2$ [650.7]: C 60.91, H 4.49, N 8.61, S 9.85; Found: C 60.25, H 4.84, N 8.17, S 8.56.

3.6. *N,N,N-Trimethyl-3-(10H-phenothiazin-10-yl)propan-1-ammoniumtriflate* (**9**) *[16]*

Compound **9** (332 mg, 90%) was synthesized according to [16] and obtained after purification as a colorless powder, Mp 168–170 °C. R_f (methanol): 0.31.

^{1}H NMR (300 MHz, methanol-d$_4$): δ 2.25 (m$_c$, 2 H). 3.01 (s, 9 H), 3.45 (m$_c$, 2 H), 4.11 (t, ^3J = 6.3 Hz, 2 H), 6.99 (m$_c$, 2 H), 7.06 (dd, ^3J = 8.1 Hz, ^4J = 0.9 Hz, 2 H), 7.19 (dd, ^3J = 7.6 Hz, ^4J = 1.6 Hz, 2 H), 7.25 (m$_c$, 2 H). ^{13}C NMR (125 MHz, methanol-d$_4$): δ 21.8 (CH_2), 44.7 (CH_2), 53.5 (CH_3), 65.7 (CH_2), 117.3 (CH), 124.3 (CH), 127.4 (C_{quat}), 128.6 (CH), 128.8 (CH), 146.2 (C_{quat}). IR $\tilde{\nu}$ [cm^{-1}]: 3065 (vw), 3042 (vw), 2965 (vw), 2868 (vw), 1593 (w), 1570 (w), 1485 (m), 1479 (m), 1460 (s), 1445 (m), 1422 (w), 1400 (w), 1339 (w), 1254 (vs), 1221 (s), 1200 (w), 1157 (s), 1140 (s), 1128 (m), 1109 (w), 1028 (s), 961 (w), 939 (w), 918 (w), 908 (w), 764 (s), 754 (w), 729 (m), 637 (s). MS (EI) m/z (%): 448 (1, [M]$^+$), 299 (10, [$C_{18}H_{23}N_2S$]$^+$) 284 (41, [$C_{17}H_{20}N_2S$]$^+$), 239 (100, [$C_{15}H_{14}NS$]$^+$), 212 (18, [$C_{13}H_{10}NS$]$^+$), 199 (50, [$C_{12}H_8NS$]$^+$). Anal. calcd. for $C_{19}H_{23}F_3N_2O_3S_2$ [448.5]: C 50.88, H 5.17, N 6.25, S 14.30; Found: C 50.83, H 5.36, N 6.09, S 14.05.

4. Conclusions

Starting from 3,7-dibromo phenothiazine, a library of 3,7-di(hetero)aryl-substituted 10-(3-trimethylammoniumpropyl)10*H*-phenothiazine salts can be prepared in five steps. As is typical for phenothiazines, all derivatives are redox systems with fully reversible first oxidations. The Hammett correlation reveals that the remote polar substituents in the consanguineous series of 3,7-di(hetero)aryl-substituted phenothiazine salts and their precursors correlate very well with the σ_p parameters. All salts and precursors can be excited in UV with a substituent tunable wavelength, and the compounds display blue to green-blue emission. The screening of the selected 3,7-di(hetero)aryl-substituted 10-(3-trimethylammoniumpropyl)10*H*-phenothiazine salts discloses for some derivatives a distinct inhibition of the pathogenic bacterial strains *M. tuberculosis*, *S. aureus*, *E. coli*, *A. baumannii*, and *K. pneumoniae*; however, the distinct inhibition of *P. aeruginosa* is not disclosed. Based on the ease of the synthetic approach, further 3,7-diaryl- and 3-aryl-substituted phenothiazines with an *N*-propyl trimethylammonium sidechain can be tested; eventually, irradiation with UV light can also occur to trigger photoinduced electron transfer. Studies focused on scrutinizing the photophysical and biological properties of these *N*-propyl trimethylammonium phenothiazine derivatives are currently underway.

Supplementary Materials: The following supporting information can be downloaded at: https://www.mdpi.com/article/10.3390/molecules29092126/s1, ^{1}H and ^{13}C NMR spectra of compounds **7** and **8** (Figures S1–S20), correlation of $E_0{}^{0/+1}$ of compounds **7** and **8** against Hammett parameters (Tables S1–S4; Figures S21–S30).

Author Contributions: The work consists of parts of the Ph.D. thesis of H.K., which was supervised by T.J.J.M. The conceptualization was outlined by T.J.J.M. Synthetic studies, analytical assignments, and electroanalytical and photophysical investigations with the compounds **7** and **8** were performed by H.K., who compiled and interpreted the obtained data. The antibacterial activity testing of compounds **8a**, **8b**, **8d**, **8e**, **8f**, and **9** was performed by L.v.G. and jointly interpreted by L.v.G. and R.K. The writing of the original draft was completed by T.J.J.M., and the review and editing was completed by H.K., L.v.G., and R.K. Project administration and funding acquisition was completed by T.J.J.M. All authors have read and agreed to the published version of the manuscript.

Funding: This research was funded by Deutsche Forschungsgemeinschaft (MU 1088/9-1 and MU 1088/13-1) and Fonds der Chemischen Industrie.

Institutional Review Board Statement: Not applicable.

Informed Consent Statement: Not applicable.

Data Availability Statement: All the data in this work are included in the manuscript and the Supplementary Materials.

Acknowledgments: The authors thank Burak Kürsat-Menekse for weighing the compounds **8a**, **8b**, **8d**, **8e**, **8f**, and **9** for the preparation of their stock solutions.

Conflicts of Interest: The authors declare no conflicts of interest.

References

1. Ohlow, M.J.; Moosmann, B. Foundation review: Phenothiazine: The seven lives of pharmacology's first lead structure. *Drug Discov. Today* **2011**, *16*, 119–131. [CrossRef] [PubMed]
2. Revoju, S.; Matuhina, A.; Canil, L.; Salonen, H.; Hiltunen, A.; Abate, A.; Vivo, P. Structure-induced optoelectronic properties of phenothiazine-based materials. *J. Mater. Chem. C* **2020**, *8*, 15486–15506. [CrossRef]
3. Khan, F.; Misra, R. Recent advances in the development of phenothiazine and its fluorescent derivatives for optoelectronic applications. *J. Mater. Chem. C* **2023**, *11*, 2786–2825. [CrossRef]
4. Huang, Z.-S.; Meier, H.; Cao, D. Phenothiazine-based dyes for efficient dye-sensitized solar cells. *J. Mater. Chem. C* **2016**, *4*, 2404–2426. [CrossRef]
5. Zhou, Z.; Franz, A.W.; Hartmann, M.; Seifert, A.; Müller, T.J.J.; Thiel, W.R. Novel Organic/Inorganic Hybrid Materials by Covalent Anchoring of Phenothiazines on MCM-41. *Chem. Mater.* **2008**, *20*, 4986–4992. [CrossRef]
6. Jaszczyszyn, A.; Gąsiorowski, K.; Świątek, P.; Malinka, W.; Cieślik-Boczula, K.; Petrus, J.; Czarnik-Matusewicz, B. Chemical structure of phenothiazines and their biological activity. *Pharmacol. Rep.* **2012**, *64*, 16–23. [CrossRef] [PubMed]
7. Pluta, K.; Morak-Młodawska, B.; Jeleń, M. Recent progress in biological activities of synthesized phenothiazines. *Eur. J. Med. Chem.* **2011**, *46*, 3179–3189. [CrossRef] [PubMed]
8. Lagadinou, M.; Onisor, M.O.; Rigas, A.; Musetescu, D.-V.; Gkentzi, D.; Assimakopoulos, S.F.; Panos, G.; Marangos, M. Antimicrobial Properties on Non-Antibiotic Drugs in the Era of Increased Bacterial Resistance. *Antibiotics* **2020**, *9*, 107. [CrossRef] [PubMed]
9. Zhou, W.; Jiang, X.; Zhen, X. Development of organic photosensitizers for antimicrobial photodynamic therapy. *Biomater. Sci.* **2023**, *11*, 5108–5128. [CrossRef]
10. González-González, A.; Vazquez-Jimenez, L.K.; Paz-González, A.D.; Bolognesi, M.L.; Rivera, G. Recent Advances in the Medicinal Chemistry of Phenothiazines, New Anticancer and Antiprotozoal Agents. *Curr. Med. Chem.* **2021**, *28*, 7910–7936. [CrossRef]
11. Posso, M.C.; Domingues, F.C.; Ferreira, S.; Silvestre, S. Development of Phenothiazine Hybrids with Potential Medicinal Interest: A Review. *Molecules* **2022**, *27*, 276. [CrossRef] [PubMed]
12. Rácz, B.; Spengler, G. Repurposing Antidepressants and Phenothiazine Antipsychotics as Efflux Pump Inhibitors in Cancer and Infectious Diseases. *Antibiotics* **2023**, *12*, 137. [CrossRef] [PubMed]
13. Boyd-Kimball, D.; Gonczy, K.; Lewis, B.; Mason, T.; Siliko, N.; Wolfe, J. Classics in Chemical Neuroscience: Chlorpromazine. *ACS Chem. Neurosci.* **2019**, *10*, 79–88. [CrossRef] [PubMed]
14. Kaur, B.; Gupta, J.; Sharma, S.; Sharma, D.; Sharma, S. Focused review on dual inhibition of quorum sensing and efflux pumps: A potential way to combat multi drug resistant Staphylococcus aureus infections. *Int. J. Biol. Macromol.* **2021**, *190*, 33–43. [CrossRef] [PubMed]
15. Blair, J.M.A.; Webber, M.A.; Baylay, A.J.; Ogbolu, D.O.; Piddock, L.J.V. Molecular mechanisms of antibiotic resistance. *Nat. Rev. Microbiol.* **2015**, *13*, 42–51. [CrossRef] [PubMed]
16. Schäfgen, B.; Khelwati, H.; Bechtel, D.F.; DeCuyper, A.; Schüssler, A.; Neuba, A.; Pierik, A.J.; Ernst, S.; Müller, T.J.J.; Thiel, W.R. Phenothiazine electrophores immobilized on periodic mesoporous organosilicas by ion exchange. *New J. Chem.* **2019**, *43*, 16396–16410. [CrossRef]
17. Rednic, M.I.; Szima, S.; Bogdan, E.; Hădade, N.D.; Terec, A.; Grosu, I. Podands with 3,7,10-trisubstituted phenothiazine units: Synthesis and structural analysis. *Rev. Roum. Chim.* **2015**, *60*, 637–642.
18. Saavedra, J.Z.; Resendez, A.; Rovira, A.; Eagon, S.; Haddenham, D.; Singaram, B. Reaction of $InCl_3$ with Various Reducing Agents: InCl3–NaBH4-Mediated Reduction of Aromatic and Aliphatic Nitriles to Primary Amines. *J. Org. Chem.* **2012**, *77*, 221–228. [CrossRef]
19. Sailer, M.; Nonnenmacher, M.; Oeser, T.; Müller, T.J.J. Synthesis and Electronic Properties of 3-Acceptor-Substituted and 3,7-Bisacceptor-Substituted Phenothiazines. *Eur. J. Org. Chem.* **2006**, *2006*, 423–435. [CrossRef]
20. Mayer, L.; Müller, T.J.J. 3,10-Diaryl Phenothiazines—One-pot Synthesis and Conformational Tuning of Ground and Excited State Electronics. *Eur. J. Org. Chem.* **2021**, *2021*, 3516–3527. [CrossRef]
21. Krämer, C.S.; Zeitler, K.; Müller, T.J.J. Synthesis of (Hetero)Aryl Bridged and Directly Linked Redox Active Phenothiazinyl Dyads and Triads. *Tetrahedron Lett.* **2001**, *42*, 8619–8624. [CrossRef]
22. Sailer, M.; Franz, A.W.; Müller, T.J.J. Synthesis and Electronic Properties of Monodisperse Oligophenothiazines. *Chem. Eur. J.* **2008**, *14*, 2602–2614. [CrossRef] [PubMed]
23. Hansch, C.; Leo, A.; Taft, R.W. A survey of Hammett substituent constants and resonance and field parameters. *Chem. Rev.* **1991**, *91*, 165–195. [CrossRef]

24. Krämer, C.S.; Müller, T.J.J. Synthesis and Electronic Properties of Alkynylated Phenothiazines. *Eur. J. Org. Chem.* **2003**, *2003*, 3534–3548. [CrossRef]
25. Hauck, M.; Stolte, M.; Schönhaber, J.; Kuball, H.-G.; Müller, T.J.J. Synthesis, Electronic and Electrooptical Properties of Emissive Solvatochromic Phenothiazinyl Merocyanine Dyes. *Chem. Eur. J.* **2011**, *17*, 9984–9998. [CrossRef] [PubMed]
26. May, L.; Müller, T.J.J. Widely Electronically Tunable 2,6-Disubstituted Dithieno [1,4]thiazines—Electron-Rich Fluorophores up to Intense NIR Emission. *Chem. Eur. J.* **2020**, *26*, 12978–12986. [CrossRef]
27. Yang, L.; Feng, J.-K.; Ren, A.-M. Theoretical Study on Electronic Structure and Optical Properties of Phenothiazine-Containing Conjugated Oligomers and Polymers. *J. Org. Chem.* **2005**, *70*, 5987–5996. [CrossRef]
28. Zanello, P. *Ferrocenes*; Togni, A., Hayashi, T., Eds.; VCH: Weinheim, Germany, 1995; pp. 317–430.
29. CLSI. *Methods for Dilution Antimicrobial Susceptibility Tests for Bacteria That Grow Aerobically; Approved Standard*, 10th ed.; CLSI Document M07-A10; Clinical and Laboratory Standards Institute: Wayne, PA, USA, 2015.

molecules

MDPI

Article

New Library of Iodo-Quinoline Derivatives Obtained by an Alternative Synthetic Pathway and Their Antimicrobial Activity

Cristina Maria Al-Matarneh [1,2,*], Alina Nicolescu [3], Ioana Cristina Marinaş [2], Mădalina Diana Găboreanu [2], Sergiu Shova [4], Andrei Dascălu [1], Mihaela Silion [5] and Mariana Pinteală [1]

1 Center of Advanced Research in Bionanoconjugates and Biopolymers, "Petru Poni" Institute of Macromolecular Chemistry of Romanian Academy, 41A Grigore Ghica Voda Alley, 700487 Iasi, Romania; idascalu@icmpp.ro (A.D.); pinteala@icmpp.ro (M.P.)

2 Research Institute of the University of Bucharest—ICUB, 91-95 Spl. Independentei, 050095 Bucharest, Romania; ioana.cristina.marinas@gmail.com (I.C.M.); gaboreanu.diana.madalina@s.bio.unibuc.ro (M.D.G.)

3 NMR Laboratory "Petru Poni" Institute of Macromolecular Chemistry of Romanian Academy, 41A Grigore Ghica Voda Alley, 700487 Iasi, Romania; alina@icmpp.ro

4 Department of Inorganic Polymers "Petru Poni" Institute of Macromolecular Chemistry of Romanian Academy, 41A Grigore Ghica Voda Alley, 700487 Iasi, Romania; shova@icmpp.ro

5 Physics of Polymers and Polymeric Materials Department, "Petru Poni" Institute of Macromolecular Chemistry, 41A Grigore Ghica Voda Alley, 700487 Iasi, Romania; silion.mihaela@icmpp.ro

* Correspondence: almatarneh.cristina@icmpp.ro

Abstract: 6-Iodo-substituted carboxy-quinolines were obtained using a one-pot, three-component method with trifluoroacetic acid as a catalyst under acidic conditions. Iodo-aniline, pyruvic acid and 22 phenyl-substituted aldehydes (we varied the type and number of radicals) or O-heterocycles, resulting in different electronic effects, were the starting components. This approach offers advantages such as rapid response times, cost-effective catalysts, high product yields and efficient purification procedures. A comprehensive investigation was conducted to examine the impact of aldehyde structure on the synthesis pathway. A library of compounds was obtained and characterized by FT-IR, MS, ^{1}H NMR and ^{13}C NMR spectroscopy and single-ray crystal diffractometry. Their antimicrobial activity against *S. epidermidis*, *K. pneumonie* and *C. parapsilosis* was tested in vitro. The effect of iodo-quinoline derivatives on microbial adhesion, the initial stage of microbial biofilm development, was also investigated. This study suggests that carboxy-quinoline derivatives bearing an iodine atom are interesting scaffolds for the development of novel antimicrobial agents.

Keywords: iodo-quinolines; Doebner synthesis; antimicrobial activity; *S. epidermidis*; *K. pneumonie*; *C. parapsilosis*

1. Introduction

Since the majority of therapeutic medications are generated from heterocyclic structures, the synthesis of biologically active heterocyclic scaffolds has continuously drawn great interest in medicinal chemistry. This feature was revealed by Vitaku et al. in a remarkable review in 2014 [1], which showed that 85% of FDA-approved small-molecule drugs contain nitrogen, with 59% of them including at least one azaheterocycle. These are very high percentages, far higher than the effect figures previously revealed for fluorine (15%) and sulfur (26%). It is interesting to note that although the average amount of nitrogen atoms per medication for all small-molecule pharmaceuticals is 2.3 N/drug, it is 35% higher for drugs that include a nitrogen heterocycle (3.1 N/drug). In this context, the quinoline nucleus is present in a wide variety of synthetic and natural azaheterocyclic products with intriguing biological characteristics [2]. Although quinoline itself possesses few applications, many of its derivatives are useful in a variety of domains, including pharmaceuticals and advanced functional materials, one example being fluorescent probes containing a quinoline-based core [3,4]. Quinoline-containing compounds have a well-established pharmacological value, starting with the famous chloroquine, used for decades to control and eradicate malaria. Other currently available drugs containing these privileged moieties are antimalarials (quinine, quinidine, mefloquine, amodiaquine, primaquine and so on), antiviral (saquinavir), antibacterial (fluoroquinolones such as ciprofloxacin, sparfloxacin, gatifloxacin and so on), antifungal–antiprotozoal (clioquinol) or anthelmintic (oxamn and vesnarinone) [5]. Since the structural change of a favored moiety with therapeutic ability significantly alters its therapeutic value, the development of novel synthetic medicines based on the quinoline scaffold continues to be an active research topic.

The need to discover novel antimicrobial agents with better properties and fewer side effects is a continuous preoccupation in the medical community. When it comes to different Gram-positive and Gram-negative bacteria species, the majority of quinoline compounds show good antimicrobial action, determined by the substitution of the heterocyclic pyridine ring rather than the aromatic moiety [6]; some examples in this sense are presented in Figure 1. Unfortunately, the widespread overuse of quinoline-based drugs led to an increased occurrence of antimicrobial resistance. As a result, several well-established synthetic techniques for creating novel, appropriately substituted quinoline scaffolds were developed over time: Combes, Skraup, Döbner–Von Miller, Conrad–Limpach, Pfitzinger, Friedländer and Povarov syntheses or the homogeneous metal-catalyzed hetero-annulation of acyclic precursors [7].

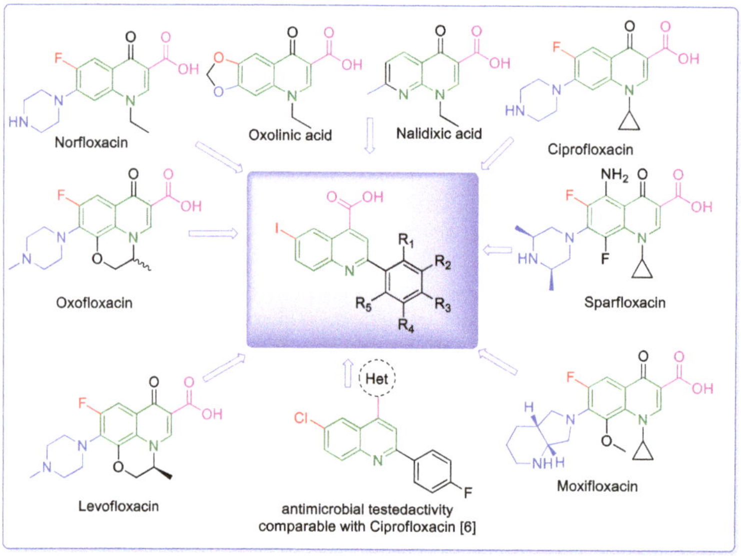

Figure 1. Rational design used in this study to obtain iodine carboxy-quinolines.

One of the most significant groups of quinoline derivatives is quinoline-4-carboxylic acids, which are used as active ingredients in commercial antioxidants and have a range of therapeutic uses [8]. In order to obtain physiologically active carboxy-quinolines, the widely used reactions are Doebner synthesis (uses activated substituted amines, aldehydes and pyruvic acid) and Pfitzinger condensation (the reaction of isatins with α-methylidene carbonyl compounds) [9]. These classic synthetic approaches suffer from limited sources of precursors, harsh reaction conditions, low yields or selectivity and difficult operations. Thus, modified approaches of these classical syntheses are still of great value, especially if they can be more efficient and easy to operate at a lower cost.

In light of this, keeping with our quest for physiologically active hybrid heterocyclic molecules [10,11], as part of our concern for microbial active compounds [12–15], we present the synthesis of a library of novel iodine carboxy-quinoline derivatives as well as the resulting by-products. For this series of compounds, we looked at the influence of the electronic effects of the starting components and structure, and we evaluated the in vitro antibacterial activity.

2. Results and Discussion

2.1. Chemistry

Povidone-iodine is a topical antiseptic medication used for wound therapy and infection prevention. It is a chemical compound of polyvinylpyrrolidone and triiodide [16]. Carboxy-quinolines substituted in position 5 with withdrawing radicals are mainly obtained through the Pfitzinger condensation of an appropriate isatin and α-methylidene carbonyl compounds in the presence of potassium hydroxide in aqueous ethanol under reflux followed by neutralization. In our previous study [16], we found that using the Doebner reaction conditions (ethanol, TFA as a catalyst and reflux for 12 h), we obtained pyrrole-2-one derivatives along with a few by-products. In this investigation, we changed the solvent conditions with acetic acid while keeping the other conditions constant. As a result, we successfully obtained quinoline derivatives. In addition to the desired quinoline derivatives, we separated from the reaction mixtures either pyrrole-2-ones, furan derivatives or methyl carboxy-quinoline as the major product, as presented in Figure 2. Based on these findings, we concluded that the main factors influencing the reaction pathways were the electronic effects and position of substituents in the starting aldehydes [17].

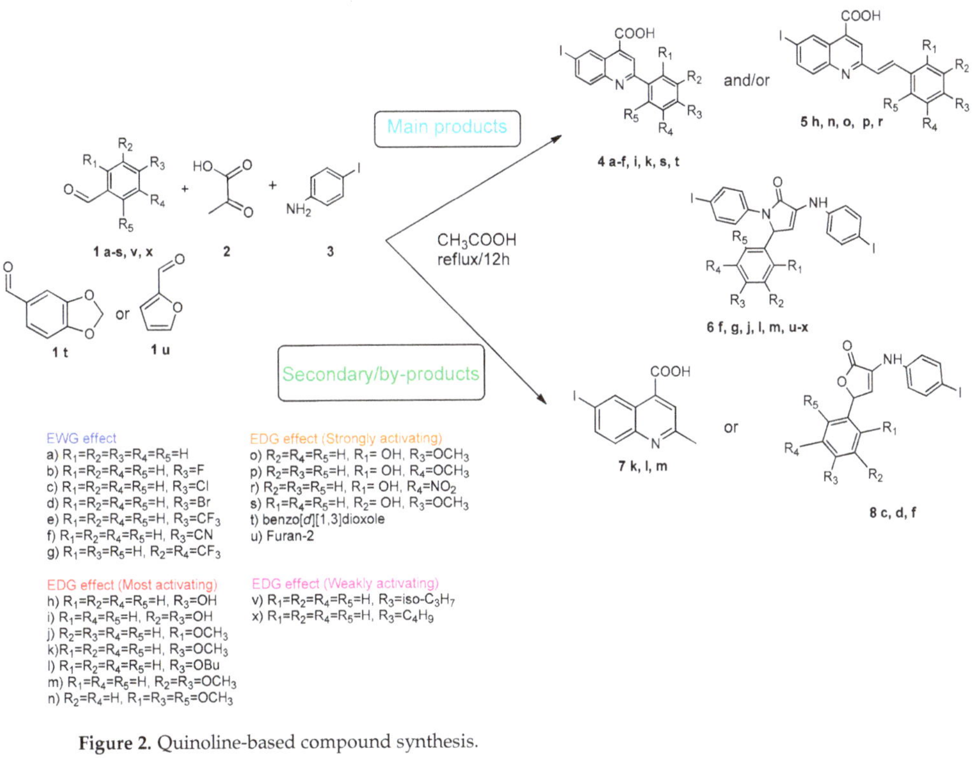

Figure 2. Quinoline-based compound synthesis.

Therefore, in the presence of an electron-withdrawing group at the *para* position of the aldehyde reactant (**1c**, **d**, **f**), the main product generated is carboxy-quinolines (**4a–f**), along with **6f** or/and **8c**, **d**, **f** as by-products. These additional structures are formed when the substituents are strongly deactivating, such as halogen and cyan groups. By introducing an additional electron-withdrawing group and repositioning them to the *meta*, *meta'* locations, the pyrrole-2-one derivative **6g** was exclusively obtained. Moving to the electron-donating group (weakly activating in the case of aldehydes **1v** and **1x**), we separated the same pyrrole-2-one derivatives (**6v** and **6x**). In the case of the strongly activating electron-donating group, we obtained different results depending on the position and/or combination of substituents. Therefore, carboxy-quinolines were produced when the methoxy group was the lone substituent in the *para* position (**4k**) or when the *para* position was occupied by the hydroxy or methoxy group and the *meta* position was occupied by the hydroxy group (**4s**). When a sterically hindered group was used such as butoxy that occupied the *para* position, a pyrrole-2-one derivative and a quinoline methyl substituted as a secondary product were formed. Once the **1m** reactant, containing two methoxy groups, was applied in the reaction, we did not achieve the intended formation of quinoline, but instead, the pyrrole derivative **6m** was produced. When the number of methoxy substituents in the aldehyde reactant is increased to 3 (**1n**), it leads to carboxy-quinoline derivatives (**5n**) containing an ethylenic bridge between the quinolinic and aldehyde cores. Similar compounds (**5o**, **5p**, **5r**) were formed when the initial aldehyde had two substituents. During a particular experiment, the replacement of the *para* group with a hydroxyl group, which is a powerful electron donor, led to the formation of a derivative of the same type (**5h**). For oxygen heterocycle aldehydes **1t** and **1u**, we noticed that the inclusion of a phenyl core led to the stability of the structure, resulting in the formation of **4t** quinoline. Specifically, derivative **1u** mostly produced the chemical pyrrole-2-one **6u**. A schematic representation of the electronic effect on the final compound's formation is presented in Figure 3, and a detailed table with the obtained products and by-products is presented in Table 1.

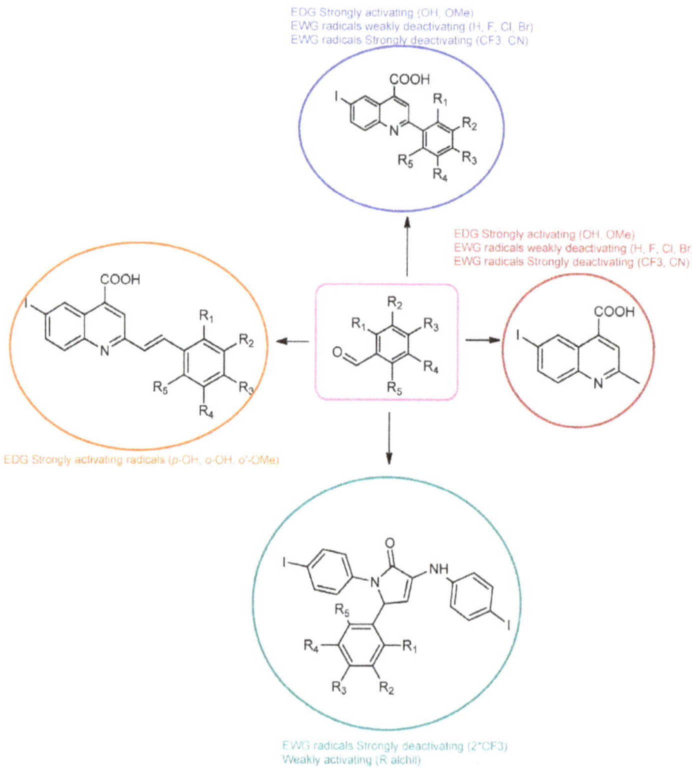

Figure 3. Reaction products depending on aldehyde substituents.

Table 1. Compressive obtained structures depending on the starting materials.

1	4	5	6	7	8	1	4	5	6	7	8	1	4	5	6	7	8
a	x					h		x				o		x			
b	x					i	x					p		x			
c	x				x	j			x			r		x			
d	x				x	k	x			x		s	x				
e	x					l			x	x		t	x				
f	x		x		x	m			x	x		u			x		
g			x			n		x				v			x		
												x			x		

1: starting aldehyde material; 4: quinoline main structure; 5: ethylene-bridged quinolinic structure; 6: pyrrole-2-one structure; 7: methyl quinoline by-product; 8: furan-2-one by-product.

The structures of the newly synthesized compounds have been established by employing spectral (NMR, IR) mass spectrometry and X-ray diffraction studies (details are presented in the Section 3 and Figures S1–S10 from the Supplementary Information).

Details on proton and proton–carbon spin systems were acquired using homo- and heteronuclear bidimensional correlation experiments such as H, H-COSY, H,C-HSQC and H,C-HMBC. In the ^{1}H-NMR spectra, the formation of the carboxy-quinoline core is indicated by the presence of four signals within the range of 7.80–9.20 ppm, with splitting patterns dependent on the proton–proton scalar couplings: doublet at 7.90 ppm, doublet of doublets at 8.10 ppm, singlet at 8.50 ppm and another doublet at 9.14 ppm. The proton of the carboxylic group presents a very broad resonance signal centered at around 14 ppm. Quinoline's aromatic substituent has proton resonance signals in the same region, their number and multiplicity being determined by the structure of the starting benzaldehyde. Thus, the presence of *para*-substituted benzaldehyde moieties in the reaction products **4b**–**f** and **4k** are recognized after the two doublets with a roof-effect pattern, whereas the moieties of disubstituted benzaldehydes have resonance signals as two doublets with 2 and 8 Hz coupling constants and a doublet of doublets (**4i** and **4s**). In the case of fluorine-substituted benzaldehyde moieties (**4b**), the additional proton–fluorine couplings alter the splitting pattern, generating a triplet and a doublet of doublets. The ethylenic bridge identified in quinoline derivative **5** was deduced from the presence in the proton spectra of two doublets with 16 Hz coupling constants, resonating in the interval 7.30–8.15 ppm. Both the coupling constants and chemical shift values are in agreement with this proposed vinylic group, indicating a *trans* proton configuration. The corresponding carbon atoms were identified in the 120–130 ppm interval in ^{13}C-NMR spectra, where double-bond carbon atoms usually resonate. Three-bond proton–carbon correlations identified in the HMBC spectrum support the covalent link of the ethylenic group with carboxy-quinoline and benzaldehyde cores. Both vinylic protons present correlation signals in the long-range HMBC spectrum with either quinoline's CH-3 or benzaldehyde's C-14 carbon atoms, as exemplified in Supporting Information Figure S4 for derivative **5o**.

The formation of side products with pyrrole-2-one (**6**) or furan-2-one (**8**) cores was readily seen in the proton NMR spectra. In the case of these derivatives, the carboxy-quinoline characteristic ^{1}H-NMR pattern was replaced by two doublets with 3 Hz coupling constants, resonating at around 6.0–7.0 ppm, previously assigned by us as fast NMR indicators for the formation of pyrrole-2-one or furan-2-one cores [17]. The rest of the proton signals present in the spectra of side products were also in good agreement with our previous NMR description of a series of iodo-dihydro-pyrrole-2-one derivatives [17].

The presence in the proton spectra of a carboxy-quinoline NMR pattern, with a new singlet at 2.70 ppm in the absence of benzaldehyde characteristic signals, clearly indicated a carboxy-quinoline substituted only with methyl, as exemplified in Figure S5 from the Supporting Information for compound **7**.

The unambiguous proton and carbon signal assignments for all new compounds, obtained from bidimensional correlation experiments, are presented in the Section 3.

The IR spectra of the obtained compounds show absorption bands specific for both aromatic and aliphatic C-H bonds, indicating the presence of the N-heterocycles (Section 3 and Figures S1–S10 from Supporting Information). The bands above 3300 cm^{-1} can be attributed to NH bonds and OH groups depending on the structure of the molecule being studied, while the bands related to C-N stretching bonds are in the range of 1335–1240 cm^{-1} and the bands specific to the absorption of ketone carbonyl groups are in the range of 1700–1650 cm^{-1}. All the other observed bands are in good agreement with the proposed structure [17].

The interaction between iodo-aniline, aldehyde derivatives and pyruvic acid with the formation of compounds **4**, **5**, **6**, **7** and **8** was also confirmed by MALDI-MS analysis (Section 3 and Figures S1–S10 from Supporting Information). Only positive ions were studied because the molecular ion only forms in the positive mode. By using the aforementioned method, we were able to achieve strong correlations between estimated mass and measured mass for each molecule, confirming all posited structures.

The structure and chemical composition of compounds **4b**, **c**, **d** and **e** were also confirmed by the single-crystal X-ray diffraction method (full information can be found in Tables S1–S5 from the Supporting Information). Accordingly, it was demonstrated that the compounds have a molecular crystal structure comprising neutral units, as illustrated in Figure 4 (**4d** (a), **4e** (b), **4b** (c) and **4c** (d)). It is to be noted that compounds **4c** and **4d**, which crystallize in the P21/n space group with very close unit cell parameters, are essentially planar, while for **4b** and **4e,** the phenyl and quinoline nuclei are rotated by 20.9(2)° and 22.5(1)°, respectively. This is caused not by the presence of steric effects but due to the peculiarities of the crystal structure packing. Further analysis has shown that the crystal packing of **4c** and **4d** is also similar, being driven by the presence of C-H···O, π-π stacking and halogen intermolecular contacts, which is illustrated in Figure S6.

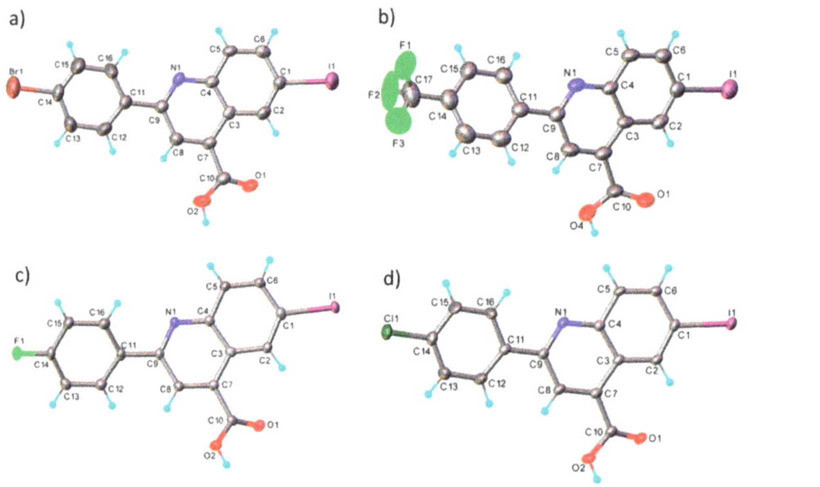

Figure 4. X-ray molecular structure of **4d** (**a**), **4e** (**b**), **4b** (**c**) and **4c** (**d**) with atom labeling and thermal ellipsoids at 50% level.

In the crystals of **4b** and **4c,** the asymmetric units are assembled to form a dense three-dimensional supramolecular network. As an example, a view of the crystal packing along the *b*-axis for **4d** is shown in Figure S7. The results of the X-ray diffraction study for compounds **4b** and **4e** are shown in Figure S8. It is to be noted that, as a whole, the system of intermolecular interactions in crystals **4e** and **4b** resembles well that observed for the above-mentioned **4d** and **4c**. As a result, the crystal structure of **4e** and **4b** is also characterized as a 3D supramolecular architecture, as depicted in Figure S9.

Important data about purity are revealed by powder X-ray diffraction pattern (PXRD) analysis of the produced substances. The PXRD of the synthesized compounds **4d** and **4e** are shown in Figure 5 and for compounds **4b** and **4c** in Figure S10. The diffraction peaks of the synthesized compounds are in good agreement with the simulated data.

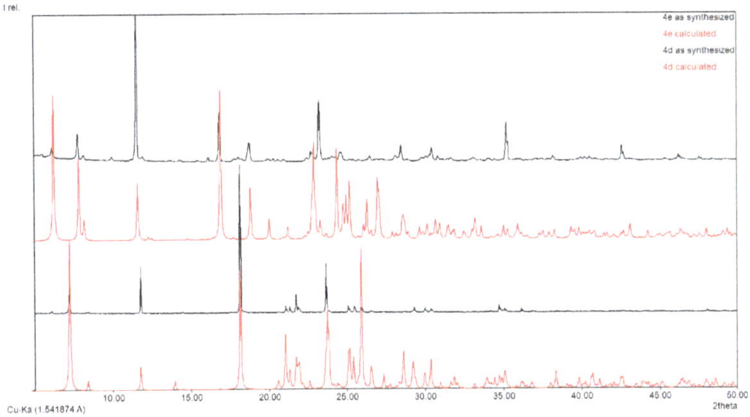

Figure 5. Powder XRD of compounds **4d** and **4e**.

2.2. Antimicrobial Activity

The most commonly used quinolones (ciprofloxacin, norfloxacin, oxofloxacin and levofloxacin) contain in their chemical structures, in addition to a quinolone center, a carboxyl group and a fluorine atom in position 5. In this study, we kept the carboxyl group in our structures and replaced fluorine with iodine (in the same position) in order to obtain better antimicrobial properties, as iodine insertion in organic molecules increases the antimicrobial activity against *S. aureus* [18–20].

Minimum inhibitory concentration (MIC) values were determined by the binary microdilution method. The results of the iodo-quinoline derivatives were compared with the basic structure and the solvent used (Table 2). Based on the information provided in Table 1, all compounds exhibited antibacterial effects against *S. epidermidis*, had no effect on *K. pneumoniae* and showed varying degrees of antifungal activity against *C. parapsilosis*. It is evident that the compounds exhibit varying levels of antibacterial and antifungal action based on the radicals linked to the basic structure.

Table 2. Antimicrobial activity is expressed as MIC (mg/mL), MBC (mg/mL) and MBEC-(mg/mL).

	S. epidermidis			*K. pneumoniae*			*C. parapsilosis*		
	MIC (mg/mL)	MBC (mg/mL)	MICMA (mg/mL)	MIC (mg/mL)	MBC (mg/mL)	MICMA (mg/mL)	MIC (mg/mL)	MBC (mg/mL)	MICMA (mg/mL)
4d	0.063	≤0.625	0.063	1.25	5	1.25	0.3125	2.5	0.3125
4a	0.625	0.625	0.3125	5	5	0.625	0.625	2.5	0.625
4b	0.625	0.625	0.3125	1.25	5	1.25	0.625	1.25	0.625
4e	0.25	≤0.625	0.25	2.5	5	1.25	0.625	2.5	0.625
4f	0.625	0.625	0.625	1.25	5	1.25	0.625	1.25	0.625
4i	1.25	2.5	1.25	1.25	5	1.25	1.25	1.25	1.25
4c	0.063	≤0.625	0.063	1.25	5	1.25	0.3125	≤0.625	0.3125
4s	1.25	2.5	1.25	1.25	5	1.25	0.625	0.625	0.625
4t	0.25	≤0.625	0.25	1.25	5	1.25	0.3125	≤0.625	≤0.15625
7	0.625	1.25	1.25	1.25	5	1.25	1.25	1.25	0.625
5r	0.3125	5	0.3125	1.25	5	1.25	0.625	5	0.625
5o	1.25	2.5	2.5	1.25	5	1.25	10	5	1.25

Table 2. *Cont.*

	S. epidermidis			*K. pneumoniae*			*C. parapsilosis*		
	MIC (mg/mL)	MBC (mg/mL)	MICMA (mg/mL)	MIC (mg/mL)	MBC (mg/mL)	MICMA (mg/mL)	MIC (mg/mL)	MBC (mg/mL)	MICMA (mg/mL)
5p	0.625	2.5	1.25	1.25	5	5	0.625	5	0.625
DMSO	2.5	2.5	2.5	1.25	5	1.25	1.25	1.25	1.25

MIC: minimum inhibitory concentration at which bacterial growth is completely inhibited; MBC: minimum concentration of drug that kills 99.9% of the test microorganisms in the original inoculum; MBEC: minimal concentration of the drug that kills 99.9% of biofilm-embedded bacteria; MICMA: minimum inhibitory concentrations for microbial adhesion.

In order to evaluate the antimicrobial activity improvement, the statistical analysis was carried out following the general structure, and the IC50 (percent inhibition is 50%) was determined for each variant (Figure 6). Therefore, the pattern of antimicrobial activity based on the IC50 against Gram-positive bacteria (*S. epidermidis*, Figure 6) may be observed as follows: in the quinoline series, the order of compounds from higher to lowest reactivity is **4d** > **4c** > **4e** > **4t** > **4b** > **4s** > **4f** > **7** > **4a** > **4i** > **DMSO** and in styryl quinoline series is **5r** > **5p** > **5o** > **DMSO**. The presence of the -C_6H_4Br group in the basic structure exhibited the highest level of activity towards the basic one, whereas the presence of the -C_6H_4Cl group in the same initial structure showed a somewhat lower level of activity. The initial structure (**7**) and the basic one (**4a**) exhibit notably higher efficiency against *S. epidermidis* than DMSO. Derivatives from the basic structure (**4a**) show significantly higher activity compared to the **5o, 5p** and **5r** variants.

Regarding Gram-negative bacteria such as *K. pneumoniae*, it can be concluded that the MIC values are determined by DMSO (Table 2). In some cases, DMSO even enhanced cell proliferation (**4e** and **4a**), resulting in higher values compared to those attributed to the solvent. Analyzing the IC50 values (Figure 6), it is evident that the basic structure exhibited a substantially lower IC50 value in comparison to the majority of the derivatives. This observation indicates that the antibacterial effectiveness of quinoline derivatives was ineffective against Gram-negative bacteria, perhaps due to the distinctive outer membrane of these cells. Resistance may arise due to mutations in porins or other constituents, alterations in the hydrophobic nature of the outer membrane or other modifications induced in Gram-negative bacteria. Gram-negative bacteria exhibit greater resistance compared to Gram-positive bacteria due to the absence of a critical membrane in Gram-positive bacteria [21]. The pattern of antibiotic efficacy against *K. pneumoniae*, as measured by IC50, was as follows: for the quinoline series, it is **4a** > **4s** > **4b** > **4i** > **7** > **4f** > **4t** > **DMSO** > **4d** > **4e** > **4c**, and for the styryl quinoline series, it is **5r** > **DMSO** > **5o** > **5p**. By comparing the IC50, it can be seen that the IC50 value for the basic structure was significantly lower compared to most of the derivatives. This aspect suggests that the antibacterial activity of quinoline derivatives was not active on Gram-negative bacteria, probably due to the specific outer membrane of these cells. Resistance can result from mutations in porins or other components, changes in the hydrophobicity of the outer membrane or other modifications triggered in Gram-negative bacteria. Gram-negative bacteria are more resistant than Gram-positive bacteria because Gram-positive bacteria lack this crucial membrane [22]. The trend of antimicrobial activity against *K. pneumoniae*, according to IC50, was as follows: **4a** > **4s** > **4b** > **4i** > **7** > **4f** > **4t** > **DMSO** > **4d** > **4e** > **4c** for quinoline series and **5r**> **DMSO** > **5o** > **5p** for styryl quinoline series.

The antifungal activity against the fungus *C. parapsilosis* (Figure 6), as measured by IC50, showed no significant variations between the derivatives and the original compound. The antifungal activity exhibited a significant difference compared to the solvent control ($p < 0.0001$, except for **4i** ($p > 0.05$)). The trend observed for the quinoline series was **4d** > **4c** > **4a** > **4e** > **4t** > **7** > **4f** > **4s** > DMSO > **4i**, and for the styryl quinoline series, it was **5s** > **5p** > DMSO > **5r.**

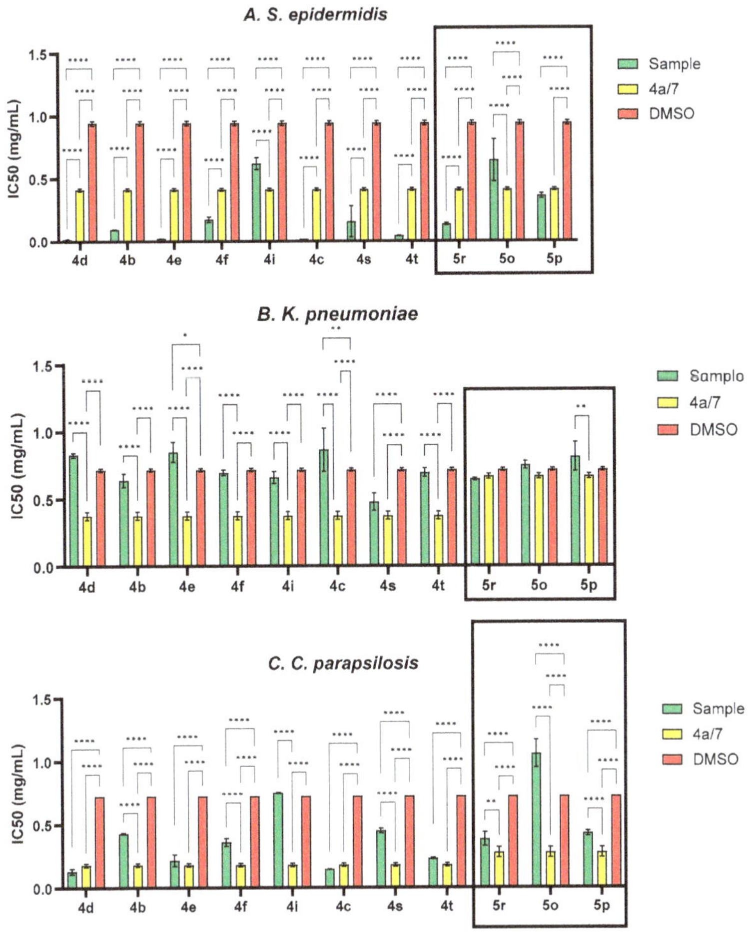

Figure 6. IC50 values (mg/mL) for each compound tested related to the general structure and solvent control for *S. epidermidis*, *K. pneumoniae* and *C. parapsilosis* strains (* $p < 0.05$, ** $p < 0.01$, **** $p < 0.0001$).

The compounds that presented an MBC value lower than the solvent were considered active. The MBC value is usually greater than or equal to the MIC value. From Table 2, it can be seen that in the case of *S. epidermidis*, the variants **4d**, **4a**, **4b**, **4e**, **4f**, **4c**, **4t** and **7** have a better microbicidal effect than the solvent used, the most active being **4s**, **4c**, **4e** and **4d**. In acute wounds, *S. epidermidis* is the most common strain [23]. Furthermore, *S. epidermidis* is not just a skin contaminant; its peptides contribute to the growth of pathogenic bacteria and might be a source of genes that confer antibiotic resistance [21]. Gram-negative bacteria were more frequently identified from chronic wounds than Gram-positive bacteria [24]. It is advised to utilize these compounds as active principles for formulations intended to treat acute wounds because they do not exhibit bactericidal action on Gram-negative bacteria, like the *K. pneumoniae* strain. According to Dowd et al. [25], the two most common species of yeast found in wounds were *C. albicans* and *C. parapsilosis*. We evaluated only the *C. parapsilosis* strain in this investigation to compare the antifungal activity of all iodo-quinoline derivatives. Table 2 reveals that only the variants **4c**, **4s** and **4t** have a fungicidal impact against *C. parapsilosis*; the rest of the compounds' action is given by DMSO unless they have only fungistatic activity.

Self-preservation has evolved in all microbes through parasitic connections in the form of biofilms. Biofilms are made up of a complex protective glycocalyx that bacterial colonies generate to defend themselves from host defense and antimicrobial treatment [26]. Biofilms are found in chronic and acute wounds, where they provide a risk of delayed infection. This process can also induce persistent and recurring skin infections [27].

The development of novel compounds capable of inhibiting biofilm formation might be one of the solutions to both the rise in antibiotic resistance and the development of hard-to-heal wounds [28]. In this regard, the effect of iodo-quinoline derivatives on microbial adhesion, the initial stage of microbial biofilm development, was investigated. The values of the minimum inhibitory concentrations for microbial adhesion (MICMA) obtained are shown in Table 2. Compounds with MBEC values less than MIC are deemed active on microbial adhesion. The majority of the substances prevent the adherence of *S. epidermidis* cells. The compounds with greater activity related to both the basic structure (**4a**) and DMSO include **4d**, **4t**, **4e** and **4c**, while the active derivative from the **7** structure was only the **5r** variant. In the case of the **4a** base structure, a decrease in microbial adherence was also seen for the *K. pneumoniae* strain. The compounds with greater activity against *C. parapsilosis* cell adherence related to both the basic structures (**4a** and **7**) and DMSO include **4d**, **4t** and **4c**. A structure-antimicrobial efficiency diagram is presented in Figure 7.

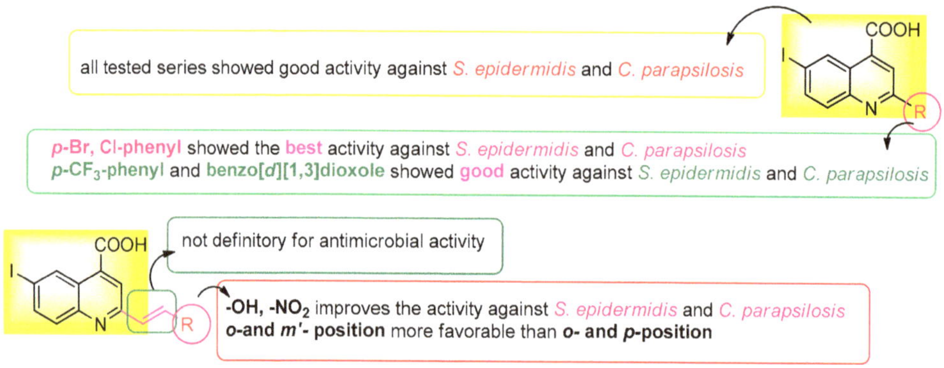

Figure 7. Antimicrobial efficiency diagram of the tested scaffolds against the three tested strains.

3. Materials and Methods

Analytical thin-layer chromatography was performed with commercial silica gel plates 60 F_{254} (Merck Darmstadt, Germany) and visualized with UV light (λ_{max} = 254 or 365 nm).

The ^{1}H and ^{13}C NMR spectra recorded for this study were obtained for solutions in DMSO-d6 at 298 K on Bruker Avance NEO 400 and 600 MHz spectrometers, equipped with a 5 mm four nuclei direct detection z-gradient probe (H,C,F,Si-QNP) and 5 mm inverse detection multinuclear z-gradient probe, respectively, with solvent peaks as reference. All chemical shifts were reported in ppm (δ), and coupling constants (*J*) were reported in Hertz (Hz). All the assignments were confirmed by information from proton–proton and proton–carbon two-dimensional NMR correlation experiments (COSY, HSQC and HMBC).

IR spectra were recorded on a Shimadzu IRTracer-100 instrument (Shimadzu U.S.A. Manufacturing, Inc., Canby, OR, USA). The melting point of the compounds was measured on MEL-TEMP capillary melting point apparatus from ambient temperature up to 400 °C. All commercially available products were used without further purification unless otherwise specified.

Mass spectra were acquired on a Bruker RapifleX MALDI-TOF/TOF (Bruker Daltonics, Bremen, Germany) equipped with a Smartbeam 3D laser. The FlexControl Version 4.0 and FlexAnalysis Version 4.0 software (Bruker, Bremen, Germany) were used to control the instrument and process the MS spectra.

The samples were initially dissolved in DMSO and subsequently diluted 10-fold in methanol. For the MALDI matrix solutions, 20 mg of α-cyano-4-hydroxycinnamic acid (HCCA) was dissolved in 1 mL of methanol. Subsequently, the MALDI matrix solution and the sample solution were mixed in a 2:1 ratio, and finally, 1 μL from each resulting solution was deposited onto the MALDI target and dried at room temperature prior to MALDI-MS analysis. Mass calibration of MALDI-TOF/TOF-MS was carried out using the peptide mixture standard solution (Bruker Daltonics, Bremen, Germany).

FlexControl 4.0 was used to optimize and acquire data using the following parameters: positive ion polarity in reflector mode, mass scan range *m/z* 100–1600 Da), digitizer 1.25 GHz, detector voltage 2117 V, 1000 shots per pixel and 5 kHz laser frequency. The laser power was set at 60% to 80% of the maximum and 1000 laser shots were accumulated for each spectrum.

3.1. General Procedure for Synthesis of Compounds **4**, **5**, **6**, **7** *and* **8**

Aldehydes **1a–v** (1 mmol) were solubilized in a minimum amount of acetic acid. A mixture containing pyruvic acid (1,5 mmoli) and TFA (20 μL) as a catalyst was then added to acetic acid and stirred for 10 min. Finally, iodo-aniline (1 mmol) was dissolved in a minimum amount of acetic acid and added and the resulting mixture was left to reflux for 12 h. The required products were obtained by filtering out the resulting suspension and washing the solid with ethanol. Dichloromethane and ethanol were used to facilitate recrystallization.

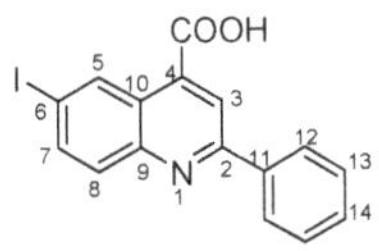

6-iodo-2-phenylquinoline-4-carboxylic acid 4a: crystallized from acetic acid; brown powder; 79% yield; mp 282–285 °C; IR ATR ν(cm^{-1}): 3650, 3321, 3271, 3074, 2982, 2971, 2886, 1690, 1653, 1597, 1541, 1512, 1482, 1443, 1375, 1328, 1259, 1233, 1154, 1089, 1021, 952, 864, 764, 677, 587.

^{1}H NMR (600.1 MHz, DMSO-d_6, δ (ppm)): 7.56 (t, ^{3}J = 7 Hz, 1H, H-14), 7.59 (t, ^{3}J = 7 Hz, 2H, H-13), 7.95 (d, ^{3}J = 9 Hz, 1H, H-8), 8.13 (dd, ^{3}J = 9 Hz, ^{4}J = 2 Hz, 1H, H-7), 8.30 (d, ^{3}J = 7 Hz, 2H, H-12), 8.52 (s, 1H, H-3), 9.14 (d, ^{4}J = 2 Hz, 1H, H-5), 14.03 (bs, 1H, OH).

^{13}C NMR (150.9 MHz, DMSO-d_6, δ (ppm)): 94.8 (C-6), 120.2 (CH-3), 125.1 (C-10), 127.2 (CH-12), 129.0 (CH-13), 130.2 (CH-14), 131.6 (CH-8), 133.9 (CH-5), 135.9 (C-4), 137.5 (C-11), 138.5 (CH-7), 147.3 (C-9), 156.4 (C-2), 167.1 (COOH).

HRMS (MALDI-TOF/TOF) *m/z*: $[M + H]^+$ Calcd for $C_{16}H_{11}INO_2$ 375.9834; found 375.9827.

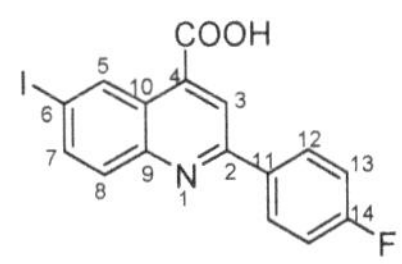

2-(4-fluorophenyl)-6-iodoquinoline-4-carboxylic acid 4b: crystallized from acetic acid; yellow powder; 82% yield; mp 131–132 °C; IR ATR ν(cm^{-1}): 3649, 3272, 3066, 2982, 2971, 2884, 1684, 1653, 1594, 1540, 1507, 1481, 1373, 1328, 1259, 1233, 1155, 1088, 1020, 957, 940, 828, 677, 585, 507.

^{1}H NMR (600.1 MHz, DMSO-d_6, δ (ppm)): 7.40 (t, $^3J_{H,H}$ = $^3J_{H,F}$ = 9 Hz, 2H, H-13), 7.92 (d, ^{3}J = 9 Hz, 1H,H-8), 8.11 (dd, ^{3}J = 9 Hz, ^{4}J = 2 Hz, 1H, H-7), 8.36 (dd, $^3J_{H,H}$ = 9 Hz, $^4J_{H,F}$ = 5 Hz, 2H, H-12), 8.50 (s, 1H, H-3), 9.12 (d, ^{4}J = 2 Hz, 1H, H-5), 14.14 (bs, 1H, OH).

^{1}H NMR (400.1 MHz, DMSO-d$_6$, δ (ppm)): 3.86 (s, 3H, OCH_3), 7.12 (d, ^{3}J = 9 Hz, 2H, H-13), 7.88 (d, ^{3}J = 9 Hz, 1H, H-8), 8.08 (dd, ^{3}J = 9 Hz, ^{4}J = 2 Hz, 1H, H-7), 8.27 (d, ^{3}J = 9 Hz, 2H, H-12), 8.46 (s, 1H, H-3), 9.09 (d, ^{4}J = 2 Hz, 1H, H-5), 14.05 (bs, 1H, OH).

^{13}C NMR (100.6 MHz, DMSO-d$_6$, δ (ppm)): 55.3 (OCH_3), 94.0 (C-6), 114.4 (CH-13), 119.7 (CH-3), 124.8 (C-10), 128.8 (CH-12), 129.9 (C-11), 130.7 (CH-8), 133.9 (CH-5), 135.6 (C-4), 138.4 (CH-7), 147.3 (C-9), 155.2 (C-2), 161.1 (C-14), 167.01 (COOH).

HRMS (MALDI-TOF/TOF) *m/z*: [M + H]$^+$ Calcd for $C_{17}H_{10}IN_2O_2$ 405.9940; found 405.9960.

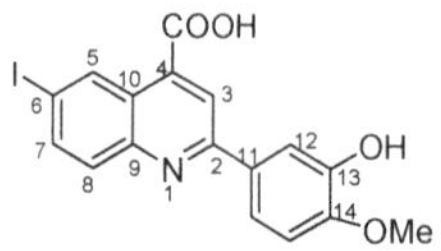

2-(3-hydroxy-4-methoxyphenyl)-6-iodoquinoline-4-carboxylic acid 4s: crystallized from acetic acid; orange powder; 85% yield; mp 292–295 °C; IR ATR ν(cm^{-1}): 3648, 3090, 2982, 2970, 2883, 1717, 1638, 1598, 1571, 1516, 1396, 1380, 1304, 1263, 1222, 1146, 1077, 1019, 952, 867, 809, 797, 770, 581, 524, 509.

^{1}H NMR (400.1 MHz, DMSO-d$_6$, δ (ppm)): 3.87 (s, 3H, CH_3), 7.09 (d, ^{3}J = 8 Hz, 1H, H-15), 7.72 (dd, ^{3}J = 8 Hz, ^{4}J = 2 Hz, 1H, H-16), 7.81 (d, ^{4}J = 2 Hz, 1H, H-12), 7.87 (d, ^{3}J = 9 Hz, 1H, H-8), 8.08 (dd, ^{3}J = 9 Hz, ^{4}J = 2 Hz, 1H, H-7), 8.41 (s, 1H, H-3), 9.11 (d, ^{4}J = 1 Hz, 1H, H-5), 9.36 (bs, 1H, OH), 13.97 (bs, 1H, OH).

^{13}C NMR (100.6 MHz, DMSO-d$_6$, δ (ppm)): 55.6 (OCH_3), 93.9 (C-6), 112.1 (CH-12), 113.9 (CH-15), 118.9 (CH-16), 119.8 (CH-3), 124.9 (C-10), 130.2 (C-11), 131.3 (CH-8), 133.9 (CH-5), 135.4 (C-4), 138.4 (CH-7), 146.9 (C-13), 147.3 (C-9), 149.9 (C-14), 156.1 (C-2), 167.1 (COOH).

HRMS (MALDI-TOF/TOF) *m/z*: [M + H]$^+$ Calcd for $C_{17}H_{13}INO_4$ 421.9889; found 421.9896.

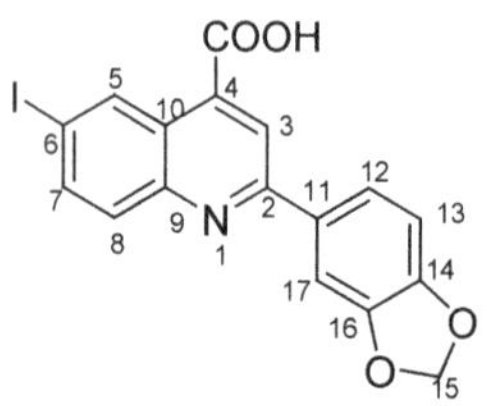

2-(benzo[d][1,3]dioxol-5-yl)-6-iodoquinoline-4-carboxylic acid 4t: crystallized from acetic acid; brown powder; 77% yield; mp 205–206 °C; IR ATR ν(cm^{-1}): 3660, 2978, 2889, 1744, 1604, 1580, 1505, 1452, 1372,1352, 1239, 1152, 1114, 1033, 941, 867, 825, 809, 603, 519.

^{1}H NMR (400.1 MHz, DMSO-d$_6$, δ (ppm)): 6.15 (s, 2H, CH_2), 7.09 (d, ^{3}J = 8 Hz, 1H, H-13), 7.83–7.86 (m, 2H, H-12 and H-17), 7.88 (d, ^{3}J = 9 Hz, 1H, H-8), 8.08 (dd, ^{3}J = 9 Hz, ^{4}J = 2 Hz, 1H, H-7), 8.44 (s, 1H, H-3), 9.08 (d, ^{4}J = 1 Hz, 1H, H-5), 9.36 (bs, 1H, OH), 13.06 (bs, 1H, OH).

^{13}C NMR (100.6 MHz, DMSO-d$_6$, δ (ppm)): 94.2 (C-6), 101.6 (CH_2), 106.9 (CH-17), 108.6 (CH-13), 119.8 (CH-3), 122.0 (CH-12), 124.9 (C-10), 131.4 (C-11), 131.8 (CH-8), 133.8 (CH-5), 135.8 (C-4), 138.4 (CH-7), 147.1 (C-9), 148.2 (C-14), 149.2 (C-16), 155.7 (C-2), 167.1 (COOH).

HRMS (MALDI-TOF/TOF) *m/z*: [M + H]$^+$ Calcd for $C_{17}H_{11}INO_4$ 419.9733; found 419.9708.

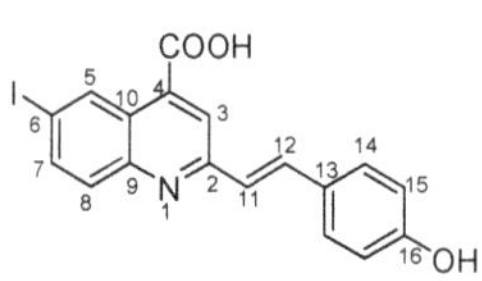

(*E*)-2-(4-hydroxystyryl)-6-iodoquinoline-4-carboxylic acid 5h: crystallized from acetic acid; yellow powder; 57% yield; mp 221–223 °C; IR ATR ν(cm^{-1}): 3314, 3024, 2970, 2953, 2922, 2853, 1740, 1666, 1646, 1582, 1534, 1487, 1415, 1386, 1231, 1217, 1162, 1063, 1008, 916, 828, 814, 792, 777, 742, 633, 620, 543.

^{1}H NMR (400.1 MHz, DMSO-d$_6$, δ (ppm)): 6.84 (d, ^{3}J = 8 Hz, 2H, H-15), 7.33 (d, ^{3}J = 16 Hz, 1H, H-11), 7.62 (d, ^{3}J = 8 Hz, 2H, H-14), 7.82 (d, ^{3}J = 9 Hz, 1H, H-8), 7.84 (d, ^{3}J = 16 Hz, 1H, H-12), 8.05 (dd, ^{3}J = 9 Hz, ^{4}J = 2 Hz, 1H, H-7), 8.24 (s, 1H, H-3), 9.08 (d, ^{4}J = 1 Hz, 1H, H-5), 9.85 (bs, 1H, OH), 14.04 (bs, 1H, OH).

^{13}C NMR (100.6 MHz, DMSO-d$_6$, δ (ppm)): 93.6 (C-6), 115.7 (CH-15), 121.6 (CH-3), 124.2 (CH-11), 124.9 (C-10), 127.0 (C-13), 129.2 (CH-14), 130.9 (CH-8), 133.9 (CH-5), 134.9 (C-4), 135.6 (CH-12), 138.2 (CH-7), 147.4 (C-9), 156.4 (C-2), 158.6 (C-16), 167.2 (COOH).

HRMS (MALDI-TOF/TOF) *m/z*: [M + H]$^+$ Calcd for $C_{18}H_{13}INO_3$ 417.9940; found 417.9931.

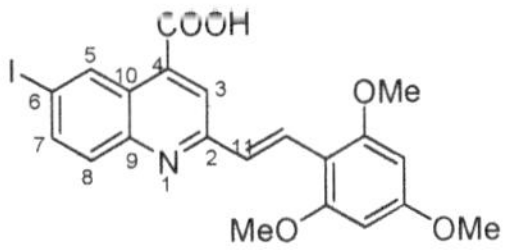

(*E*)-6-iodo-2-(2,4,6-trimethoxystyryl)quinoline-4-carboxylic acid 5n: crystallized from acetic acid; yellow powder; 47% yield; mp 328–329 °C; IR ATR ν(cm^{-1}): 3649, 2982, 2971, 2891, 1560, 1380, 1253, 1154, 1073, 967, 809, 519.

^{1}H NMR (400.1 MHz, DMSO-d$_6$, δ (ppm)): 3.87 (s, 3H, OCH_3), 3.94 (s, 6H, 2xOCH_3), 6.35 (s, 2H, H-15), 7.67 (d, ^{3}J = 16 Hz, 1H, H-11), 7.82 (d, ^{3}J = 9 Hz, 1H, H-8), 8.06 (dd, ^{3}J = 9 Hz, ^{4}J = 2 Hz, 1H, H-7), 8.11 (s, 1H, H-3), 8.14 (d, ^{3}J = 16 Hz, 1H, H-12), 9.07 (d, ^{4}J = 1 Hz, 1H, H-5), 13.08 (bs, 1H, OH).

^{13}C NMR (100.6 MHz, DMSO-d$_6$, δ (ppm)): 55.4 (OCH_3), 55.9 (2xOCH_3), 91.0 (CH-15), 93.3 (C-6), 105.9 (C-13), 121.9 (CH-3), 124.9 (C-10), 126.6 (CH-12), 126.9 (CH-11), 130.9 (CH-8), 133.9 (CH-5), 134.7 (C-4), 138.2 (CH-7), 147.4 (C-9), 157.5 (C-2), 160.2 (C-14), 161.7 (C-16), 167.1 (COOH).

HRMS (MALDI-TOF/TOF) *m/z*: [M + H]$^+$ Calcd for $C_{21}H_{19}INO_5$ 492.0308; found 492.0335.

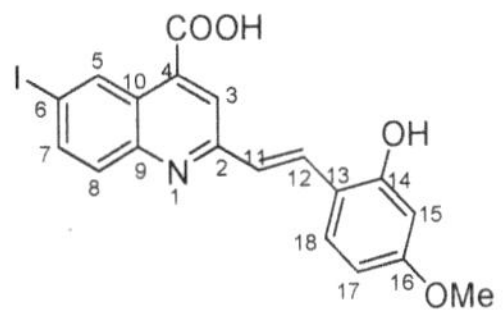

(*E*)-2-(2-hydroxy-4-methoxystyryl)-6-iodoquinoline-4-carboxylic acid 5o: crystallized from acetic acid; yellow powder; 59% yield; mp 226–227 °C; IR ATR ν(cm^{-1}): 3649, 2982, 2970, 2889, 1700, 1576, 1558, 1382, 1255, 1223, 1147, 1074, 968, 95, 809, 737, 668.

^{1}H NMR (400.1 MHz, DMSO-d$_6$, δ (ppm)): 3.76 (s, 3H, OCH_3), 6.50–6.51 (m, 2H, H-15and H-17), 7.41 (d, ^{3}J = 16 Hz, 1H, H-11), 7.66 (d, ^{3}J = 9 Hz, 1H, H-18), 7.82 (d, ^{3}J = 9 Hz, 1H, H-8), 8.05 (dd, ^{3}J = 9 Hz, ^{4}J = 2 Hz, 1H, H-7), 8.05 (d, ^{3}J = 16 Hz, 1H, H-12), 8.17 (s, 1H, H-3), 9.08 (d, ^{4}J = 2 Hz, 1H, H-5), 10.20 (bs, 1H, OH), 13.74 (bs, 1H, OH).

^{13}C NMR (150.9 MHz, DMSO-d$_6$, δ (ppm)): 55.0 (OCH_3), 93.5 (C-6), 101.2 (CH-15), 105.9 (CH-17), 115.8 (C-13), 121.7 (CH-3), 124.4 (CH-11), 124.9 (C-10), 128.8 (CH-18), 130.8 (CH-12), 131.1 (CH-8), 133.9 (CH-5), 134.7 (C-4), 138.2 (CH-7), 147.5 (C-9), 156.8 (C-2), 157.4 (C-13), 161.1 (C-16), 167.1 (COOH).

HRMS (MALDI-TOF/TOF) *m/z*: [M + H]$^+$ Calcd for $C_{19}H_{15}INO_4$ 448.0046; found 448.0029.

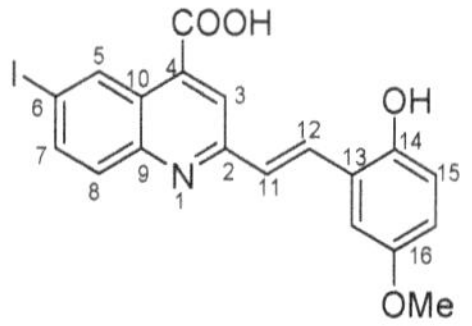

(*E*)-2-(2-hydroxy-5-methoxystyryl)-6-iodoquinoline-4-carboxylic acid 5p: crystallized from acetic acid; brown powder; 74% yield; mp 324–325 °C; IR ATR ν(cm^{-1}): 3648, 2981, 2971, 2891, 1653, 1593, 1559, 1383, 1262, 1225, 1150, 1075, 1017, 966, 949, 865, 807, 579.

^{1}H NMR (600.1 MHz, DMSO-d_6, δ (ppm)): 3.77 (s, 3H,OCH_3), 6.82 (dd, ^{3}J = 9 Hz, ^{4}J = 3 Hz, 1H, H-16), 6.87 (d, ^{3}J = 9 Hz, 1H, H-15), 7.30 (d, ^{3}J = 3 Hz, 1H, H-18), 7.58 (d, ^{3}J = 16 Hz, 1H, H-11), 7.85 (d, ^{3}J = 9 Hz, 1H, H-8), 8.07 (dd, ^{3}J = 9 Hz, ^{4}J = 2 Hz, 1H, H-7), 8.10 (d, ^{3}J = 16 Hz, 1H, H-12), 8.21 (s, 1H, H-3), 9.12 (d, ^{4}J = 2 Hz, 1H, H-5), 9.62 (bs, 1H, OH), 14.04 (bs, 1H, OH).

^{13}C NMR (150.9 MHz, DMSO-d_6, δ (ppm)): 55.4 (OCH_3), 93.9 (C-6), 110.9 (CH-18), 116.9 (CH-16), 117.0 (CH-15), 122.0 (CH-3), 122.9 (C-13), 125.1 (C-10), 127.0 (CH-11), 130.6 (CH-12), 131.1 (CH-8), 134.0 (C-4), 134.8 (CH-5), 138.3 (CH-7), 147.5(C-9), 150.1 (C-14), 152.3 (C-17), 156.4 (C-2), 167.0 (COOH).

HRMS (MALDI-TOF/TOF) *m*/*z*: $[M + H]^+$ Calcd for $C_{19}H_{15}INO_4$ 448.0046; found 448.0032.

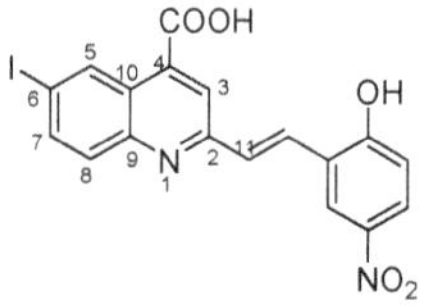

(*E*)-2-(2-hydroxy-5-nitrostyryl)-6-iodoquinoline-4-carboxylic acid 5r: crystallized from acetic acid; yellow powder; 88% yield; mp 328–330 °C; IR ATR ν(cm^{-1}): 3650, 3114, 3084, 2982, 2971, 2883, 1700, 1683, 1654, 1594, 1559, 1483, 1381, 1339, 1266, 1147, 969, 943, 826, 748, 691, 628.

^{1}H NMR (600.1 MHz, DMSO-d_6, δ (ppm)): 7.13 (d, ^{3}J = 9 Hz, 1H, H-15), 7.81 (d, ^{3}J = 16 Hz, 1H, H-11), 7.88 (d, ^{3}J = 9 Hz, 1H, H-8), 8.09 (dd, ^{3}J = 9 Hz, ^{4}J = 2 Hz, 1H, H-7), 8.13 (dd, ^{3}J = 9 Hz, ^{4}J = 3 Hz, 1H, H-16), 8.13 (d, ^{3}J = 16 Hz, 1H, H-12), 8.30 (s, 1H, H-3), 8.64 (d, ^{4}J = 3 Hz, 1H, H-18), 9.13 (d, ^{4}J = 2 Hz, 1H, H-5), 11.76 (bs, 1H, OH), 14.06 (bs, 1H, OH).

^{13}C NMR (150.9 MHz, DMSO-d_6, δ (ppm)): 94.5 (C-6), 116.4 (CH-15), 122.5 (CH-3), 123.5 (C-13), 123.7 (CH-18), 125.3 (C-10), 125.5 (CH-16), 128.6 (CH-12), 129.8 (CH-11), 131.2 (CH-8), 133.9 (CH-5), 135.0 (C-4), 138.4 (CH-7), 140.0 (C-17), 147.4 (C-9), 155.8 (C-2), 161.8 (C-14), 167.0 (COOH).

HRMS (MALDI-TOF/TOF) *m*/*z*: $[M + H]^+$ Calcd for $C_{18}H_{12}IN_2O_5$ 462.9791; found 462.9799.

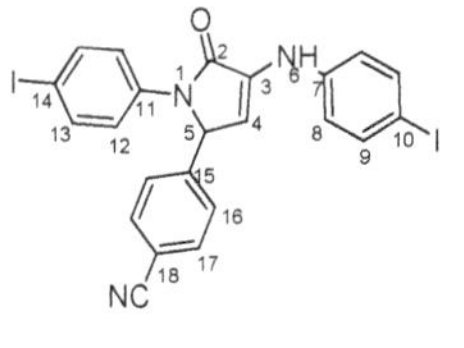

4-(1-(4-iodophenyl)-4-((4-iodophenyl)amino)-5-oxo-2,5-dihydro-1H-pyrrole-2-yl)benzonitrile 6f: [13]

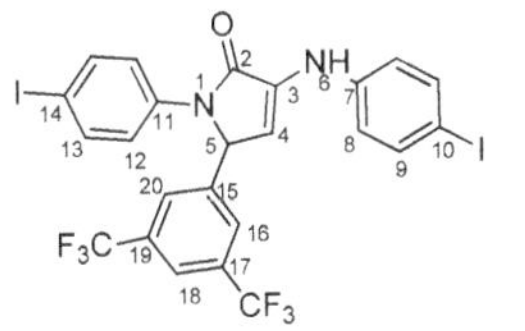

5-(3,5-bis(trifluoromethyl)phenyl)-1-(4-iodophenyl)-3-((4-iodophenyl)amino)-1H-pyrrole-2(5H)-one 6g: [13]

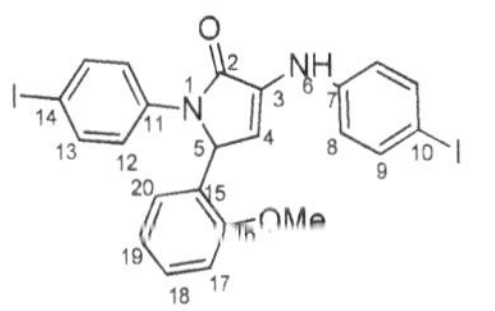

1-(4-iodophenyl)-3-((4-iodophenyl)amino)-5-(2-methoxyphenyl)-1H-pyrrole-2(5H)-one 6j: [13]

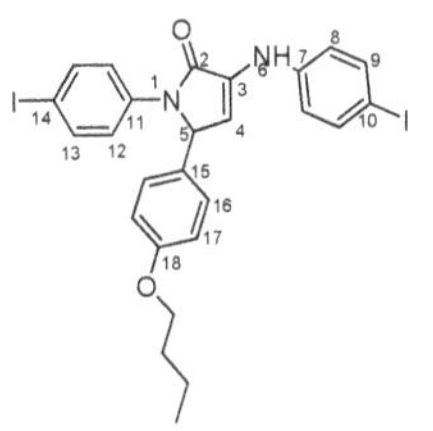

5-(4-butoxyphenyl)-1-(4-iodophenyl)-3-((4-iodophenyl)amino)-1H-pyrrole-2(5H)-one 6l: crystallized from acetic acid; black powder; 53% yield; mp 229–230 °C; IR ATR ν(cm^{-1}): 3117, 3077, 3027, 2944, 1738, 1595, 1484, 1434, 1369, 1284, 1230, 1217, 1174, 1143, 840, 827, 668, 628, 529.

^{1}H NMR (400.1 MHz, DMSO-d$_6$, δ (ppm)): 0.90 (t, ^{3}J = 7 Hz, 3H,CH$_3$), 1.39 (sextet, ^{3}J = 7 Hz, 2H, CH$_2$), 1.64 (quintet, ^{3}J = 7 Hz, 2H,CH$_2$), 3.88 (t, ^{3}J = 7 Hz, 2H, OCH$_2$), 5.98 (d, ^{3}J = 2 Hz, 1H, H-5), 6.33 (d, ^{3}J = 2 Hz, 1H, H-4), 6.83 (d, ^{3}J = 8 Hz, 2H, H-17), 7.14 (d, ^{3}J = 8 Hz, 4H, H-8 and H-16), 7.45 (d, ^{3}J = 9 Hz, 2H, H-12), 7.54 (d, ^{3}J = 9 Hz, 2H, H-9), 7.66 (d, ^{3}J = 9 Hz, 2H, H-13), 8.32 (s, 1H, NH).

^{13}C NMR (100.6 MHz, DMSO-d$_6$, δ (ppm)): 13.6 (CH$_3$), 18.7 (CH$_2$-21), 30.6 (CH$_2$-20), 61.7 (CH-5), 67.0 (CH$_2$-19), 82.4 (C-10), 88.9 (C-14), 111.3 (CH-4), 114.6 (CH-17), 119.1 (CH-8), 123.5 (CH-12), 128.0 (CH-16), 128.7 (C-15), 131.3 (C-3), 136.8 (C-11), 137.3 (CH-13), 137.4 (CH-9), 141.8 (C-7), 158.3 (C-18), 166.2 (CO-2).

HRMS (MALDI-TOF/TOF) *m/z*: [M + H]$^+$ Calcd for $C_{26}H_{25}I_2N_2O_2$ 651.0005; found 651.0027.

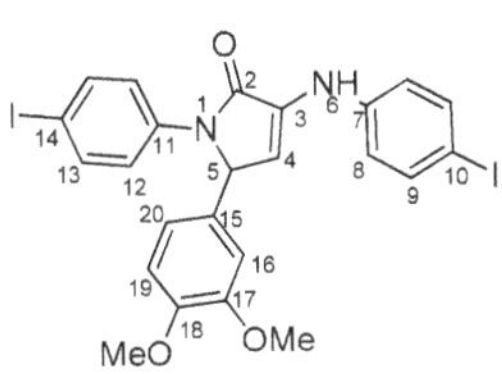

5-(3,4-dimethoxyphenyl)-1-(4-iodophenyl)-3-((4-iodophenyl)amino)-1H-pyrrole-2(5H)-one 6m: [13]

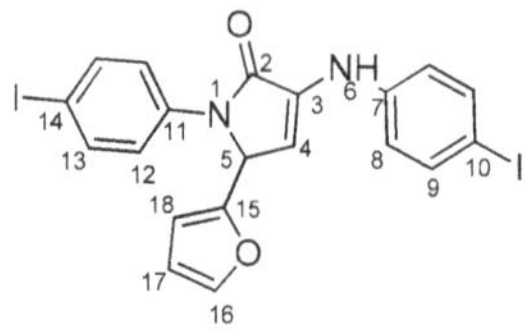

5-(furan-2-yl)-1-(4-iodophenyl)- 3-((4-iodophenyl)amino)-1H-pyrrole-2(5H)-one 6u: [13]

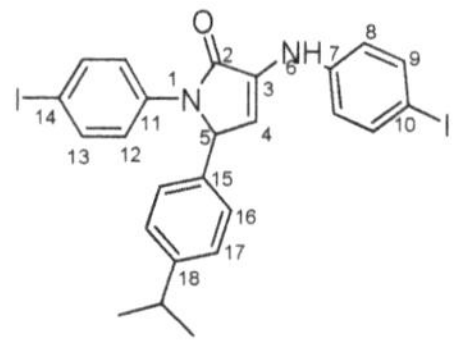

1-(4-iodophenyl)-3-((4-iodophenyl)amino)-5-(4-isopropylphenyl)-1H-pyrrole-2(5H)-one 6v: crystallized from acetic acid; black powder; 53% yield; mp 197–200 °C; IR ATR ν(cm^{-1}): 3649, 3315, 2982, 2971, 2883, 1666, 1648, 1584, 1533, 1487, 1456, 1417, 1390, 1264, 1162, 1060, 1008, 957, 813, 792, 776, 638, 560, 507.

^{1}H NMR (600.1 MHz, DMSO-d$_6$, δ (ppm)): 1.13 (d, ^{3}J = 7 Hz, 3H, CH$_3$), 1.14 (d, ^{3}J = 7 Hz, 3H, CH$_3$), 2.81 (septet, ^{3}J = 7 Hz, 1H, CH), 6.02 (d, ^{3}J = 3 Hz, 1H, H-5), 6.36 (d, ^{3}J = 3 Hz, 1H, H-4), 7.14 (d, ^{3}J = 9 Hz, 2H, H-8), 7.17 (bs, 4H, H-16 and H-17), 7.47 (d, ^{3}J = 9 Hz, 2H, H-12), 7.53 (d, ^{3}J = 9 Hz, 2H, H-9), 7.68 (d, ^{3}J = 9 Hz, 2H, H-13), 8.31 (s, 1H, NH).

^{13}C NMR (150.9 MHz, DMSO-d$_6$, δ (ppm)): 23.7 (2xCH$_3$), 32.9 (CH), 61.9 (CH-5), 82.5 (C-10), 88.9 (C-14), 111.3 (CH-4), 119.1 (CH-8), 123.4 (CH-12), 127.7 and 127.8 (CH-16 and CH-17), 131.2 (C-3), 134.7 (C-15), 136.9 (C-11), 137.4 (CH-9 and CH-13), 141.8 (C-7), 147.9 (C-18), 166.3 (CO-2).

HRMS (MALDI-TOF/TOF) *m*/*z*: [M + H]$^+$ Calcd for $C_{25}H_{23}I_2N_2O$ 620.9909; found 620.9929.

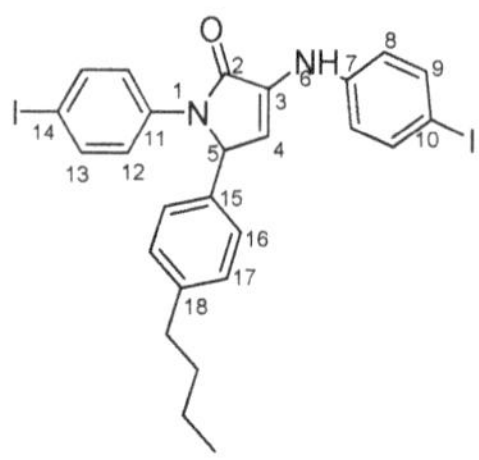

5-(4-butylphenyl)-1-(4-iodophenyl)-3-((4-iodophenyl)amino)-1H-pyrrole-2(5H)-one 6x: crystallized from acetic acid; black powder; 53% yield; mp 211–212 °C; IR ATR ν(cm^{-1}): 3313, 3024, 2956, 2927, 2854, 1739, 1665, 1646, 1585, 1535, 1487, 1416, 1388, 1229, 1218, 1164, 1063, 1009, 914, 828, 813, 793, 776, 741, 636, 621 544.

^{1}H NMR (600.1 MHz, DMSO-d$_6$, δ (ppm)): 0.86 (t, ^{3}J = 7 Hz, 3H, CH$_3$), 1.26 (sextet, ^{3}J = 7 Hz, 2H, CH$_2$), 1.48 (quintet, ^{3}J = 7 Hz, 2H, CH$_2$), 2.47–2.49 (m, 2H, CH$_2$ overlapped with DMSO), 6.00 (d, ^{3}J = 2 Hz, 1H, H-5), 6.36 (d, ^{3}J = 2 Hz, 1H, H-4), 7.09–7.15 (m, 6H, H-8, H-16, H-17), 7.45 (d, ^{3}J = 9 Hz, 2H, H-12), 7.54 (d, ^{3}J = 9 Hz, 2H, H-9), 7.66 (d, ^{3}J = 9 Hz, 2H, H-13), 8.32 (s, 1H, NH).

^{13}C NMR (100.6 MHz, DMSO-d$_6$, δ (ppm)): 13.7 (CH$_3$), 21.7 (CH$_2$-21), 32.9 (CH$_2$-20), 34.4 (CH-19), 62.0 (CH-5), 82.5 (C-10), 88.9 (C-14), 111.2 (CH-4), 119.1 (CH-8), 123.4 (CH-12), 126.6 (CH-16), 128.7 (CH-17), 131.2 (C-3), 134.5 (C-15), 136.8 (C-11), 137.3 (CH-13), 137.4 (CH-9), 141.8 (C-7), 142.0 (C-18), 166.3 (CO-2).

HRMS (MALDI-TOF/TOF) m/z: $[M + H]^+$ Calcd for $C_{26}H_{25}I_2N_2O$ 635.0056; found 635.0032.

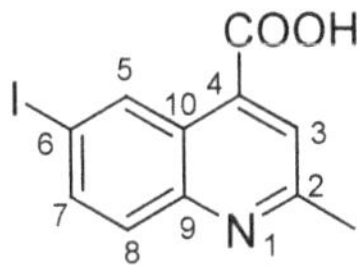

6-iodo-2-methylquinoline-4-carboxylic acid 7: crystallized from acetic acid; brown powder; 37% yield; mp 289–290 °C; IR ATR ν(cm^{-1}): 3656, 3083, 2979, 2887, 1714, 1603, 1579, 1504, 1454, 1377, 1351, 1240, 1152, 1113, 1032, 937, 867, 810, 604, 519.

^{1}H NMR (600.1 MHz, DMSO-d_6, δ (ppm)): 2.71 (s, 3H, CH_3), 7.79 (d, ^{3}J = 9 Hz, 1H, H-8), 7.89 (s, 1H, H-3), 8.05 (dd, ^{3}J = 9 Hz, ^{4}J = 2 Hz, 1H, H-7), 9.10 (d, ^{4}J = 2 Hz, 1H, H-5), 14.00 (bs, 1H, OH).

^{13}C NMR (150.9 MHz, DMSO-d_6, δ (ppm)): 24.7 (CH_3), 93.7 (C-6), 123.8 (CH-3), 124.5 (C-10), 130.8 (CH-8), 133.9 (CH-5), 134.4 (C-4), 137.9 (CH-7), 147.0 (C-9), 159.6 (C-2), 167.1 (COOH).

HRMS (MALDI-TOF/TOF) m/z: $[M + H]^+$ Calcd for $C_{11}H_9INO_2$ 313.9678; found 313.9664.

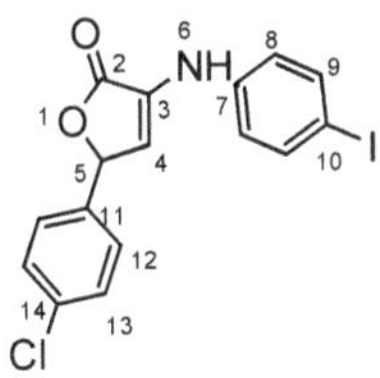

5-(4-chlorophenyl)-3-((4-iodophenyl)amino)furan-2(5H)-one 8c: crystallized from acetic acid; yellow powder; 27% yield; mp 161–162 °C; IR ATR ν(cm^{-1}): 3460, 3304, 3020, 2968, 1741, 1575, 1535, 1444, 1371, 1222, 1092, 912, 802, 528.

^{1}H NMR (400.1 MHz, DMSO-d_6, δ (ppm)): 6.23 (d, ^{3}J = 2 Hz, 1H, H-5), 6.82 (d, ^{3}J = 2 Hz, 1H, H-4), 7.15 (d, ^{3}J = 9 Hz, 2H, H-8), 7.42 (d, ^{3}J = 9 Hz, 2H, H-12), 7.50 (d, ^{3}J = 9 Hz, 2H, H-13), 7.57 (d, ^{3}J = 9 Hz, 2H, H-9), 8.59 (s, 1H,NH).

HRMS (MALDI-TOF/TOF) m/z: $[M + H]^+$ Calcd for $C_{16}H_{12}ClINO_2$ 411.9601, found 411.9584.

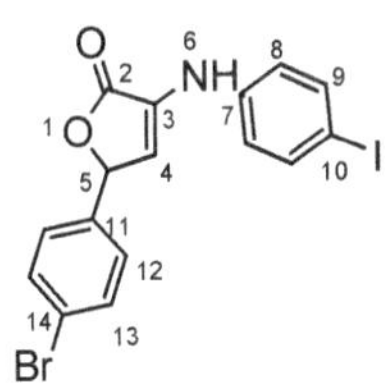

5-(4-bromophenyl)-3-((4-iodophenyl)amino)furan-2(5H)-one 8d: crystallized from acetic acid; brown powder; 31% yield; mp 182–184 °C; IR ATR ν(cm^{-1}): 3463, 3071, 3031, 2942, 2360, 2245, 1740, 1722, 1588, 1366, 1226, 1218, 1205, 1173, 899, 823, 787, 674, 640, 569, 544, 519.

^{1}H NMR (400.1 MHz, DMSO-d_6, δ (ppm)): 6.21 (d, ^{3}J = 2 Hz, 1H, H-5), 6.82 (d, ^{3}J = 2 Hz, 1H, H-4), 7.14 (d, ^{3}J = 9 Hz, 2H, H-8), 7.35 (d, ^{3}J = 9 Hz, 2H, H-12), 7.57 (d, ^{3}J = 9 Hz, 2H, H-13), 7.63 (d, ^{3}J = 9 Hz, 2H, H-9), 8.60 (s, 1H, NH).

HRMS (MALDI-TOF/TOF) m/z: $[M + H]^+$ Calcd for $C_{16}H_{12}BrINO_2$ 455.9096, found 455.9078.

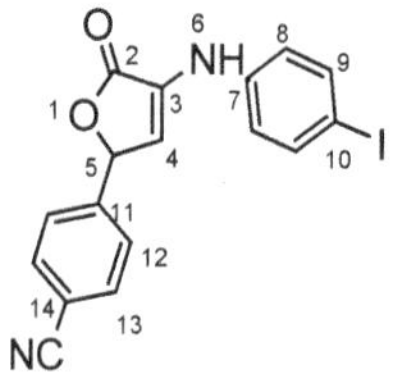

4-(4-((4-iodophenyl)amino)-5-oxo-2,5-dihydrofuran-2-yl)benzonitrile 8f: crystallized from acetic acid; brown powder; 29% yield; mp 184–185 °C; IR ATR ν(cm^{-1}): 3463, 3071, 3031, 2942, 2360, 2245, 1740, 1722, 1588, 1366, 1226, 1218, 1205, 1173, 899, 823, 787, 674, 640, 569, 544, 519.

^{1}H NMR (400.1 MHz, DMSO-d$_6$, δ (ppm)): 6.32 (d, ^{3}J = 2 Hz, 1H, H-5), 6.86 (d, ^{3}J = 2 Hz, 1H, H-4), 7.14 (d, ^{3}J = 9 Hz, 2H, H-8), 7.57 (d, ^{3}J = 9 Hz, 2H, H-9), 7.60 (d, ^{3}J = 9 Hz, 2H, H-12), 7.90 (d, ^{3}J = 9 Hz, 2H, H-13), 8.63 (s, 1H, NH).

HRMS (MALDI-TOF/TOF) *m/z*: [M + H]$^+$ Calcd for $C_{17}H_{12}IN_2O_2$ 402.9943; found 402.9942.

3.2. X-ray Crystallography

Single-crystal X-ray diffraction data for **4d** and **4e** were collected on an Oxford-Diffraction XCALIBUR Eos CCD diffractometer with graphite monochromated Mo Kα radiation, while **4b** and **4c** were collected on an XtaLAB Synergy, Dualflex, HyPix diffractometer using Cu Kα radiation. The unit cell determination and data integration were carried out using the CrysAlisPro package from Oxford Diffraction [29]. The multi-scan correction for absorption was applied. The structures were solved with the program SHELXT using the intrinsic phasing method and refined by the full-matrix least-squares method on F^2 with SHELXL [30,31]. Olex2 was used as an interface to the SHELX programs [32]. Non-hydrogen atoms were refined anisotropically. Hydrogen atoms were added in idealized positions and refined using a riding model. Selected crystallographic data and structure refinement details are provided in Table S1 and the corresponding CIF files. The supplementary crystallographic data can be obtained free of charge via www.ccdc.cam.ac.uk (or from the Cambridge Crystallographic Data Centre, 12 Union Road, Cambridge CB2 1EZ, UK; fax: (+44) 1223–336-033; or deposit@ccdc.ca.ac.uk).

3.3. Powder X-ray

X-ray diffraction analysis was performed in a Rigaku Miniflex 600 diffractometer using CuKα emission in the angular range 2–50 °(2θ) with a scanning step of 0.01° and a recording rate of 2 °/min.

3.3.1. Antimicrobial Activity

Microbial Strains

The antimicrobial activity was performed on reference microbial strains belonging to Gram-positive bacteria (*Staphylococcus epidermidis* ATCC 12228), Gram-negative bacteria (*Klebsiella pneumoniae* ATCC 13368) and yeast (*Candida parapsilosis* ATCC 22019).

3.3.2. Quantitative Evaluation of Antimicrobial Activity

In a 96-well plate, quantitative analysis was carried out using the serial binary microdilution technique in liquid media (Tryptone Soy Broth for bacteria and Sabouraud for yeast). Stock solutions of 10 mg/Ml in DMSO were prepared, with concentrations ranging from 5 to 0.016 mg/Ml. Simultaneously, under the same working circumstances, repeated dilutions with DMSO were performed to obtain the negative control. Each well received 10 μL of microbial solution adjusted to 1.5 × 10^8 CFU/Ml from cultures cultivated for 18 to 24 h. MIC values were calculated both macroscopically, as the final concentration at which no microbial growth was seen, and spectrophotometrically. The absorbance of the microbial cultures was measured at 620 nm using a Molecular Devices FlexStation 3 UV-Vis spectrophotometer (San Jose, CA, USA). For each concentration, a distinct blank was created for each sample. The data were analyzed using the Prism GraphPad 9.0 software's log (inhibitory) vs. response analysis function—Variable Slope (four parameters)—in order to generate the IC50 (the sample concentration that inhibits 50% microbiological viability from a microbial inoculum). To measure the minimal microbicidal concentrations (MMC), 5 uL of each well was spotted on a solid medium. The plates were incubated at 37 °C for 20–24 h. The MMC was defined as the last concentration at which no microbial colony development was detected.

3.3.3. The Effect on Microbial Adhesion

Following the quantitative investigation of antimicrobial activity, microbial adherence was assessed using the slime technique following fixation with methanol and staining with crystal violet (0.1%). At 490 nm, the absorbance of biological material suspended in 33% acetic acid was determined.

3.3.4. Statistical Analysis

The data were presented as means ± standard deviation (SD) (as established by duplicate analysis). GraphPad Prism 10v was used for statistical analysis. For comparison of the IC50, general structure and solvent utilized, data were analyzed using standard two-way ANOVA with a two-step increase for multiple comparisons (Tukey). The significance threshold was chosen at $p < 0.05$.

4. Conclusions

This study presents the design and synthesis of a novel family of carboxy-quinolines containing an iodine atom. The compounds were produced utilizing a one-pot, two-component technique, with trifluoroacetic acid serving as an efficient catalyst. The applicable method is an advancement of the Doebner reaction, which involves beginning with aniline and withdrawing substituents to produce desired quinolines. This technology offers many benefits, including reduced reaction times, a simplified purification process, the use of cost-effective catalysts and high product yields.

A complete investigation regarding the influence of the starting aldehyde was made. The obtained compounds were characterized by physicochemical techniques (^{1}H-NMR, ^{13}C-NMR, MS, T-IR, melting point and monocrystal and powder X-ray in some cases) in order to confirm the structure and the synthesis procedure.

The antibacterial activity of the new products was investigated and all tested series showed good activity against *S. epidermidis* and *C. parapsilosis*. The best antimicrobial activity was found for the compounds containing *p*-Br and Cl-phenyl substituents. The derivatives with *p*-CF_3-phenyl and benzo[d][1,3]dioxole moieties also showed good activity against *S. epidermidis* and *C. parapsilosis*. These compounds inhibited the adhesion of *S. epidermidis* and *C. parapsilosis* cells. Due to the hydrophobic nature of these compounds, they can easily penetrate the *Stratum corneum*, thus increasing their bioavailability for skin infection. In this context, we propose to continue the research to find new strategies to prevent the formation of chronic wounds due to microbial biofilm formation.

Supplementary Materials: The following supporting information can be downloaded at: https://www.mdpi.com/article/10.3390/molecules29040772/s1, Figure S1.A. The 1H-NMR spectrum corresponding to compound **4c**, recorded in DMSO-d6, at 400.1 MHz; Figure S1.B. The 13C-NMR spectrum corresponding to compound **4c**, recorded in DMSO-d6, at 100.6 MHz; Figure S1.C. Detailed region of the H,C-HSQC spectrum corresponding to compound **4c**, showing the correlation signals for protonated carbons; Figure S1.D. The H,C-HMBC spectrum corresponding to compound **4c**, showing 2 or 3 bond correlation signals between protons and carbons, used mainly to assign quaternary carbons; Figure S1.E. MALDI-MS spectra of compound **4c**; Figure S1.F. FT-IR spectrum of compound **4c**; Figure S2.A. The 1H-NMR spectrum corresponding to compound **4t**, recorded in DMSO-d6, at 400.1 MHz; Figure S2.B. The 13C-NMR spectrum corresponding to compound

4t, recorded in DMSO-d6, at 100.6 MHz; Figure S2.C. MALDI-MS spectrum of the compound **4t**; Figure S2.D. FT-IR spectrum of compound **4t**; Figure S3.A. The 1H-NMR spectrum corresponding to compound **8c**, recorded in DMSO-d6, at 400.1 MHz; Figure S3.B. MALDI-MS spectra of compound **8c**; Figure S3.C. FT-IR spectrum of compound **8c**; Figure S4.A. The 1H-NMR spectrum corresponding to compound **5n**, recorded in DMSO-d6, at 400.1 MHz; Figure S4.B. The 13C-NMR spectrum corresponding to compound **5n**, recorded in DMSO-d6, at 100.6 MHz; Figure S4.C. The H,C-HMBC spectrum corresponding to compound **5n**, recorded in DMSO-d6, showing correlation signals between vinylic protons and either quinoline's CH-3 or benzaldehyde's C-14 carbon atoms; Figure S4.D. Maldi-MS spectrum corresponding to compound **5n**; Figure S4.E. FT-IR spectrum of compound **5n**; Figure S5.A. The 1H-NMR spectrum corresponding to compound **7**, recorded in DMSO-d6, at 600.1 MHz; Figure S5.B. The 13C-NMR spectrum corresponding to compound **7**, recorded in DMSO-d6, at 150.9 MHz; Figure S5.C. Maldi-MS spectrum corresponding to compound **7**; Figure S5.D. FT-IR spectrum of compound **7**; Figure S6. View of the asymmetric unit (large atoms and bonds radii) showing its interaction with adjacent molecules for **4c** (a) and **4d** (b); Figure S7. Crystal packing viewed along b-axis for **4d**; Figure S8. View of the asymmetric unit (large atoms and bonds radii) showing its interaction with adjacent molecules for **4b** (a) and **4e** (b); Figure S9. Crystal packing viewed along b-axis for compounds **4b** (a) and **4e** (b); Figure S10. Powder XRD of compounds **4b** and **4c**; Table S1. Crystal data and details of structure refinement for **4b**, **4c**, **4d** and **4e**; Table S2. Bond distances and angles for **4b**; Table S3. Bond distances and angles for **4c**; Table S4. Bond distances and angles for **4d**; Table S5. Bond distances and angles for **4e**.

Author Contributions: Conceptualization, C.M.A.-M.; methodology, A.N. and I.C.M.; validation, A.N. and I.C.M.; investigation, C.M.A.-M., A.N., I.C.M., M.D.G., S.S., A.D., M.S. and M.P.; data curation, C.M.A.-M., A.N., M.P. and I.C.M.; writing—original draft preparation, C.M.A.-M., A.N., I.C.M., S.S. and M.P.; project administration, C.M.A.-M. All authors have read and agreed to the published version of the manuscript.

Funding: This research received no external funding.

Data Availability Statement: The data presented in this study are available in article and Supplementary Materials.

Acknowledgments: We acknowledge the support provided by the ICUB Fellowship for Young Researchers (A.C.M., Contract no. 26260/5 December 2022). We acknowledge the support provided by the European Union's Horizon Europe research and innovation program under grant agreement No. 101086667, project BioMat4CAST (BioMat4CAST—"Petru Poni" Institute of Macromolecular Chemistry Multi-Scale In Silico Laboratory for Complex and Smart Biomaterials).

Conflicts of Interest: The authors declare no conflicts of interest.

References

1. Vitaku, E.; Smith, D.T.; Njardarson, J.T. Analysis of the Structural Diversity, Substitution Patterns, and Frequency of Nitrogen Heterocycles among U.S. FDA Approved Pharmaceuticals. *J. Med. Chem.* **2014**, *57*, 10257–10274. [CrossRef]
2. Islam, R.; Hossain, M.I.; Okamoto, Y.; Nagamatsu, T.; Anraku, K.; Okawara, T. Facile Synthesis of 2-Phenylquinoline-4-Carboxamide Derivatives with Variant Structural Features. *Heterocycles* **2014**, *89*, 693–708. [CrossRef]
3. Jun, J.V.; Petersson, E.J.; Chenoweth, D.M. Rational Design and Facile Synthesis of a Highly Tunable Quinoline-Based Fluorescent Small-Molecule Scaffold for Live Cell Imaging. *J. Am. Chem. Soc.* **2018**, *140*, 9486–9493. [CrossRef] [PubMed]
4. Rouffet, M.; de Oliveira, C.A.F.; Udi, Y.; Agrawal, A.; Sagi, I.; McCammon, J.A.; Cohen, S.M. From Sensors to Silencers: Quinoline- and Benzimidazole-Sulfonamides as Inhibitors for Zinc Proteases. *J. Am. Chem. Soc.* **2010**, *132*, 8232–8233. [CrossRef] [PubMed]
5. Afzal, O.; Kumar, S.; Haider, M.R.; Ali, M.R.; Kumar, R.; Jaggi, M.; Bawa, S. A Review on Anticancer Potential of Bioactive Heterocycle Quinoline. *Eur. J. Med. Chem.* **2015**, *97*, 871–910. [CrossRef] [PubMed]
6. Narwal, S.; Kumar, S.; Verma, P.K. Synthesis and Therapeutic Potential of Quinoline Derivatives. *Res. Chem. Intermed.* **2017**, *43*, 2765–2798. [CrossRef]
7. Wang, Y.; Xin, X.; Liang, Y.; Lin, Y.; Zhang, R.; Dong, D. A Facile and Efficient One-Pot Synthesis of Substituted Quinolines from A-Arylamino Ketones Under Vilsmeier Conditions. *Eur. J. Org. Chem.* **2009**, *2009*, 4165–4169. [CrossRef]
8. Michael, J.P. Indolizidine and Quinolizidine Alkaloids. *Nat. Prod. Rep.* **2005**, *22*, 603–626. [CrossRef] [PubMed]
9. Lv, Q.; Fang, L.; Wang, P.; Lu, C.; Yan, F. A Simple One-Pot Synthesis of Quinoline-4-Carboxylic Acid Derivatives by Pfitzinger Reaction of Isatin with Ketones in Water. *Monatshefte Fur. Chem.* **2013**, *144*, 391–394. [CrossRef]
10. Al Matarneh, C.M.; Amarandi, R.M.; Craciun, A.M.; Mangalagiu, I.I.; Zbancioc, G.; Danac, R. Design, Synthesis, Molecular Modelling and Anticancer Activities of New Fused Phenanthrolines. *Molecules* **2020**, *25*, 527. [CrossRef]

11. Al Matarneh, C.M.; Ciobanu, C.I.; Apostu, M.O.; Mangalagiu, I.I.; Danac, R. Cycloaddition versus Amidation in Reactions of 2-Amino-2-Oxoethyl-Phenanthrolinium Ylides to Activated Alkynes and Alkenes. *Comptes Rendus Chim.* **2018**, *21*, 1–8. [CrossRef]
12. Al Matarneh, C.M.; Sardaru, M.C.; Apostu, M.O.; Rosca, I.; Ciobanu, C.I.; Mangalagiu, I.I.; Danac, R. Synthesis and Antibacterial Evaluation of New Pyrrolo[3′,4′:3,4]Pyrrolo[1,2-a]Quinoline and Pyrrolo[3′,4′:3,4]Pyrrolo[2,1 − a]Isoquinoline Derivatives. *Stud. Univ. Babeș-Bolyai Chem.* **2019**, *64*, 67–80. [CrossRef]
13. Al-Matarneh, C.M.; Nicolescu, A.; Marinas, I.C.; Chifiriuc, M.C.; Shova, S.; Silion, M.; Pinteala, M. Novel Antimicrobial Iodo-Dihydro-Pyrrole-2-One Compounds. *Future Med. Chem.* **2023**, *15*, 1369–1391. [CrossRef]
14. Al Matarneh, C.; Ciobanu, C.I.; Mangalagiu, V.; Zbancioc, G.; Danac, R. Microwave Assisted Synthesis of Six Member Ring Azaheterocycles and Their Antimycobacterial and Anticancer Evaluation. *Rev. Chim.* **2020**, *71*, 287–293. [CrossRef]
15. Al-Matarneh, C.; Rosca, I.; Shova, S.; Danac, R. Synthesis and Properties of New Fused Pyrrolo-1,10-Phenanthroline Type Derivatives. *J. Serbian Chem. Soc.* **2021**, *86*, 901–915. [CrossRef]
16. Betadine, Kenalog, Marbeta, Marcaine, P-Care M, P-Care MG, P-Care, P-Care X, Summer's Eve Medicated. Available online: https://go.drugbank.com/drugs/DB06812 (accessed on 18 January 2024).
17. Zarghi, A.; Ghodsi, R. Bioorganic & Medicinal Chemistry Design, Synthesis, and Biological Evaluation of Ketoprofen Analogs as Potent Cyclooxygenase-2 Inhibitors. *Bioorg. Med. Chem.* **2010**, *18*, 5855–5860. [CrossRef]
18. Sareena, C.; Tharamel, S. Microbial Pathogenesis Identification of 4-Diphenylamino 3-Iodo Coumarin as a Potent Inhibitor of DNA Gyrase B of S. Aureus. *Microb. Pathog.* **2020**, *147*, 104387. [CrossRef] [PubMed]
19. Zander, J.; Besier, S.; Faetke, S.; Saum, S.H.; Müller, V.; Wichelhaus, T.A. International Journal of Antimicrobial Agents Antimicrobial Activities of Trimethoprim/Sulfamethoxazole, 5-Iodo-2-Deoxyuridine and Rifampicin against Staphylococcus Aureus. *Int. J. Antimicrob. Agents* **2010**, *36*, 562–565. [CrossRef]
20. Tonoyan, L.; Friel, R.; Flaherty, V.O. Mutation Rate and Ef Fl Ux Response of Bacteria Exposed to a Novel Antimicrobial Iodo-Thiocyanate Complex. *J. Glob. Antimicrob. Resist.* **2020**, *22*, 13–17. [CrossRef]
21. Kaftandzieva, A.; Kostovski, M.; Mehmeti, B.; Mirchevska, G. The Most Common Bacterial Isolates from Wound Samples—A Three-Year Study. *Arch. Public Health* **2021**, *13*, 77–90. [CrossRef]
22. Breijyeh, Z.; Jubeh, B.; Karaman, R. Resistance of Gram-Negative Bacteria to Current Antibacterial Agents and Approaches to Resolve It. *Molecules* **2020**, *25*, 1340. [CrossRef]
23. Rahim, K.; Saleha, S.; Zhu, X.; Huo, L.; Basit, A.; Franco, O.L. Bacterial Contribution in Chronicity of Wounds. *Microb. Ecol.* **2017**, *73*, 710–721. [CrossRef]
24. Brackman, G.; De Meyer, L.; Nelis, H.J.; Coenye, T. Biofilm Inhibitory and Eradicating Activity of Wound Care Products against Staphylococcus Aureus and Staphylococcus Epidermidis Biofilms in an in Vitro Chronic Wound Model. *J. Appl. Microbiol.* **2013**, *114*, 1833–1842. [CrossRef]
25. Dowd, S.E.; Delton Hanson, J.; Rees, E.; Wolcott, R.D.; Zischau, A.M.; Sun, Y.; White, J.; Smith, D.M.; Kennedy, J.; Jones, C.E. Survey of Fungi and Yeast in Polymicrobial Infections in Chronic Wounds. *J. Wound Care* **2011**, *20*, 40–47. [CrossRef] [PubMed]
26. Leaper, D.; Assadian, O.; Edmiston, C.E. Approach to Chronic Wound Infections. *Br. J. Dermatol.* **2015**, *173*, 351–358. [CrossRef] [PubMed]
27. Garg, S.S.; Dubey, R.; Sharma, S.; Vyas, A.; Gupta, J. Biological Macromolecules-Based Nanoformulation in Improving Wound Healing and Bacterial Biofilm-Associated Infection: A Review. *Int. J. Biol. Macromol.* **2023**, *247*, 125636. [CrossRef] [PubMed]
28. Dofe, V.S.; Sarkate, A.P.; Kathwate, S.H.; Gill, C.H. Synthesis, Antimicrobial Activity and Anti-Biofilm Activity of Novel Tetrazole Derivatives. *Heterocycl. Commun.* **2017**, *23*, 325–330. [CrossRef]
29. Rigaku Corporation. *CrysAlisPro Single Crystal X-ray Diffraction*; Rigaku Corporation: Tokyo, Japan, 2015.
30. Sheldrick, G.M. SHELXT—Integrated Space-Group and Crystal-Structure Determination. *Acta Crystallogr. Sec. A Found Crystallogr.* **2015**, *71*, 3–8. [CrossRef] [PubMed]
31. Sheldrick, G.M. Crystal Structure Refinement with SHELXL. *Acta Crystallogr. Sec. C Struct. Chem.* **2015**, *71*, 3–8. [CrossRef]
32. Dolomanov, O.V.; Bourhis, L.J.; Gildea, R.J.; Howard, J.A.K.; Puschmann, H. OLEX2: A Complete Structure Solution, Refinement and Analysis Program. *J. Appl. Crystallogr.* **2009**, *42*, 339–341. [CrossRef]

molecules

MDPI

Article

A Convenient Synthesis of Novel Isoxazolidine and Isoxazole Isoquinolinones Fused Hybrids

Konstantinos A. Ouzounthanasis [1], Stergios R. Rizos [2] and Alexandros E. Koumbis [1,*]

[1] Laboratory of Organic Chemistry, Department of Chemistry, Aristotle University of Thessaloniki, 54124 Thessaloniki, Greece

[2] Department of Chemistry and Chemical Biology, Harvard University, 12 Oxford St., Cambridge, MA 02138, USA; stergiosrizos@fas.harvard.edu

* Correspondence: akoumbis@chem.auth.gr

Abstract: Isoxazolidine, isoxazole, and isoquinolinone rings are present in the structure of several natural products and/or pharmaceutically interesting compounds. In this work, facile and efficient pathways have been developed for the preparation of fused frameworks bearing those heterocycles. The successful approaches for both isoxazolidine/isoquinolinone and isoxazole/isoquinolinone hybrid syntheses relied initially on 1,3-dipolar cycloadditions of nitrones and nitrile oxides to indenone and 2-propargylbenzamide, respectively. The construction of the isoquinolinone lactam system followed by performing a selective Schmidt reaction for isoxazolidine derivatives (two steps overall), whereas the isoxazole lactams were reached via an Ullmann-type cyclisation (three steps overall). Key observations were made regarding the stereo- and regioselectivities of the reactions employed, and small libraries of the targeted hybrids were prepared, demonstrating the general applicability of these strategies.

Keywords: isoxazole; isoxazolidine; isoquinolinone; indanone; nitrone; nitrile oxide; 1,3-dipolar cycloaddition; Schmidt reaction; Ullmann reaction

1. Introduction

The 1,3-dipolar cycloaddition reaction (Huisgen cycloaddition) is established as a powerful tool for constructing five-membered heterocycles [1]. Among the various heterocyclic rings that are easily accessible via this strategy, isoxazolidines and isoxazoles, the cycloadducts obtained when nitrones or nitrile oxides are employed as 1,3-dipoles in reactions with alkenes and alkynes, respectively, are of great importance [2,3]. These scaffolds are present in the structure of several natural products, and in many instances, they were proved to be crucial for the observed biological activity [4] (Figure 1). A plethora of drugs and other compounds of interest to the pharmaceutical industry also contain those five-membered heterocycles [5,6], thus increasing the synthetic value of the 1,3-dipolar cycloaddition approach not only for the ease of preparing heterocyclic compounds but also for inserting high structural complexity in a straightforward manner.

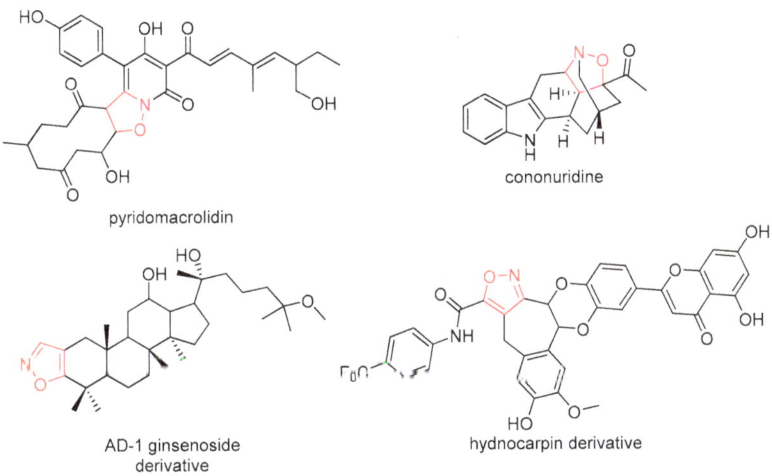

Figure 1. Representative examples of natural products and bioactive derivatives bearing the isoxazolidine or isoxazole moiety [7–10].

Another structural feature commonly appearing in biological active compounds, including either natural products or synthetic drugs, is the isoquinolinone or tetrahydroisoquinolinone moiety [11] (Figure 2). Such ring systems are often found in several pharmacophores, and positive results of Structure–Activity Relationship (SAR) studies support the design of new compounds bearing the aforementioned scaffold [12]. A convenient approach for the construction of the benzolactamic backbone present in those ring systems is the Schmidt reaction [13,14]. Apparently, the harsh acidic/thermal conditions required for the Schmidt protocol could be considered a drawback for the range of desired potential substrates. However, proper modifications may lead to a larger substrate pool being tolerated.

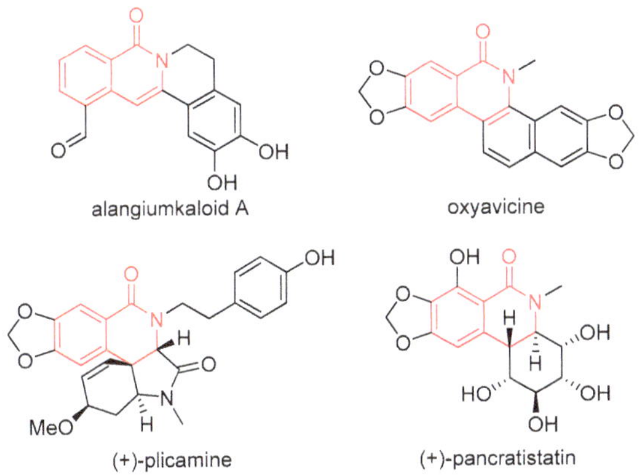

Figure 2. Isoquinolinone or tetrahydroisoquinolinone scaffold in some natural products [15–18].

The facile preparation of libraries of novel hybrid compounds combining the two interesting structural features (isoxazolidine or isoxazole heterocycle and (tetra)isoquinolinone ring system) seems to be an appealing and challenging goal. A versatile synthetic methodology could provide potent lead compounds against various biological targets (enzyme inhibition, antifungal activity, etc.). Therefore, development of a concise synthetic sequence that simultaneously maintains a high level of versatility was the task we decided to address. Moreover, observations of the chemical behaviour and reactivity of both the heterocyclic intermediates and the targeted compounds could extract valuable information for the development of future synthetic routes towards other related compounds, so as to avoid unforeseen complications and the formation of unwanted side products in newly designed syntheses.

According to our original design, the employment of two key reactions was envisioned for the construction of the main core of the targeted compounds (Figure 3). The introduction of the amide functional group was planned via a Schmidt reaction to a ketone isoxazolidine or isoxazoline cycloadduct. The latter can be prepared utilising an 1,3-dipolar cycloaddition reaction between indenone, serving as the dipolarophile partner, and a nitrone or nitrile oxide 1,3-dipole. In the case of isoxazoline/isoquinolinones, a final oxidation step is required to obtain the target isoxazole derivatives. Both 1,3-dipoles and the dipolarophile can be accessed through known synthetic methodologies from commercially available reagents (Scheme 1).

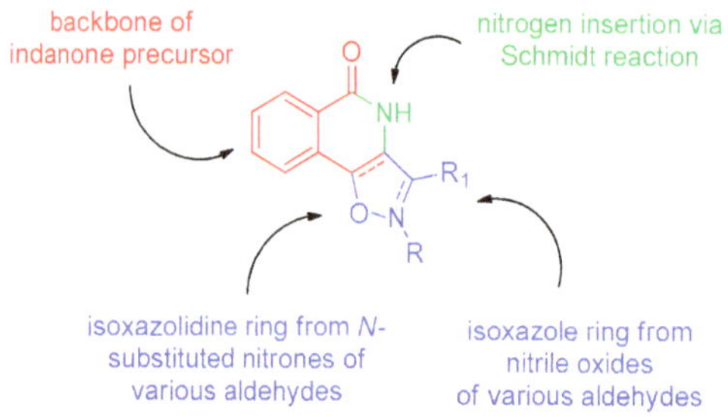

Figure 3. Design of targeted compounds.

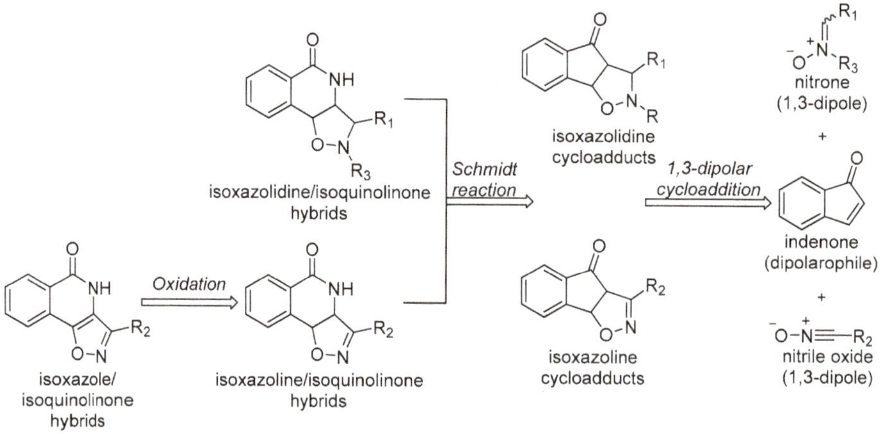

Scheme 1. Retrosynthetic plan for the targeted hybrids.

2. Results and Discussion

2.1. Synthesis of the Starting Materials

Due to its instability, indenone (**3**) was prepared from 3-bromoindanone (**2**) upon treatment with Et_3N, each time prior to its use [19]. Bromide **2** was prepared from 1-indanone (**1**) at a large scale (up to 10 g) following a known procedure [20] and was kept in the freezer for months without any sign of decomposition (Scheme 2).

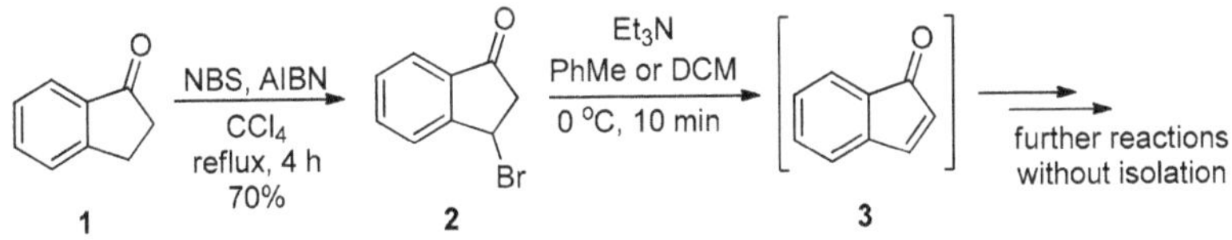

Scheme 2. Preparation of dipolarophile **3**.

For the first batch of 1,3-dipoles, we chose to proceed with *N*-benzyl nitrones of various commercially available aldehydes, consisting mostly of benzaldehydes. A modified procedure of a typical one reported in the literature [21] was followed, and, in total, 14 nitrones [22–27] were synthesised (Scheme 3A). In a similar manner, a variety of aldehydes were transformed to the corresponding oximes (Scheme 3B), following a known general protocol [28]. These oximes [29–37] served as the precursors of the required nitrile oxides.

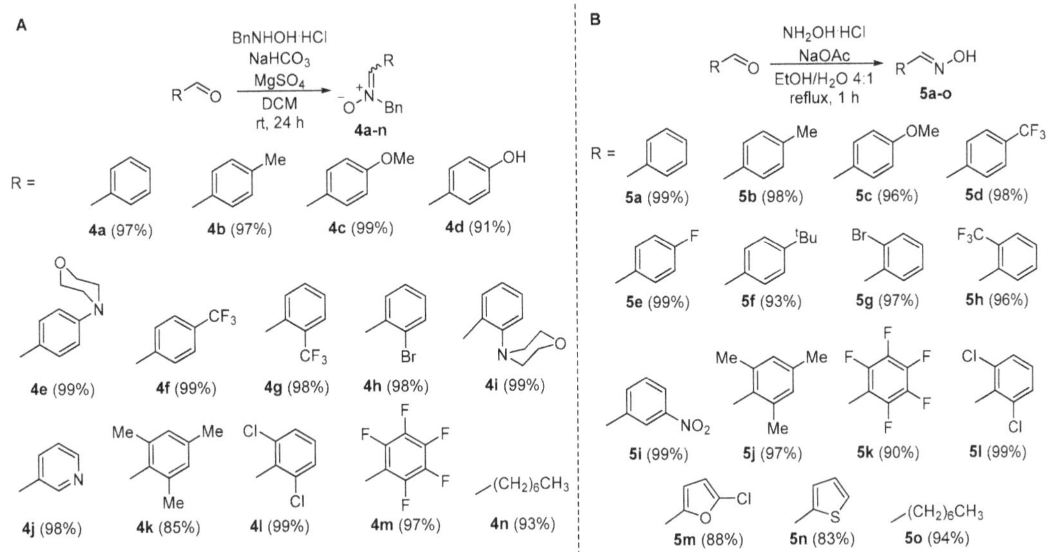

Scheme 3. (**A**). Preparation of nitrones **4**. (**B**). Preparation of nitrile oxide precursors **5**.

2.2. *Synthesis of Isoxazolidine/Isoquinolinone Hybrids*

2.2.1. Optimisation Studies for the 1,3-dipolar Cycloaddition Reaction between Indenone and Nitrones

In principle, four cycloadducts are to be expected as plausible products from the 1,3-dipolar cycloaddition reaction between a nitrone and a cyclic alkene. The *E*/*Z*-nitrone isomerisation in combination with the competing effects of secondary interactions, such as pi stacking and steric hindrance from bulky groups, can play a crucial role, thus rendering it rather hard to predict the stereoselective outcome (Scheme 4). Additionally, the issue of regioselectivity (regioisomers not shown) may complicate the situation even more. However, we presumed that in our case only one regio-orientation would be favoured due to the α,β-unsaturated system present in indenone.

To identify the preference towards the two competing stereoisomeric cycloadducts, several conditions were examined for the model reaction between *N*-benzyl nitrone of benzaldehyde (**4a**) and the in situ prepared indenone (**3**) (Table 1). Initially, simply refluxing the two partners in toluene afforded only two cycloadducts, **6a** and **7a**, with a good total yield (70%) (entry 1), whereas at lower temperatures the reaction seemed to be sluggish, regardless of the solvent used (only representative entries 1–3 are given in Table 1). The two adducts were successfully separated with column chromatography. 2D NMR spectroscopy (Figure 4) revealed that both cycloadducts emerged from the same regio orientation of the nitrone and were stereoisomers in regard to the relative geometry of the nitrone (*E*/*Z* isomerisation). Next, we settled with conducting a thorough screening in an attempt to determine conditions that may lead to the selective preparation of each of the two cycloadducts and further improve the yield. Metal triflates are widely used in such reactions with nitrones to control stereo- and regioselectivity because a lot of those compounds are commercially available [38–40]. Therefore, besides modifying the solvent, temperature, and reaction time, we also checked the influence of a reasonable number of metal triflates in substoichiometrical amounts as additives in order to investigate whether one of the cycloadducts is favoured over the other. Although we have not managed to establish a completely selective protocol, it is worth mentioning that $Zn(OTf)_2$ (entries 7 and 8) led to the best ratio (almost 2:1 in favour of *endo*-adduct **6a**) and an overall yield of 80%. The best results in terms of the overall yield of the reaction (90%) were obtained using AgOTf as the additive, but without a significant effect on selectivity regarding the formation of the two cycloadducts (**6a**/**7a** 1.25:1, entry 12). The silver salt counteranion proved to be of little importance because similar results for the stereoselectivity and the yield were observed with all silver salts employed (entries 12, 18, and 19). In an attempt to reverse the regioselectivity by taking advantage of their high affinity towards oxygen, titanium, and tin, Lewis acids were used (entries 16 and 17), but with no success. In general, heating the reaction mixture at 80 °C gave optimum results both in terms of productivity and time of completion.

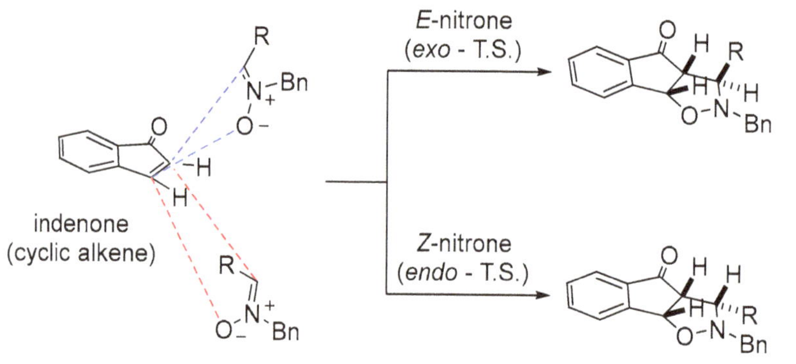

Scheme 4. The two possible stereoisomeric adducts for the reaction of indenone **3** and a nitrone.

Table 1. Optimisation studies for the 1,3-dipolar cycloaddition reaction of **3** with nitrone **4a**.

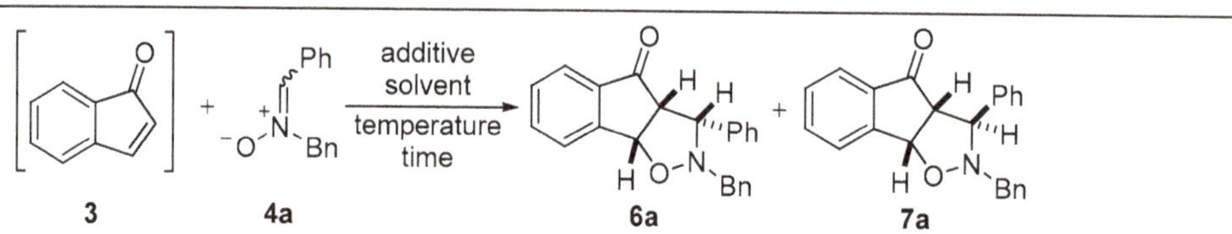

Entry	3 (equiv.)	4a [1] (equiv.)	Solvent [2]	Temp. (°C)	Additive (0.25 equiv.)	Time (h)	6a [5]	7a [5]	Yield [5] (comb.)
1	1	1	PhMe	110	-	24	40%	30%	70%
2	1	1	DCM	25	-	72	18%	7%	25%
3	1	1	none	80	-	2	32.5%	15.5%	48%

Table 1. *Cont.*

Entry	3 (equiv.)	4a [1] (equiv.)	Solvent [2]	Temp. (°C)	Additive (0.25 equiv.)	Time (h)	6a [5]	7a [5]	Yield [5] (comb.)
4	1	1	DCM	25	$Cu(OTf)_2$	24	10%	<1%	10%
5	1	1	PhMe	40	$Sc(OTf)_3$	72	43%	20%	63%
6	1	1	PhMe	40	$Mg(OTf)_2$	72	39.5%	25.5%	65%
7	1.2	1	PhMe	60	$Zn(OTf)_2$	24	50%	28%	78.5%
8	1.2	1	PhMe	60	$Zn(OTf)_2$	48	54.5%	25.5%	80%
9	1.2	1	PhMe	60	$In(OTf)_3$	48	35%	20.5%	55.5%
10	1.2	1	PhMe	60	AgOTf	72	45.5%	36%	81.5%
11 [3]	1.2	1	PhMe	80	$Zn(OTf)_2$	24	50%	34%	84%
12	1.2	1	PhMe	80	AgOTf	24	50%	40%	90%
13	1.2	1	PhMe	80	$Cu(OTf)_2$	24	31%	19%	50%
14 [4]	1.2	1	PhMe	80	$Zn(OTf)_2$	3	34.5%	18.5%	53%
15 [4]	1.2	1	PhMe	100	AgOTf	1	46%	34%	80%
16 [3]	1	1	PhMe	25	$SnCl_4$	24	-	-	-
17 [3]	1	1	PhMe	25	$TiCl_4$	24	-	-	-
18	1.2	1	PhMe	80	Ag_2CO_3	24	45%	35.5%	80.5%
19	1.2	1	PhMe	80	Ag_2O	24	45%	35%	80%

[1] 1 mmole of **4a** used; [2] 7 mL/mmole of **4a**; [3] 1 equiv. of the additive was used; [4] the reaction was carried out under μW irradiation; [5] yield after purification through flash column chromatography based on the amount of nitrone used.

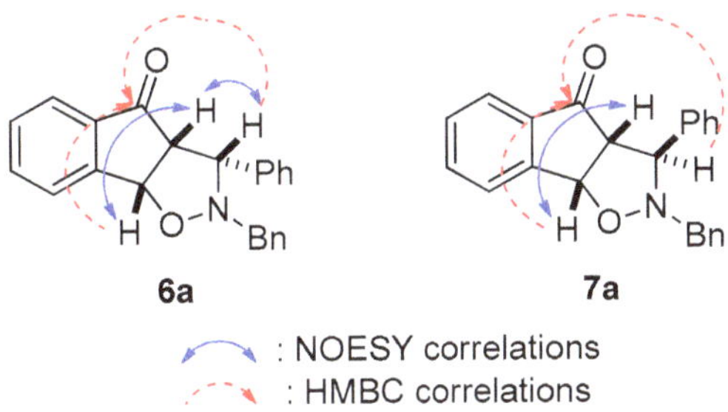

Figure 4. Determining the structure of cycloadducts **6a** and **7a** using 2D NMR spectra.

2.2.2. Synthesis of Diverse Isoxazolidines

We then applied our optimised conditions (entry 12, Table 1) by using as dipoles the *N*-benzyl nitrones **4** shown in Scheme 3A (Table 2). For almost every nitrone examined, the *endo* cycloadduct **6** was the predominant one, in accordance with the optimisation studies. However, in two cases (entries 13 and 14), the *exo* cycloadduct was the main product of the reaction. It should be noted that for the *o*-substituted phenyl nitrones **4g** and **4h** (entries 7 and 8), the formation of the corresponding regioisomers **8** was also verified. 2D NMR studies were again used to prove that the latter originate from an *exo*-T.S. pathway (in a way similar to the analysis shown in Figure 4).

2.2.3. Optimisation of the Schmidt Reaction of Isoxazolidine Adducts

The Schmidt reaction represents a convenient method for the conversion of cyclic ketones to lactams. The nitrogen migration depends strongly on the acidic medium that the reaction is carried in (H_2SO_4, PPA, HCl) [13], while the respective substrate holds a decisive role as well (electronic effects of existing substituents may alter the outcome) [41–43].

Table 2. Synthesis of isoxazolidine cycloadducts using various nitrones.

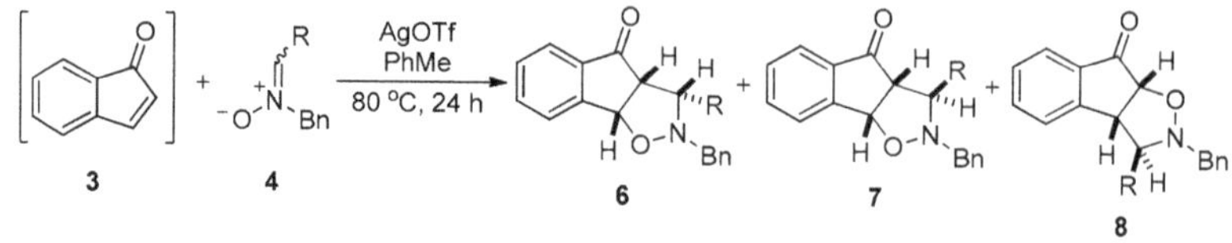

Entry	R =	*endo* Cycloadduct	*exo* Cycloadduct	Regio Cycloadduct	Yield [1] (comb.)
1	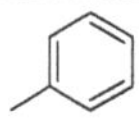	**6a** (50%)	**7a** (40%)	-	90%
2	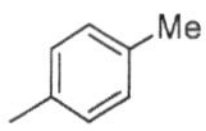	**6b** (40%)	**7b** (37%)	-	77%
3	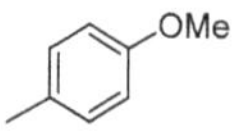	**6c** (47%)	**7c** (24%)	-	71%
4	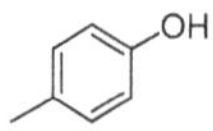	**6d** (45%)	**7d** (26%)	-	71%
5	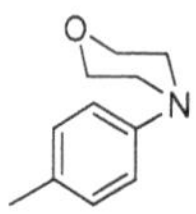	**6e** (43%)	**7e** (26%)	-	69%
6	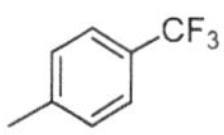	**6f** (54%)	**7f** (24%)	-	78%
7	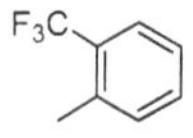	**6g** (31%)	**7g** (24%)	**8g** (16%)	71%
8	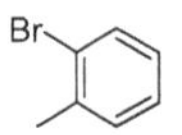	**6h** (28%)	**7h** (20%)	**8h** (23%)	71%
9	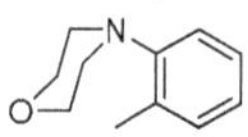	**6i** (42%)	**7i** (34%)	-	76%
10	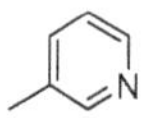	**6j** (55%)	**7j** (27%)	-	82%
11	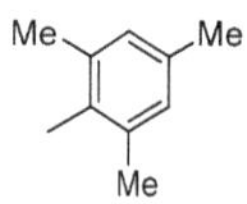	**6k** (32%)	**7k** (20%)	-	52%
12	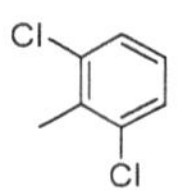	**6l** (40%)	-	-	40%
13	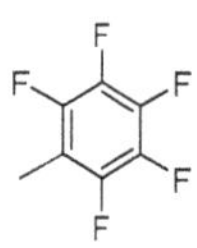	**6m** (18%)	**7m** (37%)	-	55%
14	$(CH_2)_6CH_3$	**6n** (21%)	**7n** (55%)	-	76%

[1] yield after purification through flash column chromatography based on the amount of nitrone used.

To investigate the optimal conditions for the transformation of our ketone cycloadducts (**6** and **7**) to lactams (tetrahydroisoquinolinones), a number of experiments were conducted using *endo* cycloadduct **6a** as the model substrate (Table 3). Interestingly enough, many of the more commonly used acids (HCl, H_2SO_4) failed to give any reaction, or the yield was meagre. Trifluoroacetic acid (TFA) and the extremely powerful triflic acid (TfOH) also failed to afford any product, and the starting material was recovered intact in both cases. The first encouraging result was obtained using methanesulfonic acid (MsOH), as shown in entry 9. Ultimately, proper modifications led to a protocol that furnished the desired tetrahydroisoquinolinone **9a**, with a very good yield (entry 11). 2D NMR studies (Figure 5) confirmed the nitrogen migration to the sp^3 (alkyl) carbon and retention of the relative stereochemistry of protons of the starting adduct **6a** in hybrid **9a**. Additionally, the formation of the tetrazole by-product **10a** was observed. In order to test whether amide **9a** could be formed exclusively, the Schmidt reaction was also carried out at a lower temperature (0 °C) in DCM or $CHCl_3$, but the starting isoxazolidine **6a** remained intact (experiments not shown in Table 3).

Table 3. Investigation of the Schmidt reaction conditions on **6a**.

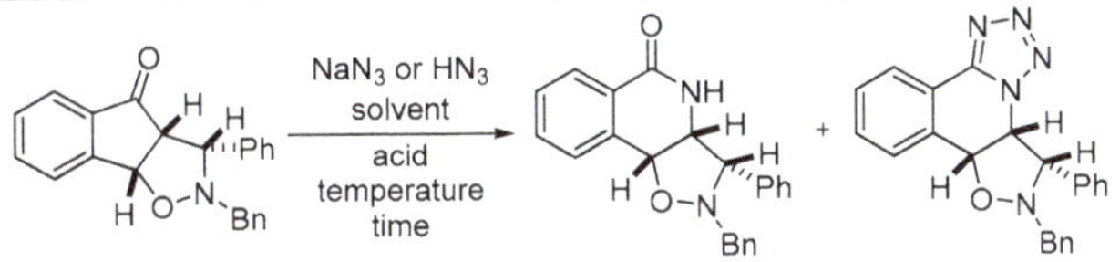

Entry	Acid	Reagent	Solvent [4]	Temp. (°C)	Time (h)	9a [5]	10a [5]	Yield [5] (comb.)
1 [1]	aq. HCl (37%)	NaN_3	PhMe	25	24	-	-	rsm
2 [1]	aq. HCl (37%)	NaN_3	PhMe	50	24	-	-	rsm
3 [2]	aq. HCl (37%)	NaN_3	-	80	24	-	-	rsm
4 [2]	TFA	NaN_3	-	25	24	-	-	rsm
5 [2]	TFA	NaN_3	-	72	24	-	-	rsm
6 [1]	TfOH	NaN_3	PhMe	25	24	-	-	rsm
7 [2]	TfOH	NaN_3	-	25	24	-	-	rsm
8 [1]	H_2SO_4	NaN_3	PhMe	80	24	15%	-	15%
9 [1]	MsOH	NaN_3	PhMe	25	24	<10%	-	~10% + rsm
10 [3]	MsOH	HN_3	PhMe	25	48	76%	13%	89%
11 [3]	MsOH	HN_3	DCM	25	48	78%	14%	92%

[1] 0.1 mmoles of **6a**, 3 equiv. NaN_3, 0.1 equiv. of the acid; [2] 0.1 mmoles of **6a**, 3 equiv. NaN_3, acid is used as the solvent; [3] 0.1 mmoles of **6a**, 2.5M HN_3 in solvent, ratio of solvent/MsOH 2:1; [4] 20 mL/mmole of **6a**; [5] yield after purification through flash column chromatography; rsm = recovered starting material.

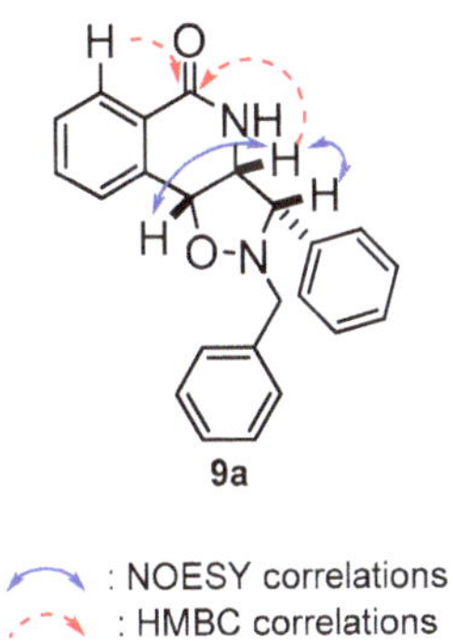

Figure 5. Determining the structure of lactam **9a** using 2D NMR spectra.

It is known [44,45] that this type of tetrazole by-product may be derived via an alternative pathway of the Schmidt reaction, where the initially formed nitrilium ion is quenched by hydrazoic acid when the latter is present at high concentrations. To exclude the possibility of tetrazole ring formation via a post-amidation reaction, we left pure lactam **9a** to react under our Schmidt protocol conditions, but the latter remained intact. Applying other methods that transform amides to tetrazoles at higher temperatures [46,47] led to the consumption of starting lactam and the formation of parent isoquinolinone **11** and the original nitrone **4a** via an unexpected *retro* 1,3-dipolar cycloaddition reaction (Scheme 5). As it was further concluded, the same *retro* cycloaddition occurs when lactam **9a** is simply heated above 80 °C in an appropriate solvent.

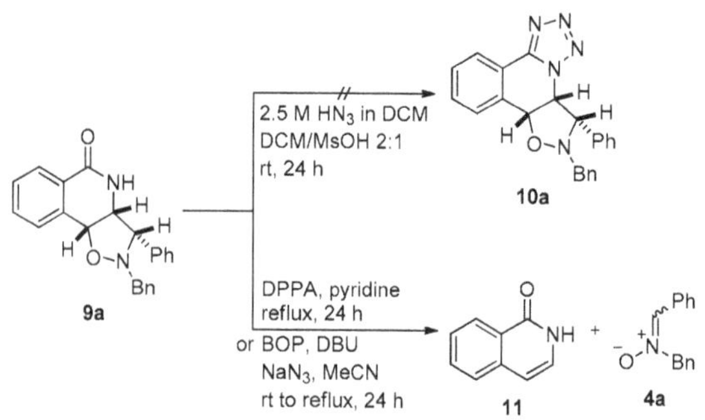

Scheme 5. Attempting tetrazole formation from lactam **9a**.

2.2.4. Synthesis of the Targeted Isoxazolidine/Tetrahydroisoquinolinone Hybrids

Having gathered critical insight, we proceeded to prepare a number of isozaxolidine hybrids. The optimised protocol for the Schmidt reaction was applied to both *endo* and *exo* cycloadducts, and the results are summarised in Tables 4 and 5, respectively. Regardless of the stereoisomer used, this protocol furnished (in most cases, uneventfully) the desired tetrahydroisoquinolinones (**9** and **12**), although some rather distinct differences between the two can be underlined. *Endo* cycloadducts **6** gave the lactams in moderate to very good yields, whereas the *exo* ones (**7**) furnished the analogous lactams in excellent yields in almost every case. The reaction time was also notably shorter for *exo* cycloadducts compared to the *endo* ones. Furthermore, the tetrazole by-product of *exo* analogs was formed in almost every instance. Presumably, the difference in the conformation of the five-membered heterocyclic ring holds the answer to this overall quite different behaviour, but a complete explanation cannot be deduced from the existing data.

Table 4. Synthesis of tetrahydroisoquinolinones **9**.

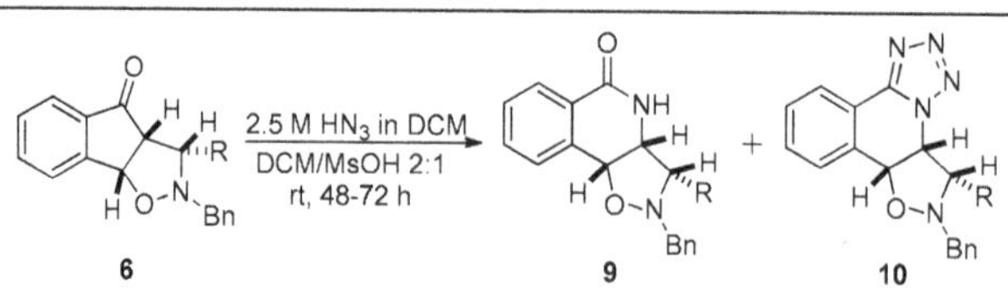

Entry	R =	Lactam [1]	Tetrazole [1]	Combined Yield [1]
1	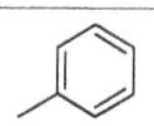	**9a** (78%)	**10a** (14%)	89%
2	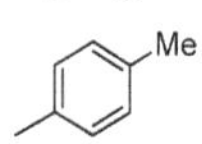	**9b** (59%)	**10b** (19%)	78%

Table 4. *Cont.*

Entry	R =	Lactam [1]	Tetrazole [1]	Combined Yield [1]
3	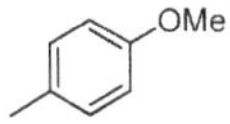	**9c** (62%)	-	62%
4	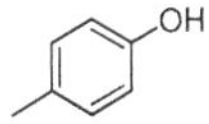	**9d** (54%)	-	54%
5	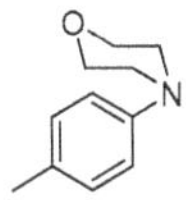	**9e** (81%)	-	81%
6	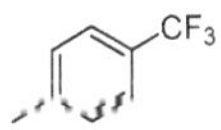	**9f** (70%)	-	70%
7	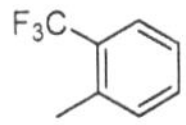	**9g** (65%)	**10g** (12%)	77%
8	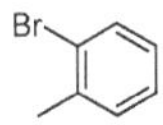	**9h** (65%)	**10h** (17%)	82%
9	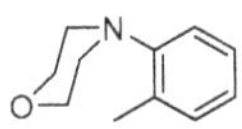	**9i** (51%)	-	51%
10	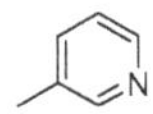	**9j** (53%)	-	53%
11	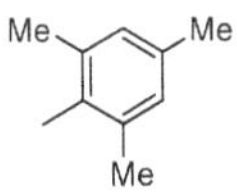	**9k** (90%)	-	90%
12	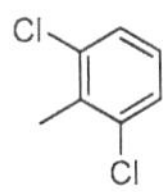	**9l** (72%)	**10l** (13%)	85%
13	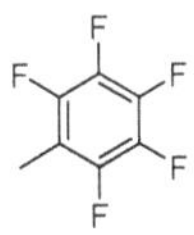	**9m** (65%)	-	65%
14	$(CH_2)_6CH_3$	**9n** (69%)	**10n** (11%)	80%

[1] yield after purification through flash column chromatography.

Table 5. Synthesis of tetrahydroisoquinolinones **12**.

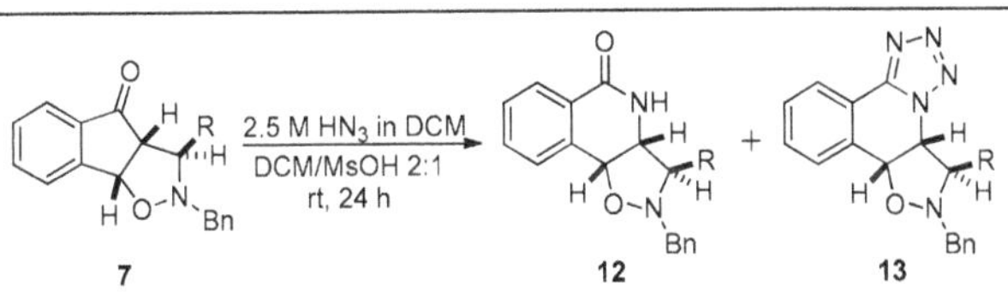

Entry	R =	Lactam [1]	Tetrazole [1]	Combined Yield [1]
1	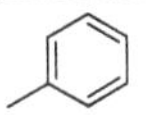	**12a** (82%)	**13a** (14%)	96%
2	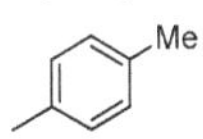	**12b** (72%)	**13b** (14%)	86%

Table 5. *Cont.*

Entry	R =	Lactam [1]	Tetrazole [1]	Combined Yield [1]
3	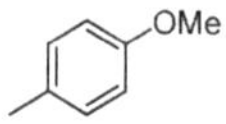	**12c** (83%)	**13c** (14%)	97%
4	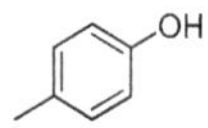	**12d** (75%)	-	75%
5	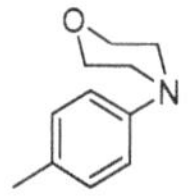	**12e** (84%)	**13e** (14%)	98%
6	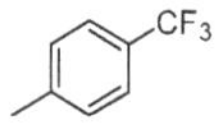	**12f** (85%)	**13f** (14%)	99%
7	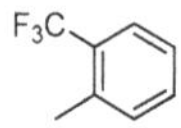	**12g** (82%)	**13g** (14%)	96%
8	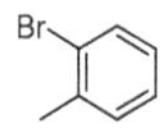	**12h** (87%)	**13h** (12%)	99%
9	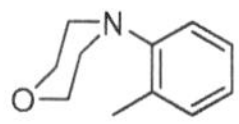	**12i** (84%)	**13i** (8%)	92%
10	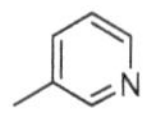	**12j** (64%)	-	64%
11	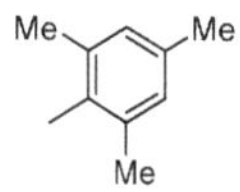	-	-	decomposed
13	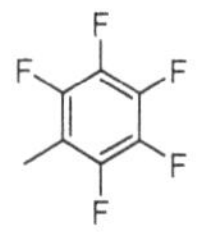	**12m** (81%)	**13m** (12%)	93%
14	$(CH_2)_6CH_3$	**12n** (90%)	**13n** (9%)	99%

[1] yield after purification through flash column chromatography.

2.3. *Synthesis of Isoxazole/Isoquinolinone Hybrids*

2.3.1. Model 1,3-dipolar Cycloaddition Reaction between Indenone and Benzonitrile Oxide and the Following Schmidt Reaction

The 1,3-dipolar cycloaddition reaction between indenone **3** and nitrile oxides was planned as a one-pot procedure. Treating oximes **6** with NCS in DMF furnished the corresponding hydroxamoyl chlorides **14** [48], which were then used without further purification. In the model reaction, one equivalent of Et_3N was added to the solution of the in situ prepared indenone **3** before the dropwise addition of phenylhydroxamoyl chloride (Scheme 6). This reaction smoothly furnished the two regioisomeric adducts (**15** and **16**), which were isolated and characterised. Once again, the major regioisomer **15** was the one favoured due to the α,β-unsaturated system present in the dipolarophile.

Isoxazoline cycloadduct **15** was then submitted to the previously optimised Schmidt reaction protocol (i.e., entry 11, Table 3). To our surprise, this reaction did not furnish the desired lactam, but instead nitrile **17** was isolated as the main product. Traces of the corresponding isoxazole/benzamide **18a** were also identified in the reaction mixture (Scheme 7). Alternatively, heating of **15** in the presence of sodium azide (NaN_3) and concentrated sulfuric acid gave benzamide **18a** as the major product. Obviously, the latter derived upon hydrolysis of the initially formed nitrile **17**.

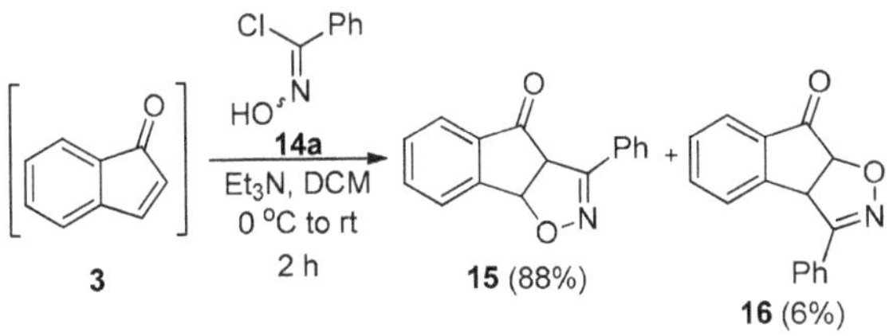

Scheme 6. 1,3-dipolar cycloaddition between indenone **3** and benzonitrile oxide **14a**.

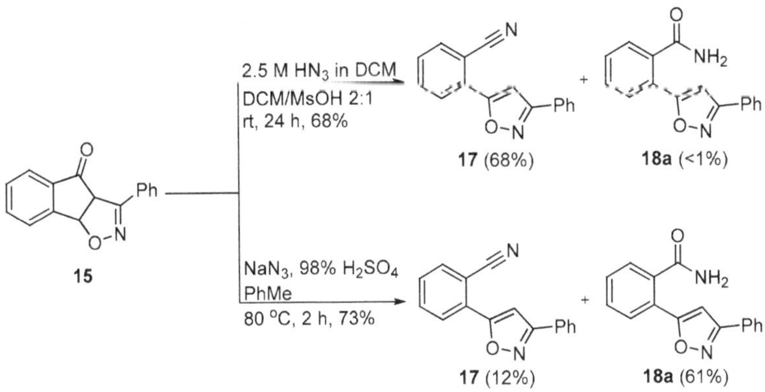

Scheme 7. Schmidt reaction on ketone isoxazoline **15**.

Both the Schmidt reaction and the Beckmann rearrangement share the same mechanistic pathway once the nitrogen migration occurs. Nitriles are common side products generated from the Beckmann rearrangement, and when those are the major products, the transformation is named "Beckmann fragmentation" instead [49]. We assume that in our case, this fragmentation is greatly favoured because the intermediate nitrilium ion **20** readily collapses to the significantly more stable aromatic isoxazole **17** (Scheme 8).

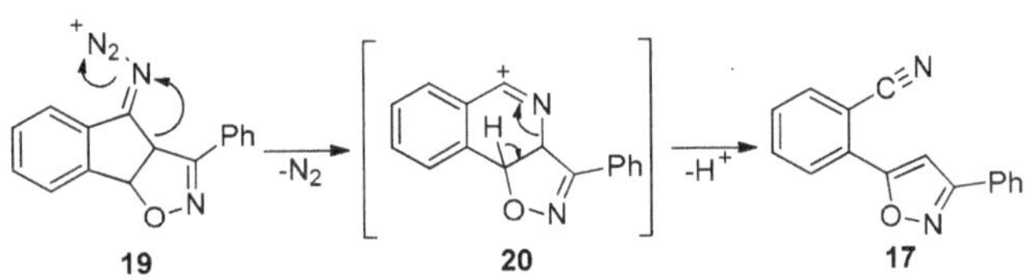

Scheme 8. Nitrile/isoxazole formation as a result of a Beckmann fragmentation.

2.3.2. Revision of the Synthetic Plan

The unforeseen outcome of the Schmidt reaction on isoxazoline adduct **15** made us reconsider our synthetic approach towards the targeted isoxazole/isoquinolinone hybrids. Interestingly, derivative **17a** attracted our attention because it incorporates the isoxazole/benzamide framework, which is ultimately present in our targeted hybrid compounds. Thus, a revised short, synthetic route involving isoxazole/benzamide derivatives was designed. According to this plan, cyclisation of the appropriate 4-iodo isoxazole intermediates via a copper-catalysed Ullmann-type reaction could afford the desired isoxazole/isoquinolinone hybrids. The required iodo derivatives could be prepared via an iodination reaction of the corresponding isoxazole/benzamides, which simply represent the adducts of 1,3-dipolar cycloadditions between 2-propargylbenzamide, now serving as the new dipolarophile, and various in situ prepared nitrile oxides (Scheme 9).

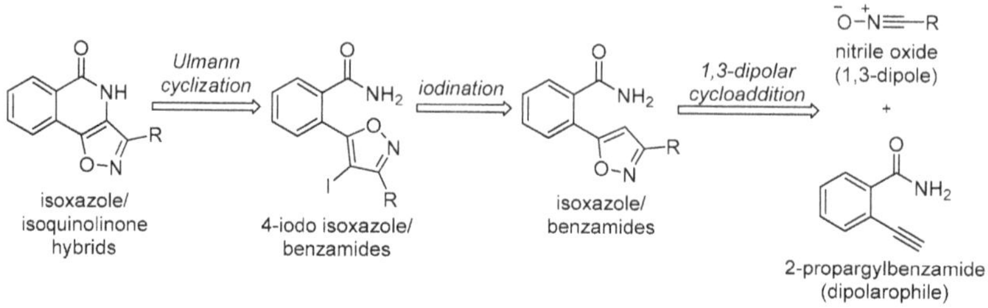

Scheme 9. Revised retrosynthetic plan towards isoxazole/isoquinolinone hybrids.

2.3.3. Isoxazole/Benzamides via an 1,3-dipolar Cycloaddition Reaction

A Sonogashira cross coupling reaction to commercially available 2-iodobenzamide and subsequent cleavage of the TMS group, following a typical procedure, afforded 2-propargylbenzamide **21** [50]. A series of in situ generated nitrile oxides (see Scheme 6) reacted with propargyl dipolarophile **21** to give isoxazole/benzamides **18** as single products and in very good yields (Table 6).

Table 6. Synthesis of isoxazole/benzamides **18**.

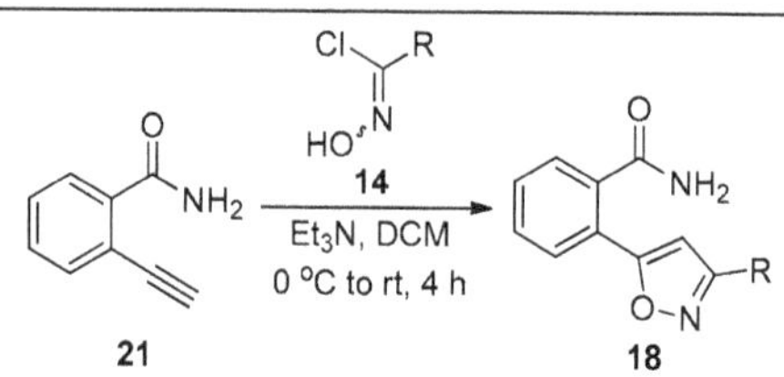

Entry	R =	Yield [1]	Entry	R =	Yield [1]
1	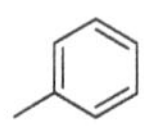	**18a** (84%)	9	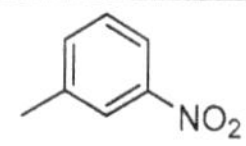	**18i** (83%)
2	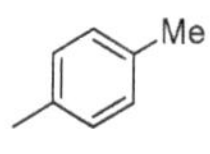	**18b** (89%)	10 [2]	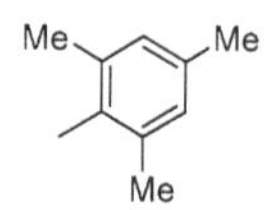	**18j** (90%)
3	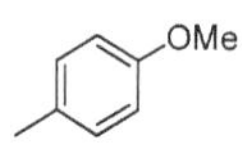	**18c** (87%)	11 [3]	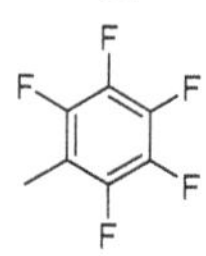	**18k** (84%)
4	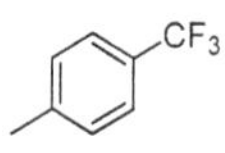	**18d** (86%)	12 [2]	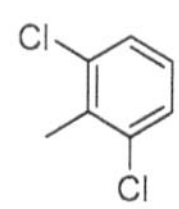	**18l** (93%)
5	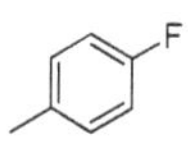	**18e** (88%)	13	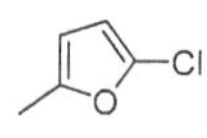	**18m** (80%)
6	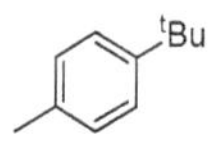	**18f** (91%)	14	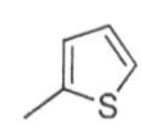	**18n** (79%)

Table 6. *Cont.*

Entry	R =	Yield [1]	Entry	R =	Yield [1]
7	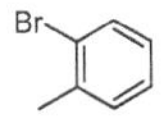	**18g** (92%)	15	$(CH_2)_6CH_3$	**18o** (92%)
8	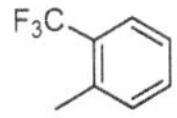	**18h** (89%)			

[1] yield after purification through flash column chromatography; [2] PhMe was used instead of DCM and the reaction mixture was refluxed; [3] the reaction mixture was heated to 40 °C.

2.3.4. Iodination of the Isoxazole Ring

The isoxazole heterocycle is prone to undergoing various transformations, including halogenation. The protocol for the iodination of the 4-position of the isoxazole ring of our derivatives **18** involved treatment with NIS (*N*-iodosuccinimide) in trifluoroacetic acid under microwave irradiation [51]. This afforded the 4-iodoisoxazoles **22** in excellent yields for most of the substrates (Table 7). Exceptions were the analogs bearing the mesityl, furanyl, and thiophene groups (**18j**, **18m**, and **18n**, respectively), which failed to exclusively give the desired products. Instead, complicated mixtures of mono- and polyiodinated products were obtained, regardless of the equivalents of NIS used. It is also worth mentioning that two equivalents of NIS were used for isoxazole **18c**, due to its facile concomitant iodination on the electronically rich p-methoxyphenyl ring.

Table 7. Iodination of isoxazole derivatives **18**.

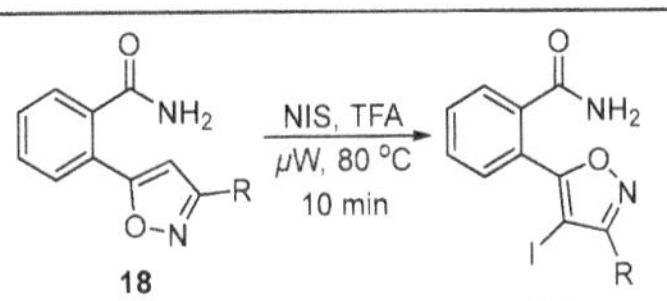

18 **22**

Entry	R =	Yield [1]	Entry	R =	Yield [1]
1	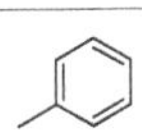	**22a** (97%)	9 [3]	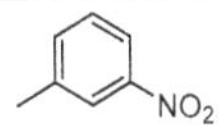	**22i** (99%)
2	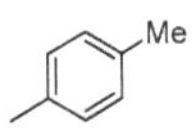	**22b** (99%)	10	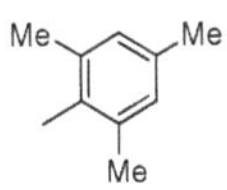	-
3 [2]	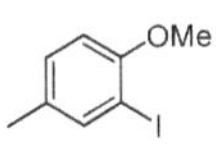	**22c** (88%)	11	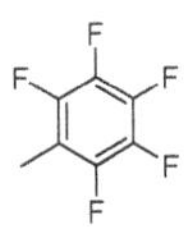	**22k** (99%)
4	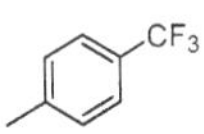	**22d** (99%)	12	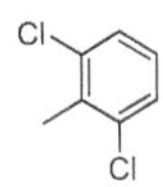	**22l** (98%)
5	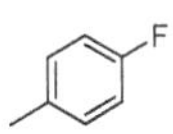	**22e** (97%)	13	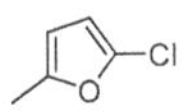	-
6	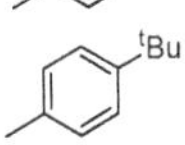	**22f** (97%)	14	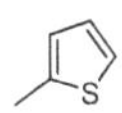	-
7	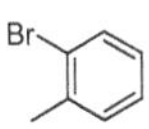	**22g** (99%)	15	$(CH_2)_6CH_3$	**22o** (98%)

Table 7. *Cont.*

Entry	R =	Yield [1]	Entry	R =	Yield [1]
8	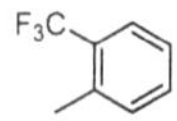	**22h** (99%)			

[1] yield after purification through flash column chromatography; [2] 2 equiv. of NIS were used; [3] yield after purification through trituration with Et_2O/n-hexane.

2.3.5. Optimisation Studies for the Ullmann Reaction

The copper-catalysed C–N bond formation (Ullmann reaction) remains an appealing alternative for the Buchwald–Hartwig reaction, because the latter depends on more expensive and less stable palladium catalysts. A thorough screening to determine the best conditions (copper catalyst, base, solvent, and temperature) was conducted on iodo-isoxazole **22a** (Table 8). Copper(I) thiophene-2-carboxylate (CuTC), being a useful reagent in such reactions, gave inarguably superior results in comparison to the more commonly used copper(I) iodide (CuI) [52]. Typical inorganic bases and solvents were tested, and the best results were obtained upon using K_2CO_3 and DMF while maintaining the temperature at 80 °C (entry 10).

2.3.6. Building a Library of Isoxazole/Isoquinolinone Hybrids via the Ullmann-Type Cyclisation

The optimum conditions for Ullmann cyclisation were, thereafter, applied to the rest of the iodo-isoxazoles (Table 9). Analogs bearing a p-substituted aryl group afforded the desired isoquinolinones in very good yields (67–93%). Isoquinolinones where the aryl group is ortho-substituted were isolated in moderate yields (58–61%), whereas analogs with polysubstituted aryl groups furnished the targeted products in generally low yields (29–46%). Nitro analog **22i** failed to give the Ullmann cyclisation because it gradually decomposed during the reaction. Moreover, isoquinolinone derivative **23c** was subjected to a reductive deiodination employing 10% Pd/C to furnish isoquinolinone **24**.

Table 8. Ullmann cyclisation optimisation studies.

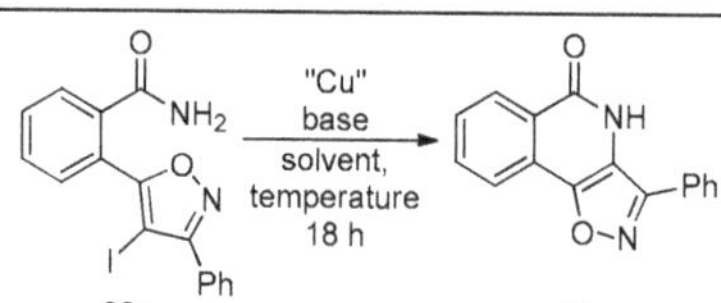

22a **23a**

Entry [1]	Copper Source (equiv.)	Base (2 equiv.)	Solvent [2]	Temp. (°C)	Yield [5]
1	CuI (0.15)	^{t}BuONa	^{t}BuOH	100	16%
2	CuI (0.15)	^{t}BuONa	1,4-dioxane	100	20%
3 [3]	CuI (0.15)	Cs_2CO_3	DMF	100	15%
4	CuI (1)	Cs_2CO_3	DMF	100	56%
5	CuTC (0.15)	K_3PO_4	DMF	100	50%
6	CuTC (0.25)	K_3PO_4	DMF	80	61%
7	CuTC (0.15)	^{t}BuONa	DMF	100	55%
8	CuTC (0.15)	Cs_2CO_3	DMF	100	56%
9	CuTC (0.25)	Cs_2CO_3	DMF	80	73%
10	CuTC (0.25)	K_2CO_3	DMF	80	78%
11	CuTC (0.25)	K_2CO_3	DMF	120	-
12	CuTC (0.25)	K_2CO_3	DMSO	80	31%
13 [4]	CuTC (0.25)	K_2CO_3	DMF	80	48%

[1] 0.1 mmoles of iodo-benzamide **22a** was used; [2] 10 mL/mmole of **22a**; [3] 1 equiv. of DMEDA was used as additive; [4] 1 equiv. of base was used; [5] yield after purification through trituration with Et_2O/n-hexane.

Table 9. Synthesis of isoquinolinones **23**.

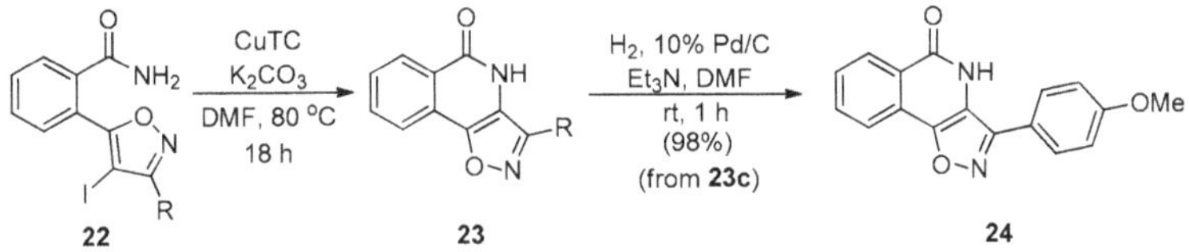

22 **23** **24**

Entry	R =	Yield [1]	Entry	R =	Yield [1]
1	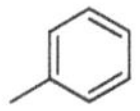	**23a** (78%)	7	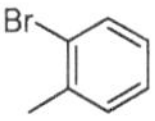	**23g** (58%)
2	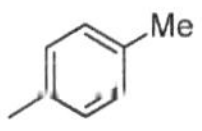	**23b** (67%)	8	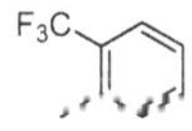	**23h** (61%)
3	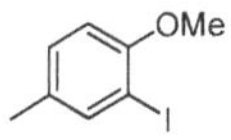	**23c** (93%)	9	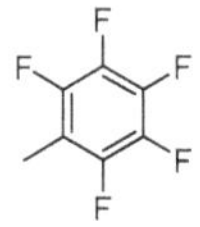	**23k** (29%)
4	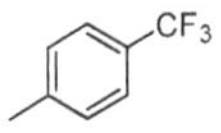	**23d** (81%)	10	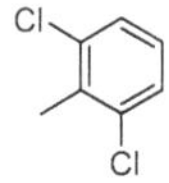	**23l** (46%)
5	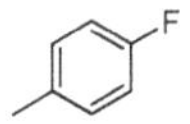	**23e** (89%)	11	$-(CH_2)_6CH_3$	**23o** (80%)
6	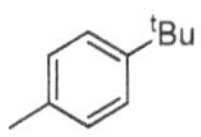	**23f** (78%)			

[1] yield after purification through recrystallisation (dissolved in 10% TFA in DCM and crystallised upon adding a few drops of MeOH).

3. Experimental Section

3.1. General Information

All anhydrous reactions were performed under an argon atmosphere using oven-dried (120 °C) or flame-dried glassware (under vacuum) with dry solvents under anhydrous conditions. THF, 1,4-dioxane, and toluene were distilled over sodium/benzophenone under an argon atmosphere into a dry Schlenk Kjeldahl storage flask containing activated molecular sieves (4 Å), and they were allowed to stand for at least for 24 h. DCM was distilled over calcium hydride (CaH_2) before use. Carbon tetrachloride was distilled over phosphorus pentoxide before use. All reactions requiring high temperatures were conducted using silicon oil baths as the heating medium. Flash column chromatography was performed by employing silica gel 60 (40–63 µm, Merck). Reactions were monitored through TLC using 0.25 mm silica gel 60 F_{254} plates purchased from Merck. TLC plates were visualised through exposure to ultraviolet light (UV) and/or exposure to an acidic solution of *p*-anisaldehyde or a solution of ninhydrin stain, followed by heating with a heat gun (400 °C). All commercially available reagents and solvents were purchased from Fluorochem, Sigma-Aldrich & Merck, Fischer Scientific, and TCI Chemicals and used as such. Molecular sieves (3 Å and 4 Å) were dried under a high vacuum by being heated with a propane torch in a round-bottom flask for 1–2 min, and the procedure was repeated 2–3 times. Celite®® 545 was purchased from Fluorochem. ^1H, ^{13}C, ^{19}F, and 2D NMR spectra were recorded with an Agilent-500/54 spectrometer. Unless otherwise stated, all NMR spectra were recorded at 25 °C. Proton chemical shifts are reported in parts per million (δ scale) and are calibrated relative to a residual nondeuterated solvent as an internal reference ($CDCl_3$: δ 7.26, DMSO-d_6: δ 2.5 ppm). Carbon chemical shifts are reported in parts per million (δ scale) and are referenced from the central peak of the carbon resonance of the solvent ($CDCl_3$: 77.00, DMSO-d_6: 40.00 ppm). Infrared (IR) data were recorded in a scan range from 400 to 4000 cm^{-1} on a Thermo Scientific Nicolet 6700 FT-IR spectrometer equipped with a diamond attenuated total reflection (ATR) stage. HRMS data were acquired using an Agilent 6540 HRMS-QTOF model equipped with a Dual AJS ESI-MS system or with a Q-TOF (Time of Flight Mass Spectrometry) Maxis Impact (Bruker Daltonics, Bremen, Germany) with ESI source and U-HPLC Thermo Dionex UltiMate 3000 RSLC (ThermoFisher Scientific, Dreieich, Germany) pump and autosampler. Melting points were determined on a A.KRÜSS Optronic Melting Point Meters KSP1N model apparatus. Reactions under microwave irradiation were carried out using a Biotage Initiator+ microwave synthesiser.

3.2. *Synthesis of 3-Bromo-2,3-dihydro-1H-inden-1-one* (**2**)

Compound **2** was prepared following a modified procedure of that reported in the literature [20]. 1-Indanone (10.04 g, 76.0 mmoles, 1.0 equiv.) was dissolved in anhydrous CCl_4 (150 mL), and then *N*-bromosuccinimide (13.52 g, 76.0 mmoles, 1.0 equiv.) and AIBN (0.125 g, 0.76 mmoles, 0.01 equiv.) were sequentially added at room temperature. The resulting suspension was refluxed for 4 h in an oil bath protected from light, allowed to cool to room temperature, and then filtered through a short pad of Celite. After evaporation of the solvent under reduced pressure, the crude mixture was purified through flash column chromatography (silica gel, *n*-hexane/Et_2O 35:1 *v*/*v*) to give **2** (11.23 g, 70% yield) as an orange oil, which solidified in the freezer. R_f = 0.43 (*n*-hexane/EtOAc 4:1 *v*/*v*). All spectroscopic data were in accordance with those reported in the literature [53].

3.3. *General Procedure for the Preparation of Nitrones* **4**

To a round bottom flask containing *N*-benzylhydroxylamine hydrochloride (2.0 equiv.), anhydrous DCM (40 mL/g) was added under an argon atmosphere. To the resulting suspension of $MgSO_4$ (2.0 equiv.), the corresponding aldehyde (1.1 equiv.) and $NaHCO_3$ (1.1 equiv.) were added in that order, and the mixture was stirred for 24 h at room temperature. The reaction mixture was then filtered through a short pad of Celite, and the solvent was evaporated under reduced pressure. The crude mixture was purified through flash column chromatography (silica gel, *n*-hexane/EtOAc 5:1 *v*/*v* to 1:1 *v*/*v*) to give the desired nitrones **4**.

3.4. *General Procedure for the Preparation of Oximes* **5**

Oximes **5** were prepared following a procedure reported previously in the literature [28].

3.5. *General Procedure for the Preparation of Isoxazolidine Cycloadducts* **6** *and* **7**

3-Bromo-1-indanone (**2**, 0.726 g, 3.44 mmoles, 1.2 equiv.) was dried azeotropically through evaporation with PhMe (3x10 mL) and dissolved in anhydrous PhMe (20 mL) under an argon atmosphere. The mixture was placed in an ice bath, and anhydrous Et_3N (0.48 mL, 3.44 mmoles, 1.2 equiv.) was added dropwise. After stirring for 15 min at 0 °C, AgOTf (0.185 g, 0.72 mmoles, 0.25 equiv.) and a solution of the corresponding nitrone (2.87 mmoles, 1 equiv.) in anhydrous PhMe (3 mL) were added to the reaction mixture, which was then placed in an oil bath at 80 °C. After stirring for 24 h at that temperature, the reaction mixture was cooled to room temperature, diluted with a small amount of DCM, and filtered through a short pad of Celite. The solvent was evaporated under reduced pressure, and the crude mixture was purified through flash column chromatography (silica gel, *n*-hexane/EtOAc 20:1 *v*/*v* to 12:1 *v*/*v*) to give the corresponding cycloadducts **6** and **7** (and, in some cases, regioisomers **8**).

3.6. General Procedure for the Preparation of Hybrids 9 and 12

A solution of HN_3 (2.5 M) in DCM was prepared by dissolving NaN_3 (0.650 g, 10 mmoles) in H_2O (1 mL), adding the organic solvent (4 mL), cooling the mixture to −10 °C, and slowly adding MsOH (1 mL) over a period of 20 min under vigorous stirring. The mixture was stirred for another 10 min, and the organic layer was then separated. The starting ketone (**6** or **7**, 0.15 mmoles) was dissolved to this solution (3 mL), and the mixture was stirred for 30 min at room temperature. MsOH (1.5 mL) was added dropwise over a period of 3 h. The resulting mixture was stirred for 24–72 h at room temperature, diluted with DCM (5 mL), placed in an ice bath, and slowly quenched through the addition of a saturated aqueous Na_2CO_3 solution until effervescence ceased. The organic layer was separated, and the aqueous layer was extracted with DCM (3 × 10 mL). The organic layers were combined, washed with brine (40 mL), and dried over anhydrous Na_2SO_4. After evaporation of the solvent under reduced pressure, the crude mixture was purified through flash column chromatography (silica gel, *n*-hexane/EtOAc 5:1 to 1:1 *v*/*v*) to give the desired isoxazolidine/tetrahydroisoquinolinone hybrids **9** and **12** (and tetrazole by-products **10** and **13**, respectively).

3.7. Synthesis of Isoxazoline Cycloadducts 15 and 16

3-Bromo-1-indanone (**2**, 1.00 g, 4.74 mmoles, 1 equiv.) was dissolved in DCM (50 mL), and the mixture was placed in an ice bath. Et_3N (0.79 mL, 5.69 mmoles, 1.2 equiv.) was added dropwise, and the yellow solution was stirred for 15 min at 0 °C. A DCM solution (100 mL) of phenylhydroxamoyl chloride **14a** (0.811 g, 5.21 mmoles, 1.1 equiv.) was then added via a pressure-equalising dropping funnel over a period of 1 h, while the temperature was maintained at 0 °C. After the addition, the reaction mixture was left stirring for another 1 h, and the reaction temperature was gradually allowed to reach 20 °C. The reaction mixture was quenched through the addition of aqueous semi-saturated aqueous NaCl solution (150 mL), the organic layer was separated, and the aqueous layer was extracted with DCM (3 × 150 mL). The organic layers were combined, washed with brine (500 mL), and dried over anhydrous Na_2SO_4. After evaporation of the solvent under reduced pressure, the crude mixture was purified through flash column chromatography (silica gel, *n*-hexane/EtOAc 12:1 to 10:1 *v*/*v*) to give cycloadducts **15** (1.04 g, 88% yield) and **16** (0.071 g, 6% yield).

3-Phenyl-3a,8b-dihydro-4H-indeno [2,1-d]isoxazol-4-one (15): white solid; m.p. 153–154 °C; R_f = 0.31 (*n*–hexane/EtOAc 4:1 *v*/*v*); ^1H NMR (500 MHz, $CDCl_3$): δ = 8.03–7.96 (m, 2H), 7.82 (d, *J* = 7.7 Hz, 1H), 7.79–7.73 (m, 2H), 7.55 (t, *J* = 7.4 Hz, 1H), 7.46–7.41 (m, 3H), 6.31 (d, *J* = 8.3 Hz, 1H), 4.75 (d, *J* = 8.3 Hz, 1H) ppm; ^{13}C NMR (125 MHz, $CDCl_3$): δ = 197.2, 152.6, 150.7, 136.3, 134.5, 130.6, 130.4, 128.6, 128.0, 127.9, 126.9, 124.2, 83.0, 60.6 ppm; FT-IR (neat): ν = 3057, 2941, 1723, 1604, 1590, 1334, 1269, 884, 761, 696 cm^{-1}; HRMS (ESI), *m*/*z*: $[M + Na]^+$ calcd for $C_{16}H_{11}NNaO_2^+$ 272.0682; 272.0679.

3-Phenyl-3a,8a-dihydro-8H-indeno [1,2-d]isoxazol-8-one (16): white solid; m.p. 172–173 °C; R_f = 0.18 (*n*–hexane/EtOAc 4:1 *v*/*v*); ^1H NMR (500 MHz, $CDCl_3$): δ = 7.85 (d, *J* = 7.7 Hz, 1H), 7.83–7.79 (m, 2H), 7.53 (td, *J* = 7.5, 1.3 Hz, 1H), 7.50–7.47 (m, 3H), 7.43 (t, *J* = 7.9 Hz, 2H), 5.42 (d, *J* = 8.6 Hz, 1H), 5.33 (d, *J* = 8.6 Hz, 1H) ppm; ^{13}C NMR (125 MHz, $CDCl_3$): δ = 198.7, 156.9, 150.3, 136.1, 134.4, 130.6, 129.3, 128.1, 128.0, 127.6, 126.2, 125.7, 84.7, 53.1 ppm; FT-IR (neat): ν = 3058, 2962, 1716, 1592, 1444, 1347, 1251, 881, 759, 694 cm^{-1}. HRMS (ESI), *m*/*z*: $[M + Na]^+$ calcd for $C_{16}H_{11}NNaO_2^+$ 272.0682; 272.0686.

3.8. Synthesis of Nitrile 17 and Benzamide 18a from 15

Procedure A: The solution of HN_3 (2.5 M) in DCM was prepared as described in Section 3.6. Isoxazoline adduct **15** (0.424 g, 1.7 mmoles, 1 equiv.) was dissolved in DCM (6 mL, 2.5 M HN_3 solution) and left stirring for 30 min at room temperature. MsOH (3 mL) was added dropwise over a period of 3 h. The resulting mixture was stirred for 24 h at room temperature, diluted with DCM (15 mL), placed in an ice bath, and slowly quenched through the addition of saturated aqueous Na_2CO_3 solution until effervescence ceased.

The organic layer was separated, and the aqueous layer was extracted with DCM (3 × 20 mL). The organic layers were combined, washed with brine (80 mL), and dried over anhydrous Na_2SO_4. After evaporation of the solvent under reduced pressure, the crude mixture was purified through flash column chromatography (silica gel, *n*-hexane/EtOAc 10:1 to 1:1 *v*/*v*) to give nitrile **17** (0.247 g, 68% yield) and benzamide **18a** (4 mg, 1%).

Procedure B: Cycloadduct **15** (0.112 g, 0.45 mmoles, 1 equiv.) was dissolved in PhMe (5 mL), and the clear solution was placed in an ice bath. NaN_3 (0.088 g, 1.35 mmoles, 3 equiv.) and H_2SO_4 (98%, 3 µL, 0.045 mmoles, 0.1 equiv.) were sequentially added at 0 °C, and the reaction mixture was then placed in an oil bath at 80 °C. After stirring for 2 h, the mixture was allowed to cool to toom temperature, placed in an ice bath, diluted with a small amount of EtOAc, and slowly quenched through the addition of saturated aqueous Na_2CO_3 solution until effervescence ceased. The organic layer was separated, and the aqueous layer was extracted with EtOAc (3 × 10 mL). The organic layers were combined, washed with brine (30 mL), and dried over anhydrous Na_2SO_4. After evaporation of the solvent under reduced pressure, the crude mixture was purified through flash column chromatography (silica gel, *n*-hexane/EtOAc 10:1 to 1:1 *v*/*v*) to give nitrile **17** (0.013 g, 12% yield) and benzamide **18a** (0.081 g, 61%).

2-(3-phenylisoxazol-5-yl)benzonitrile (**17**): white solid; R_f = 0.53 (*n*-hexane/EtOAc 3:1 *v*/*v*). All spectroscopic data are in accordance with those reported in the literature [54].

2-(3-phenylisoxazol-5-yl)benzamide (**18a**): white solid; m.p. 159–160 °C; R_f = 0.11 (*n*-hexane/EtOAc 1:1 *v*/*v*); ^{1}H NMR (500 MHz, DMSO-d_6): δ = 7.99 (br s, 1H), 7.89 (dd, *J* = 7.8, 1.8 Hz, 2H), 7.83–7.79 (m, 1H), 7.63–7.50 (m, 7H), 7.25 (s, 1H) ppm; ^{13}C NMR (125 MHz, DMSO-d_6): δ = 170.6, 169.8, 162.6, 137.5, 130.8, 130.7, 130.0, 129.7, 129.0, 128.9, 128.3, 127.0, 124.7, 101.1 ppm. FT-IR (neat): ν = 3332, 3158, 2804, 1663, 1623, 1464, 1401, 1137, 950, 758, 690 cm^{-1}. HRMS (ESI), *m*/*z*: $[M + Na]^+$ calcd for $C_{16}H_{12}N_2NaO_2^+$ 287.0791; found 287.0795.

3.9. *Synthesis of 2-Propargylbenzamide* (**21**)

Compound **21** was prepared following a procedure reported previously in the literature [51].

3.10. *General Procedure for the Preparation of Hydroxamoyl Chlorides* **14**

Hydroxamoyl chlorides **14** were prepared following a procedure reported previously in the literature [49].

3.11. *General Procedure for the Preparation of Isoxazole/Benzamides* **18**

2-propargyl benzamide (**21,** (0.164 g, 1.13 mmoles) was dissolved in DCM (6 mL) and placed in an ice bath. Et_3N (0.19 mL, 1.36 mmoles, 1.2 equiv.) was added, and the reaction mixture was stirred for 10 min at 0 °C. A DCM solution (4 mL) of hydroxamoyl chloride (**14,** 1.36 mmoles, 1.2 equiv.) was then added dropwise over a period of 15 min, and the mixture was left stirring at 0 °C for 1 h. The reaction mixture was gradually allowed to reach room temperature and stirred for another 4 h before it was quenched through the addition of semi-saturated aqueous NaCl solution (20 mL). The organic layer was separated, and the aqueous layer was extracted with DCM (3 × 15 mL). The organic layers were combined, washed with brine (60 mL), and dried over anhydrous Na_2SO_4. After evaporation of the solvent under reduced pressure, the crude mixture was purified through flash column chromatography (silica gel, *n*-hexane/EtOAc 5:1 to 1:2 *v*/*v*) to afford the corresponding isoxazole/benzamides **18**.

3.12. *General Procedure for the Preparation of 4-Iodo Isoxazoles* **22**

To a proper heavy-wall microwave reaction vial containing an isoxazole/benzamide **18** (0.58 mmoles), TFA (2 mL) and *N*-iodosuccinimide (0.137 g, 0.61 mmoles, 1.05 equiv.) were added at room temperature. The vial was then sealed and stirred under microwave irradiation at 80 °C for 10 min. The light purple solution was diluted with DCM (10 mL), placed in an ice bath, and neutralised with a saturated, aqueous $NaHCO_3$ solution. The organic layer was separated, and the aqueous layer was extracted with DCM (2 × 15 mL). The organic layers were combined, washed with brine (40 mL), and dried over anhydrous Na_2SO_4. After evaporation of the solvent under reduced pressure, the crude mixture was purified through flash column chromatography (silica gel, *n*-hexane/EtOAc 5:1 to 1:2 *v*/*v*) to give the corresponding 4-iodo isoxazole/benzamides **22**.

3.13. *General Procedure for the Preparation of Isoxazole/Isoquinolinones* 23

DMF (3 mL) was added to a heavy-wall sealed tube containing a 4-iodo isoxazole/benzamide **22** (0.29 mmoles) at room temperature. K_2CO_3 (0.08 g, 0.58 mmoles, 2 equiv.) and CuTC (0.014 g, 0.072 mmoles, 0.25 equiv.) were sequentially added, and the light green suspension was flushed with argon for 2–3 min. The tube was tightly sealed, and the mixture was stirred at 80 °C for 24 h. The dark greenish/brown suspension was diluted with EtOAc (5 mL) and quenched with the addition of an aqueous NH_4OH-NH_4Cl solution (10 mL). The organic layer was separated, and the aqueous layer was extracted with EtOAc (3 × 10 mL). The organic layers were combined, washed with brine (40 mL), and dried over anhydrous Na_2SO_4. After evaporation of the solvent under reduced pressure, the resulting solid residue was purified through recrystallisation (dissolved in 10% TFA in DCM and crystallised upon adding a few drops of MeOH) to afford the corresponding isoxazole/isoquinolinones **23**.

3.14. *Synthesis of Isoquinolinone* 24 *via Reductive Deiodination*

DMF (1 mL) was added to a Schlenk tube containing isoquinolinone **23c** (21 mg, 0.051 mmoles) at room temperature. Then, 10% Pd/C (3 mg) and a couple of drops of Et_3N were sequentially added, and the reaction vessel was purged with H_2 and placed under an atmosphere of H_2. After stirring for 1 h at room temperature, the resultant slurry was diluted with EtOAc (10 mL), filtered through a short a pad of Celite, and semi-saturated, aqueous NaCl solution (15 mL) was added. The organic layer was separated, and the aqueous layer was extracted with EtOAc (3 × 10 mL). The organic layers were combined, washed with brine (40 mL), and dried over anhydrous Na_2SO_4. After evaporation of the solvent under reduced pressure, the resulting solid residue was purified through recrystallisation (dissolved in 10% TFA in DCM and crystallised upon adding a few drops of MeOH) to give pure **24** (15 mg, 98% yield).

3-(4-methoxyphenyl)isoxazolo [4,5-c]isoquinolin-5(4H)-one (**24**): white solid; m.p. > 250 °C (decomposed); R_f = 0.4 (*n*-hexane/EtOAc 1:1 *v*/*v*); ^{1}H NMR (500 MHz, 4% TFA-*d* in $CDCl_3$): δ = 8.54 (d, *J* = 8.1 Hz, 1H), 8.19 (d, *J* = 7.9 Hz, 1H), 7.99 (t, *J* = 7.6 Hz, 1H), 7.83–7.77 (m, 3H), 7.11 (d, *J* = 8.3 Hz, 2H), 3.91 (s, 3H) ppm; ^{13}C NMR (125 MHz, 4% TFA-*d* in $CDCl_3$): δ = 164.4, 161.9, 152.2, 151.4, 135.0, 130.3, 129.4, 129.3, 125.6, 123.9, 121.6, 117.8, 117.3, 115.1, 55.5 ppm; FT-IR (neat): ν = 3089, 2981, 1663, 1600, 1486, 1346, 1259, 1182, 826, 769 cm^{-1}. HRMS (ESI), *m*/*z*: $[M + Na]^+$ calcd for $C_{17}H_{12}N_2NaO_3^+$ 315.0740; found 315.0746.

4. Conclusions

Herein, we presented our investigation of syntheses of novel isoxazolidine and isoxazole isoquinolinone hybrids. According to our original retrosynthesis, those fused heterocyclic compounds could be prepared via an 1,3-dipolar cycloaddition of indenone with nitrones and nitrile oxides and a subsequent Schmidt reaction. The cycloaddition with nitrones was found to be regioselective and, to some extent, depending on the reaction conditions, stereoselective, thus favouring the endo-adducts. Both endo- and exo-stereoisomers were uneventfully subjected to the Schmidt reaction to give the corresponding desired lactams (isoxazolidine/isoquinolinone hybrids). Although this scenario proved successful for isoxazolidine derivatives, a Beckmann fragmentation occurred when the Schmidt reaction protocol was applied on the indenone–benzonitrile oxide adduct. Thus, an alternative approach was adopted, which first involved an 1,3-dipolar cycloaddition reaction of an appropriate alkyne with nitrile oxides, and then an Ullmann type cyclisation of the corresponding iodinated isoxazoles, to furnish the desired isoxazole/isoquinolinone hybrids.

Differentially substituted dipoles were used to obtain small libraries of each category of hybrids. The overall syntheses represent short and relatively straightforward pathways towards these new classes of compounds, which will be evaluated in the future for their biological activity.

Supplementary Materials: The following supporting information can be downloaded at: https://www.mdpi.com/article/10.3390/molecules29010091/s1. It contains data for compounds **4–10**, **12**, **13**, **18b–18o**, **22**, and **23** and copies of ^{1}H, ^{13}C, ^{19}F, and 2D NMR spectra of all new compounds.

Author Contributions: Conceptualisation, A.E.K.; methodology, K.A.O. and S.R.R.; formal analysis, K.A.O.; data curation, K.A.O.; writing—original draft preparation, K.A.O. and A.E.K.; writing—review and editing, A.E.K.; project administration, A.E.K. All authors have read and agreed to the published version of the manuscript.

Funding: This research received no external funding.

Institutional Review Board Statement: Not applicable.

Informed Consent Statement: Not applicable.

Data Availability Statement: All data supporting the findings of this study are available within the paper and within its Supplementary Materials published online.

Acknowledgments: This research work was supported by the Hellenic Foundation for Research and Innovation (HFRI) under the 3rd Call for HFRI PhD Fellowships (Fellowship Number: 5748). K.A.O. is grateful for this fellowship. The authors express their thanks to Maroula G. Kokotou, Laboratory of Chemistry, Department of Food Science and Human Nutrition, Agricultural University of Athens, for performing the HRMS analyses of several compounds.

Conflicts of Interest: The authors declare no conflicts of interest.

References

1. Breugst, M.; Reissig, H.-U. The Huisgen Reaction: Milestones of the 1,3-Dipolar Cycloaddition. *Angew. Chem. Int. Ed.* **2020**, *59*, 12293–12307. [CrossRef] [PubMed]
2. Berthet, M.; Cheviet, T.; Dujardin, G.; Parrot, I.; Martinez, J. Isoxazolidine: A Privileged Scaffold for Organic and Medicinal Chemistry. *Chem. Rev.* **2016**, *116*, 15235–15283. [CrossRef] [PubMed]
3. Sysak, A.; Obmińska-Mrukowicz, O. Isoxazole Ring as A Useful Scaffold in A Search for New Therapeutic Agents. *Eur. J. Med. Chem.* **2017**, *137*, 292–309. [CrossRef] [PubMed]
4. Wang, X.; Hu, Q.; Tang, H.; Pan, X. Isoxazole/Isoxazoline Skeleton in the Structural Modification of Natural Products: A Review. *Pharmaceuticals* **2023**, *16*, 228. [CrossRef] [PubMed]
5. Chiacchioa, M.A.; Giofrèb, S.V.; Romeob, R.; Romeob, G.; Chiacchio, U. Isoxazolidines as Biologically Active Compounds. *Curr. Org. Synth.* **2016**, *13*, 726–749. [CrossRef]
6. Perrone, M.G.; Vitale, P.; Panella, A.; Ferorelli, S.; Contino, M.; Lavecchia, A.; Scilimati, A. Isoxazole-Based-Scaffold Inhibitors Targeting Cyclooxygenases (COXs). *ChemMedChem* **2016**, *11*, 1172–1187. [CrossRef] [PubMed]
7. Takahashi, S.; Kakinuma, N.; Uchida, K.; Hashimoto, R.; Yanagisawa, T.; Nakagawa, A. Pyridovericin and Pyridomacrolidin: Novel Metabolites from Entomopathogenic Fungi, Beauveria Bassiana. *J. Antibiot.* **1998**, *51*, 596–598. [CrossRef]
8. Nge, C.-E.; Sim, K.-S.; Lim, S.-H.; Thomas, N.F.; Low, Y.-Y.; Kam, T.-S. A Hexacyclic, Iboga-Derived Monoterpenoid Indole with a Contracted Tetrahydroazepine C-Ring and Incorporation of an Isoxazolidine Moiety, a Seco-Corynanthean, an Aspidosperma-Aspidosperma Bisindole with Anticancer Properties, and the Absolute Configuration of the Pyridopyrimidine Indole Alkaloid, Vernavosine. *J. Nat. Prod.* **2016**, *79*, 2709–2717.
9. Ma, L.; Miao, D.; Lee, J.J.; Li, T.; Chen, Y.; Su, G.; Zhao, Y. Synthesis and Biological Evaluation of Heterocyclic Ring-Fused Dammarane-Type Ginsenoside Derivatives as Potential Anti-Tumor Agents. *Bioorg. Chem.* **2021**, *116*, 105365. [CrossRef]
10. Arya, J.S.; Joseph, M.M.; Sherin, D.R.; Nair, J.B.; Manojkumar, T.K.; Maiti, K.K. Exploring Mitochondria-Mediated Intrinsic Apoptosis by New Phytochemical Entities: An Explicit Observation of Cytochrome c Dynamics on Lung and Melanoma Cancer Cells. *J. Med. Chem.* **2019**, *62*, 8311–8329. [CrossRef]
11. Rao, L.B.; Sreenivasulu, C.; Kishore, D.R.; Satyanarayana, G. Trending Strategies for the Synthesis of Quinolinones and Isoquinolinones. *Tetrahedron* **2022**, *127*, 133093. [CrossRef]
12. Humphries, P.S.; Benbow, J.W.; Bonin, P.D.; Boyer, D.; Doran, S.D.; Frisbie, R.K.; Piotrowski, D.W.; Balan, G.; Bechle, B.M.; Conn, E.L.; et al. Synthesis and SAR of 1,2,3,4-Tetrahydroisoquinolin-1-ones as Novel G-Protein-Coupled Receptor 40 (GPR40) Antagonists. *Bioorg. Med. Chem. Lett.* **2009**, *19*, 2400–2403. [CrossRef] [PubMed]
13. Crosby, I.T.; Shin, J.K.; Capuano, B. The Application of the Schmidt Reaction and Beckmann Rearrangement to the Synthesis of Bicyclic Lactams: Some Mechanistic Considerations. *Aust. J. Chem.* **2010**, *63*, 211–226. [CrossRef]

14. López, L.; Selent, J.; Ortega, R.; Masaguer, C.F.; Domínguez, E.; Areias, F.; Brea, J.; Loza, M.I.; Sanz, F.; Pastor, M. Synthesis, 3D-QSAR, and Structural Modeling of Benzolactam Derivatives with Binding Affinity for the D2 and D3 Receptors. *ChemMedChem* **2010**, *5*, 1300–1317. [CrossRef] [PubMed]
15. Pailee, P.; Prachyawarakorn, V.; Mahidol, C.; Ruchirawat, S.; Kittakoop, P. Protoberberine Alkaloids and Cancer Chemopreventive Properties of Compounds from *Alangium salviifolium*. *Eur. J. Org. Chem.* **2011**, *2011*, 3809–3814. [CrossRef]
16. Arthur, H.R.; Hui, W.H.; Ng, Y.L. An Examination of the Rutaceae of Hong Kong. Part III. The Alkaloid, Avicine, from *Zanthoxylum Avicennae*. *J. Chem. Soc.* **1959**, *803*, 4007–4009. [CrossRef]
17. Zhan, G.; Zhou, J.; Liu, R.; Liu, T.; Guo, G.; Wang, J.; Xiang, M.; Xue, Y.; Luo, Z.; Zhang, Y.; et al. Galanthamine, Plicamine, and Secoplicamine Alkaloids from Zephyranthes candida and Their Anti-acetylcholinesterase and Anti-inflammatory Activities. *J. Nat. Prod.* **2016**, *79*, 760–766. [CrossRef]
18. Manpadi, M.; Kornienko, A. Total Syntheses of Pancratistatin. A Review. *Org. Prep. Proced. Int.* **2008**, *40*, 107–161. [CrossRef]
19. Hauser, F.M.; Zhou, M.; Sun, Y. Facile Synthesis of Indenones from Indanones: A New Procedure. *Synth. Commun.* **2001**, *31*, 77–80. [CrossRef]
20. Lu, X.; Schneider, U. Aza-Morita-Baylis-Hillman Reactions Catalyzed by a Cyclopropenylidene. *Chem. Commun.* **2016**, *52*, 12980–12983. [CrossRef]
21. Chandrasekhar, B.; Ahn, S.; Ryu, J.-S. Synthesis of 4-Isoxazolines through Gold(I)-Catalyzed Cyclization of Propargylic N-Hydroxylamines. *J. Org. Chem.* **2016**, *81*, 6740–6749. [CrossRef] [PubMed]
22. He, C.-T.; Han, X.-L.; Zhang, Y.-X.; Du, Z.-T.; Si, C.-M.; Wei, B.-G. Sc(OTf)$_3$ Catalyzed [3 + 2]-Cycloaddition of Nitrones with Ynones. *Org. Biomol. Chem.* **2021**, *19*, 457–466. [CrossRef] [PubMed]
23. Bortolini, o.; Mulani, I.; De Niro, A.; Maiuolo, L.; Nardi, M.; Russo, B.; Avnet, S. Efficient Synthesis of Isoxazolidine-Substituted Bisphosphonates By 1,3-Dipolar Cycloaddition Reactions. *Tetrahedron* **2011**, *67*, 5635–5641. [CrossRef]
24. Poulsen, P.H.; Vergura, S.; Monleón, A.; Jørgensen, D.K.B.; Jørgensen, K.A. Controlling Asymmetric Remote and Cascade 1,3-Dipolar Cycloaddition Reactions by Organocatalysis. *J. Am. Chem. Soc.* **2016**, *138*, 6412–6415. [CrossRef] [PubMed]
25. Delso, I.; Terejo, T. ^{1}H–^{15}N HMBC as a Valuable Tool for the Identification and Characterization of Nitrones. *Tetrahedron Lett.* **2007**, *48*, 4101–4104. [CrossRef]
26. Chakraborty, B.; Chettri, E. Synthesis of Some Novel Class of Regioselective Spiro Isoxazolidine Derivatives via 1,3-Dipolar Cycloaddition Reaction of *N*-Benzyl-*C*-fluorosubstituted Phenyl Nitrones in Ionic Liquid. *J. Heterocycl. Chem.* **2018**, *55*, 1157–1165. [CrossRef]
27. Katahara, S.; Kobayashi, S.; Fujita, K.; Matsumoto, T.; Sato, T.; Chida, N. An Iridium-Catalyzed Reductive Approach to Nitrones from *N*-Hydroxyamides. *J. Am. Chem. Soc.* **2016**, *138*, 5246–5249. [CrossRef]
28. Mukhopadhyay, S.; Batra, S. Direct Transformation of Arylamines to Aryl Halides via Sodium Nitrite and *N*-Halosuccinimide. *Chem. Eur. J.* **2018**, *24*, 14622–14626. [CrossRef]
29. Schierle, S.; Neumann, S.; Heitel, P.; Willems, S.; Kaiser, A.; Pollinger, J.; Merk, D. Design and Structural Optimization of Dual FXR/PPARδ Activators. *J. Med. Chem.* **2020**, *63*, 8369–8379. [CrossRef]
30. Di Nunno, L.; Vitale, P.; Scilimati, A.; Simonea, L.; Capitelli, F. Stereoselective Dimerization of 3-Arylisoxazoles to Cage-Shaped Bis-b-lactams *syn* 2,6-Diaryl-3,7-diazatricyclo[4.2.0.02,5]-octan-4,8-diones Induced by Hindered Lithium Amides. *Tetrahedron* **2007**, *63*, 12388–12395. [CrossRef]
31. Tambara, K.; Dan Pantos, G. Conversion of Aldoximes into Nitriles and Amides Under Mild Conditions. *Org. Biomol. Chem.* **2013**, *11*, 2466–2472. [CrossRef] [PubMed]
32. Steiger, S.A.; Li, C.; Backos, D.S.; Reigan, P.; Natale, N.R. Dimeric Isoxazolyl-1,4-Dihydropyridines Have Enhanced Binding at the Multi-Drug Resistance Transporter. *Bioorg. Med. Chem.* **2017**, *25*, 3223–3234. [CrossRef] [PubMed]
33. Yu, J.; Lu, M. Metal-Free: A Novel and Efficient Aerobic Oxidation of Primary Amines to Oximes Using *N,N′,N″*-Trihydroxyisocyanuric Acid and Acetaldoxime as Catalysts in Water. *Synlett* **2014**, *25*, 1873–1878. [CrossRef]
34. McIntosh, M.L.; Naffziger, M.R.; Ashburn, B.O.; Zakharov, L.N.; Carter, R.G. Highly Regioselective Nitrile Oxide Dipolar Cycloadditions with *ortho*-Nitrophenyl Alkynes. *Org. Biomol. Chem.* **2012**, *10*, 9204–9213. [CrossRef] [PubMed]
35. Jawalekar, A.M.; Reubsaet, E.; Rutjes, F.P.J.T.; van Delft, F.L. Synthesis of Isoxazoles by Hypervalent Iodine-Induced Cycloaddition of Nitrile Oxides to Alkynes. *Chem. Commun.* **2011**, *47*, 3198–3200. [CrossRef] [PubMed]
36. Kanemasa, S.; Matsuda, H.; Kamimurac, A.; Kakinami, T. Synthesis of Hydroximoyl Chlorides from Aldoximes and Benzyltrimethylammonium Tetrachloroiodate (BTMA ICl_4). *Tetrahedron* **2000**, *56*, 1057–1064. [CrossRef]
37. Suzuki, K.; Watanabe, T.; Murahashi, S.-I. Oxidation of Primary Amines to Oximes with Molecular Oxygen using 1,1-Diphenyl-2-picrylhydrazyl and WO_3/Al_2O_3 as Catalysts. *J. Org. Chem.* **2013**, *78*, 2301–2310. [CrossRef]
38. Palomo, C.; Oiarbide, M.; Arceo, E.; García, J.M.; López, R.; González, A.; Linden, A. Lewis Acid Catalyzed Asymmetric Cycloadditions of Nitrones: A′-Hydroxy Enones as Efficient Reaction Partners. *Angew. Chem. Int. Ed.* **2005**, *44*, 6187–6190. [CrossRef]
39. Cao, G.; Zhou, S.; Teng, D. Synthesis of Spiroisoxazolidinyl-Benzoisothiazolines by 1,3-Dipolar Cycloaddition of Benzoisothiazole-2,2-dioxide-3-ylidenes with Nitrones. *Tetrahedron* **2017**, *73*, 2329–2333. [CrossRef]
40. Śnieżek, M.; Stecko, S.; Panfil, I.; Furman, B.; Urbańczyk-Lipkowska, Z.; Chmielewski, M. Thermal and Sc(OTf)$_3$ Catalyzed 1,3-Dipolar Cycloaddition of Open-Chain Nitrones to α,β-Unsaturated Lactones: Combined Experimental and Computational Studies. *Tetrahedron Asymmetry* **2013**, *24*, 89–103. [CrossRef]

41. Minami, S.; Tomita, M.; Takamatsu, H.; Uyeo, S. The Schmidt Reaction with Some Tetralone and Indanone Derivatives. *Chem. Pharm. Bull.* **1965**, *13*, 1084–1091. [CrossRef] [PubMed]
42. Evans, D.; Lockhart, I.M. The Schmidt Reaction with Aromatic Ketones. *J. Chem. Soc.* **1965**, 4806–4812. [CrossRef]
43. Conley, R.T. Schmidt Reactions in Polyphosphoric Acid. I. Rearrangement of Ketones. *J. Org. Chem.* **1958**, *23*, 1330–1333. [CrossRef]
44. Škorić, D.Đ.; Klisurić, O.R.; Jakimov, D.S.; Sakač, M.N.; Csanádi, J.J. Synthesis of New Bile Acid-Fused Tetrazoles Using the Schmidt Reaction. *Beilstein J. Org. Chem.* **2021**, *17*, 2611–2620. [CrossRef] [PubMed]
45. Koldobskii, G.I.; Ostrovskii, V.A. Application of the Schmidt Reaction for the Preparation of Tetrazoles. *Chem. Heterocycl. Compd.* **1975**, *11*, 626–635. [CrossRef]
46. Ishihara, K.; Shioiri, T.; Matsugi, M. An Expeditious Approach to Tetrazoles from Amides Utilizing Phosphorazidates. *Org. Lett.* **2020**, *22*, 6244–6247. [CrossRef] [PubMed]
47. Wan, Z.-K.; Wacharasindhu, S.; Levins, G.C.; Lin, M.; Tabei, K.; Mansour, T.S. The Scope and Mechanism of Phosphonium-Mediated S_NAr Reactions in Heterocyclic Amides and Ureas. *J. Org. Chem.* **2007**, *72*, 10194–10210. [CrossRef]
48. Stotani, S.; Gatta, V.; Medda, F.; Padmanaban, M.; Karawajczyk, A.; Tammela, P.; Giordanetto, F.; Tzalis, D.; Collina, S. A Versatile Strategy for the Synthesis of 4,5-Dihydroxy-2,3-Pentanedione (DPD) and Related Compounds as Potential Modulators of Bacterial Quorum Sensing. *Molecules* **2018**, *23*, 2545. [CrossRef]
49. VerHaeghe, D.G.; Weber, G.S.; Pappalardo, P.A. Beckmann Fragmentation Versus Beckmann Rearrangement in Dehydronorcamphor Derivatives. *Tetrahedron Lett.* **1989**, *30*, 4041–4044. [CrossRef]
50. Brahmchari, D.; Verma, A.K.; Mehta, S. Regio- and Stereoselective Synthesis of Isoindolin-1-ones through BuLi-Mediated Iodoaminocyclization of 2-(1-Alkynyl)benzamides. *J. Org. Chem.* **2018**, *83*, 3339–3347. [CrossRef]
51. Morita, T.; Fuse, S.; Nakamura, H. Generation of an 4-Isoxazolyl Anion Species: Facile Access to Multifunctionalized Isoxazoles. *Angew. Chem. Int. Ed.* **2016**, *55*, 1–6. [CrossRef]
52. Quana, Z.-J.; Xiaa, H.-D.; Zhanga, Z.; Daa, Y.-X. Ligand-free CuTC-catalyzed *N*-arylation of Amides, Anilines and 4-aminoantipyrine: Synthesis of *N*-arylacrylamides, 4- amido-*N*-phenylbenzamides and 4-amino (*N*-phenyl)antipyrenes. *Appl. Organometal. Chem.* **2014**, *28*, 81–85. [CrossRef]
53. Cantillo, D.; de Frutos, O.; Rincon, J.A.; Mateos, C.; Kappe, C.O. A Scalable Procedure for Light-Induced Benzylic Brominations in Continuous Flow. *J. Org. Chem.* **2014**, *79*, 223–229. [CrossRef]
54. Wang, A.; Lv, P.; Liu, Y. 4,5-Dihydro-1,2,4-oxadiazole as a Single Nitrogen Transfer Reagent: Synthesis of Functionalized Isoxazoles Assisted by $Sc(OTf)_3$ or Au(I)/ $Sc(OTf)_3$ Synergistic Catalysis. *Org. Lett.* **2023**, *25*, 4377–4382. [CrossRef]

molecules

MDPI

Article

1,1,1,3,3,3-Hexafluoro-2-Propanol-Promoted Friedel–Crafts Reaction: Metal-Free Synthesis of C3-Difluoromethyl Carbinol-Containing Imidazo[1,2-*a*]pyridines at Room Temperature

Juanjuan Gao, Zhaowen Liu, Xiaohua Guo, Longhui Wu, Zhixi Chen * and Kai Yang *

College of Pharmacy, Gannan Medical University, Ganzhou 341000, China; gaoya0758@163.com (J.G.); liuzhaowenyifan@126.com (Z.L.); xhguo20041@163.com (X.G.); 13970997583@163.com (L.W.)
* Correspondence: czxb22@gmu.edu.cn (Z.C.); kai_yangyang@126.com (K.Y.); Fax: +86-0797-8169782 (K.Y.)

Abstract: A facile and efficient method has been developed for the synthesis of C3-difluoromethyl carbinol-containing imidazo[1,2-*a*]pyridines at room temperature via the HFIP-promoted Friedel–Crafts reaction of difluoroacetaldehyde ethyl hemiacetal and imidazo[1,2-*a*]pyridines. This strategy could be applied to the direct C(sp^2)-H hydroxydifluoromethylation of imidazo[1,2-*a*]pyridines and afford a series of novel difluoromethylated carbinols in good to satisfactory yields with 29 examples. Furthermore, gram-scale and synthetic transformation experiments have also been achieved, demonstrating its potential applicable value in organic synthesis. This green protocol has several advantages, including being transition metal- and oxidant-free, being carried out at room temperature, having high efficiency, and having a wide substrate scope.

Keywords: hydroxydifluoromethylation; Friedel–Crafts reaction; imidazo[1,2-*a*]pyridine; HFIP

1. Introduction

The incorporation of fluorine-containing groups within bioactive compounds is considered one of the most useful approaches to address important issues relevant to medicinal chemistry due to their unique chemical and biological properties, such as affinity, metabolic stability, lipophilicity, cell permeability, and bioavailability [1–3]. One of the more prominent examples is the CF_2H unit, which is a good hydrogen bond donor and may serve as a bioisostere for hydroxy, thiol, and amide groups, and it has additional attractive properties such as its ability to modulate lipophilicity, polarity, and conformational preferences [4]. For these reasons, CF_2H-containing compounds may be widely applied in the fields of pharmaceuticals, agrochemicals, and advanced functional materials [5–7]. Notably, difluoromethyl carbinols containing both the difluoromethyl and hydroxyl groups are prevalent in bioactive molecules, such as antitumor agents [8], antidiabetic agents [9], Gaucher disease inhibitors [10], farnesoid X receptor modulators [11], and estrogen receptor degraders [12] (Figure 1). Therefore, the exploration of a facile synthetic protocol of difluoromethylated carbinol derivatives is an undeniably important and valuable research topic in expanding the chemical space for drug discovery [13,14].

In addition, imidazo[1,2-*a*]pyridines are widely found in natural products. They are also extensively used in modern organic synthesis and pharmaceutical and materials science [15]. Imidazo[1,2-*a*]pyridine is considered to be an important scaffold due to its various biological and pharmaceutical activities, such as antiviral, antifungal, and antitumor activities. The core structure of many commercially available drugs includes alpidem, miroprofen, necopidem, olprinone, saripidem, zolpidem, and zolimidine (Figure 2) [16–21]. Therefore, the efficient and green preparation of imidazo[1,2-*a*]pyridine derivatives with various substituents has also drawn considerable attention, especially C3-functionalized imidazo[1,2-*a*]pyridine derivatives [22]. Therefore, a sustained effort is being made to develop new approaches to synthesize C3-functionalized imidazo[1,2-*a*]pyridine derivatives. However, there are no reports on the synthesis of C3-difluoromethyl carbinol-containing imidazo[1,2-*a*]pyridines.

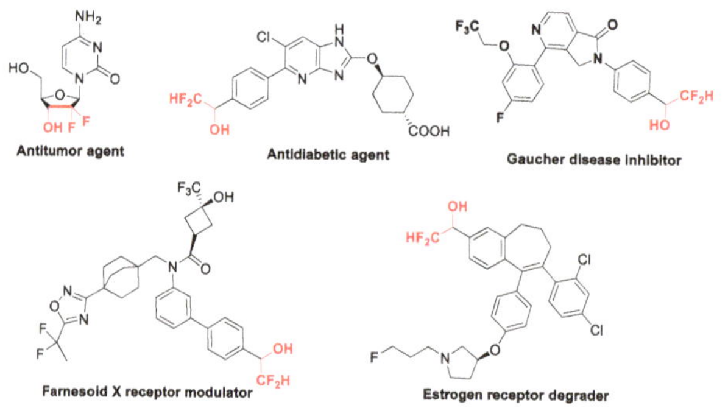

Figure 1. Representative difluoromethyl carbinol-containing biologically active molecules.

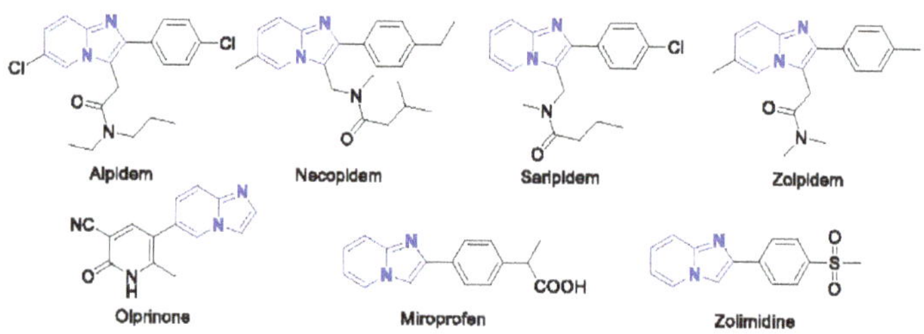

Figure 2. Representative drugs containing imidazo[1,2-*a*]pyridine scaffolds.

The strategy of C-H bond functionalization is known to be an ideal route for the preparation of diverse imidazo[1,2-*a*]pyridines as it is a straightforward, atom-economical, and synthetic step-economical method [23–25]. However, there are no reports on the C-H hydroxydifluoromethylation of imidazo[1,2-*a*]pyridines. It has been reported that 1,1,1,3,3,3-hexafluoro-2-propanol (HFIP), with a much higher polarity, increased Brønsted acidity, and strong hydrogen-bond donation [26–31], was considered to be a promising catalyst or promoter for the Friedel–Crafts reaction. For example, as early as 2012, the Naájera research group reported that HFIP serves as a promoter for the substitution reaction of allylic alcohols with nucleophiles [26]. Subsequently, HFIP was studied and applied as a catalyst or promoter in various types of Friedel–Crafts reactions. Inspired by this strategy for HFIP-promoted Friedel–Crafts alkylation, we focused our interest on developing facile methods for the synthesis of C3-difluoromethyl carbinol-containing imidazo[1,2-*a*]pyridines at room temperature via the HFIP-promoted cross-dehydrative coupling of difluoroacetaldehyde ethyl hemiacetal and imidazo[1,2-*a*]pyridines (Scheme 1). This green protocol possesses many intrinsic advantages like operational simplicity, high efficiency, atom economy, mild reaction conditions (e.g., at room temperature, transition metal- and oxidant-free, without inert gas protection), and a wide substrate scope.

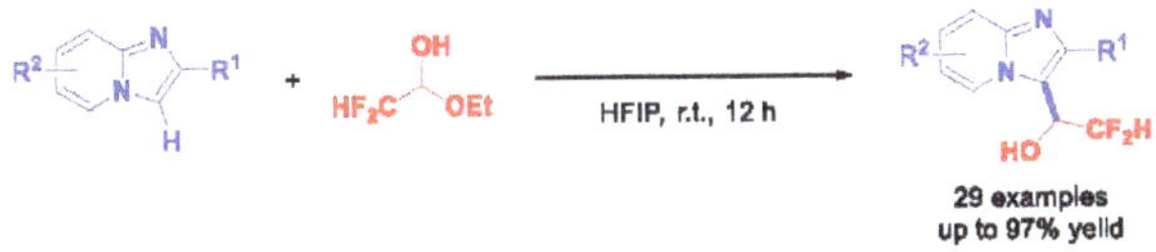

Scheme 1. The synthesis of C3-difluoromethyl carbinol-containing imidazo[1,2-*a*]pyridines.

2. Results and Discussion

2.1. Optimization of Reaction Conditions

The hydroxydifluoromethylation of imidazo[1,2-*a*]pyridine (**1a**) and commercially available difluoroacetaldehyde ethylacetal (**2a**) was selected as the model reaction for the optimization of the reaction conditions. The results are summarized in Table 1.

Table 1. Optimization of reaction conditions [a].

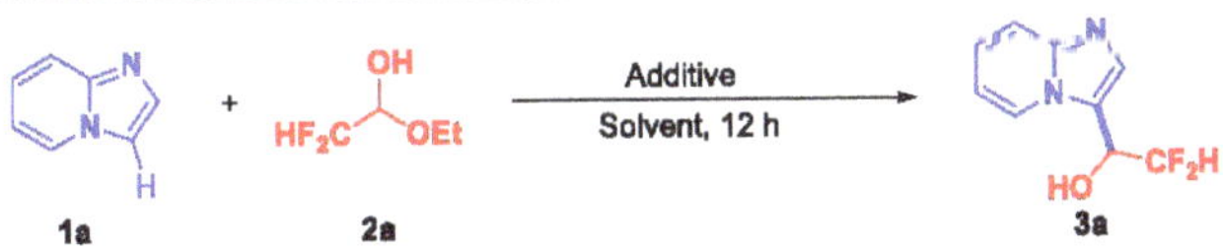

Entry	Additive (Equiv.)	Solvent	Yield (%) [b]
1	TFA (10%)	DCM	trace
2	TsOH (10%)	DCM	trace
3	$BF_3{\cdot}OEt_2$	DCM	trace
4	$Y(OTf)_3$ (10%)	DCM	trace
5	$Sc(OTf)_3$ (10%)	DCM	trace
6	HFIP (10%)	DCM	trace
7	HFIP (1.0)	DCM	50
8	HFIP (2.0)	DCM	73
9		HFIP	97

[a] Reaction conditions: **1a** (0.2 mmol), **2a** (0.4 mmol, 2.0 equiv.), solvent (1.0 mL), r.t., 12 h. [b] Isolated yield.

Initially, the effect of Brønsted acids and Lewis acids, such as trifluoroacetic acid (TFA), *p*-toluenesulfonic acid (TsOH), $BF_3{\cdot}Et_2O$, $Y(OTf)_3$, and $Sc(OTf)_3$, employed as catalysts on the reaction was investigated (Entries 1–5); it was found that only a trace amount of product **3a** was detected. HFIP was considered to be a promising catalyst or promoter considering its much higher polarity, increased Brønsted acidity, and strong hydrogen-bond donation [26–31]. Therefore, we investigated the effect of HFIP as a catalyst or promoter on this transformation. Regrettably, trace product **3a** was afforded in dichloromethane when 10 mol% HFIP was employed in the reaction, indicating that the catalytic amount of HFIP was not enough to facilitate the hydroxydifluoromethylation process (Entry 6). We attempted to increase the dosage of HFIP to 1.0 or 2.0 equivalents; product **3a** could be obtained in 50% and 73% yields, respectively (Entries 7 and 8). This result encouraged us to use HFIP as a solvent for the reaction. Excitingly, the hydroxydifluoromethylation could proceed completely to obtain **3a** in 97% yield (Entry 9). Thus, using HFIP as the solvent, the reaction of **1a** (0.2 mmol) and **2a** (0.4 mmol) was successfully achieved at room temperature for 12 h to obtain **3a** in satisfactory yield.

2.2. Scope of Reaction Substrates

Under the optimized reaction conditions, the substrate scope of this approach was examined by using different 2-substituted imidazo[1,2-*a*]pyridines. The results are summarized in Table 2.

Table 2. Substrate scope of various 2-aryl imidazo[1,2-*a*]pyridines [a, b].

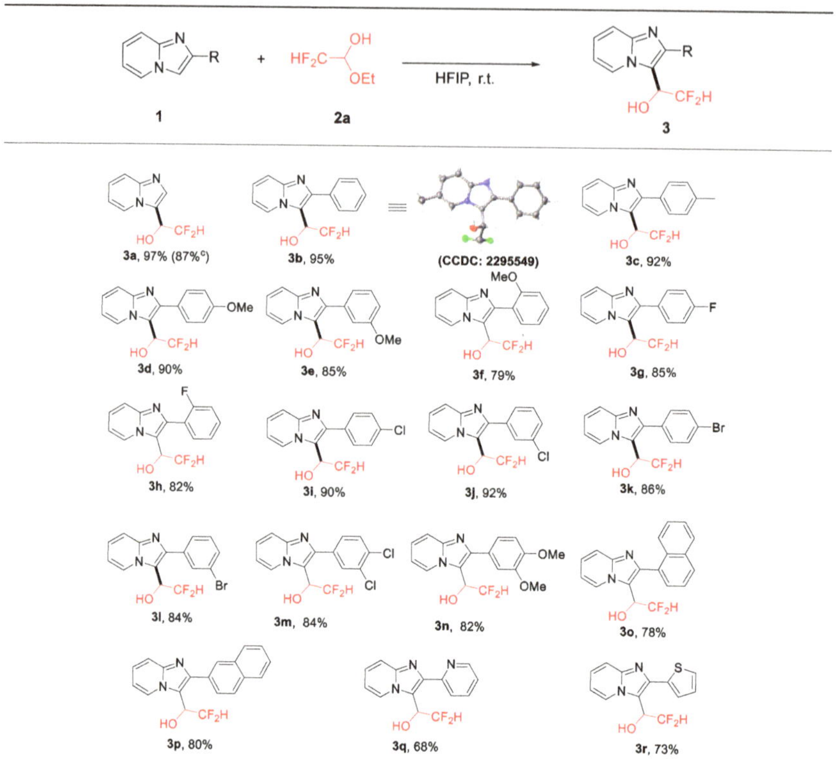

[a] Reaction conditions: **1** (0.2 mmol), **2a** (0.4 mmol, 2.0 equiv.), HFIP (1.0 mL), r.t., 12 h. [b] Isolated yield. [c] Reaction performed on a 5 mmol scale.

It was found that 2-phenyl imidazo[1,2-*a*]pyridine could react smoothly with **2a** to give the corresponding product **3b** in 95% yield. Then, a series of substituents, such as methyl, methoxyl, fluoro, chloro, and bromo groups on the C2-phenyl ring of 2-phenyl imidazo[1,2-*a*]pyridines, were well tolerated under the standard conditions. This showed that both methyl and methoxyl substituted substrates proceeded smoothly to give the corresponding products **3c–3f** in 79–92% yields. The halo-substituted substrates exhibited good reactivity in the hydroxydifluoromethylation to obtain products **3g–3l** in 82–92% yields. Dichloro-substituted and dimethoxy-substituted substrates were also subjected to smooth transformation to afford products **3m** and **3n** in 84% and 82% yields, respectively. As anticipated, 2-(naphthalen-1-yl) and 2-(naphthalen-2-yl) imidazo[1,2-*a*]pyridines underwent hydroxydifluoromethylation to give the desired products **3o** and **3p** in 78% and 80% yields, respectively. To our delight, 2-heteroaryl imidazo[1,2-*a*]pyridines were also successfully employed for the synthesis of the desired products. Using 2-(2-pyridyl)imidazo[1,2-*a*]pyridine and 2-(2-thienyl)imidazo[1,2-*a*]pyridine as substrates, corresponding products **3q** and **3r** were obtained in 68% and 73% yields, respectively. Furthermore, this conversion could be readily carried out on the 5 mmol scale to obtain the desired product **3a** in 87% yield, which proved to be easily applied to a gram-scale preparation.

To further extend the scope of this methodology, a series of imidazo[1,2-*a*]pyridines with different substituents on the pyridine ring were used to evaluate the universality of this hydroxydifluoromethylation (Table 3). As anticipated, imidazo[1,2-*a*]pyridines with an electron-donating group as well as a halogen group on the pyridine rings all worked well with **2a** and afforded the desired target molecules **3aa–3ah** in satisfactory yields (83–95%).

Trifluoromethyl, as a strong electron-withdrawing group, can also be tolerated in this conversion, giving the products in good yields (**3ai**, 89% and **3aj**, 87%). 6-Bromo-7-methyl-2-phenylimidazo[1,2-*a*]pyridine as a polysubstituted substrate was also suitable for this conversion, resulting in the product **3ak** with 90% yield.

Table 3. Substrate scope of various substituted 2-phenyl imidazo[1,2-*a*]pyridines [a, b].

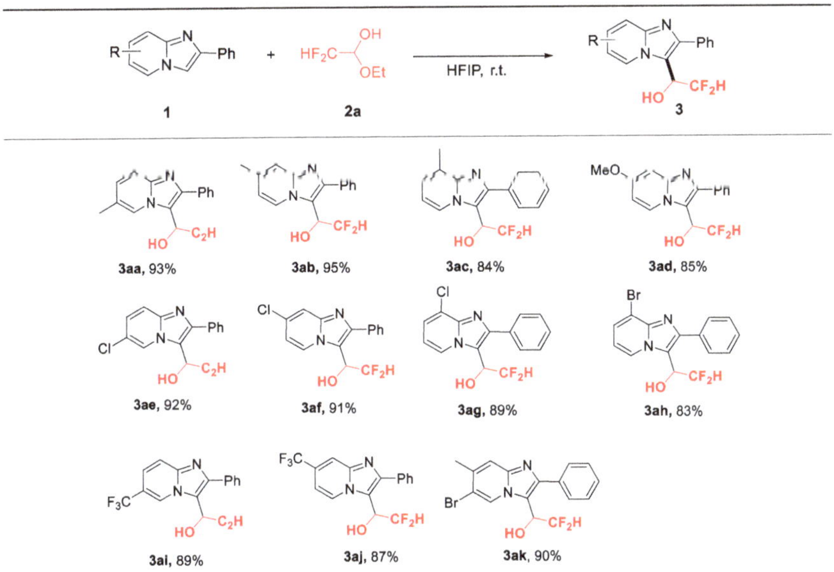

[a] Reaction conditions: **1** (0.2 mmol), **2a** (0.4 mmol, 2.0 equiv.), HFIP (1.0 mL), r.t., 12 h. [b] Isolated yield.

The structures of the desired products were confirmed by NMR (^{1}H, ^{13}C, and ^{19}F) and HM-RS data, and the structure of **3b** (CCDC 2295549) was unambiguously confirmed by single-crystal X-ray analysis, which can be seen in Supplementary Materials for details [32].

2.3. Mechanism Investigation

To gain insight into the details of the mechanism, we performed two control experiments accordingly. Adding 2.0 equivalent radical scavenger 2,2,6,6-tetramethylpiperidin-1-yloxyl (TEMPO) or BHT (butylated hydroxytoluene) to the reaction system, it was found that the corresponding compound **3a** could also be obtained in 91% and 88% isolation yields, respectively (Scheme 2). These results indicated that the reaction may not be involved in the pathway of radical participation.

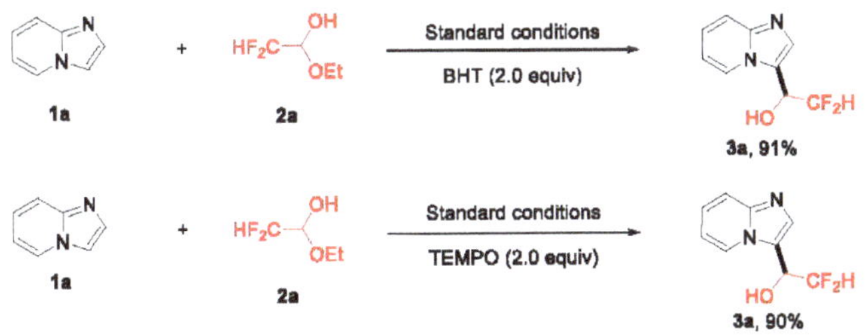

Scheme 2. Control experiments.

It has been reported that HFIP can be used to promote the generation of a C-C bond by cleavage of a C-O bond (the substrates of aromatic aldehyde hydrate, propargyl alcohol, difluoroacetaldehyde ethyl hemiacetal, etc.) via the Friedel–Crafts alkylation pathway [26–28,33–35]. Based on the properties of HFIP and the above-mentioned control experimental results, a plausible mechanism of HFIP-promoted Friedel–Crafts alkylation was put forward (Scheme 3). Initially, under the action of proton donor HFIP, the difluoroacetaldehyde ethyl hemiacetal **2a** has a tendency to remove one molecule of ethanol to generate difluoroacetaldehyde **A,** then with proton exchange with HFIP taking place and forming carbonium ions to active difluoroacetaldehyde **A**, making it more susceptible for nucleophilic addition with imidazo[1,2-*a*]pyridine **1a**, to generate the corresponding **B**. Finally, **B** undergoes the dehydrogenation process to produce the final product **3a**.

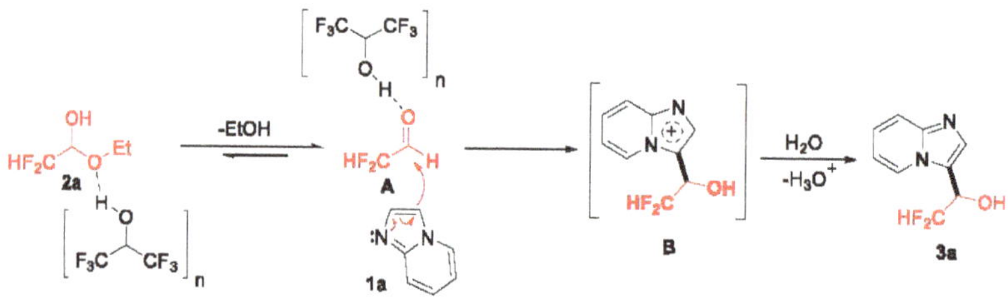

Scheme 3. A plausible reaction pathway.

2.4. Synthetic Derivatization

Moreover, the reactions are highly practical because of their wide range of applications in pharmaceutical chemistry for the production of diversified structural scaffolds and combinatorial libraries for drug discovery. The desired targets derivatized at hydroxyl positions may be promising candidates for such applications in pharmaceutical chemistry and organic synthesis. For instance, C3-difluoroacetyl imidazo[1,2-*a*]pyridines **4a** and **4b** have been prepared through Dess–Martin periodinane oxidation with **3a** and **3b** with yields of 77% and 79%, respectively (Scheme 4).

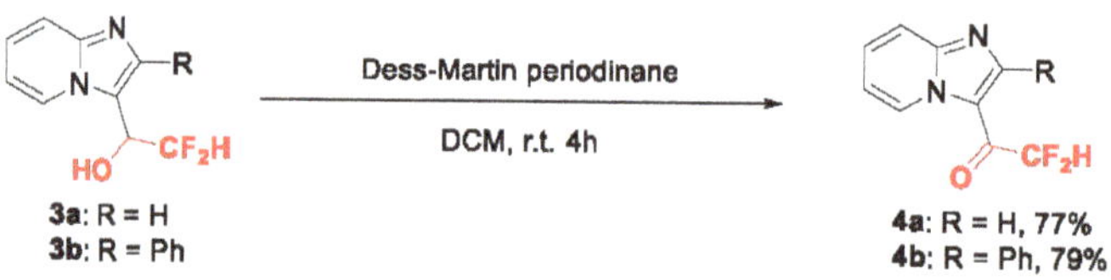

Scheme 4. Synthetic derivatization.

3. Materials and Methods

3.1. General Information

Melting point (m.p.) was performed on a Büchi Melting Point B-545 instrument without correcting. The ^{1}H, ^{13}C and ^{19}F NMR spectra were collected on a BRUKER DRX-400 spectrometer in $CDCl_3$ using tetramethylsilane (TMS) as an internal standard. High-resolution mass spectra (HRMS) were obtained with an LCMS-IT-TOF mass spectrometer. Single-crystal X-ray analysis was obtained using Bruker APEX2 Smart CCD. TLC was performed by using commercially prepared 100–400 mesh silica gel plates (GF254), and visualization was detected at 254 or 365 nm. All reagents and solvents were purchased from commercial sources and used without further purification, The 2-substituted imidazo[1,2-*a*]pyridines **1** (except for **1a**) were synthesized from 2-bromoacetophenones (or acetophenones) and various 2-aminopyridines [36,37].

3.2. *Experimental Procedure for Compounds* **3a–3ak**

The mixture of substituted imidazo[1,2-*a*]pyridine **1** (0.2 mmol, 1.0 equiv.) and difluoroacetaldehyde ethyl hemiacetal **2a** (0.30 mmol, 1.5 equiv.) in 1,1,1,3,3,3-hexafluoro-2-propanol (1.0 mL) was stirred at room temperature for 12 h. After the completion of the reaction, the reaction mixture was quenched with H_2O (15 mL) and extracted three times with ethyl acetate (3 × 15 mL). Then, the organic layer was dried over anhydrous Na_2SO_4. After the filtration and evaporation of the solvents under reduced pressure, the crude products were purified by column chromatography on silica gel to afford the desired products **3a–3ak**.

3.3. *Experimental Procedure for Compounds* **4a** *and* **4b**

Compounds **3a** or **3b** (0.1 mmol, 1.0 equiv.) and Dess–Martin periodinane (0.37 mmol, 3.7 equiv.) in DCM (1.0 mL) were stirred in a ground glass test tube at room temperature for 4 h. After monitoring the end of the reaction on TLC, the residues were purified by column chromatography on silica gel to give the pure products **4a** or **4b**.

3.4. *Characterization Data for All Products* **3a–3ak** *and* **4a–4b**

(±)-3-(2,2-Difluoro-1-hydroxy)ethyl-imidazo[1,2-a]pyridine (**3a**), white solid (38 mg, 97%); m.p. 160–162 °C; ^{1}H-NMR (400 MHz, $CDCl_3$:CD_3OD = 7:1), δ_H: 5.16–5.23 (*m*, 1H), 6.10 (*td*, *J* = 55.6, 3.2 Hz, 1H), 6.85–6.90 (*m*, 1H), 7.23–7.28 (*m*, 1H), 7.49–7.56 (*m*, 2H), 8.41 (*d*, *J* = 6.8 Hz, 1H); ^{13}C-NMR (100 MHz, $CDCl_3$:CD_3OD = 7:1), δ_C: 65.7 (*t*, *J* = 26.0 Hz), 112.7, 115.1 (*t*, *J* = 243.0 Hz), 116.9, 119.9, 125.6, 125.7, 131.8, 146.2; ^{19}F NMR (376 MHz, $CDCl_3$:CD_3OD = 7:1), δ_F, ppm: −126.14 (*d*, *J* = 285.8 Hz, 1F), 127.05 (*d*, *J* = 285.8 Hz, 1F); ESI-HRMS, *m*/*z*: Calcd for $C_9H_9F_2N_2O$ $[M + H]^+$, 199.0677, found: 199.0669.

(±)-3-(2,2-Difluoro-1-hydroxy)ethyl-2-phenyl-imidazo[1,2-a]pyridine (**3b**), white solid (52 mg, 95%); m.p. 161–163 °C; ^{1}H-NMR (400 MHz, $CDCl_3$:CD_3OD = 7:1), δ_H, ppm: 5.31–5.38 (*m*, 1H), 6.14 (*td*, *J* = 55.6, 3.2 Hz, 1H), 7.18–7.23 (*m*, 1H), 7.30–7.35 (*m*, 3H), 7.44–7.50 (*m*, 3H), 8.63 (*d*, *J* = 6.8 Hz, 1H); ^{13}C-NMR (100 MHz, $CDCl_3$:CD_3OD = 7:1), δ_C, ppm: 66.2 (*t*, *J* = 25.0 Hz), 112.3, 115.4 (*t*, *J* = 245.0 Hz), 115.6, 116.5, 125.9, 127.3, 128.3, 128.4, 128.8, 133.1, 145.4, 145.5; ^{19}F NMR (376 MHz, $CDCl_3$:CD_3OD = 7:1), δ_F, ppm: −123.70 (*d*, *J* = 282.0 Hz, 1F), −124.88 (*d*, *J* = 282.0 Hz, 1F); ESI-HRMS, *m*/*z*: Calcd for $C_{15}H_{13}F_2N_2O$ $[M + H]^+$, 275.0990, found: 275.0979.

(±)-3-(2,2-Difluoro-1-hydroxy)ethyl-2-(p-tolyl)-imidazo[1,2-a]pyridine (**3c**), white solid (53 mg, 92%); m.p. 187–189 °C; ^{1}H-NMR (400 MHz, $CDCl_3$:CD_3OD = 7:1), δ_H, ppm: 2.35 (*s*, 3H), 5.31–5.38 (*m*, 1H), 6.13 (*J* = 55.6, 3.2 Hz, 1H), 6.75–6.80 (*m*, 1H), 7.13 (*d*, *J* = 7.6 Hz, 2H), 7.16–7.21 (*m*, 1H), 7.35 (*d*, *J* = 8.0 Hz, 2H), 7.43 (*d*, *J* = 9.2 Hz, 1H), 8.61 (*d*, *J* = 6.8 Hz, 1H); ^{13}C-NMR (100 MHz, $CDCl_3$:CD_3OD = 7:1), δ_C, ppm: 21.2, 66.2 (*t*, *J* = 24.6 Hz), 112.2, 115.4 (*t*, *J* = 244.0 Hz), 116.4, 125.8, 127.3, 128.6, 128.8, 129.1, 129.3, 130.1, 138.1, 145.4; ^{19}F NMR (376 MHz, $CDCl_3$:CD_3OD = 7:1), δ_F, ppm: −123.68 (*d*, *J* = 282.0 Hz, 1F), −124.96 (*d*, *J* = 282.0 Hz, 1F); ESI-HRMS, *m*/*z*: Calcd for $C_{16}H_{15}F_2N_2O$ $[M + H]^+$, 289.1147, found: 289.1136.

(±)-3-(2,2-Difluoro-1-hydroxy)ethyl-2-(4-methoxyphenyl)-imidazo[1,2-a]pyridine (**3d**), white solid (55 mg, 90%); m.p. 178–180 °C; ^{1}H-NMR (400 MHz, $CDCl_3$:CD_3OD = 7:1), δ_H, ppm: 3.83 (*s*, 3H), 5.29–5.36 (*m*, 1H), 6.14 (*td*, *J* = 55.6, 2.4 Hz, 1H), 6.71–6.81 (*m*, 1H), 6.86–6.90 (*m*, 2H), 7.81–7.21 (*m*, 1H), 7.40–7.46 (*m*, 3H), 8.62 (*d*, *J* = 4.0 Hz, 1H); ^{13}C-NMR (100 MHz, $CDCl_3$:CD_3OD = 7:1), δ_C, ppm: 55.2, 66.2 (*t*, *J* = 25.0 Hz), 112.2, 113.9, 115.1, 115.4 (*t*, *J* = 244.0 Hz), 116.3, 115.6, 125.8, 127.2, 130.1, 145.2, 145.4, 159.6; ^{19}F NMR (376 MHz, $CDCl_3$:CD_3OD = 7:1), δ_F, ppm: −123.74 (*d*, *J* = 282.0 Hz, 1F), −124.87 (*d*, *J* = 282.0 Hz, 1F); ESI-HRMS, *m*/*z*: Calcd for $C_{16}H_{15}F_2N_2O_2$ $[M + H]^+$, 305.1096, found: 305.1081.

(±)-3-(2,2-Difluoro-1-hydroxy)ethyl-2-(3-methoxyphenyl)-imidazo[1,2-a]pyridine (**3e**), white solid (52 mg, 85%); m.p. 176–178 °C; ^{1}H-NMR (400 MHz, $CDCl_3$:CD_3OD = 7:1), δ_H, ppm: 5.34–5.42 (*m*, 1H), 6.13 (*td*, *J* = 55.6, 3.6 Hz, 1H), 6.75–6.79 (*m*, 1H), 6.84–6.87 (*m*, 1H),

7.00–7.05 (*m*, 2H), 7.14–7.23 (*m*, 2H), 7.41 (*d*, *J* = 7.2 Hz, 1H), 8.61 (*d*, *J* = 6.8 Hz, 1H); ^{13}C-NMR (100 MHz, $CDCl_3$:CD_3OD = 7:1), δ_C, ppm: 55.2, 66.3 (*t*, *J* = 26.0 Hz), 112.3, 112.9, 114.1, 114.2, 115.4 (*t*, *J* = 245.0 Hz), 115.5, 116.6, 1221.2, 125.9, 127.3, 129.5, 134.4, 145.3, 145.4, 159.4; ^{19}F NMR (376 MHz, $CDCl_3$:CD_3OD = 7:1), δ_F, ppm: −123.62 (*d*, *J* = 282.0 Hz, 1F), −124.75 (*d*, *J* = 282.0 Hz, 1F); ESI-HRMS, *m*/*z*: Calcd for $C_{16}H_{15}F_2N_2O_2$ $[M + H]^+$, 305.1096, found: 305.1081.

(±)-3-(2,2-Difluoro-1-hydroxy)ethyl-2-(2-methoxyphenyl)-imidazo[1,2-a]pyridine (**3f**), white solid (48 mg, 79%); m.p. 177–179 °C; ^{1}H-NMR (400 MHz, $CDCl_3$), δ_H, ppm: 3.56 (*s*, 1H), 5.07–5.14 (*m*, 1H), 5.91 (*td*, *J* = 55.6, 1.6 Hz, 1H), 6.70–6.77 (*m*, 2H), 6.81 (*d*, *J* = 8.4 Hz, 1H), 7.09–7.16 (*m*, 2H), 7.21–7.25 (*m*, 1H), 7.37 (*d*, *J* = 8.8 Hz, 1H), 8.57 (*d*, *J* = 6.8 Hz, 1H); ^{13}C-NMR (100 MHz, $CDCl_3$), δ_C, ppm: 55.2, 66.5 (*t*, *J* = 24.0 Hz), 111.0, 112.1, 115.7 (*t*, *J* = 246.0 Hz),, 116.6, 116.9, 120.8, 122.1, 125.4, 127.2, 130.0, 132.1, 141.3, 145.6, 156.3; ^{19}F NMR (376 MHz, $CDCl_3$), δ_F, ppm: −123.06 (*d*, *J* = 278.2 Hz, 1F), −127.67 (*d*, *J* = 278.2 Hz, 1F); ESI-HRMS, *m*/*z*: Calcd for $C_{16}H_{15}F_2N_2O_2$ $[M + H]^+$, 305.1096, found: 305.1081.

(±)-3-(2,2-Difluoro-1-hydroxy)ethyl-2-(4-fluorophenyl)-imidazo[1,2-a]pyridine 2,2- (**3g**), white solid (49 mg, 85%); m.p. 208–210 °C; ^{1}H-NMR (400 MHz, CD_3OD), δ_H, ppm: 5.28–5.35 (*m*, 1H), 6.33 (*td*, *J* = 55.6, 3.6 Hz, 1H), 6.92–6.96 (*m*, 1H), 7.18–7.24 (*m*, 2H), 7.35–7.39 (*m*, 1H), 7.55 (*d*, *J* = 9.2 Hz, 1H), 7.63–7.67 (*m*, 2H), 8.75 (*d*, *J* = 6.8 Hz, 1H); ^{13}C-NMR (100 MHz, CD_3OD), δ_C, ppm: 66.1 (*t*, *J* = 25.0 Hz), 112.3, 115.1 (*t*, *J* = 22.0 Hz), 115.7 (*t*, *J* = 243.0 Hz), 115.8, 116.1, 126.4, 127.5, 129.6 (*d*, *J* = 3.0 Hz), 130.8 (*d*, *J* = 9.0 Hz), 145.5, 144.2, 163.0 (*d*, *J* = 246.0 Hz); ^{19}F NMR (376 MHz, $CDCl_3$:CD_3OD = 7:1), δ_F, ppm: −115.40, −125.45 (*d*, *J* = 285.8 Hz, 1F), −126.28 (*d*, *J* = 282.0 Hz, 1F); ESI-HRMS, *m*/*z*: Calcd for $C_{15}H_{12}F_3N_2O$ $[M + H]^+$, 293.0896, found: 293.0885.

(±)-3-(2,2-Difluoro-1-hydroxy)ethyl-2-(2-fluorophenyl)-imidazo[1,2-a]pyridine (**3h**), white solid (48 mg, 82%); m.p. 173–175 °C; ^{1}H-NMR (400 MHz, $CDCl_3$), δ_H, ppm: 5.04–5.11 (*m*, 1H), 5.95 (*td*, *J* = 55.6, 2.0 Hz, 1H), 6.68–6.72 (*m*, 1H), 6.92–7.00 (*m*, 2H), 7.09–7.13 (*m*, 1H), 7.17–7.23 (*m*, 1H), 7.26–7.31 (*m*, 1H), 8.37 (*d*, *J* = 9.2 Hz, 1H), 8.57 (*d*, *J* = 6.0 Hz, 1H); ^{13}C-NMR (100 MHz, $CDCl_3$), δ_C, ppm: 66.4 (*t*, *J* = 24.0 Hz), 112.4, 115.4 (*t*, *J* = 246.0 Hz), 115.6, 115.8, 116.6, 120.9 (*d*, *J* = 15.0 Hz), 124.3 (*d*, *J* = 4.0 Hz), 126.0, 127.5, 130.4 (*d*, *J* = 8.0 Hz), 132.0, 138.9, 145.9, 159.4 (*d*, *J* = 245.0 Hz); ^{19}F NMR (376 MHz, $CDCl_3$), δ_F, ppm: −123.88 (*d*, *J* = 285.8 Hz, 1F), −124.66 (*d*, *J* = 285.8 Hz, 1F); ESI-HRMS, *m*/*z*: Calcd for $C_{15}H_{12}F_3N_2O$ $[M + H]^+$, 293.0896, found: 293.0885.

(±)-2-(4-Chlorophenyl)-3-(2,2-difluoro-1-hydroxy)ethyl-imidazo[1,2-a]pyridine (**3i**), white solid (55 mg, 90%); m.p. 198–200 °C; ^{1}H-NMR (400 MHz, $CDCl_3$:CD_3OD = 7:1), δ_H, ppm: 5.26–5.32 (*m*, 1H), 6.16 (*td*, *J* = 55.6, 3.2 Hz, 1H), 6.80–6.85 (*m*, 1H), 7.22–7.31 (*m*, 3H), 7.40 (*d*, *J* = 8.0 Hz, 2H), 7.46 (*d*, *J* = 9.2 Hz, 1H), 8.60 (*d*, *J* = 6.8 Hz, 1H); ^{13}C-NMR (100 MHz, $CDCl_3$:CD_3OD = 7:1), δ_C, ppm: 66.3 (*t*, *J* = 26.0 Hz), 112.6, 115.2 (*t*, *J* = 245.0 Hz), 115.6, 116.6, 126.2, 127.3, 128.7, 129.9, 131.5, 134.4, 144.2, 145.5; ^{19}F NMR (376 MHz, $CDCl_3$:CD_3OD = 7:1), δ_F, ppm: −123.74 (*d*, *J* = 282.0 Hz, 1F), −124.57 (*d*, *J* = 282.0 Hz, 1F); ESI-HRMS, *m*/*z*: Calcd for $C_{15}H_{12}ClF_2N_2O$ $[M + H]^+$, 309.0601, found: 309.0616.

(±)-2-(3-Chlorophenyl)-3-(2,2-difluoro-1-hydroxy)ethyl-imidazo[1,2-a]pyridine (**3j**), white solid (57 mg, 92%); m.p. 174–176 °C; ^{1}H-NMR (400 MHz, $CDCl_3$:CD_3OD = 7:1), δ_H, ppm: 5.28–5.26 (*m*, 1H), 6.15 (*td*, *J* = 55.6, 3.6 Hz, 1H), 6.78–6.83 (*m*, 1H), 7.20–7.23 (*m*, 1H), 7.25–7.32 (*m*, 2H), 7.35–7.38 (*m*, 1H), 7.44–7.51 (*m*, 2H), 8.63 (*d*, *J* = 6.8 Hz, 1H); ^{13}C-NMR (100 MHz, $CDCl_3$:CD_3OD = 7:1), δ_C, ppm: 66.1 (*t*, *J* = 25.0 Hz), 112.6, 115.3 (*t*, *J* = 245.0 Hz), 115.9, 116.5, 126.2, 126.9, 127.4, 128.3, 128.7, 129.7, 134.3, 134.9, 143.8, 145.5; ^{19}F NMR (376 MHz, $CDCl_3$:CD_3OD = 7:1), δ_F, ppm: −123.86 (*d*, *J* = 285.8 Hz, 1F), −124.63 (*d*, *J* = 285.8 Hz, 1F); ESI-HRMS, *m*/*z*: Calcd for $C_{15}H_{12}ClF_2N_2O$ $[M + H]^+$, 309.0601, found: 309.0616.

(±)-2-(4-Bromophenyl)-3-(2,2-difluoro-1-hydroxy)ethyl-imidazo[1,2-a]pyridine (**3k**), white solid (60 mg, 86%); m.p. 190–192 °C; ^{1}H-NMR (400 MHz, $CDCl_3$:CD_3OD = 7:1), δ_H, ppm:

5.24–5.31 (*m*, 1H), 6.16 (*td*, *J* = 55.6, 3.6 Hz, 1H), 6.80–6.85 (*m*, 1H), 7.22–7.26 (*m*, 1H), 7.32 (*d*, *J* = 8.4 Hz, 2H), 7.43–7.47 (*m*, 3H), 8.59 (*d*, *J* = 6.8 Hz, 1H); ^{13}C-NMR (100 MHz, $CDCl_3$:CD_3OD = 7:1), δ_C, ppm: 66.2 (*t*, *J* = 25.0 Hz), 112.5, 115.2 (*t*, *J* = 245.0 Hz), 115.7, 116.6, 122.7, 126.2, 127.3, 130.2, 131.6, 131.9, 144.2, 145.5; ^{19}F NMR (376 MHz, $CDCl_3$:CD_3OD = 7:1), δ_F, ppm: −123.74 (*d*, *J* = 282.0 Hz, 1F), −124.56 (*d*, *J* = 282.0 Hz, 1F); ESI-HRMS, *m*/*z*: Calcd for $C_{16}H_{15}BrF_2N_2O$ [M + H]$^+$, 353.0096, found: 353.0107.

(±)-2-(3-Bromophenyl)-3-(2,2-difluoro-1-hydroxy)ethyl-imidazo[1,2-a]pyridine (**3l**), white solid (59 mg, 84%); m.p. 186–188 °C; ^{1}H-NMR (400 MHz, $CDCl_3$:CD_3OD = 7:1), δ_H, ppm: 5.27–5.35 (*m*, 1H), 6.01–6.16 (*m*, 1H), 6.78–6.82 (*m*, 1H), 7.17–7.22 (*m*, 2H), 7.38–7.45 (*m*, 3H), 7.67 (*s*, 1H), 8.62 (*d*, *J* = 6.4 Hz, 1H); ^{13}C-NMR (100 MHz, $CDCl_3$:CD_3OD = 7:1), δ_C, ppm: 66.22 (*t*, *J* = 25.0 Hz), 112.6, 115.3 (*t*, *J* = 245.0 Hz), 115.9, 116.6, 122.5, 126.3, 127.4, 130.0, 131.3, 131.7, 135.2, 143.7, 145.6; ^{19}F NMR (376 MHz, $CDCl_3$:CD_3OD = 7:1), δ_F, ppm: −123.88 (*d*, *J* = 285.8 Hz, 1F), −124.66 (*d*, *J* = 285.8 Hz, 1F), ESI-HRMS, *m*/*z*: Calcd for $C_{15}H_{14}BrF_2N_2O$ [M + H]$^+$, 353.0096, found: 353.0107.

(±)-2-(3,4-Dichlorophenyl)-3-(2,2-difluoro-1-hydroxy)ethyl-imidazo[1,2-a]pyridine (**3m**), white solid (57 mg, 84%); m.p. 205–207 °C; ^{1}H-NMR (400 MHz, $CDCl_3$:CD_3OD = 7:1), δ_H, ppm: 5.26–5.34 (*m*, 1H), 6.18 (*td*, *J* = 55.6, 3.6 Hz, 1H), 6.83–6.87 (*m*, 1H), 7.25–7.30 (*m*, 1H), 7.38–7.53 (*m*, 3H), 7.68 (*s*, 1H), 8.63 (*d*, *J* = 6.4 Hz, 1H);^{13}C-NMR (100 MHz, $CDCl_3$:CD_3OD=7:1), δ_C, ppm: 66.2 (*t*, *J* = 26.0 Hz), 112.7, 115.2 (*t*, *J* = 245.0 Hz), 116.0, 116.6, 126.5, 127.4, 128.0, 130.5, 132.5, 132.6, 133.2, 142.9, 145.6; ^{19}F NMR (376 MHz, $CDCl_3$:CD_3OD = 7:1), δ_F, ppm: −123.85 (*d*, *J* = 285.8 Hz, 1F), −124.66 (*d*, *J* = 285.8 Hz, 1F); ESI-HRMS, *m*/*z*: Calcd for $C_{15}H_{11}Cl_2F_2N_2O$ [M + H]$^+$, 343.0211, found: 343.0213.

(±)-3-(2,2-Difluoro-1-hydroxy)ethyl-2-(3,4-dimethoxyphenyl)-imidazo[1,2-a]pyridine (**3n**), white solid (55 mg, 82%); m.p. 178–180 °C; ^{1}H-NMR (400 MHz, $CDCl_3$:CD_3OD = 7:1), δ_H, ppm: 3.86 (*s*, 3H), 3.90 (*s*, 3H), 5.35–5.42 (*m*, 1H), 6.19 (*td*, *J* = 55.6, 3.6 Hz, 1H), 6.78–6.86 (*m*, 2H), 7.04 (*d*, *J* = 8.0 Hz, 1H), 7.16 (*s*, 1H), 7.19–7.23 (*m*, 1H), 7.46 (*d*, *J* = 8.8 Hz, 1H), 8.61 (*d*, *J* = 6.8 Hz, 1H); ^{13}C-NMR (100 MHz, $CDCl_3$:CD_3OD = 7:1), δ_C, ppm: 55.7, 55.7, 66.3 (*t*, *J* = 25.0 Hz), 110.9, 112.1, 112.3, 115.1, 115.4 (*t*, *J* = 245.0 Hz), 116.3, 121.2, 125.8, 125.9, 127.2, 145.3, 145.4, 148.7, 149.0; ^{19}F NMR (376 MHz, $CDCl_3$:CD_3OD = 7:1), δ_F, ppm: −123.54 (*d*, *J* = 282.0 Hz, 1F), −124.53 (*d*, *J* = 282.0 Hz, 1F); ESI-HRMS, *m*/*z*: Calcd for $C_{17}H_{17}F_2N_2O_3$ [M + H]$^+$, 335.1202, found: 335.1195.

(±)-3-(2,2-Difluoro-1-hydroxy)ethyl-2-(naphthalen-1-yl)-imidazo[1,2-a]pyridine (**3o**), white solid (51 mg, 78%); m.p. 212–214 °C; ^{1}H-NMR (400 MHz, $CDCl_3$), δ_H, ppm: 4.68–4.76 (*m*, 1H), 5.45 (*td*, *J* = 55.6, 2.8 Hz, 1H), 6.67–6.71 (*m*, 1H), 6.94–6.99 (*m*, 2H), 7.06–7.10 (*m*, 1H), 7.15–7.19 (*m*, 1H), 7.25–7.30 (*m*, 2H), 7.41 (*d*, *J* = 8.4 Hz, 1H), 7.74 (*d*, *J* = 8.0 Hz, 1H), 8.46 (*d*, *J* = 6.8 Hz, 1H); ^{13}C-NMR (100 MHz, $CDCl_3$), δ_C, ppm: 66.0 (*t*, *J* = 25.0 Hz), 112.3, 115.2 (*t*, *J* = 245.0 Hz), 116.4, 117.2, 124.9, 125.6, 125.9, 125.9, 126.2, 127.3, 128.0, 128.3, 129.0, 130.2, 132.2, 133.4, 143.8, 145.3; ESI-HRMS, *m*/*z*: Calcd for $C_{19}H_{15}F_2N_2O$ [M + H]$^+$, 325.1147, found: 325.1154.

(±)-3-(2,2-Difluoro-1-hydroxy)ethyl-2-(naphthalen-2-yl)-imidazo[1,2-a]pyridine (**3p**), white solid (52 mg, 80%); m.p. 210–212 °C; ^{1}H-NMR (400 MHz, $CDCl_3$:CD_3OD = 7:1), δ_H, ppm: 5.38–5.42 (*m*, 1H), 6.14 (*td*, *J* = 55.6, 3.2 Hz, 1H), 6.57–6.61 (*m*, 1H), 6.96–7.01 (*m*, 1H), 7.33 (*d*, *J* = 8.8 Hz, 1H), 7.47–7.51 (*m*, 3H), 7.65 (*d*, *J* = 8.4 Hz, 1H), 7.73–7.77 (*m*, 2H), 7.85 (*s*, 1H), 8.53 (*d*, *J* = 6.8 Hz, 1H); ^{13}C-NMR (100 MHz, $CDCl_3$:CD_3OD=7:1), δ_C, ppm: 66.3 (*t*, *J* = 25.0 Hz), 112.2, 115.4 (*t*, *J* = 245.0 Hz), 115.8, 116.3, 125.8, 126.3, 126.3, 126.4, 127.2, 127.6, 127.8, 128.1, 128.3, 130.4, 132.9, 133.0, 145.2, 145.5; ^{19}F NMR (376 MHz, $CDCl_3$:CD_3OD = 7:1), δ_F, ppm: −123.52 (*d*, *J* = 285.8 Hz, 1F), −124.60 (*d*, *J* = 285.8 Hz, 1F); ESI-HRMS, *m*/*z*: Calcd for $C_{19}H_{15}F_2N_2O$ [M + H]$^+$, 325.1147, found: 325.1154.

(±)-3-(2,2-Difluoro-1-hydroxy)ethyl-2-(pyridin-2-yl)-imidazo[1,2-a]pyridine (**3q**), white solid (37 mg, 68%); m.p. 165–167 °C; ^{1}H-NMR (400 MHz, $CDCl_3$), δ_H, ppm: 5.25–5.32 (*m*, 1H), 5.87 (*td*, *J* = 55.6, 5.6 Hz, 1H), 6.78–6.82 (*m*, 1H), 7.14–7.23 (*m*, 2H), 7.56 (*d*, *J* = 9.2 Hz, 1H),

7.78–7.82 (*m*, 1H), 8.03 (*d*, *J* = 6.8 Hz, 1H), 8.38 (*d*, *J* = 8.0 Hz, 1H), 8.45 (*d*, *J* = 4.0 Hz, 1H), 9.82 (*br*, 1H); ^{13}C-NMR (100 MHz, $CDCl_3$), δ_C, ppm: 65.8 (*t*, *J* = 26.0 Hz), 113.3, 115.9 (*t*, *J* = 244.0 Hz), 118.0, 120.1, 123.0, 123.0, 124.0, 125.6, 138.4, 142.6, 145.1, 147.2, 152.6; ^{19}F NMR (376 MHz, $CDCl_3$), δ_F, ppm: −123.20 (*d*, *J* = 282.0 Hz, 1F), −127.07 (*d*, *J* = 282.0 Hz, 1F); ESI-HRMS, *m*/*z*: Calcd for $C_{14}H_{12}F_2N_3O$ [M + H]$^+$, 276.0943, found: 276.0962.

(±)-3-(2,2-Difluoro-1-hydroxy)ethyl-2-(thiophen-2-yl)-imidazo[1,2-a]pyridine (**3r**), white solid (41 mg, 73%); m.p. 163–165 °C; ^{1}H-NMR (400 MHz, $CDCl_3$:CD_3OD = 7:1), δ_H, ppm: 5.46–5.54 (*m*, 1H), 6.10 (*td*, *J* = 55.6, 3.2 Hz, 1H), 6.69–6.73 (*m*, 1H), 6.96–6.99 (*m*, 1H), 7.11–7.16 (*m*, 1H), 7.19 (*d*, *J* = 3.6 Hz, 1H), 7.27 (*d*, *J* = 5.2 Hz, 1H), 7.40 (*d*, *J* = 9.2 Hz, 1H), 8.57 (*d*, *J* = 7.2 Hz, 1H); ^{13}C-NMR (100 MHz, $CDCl_3$:CD_3OD = 7:1), δ_C, ppm: 66.4 (*t*, *J* = 25.0 Hz), 112.4, 114.9, 115.4 (*t*, *J* = 246.0 Hz), 116.4, 126.2, 126.3, 126.7, 127.4, 127.6, 135.5, 139.1, 145.7; ^{19}F NMR (376 MHz, $CDCl_3$), δ_F, ppm: −123.44 (*d*, *J* = 282.0 Hz, 1F), −127.37 (*d*, *J* = 282.0 Hz, 1F); ESI-HRMS, *m*/*z*: Calcd for $C_{13}H_{11}F_2N_2OS$ [M + H]$^+$, 281.0555, found: 281.0566.

(±)-3-(2,2-Difluoro-1-hydroxy)ethyl-6-methyl-2-phenyl-imidazo[1,2-a]pyridine (**3aa**), white solid (54 mg, 93%); m.p. 212–214 °C; ^{1}H-NMR (400 MHz, $CDCl_3$:CD_3OD = 7:1), δ_H, ppm: 2.30 (*s*, 3H), 5.28–5.36 (*m*, 1H), 6.13 (*td*, *J* = 55.6, 4.0 Hz, 1H), 7.00 (*d*, *J* = 9.2 Hz, 1H), 7.26–7.31 (*m*, 4H), 7.38–7.41 (*m*, 2H), 8.35 (s, 1H); ^{13}C-NMR (100 MHz, $CDCl_3$:CD_3OD = 7:1), δ_C, ppm: 18.4, 66.3 (*t*, *J* = 25.0 Hz), 115.1, 115.2 (*t*, *J* = 245.0 Hz), 115.8, 122.0, 124.8, 128.0, 128.4, 128.7, 129.0, 133.1, 144.5, 145.2; ^{19}F NMR (376 MHz, $CDCl_3$:CD_3OD = 7:1), δ_F, ppm: −123.27 (*d*, *J* = 282.0 Hz, 1F), −124.93 (*d*, *J* = 282.0 Hz, 1F); ESI-HRMS, *m*/*z*: Calcd for $C_{16}H_{15}F_2N_2O$ [M + H]$^+$, 289.1147, found: 289.1136.

(±)-3-(2,2-Difluoro-1-hydroxy)ethyl-7-methyl-2-phenyl-imidazo[1,2-a]pyridine (**3ab**), white solid (55 mg, 95%); m.p. 186–188 °C; ^{1}H-NMR (400 MHz, $CDCl_3$), δ_H, ppm: 2.32 (*s*, 3H), 5.26–5.34 (*m*, 1H), 6.18 (*td*, *J* = 55.6, 3.6 Hz, 1H), 7.55 (*d*, *J* = 8.8 Hz, 1H), 7.17–7.29 (*m*, 6H), 8.43 (*d*, *J* = 9.2 Hz, 1H); ^{13}C-NMR (100 MHz, $CDCl_3$), δ_C, ppm: 21.3, 66.4 (*t*, *J* = 25.0 Hz), 114.8, 115.1, 115.3 (*t*, *J* = 245.0 Hz), 126.5, 128.1, 128.4, 128.6, 132.7, 137.3, 144.8, 145.7; ^{19}F NMR (376 MHz, $CDCl_3$), δ_F, ppm: −123.73 (*d*, *J* = 282.0 Hz, 1F), −124.81 (*d*, *J* = 282.0 Hz, 1F); ESI-HRMS, *m*/*z*: Calcd for $C_{16}H_{15}F_2N_2O$ [M + H]$^+$, 289.1147, found: 289.1136.

(±)-3-(2,2-Difluoro-1-hydroxy)ethyl-8-methyl-2-phenyl-imidazo[1,2-a]pyridine (**3ac**), white solid (49 mg, 84%); m.p. 184–186 °C; ^{1}H-NMR (400 MHz, $CDCl_3$:CD_3OD = 7:1), δ_H, ppm: 2.53 (*s*, 3H), 5.21–5.28 (*m*, 1H), 6.08 (*td*, *J* = 55.6, 1.6 Hz, 1H), 6.70–6.74 (*m*, 1H), 7.03 (*d*, *J* = 6.0 Hz, 1H), 7.28–7.32 (*m*, 3H), 7.45–7.48 (*m*, 2H), 8.48 (*d*, *J* = 6.8 Hz, 1H); ^{13}C-NMR (100 MHz, $CDCl_3$:CD_3OD = 7:1), δ_C, ppm: 16.9, 66.2 (*t*, *J* = 25.0 Hz), 112.4, 115.2 (*t*, *J* = 245.0 Hz), 115.9, 124.8, 124.9, 126.7, 128.1, 128.3, 129.1, 133.3, 145.2, 145.9; ^{19}F NMR (376 MHz, $CDCl_3$:CD_3OD = 7:1), δ_F, ppm: −123.51 (*d*, *J* = 282.0 Hz, 1F), −124.57 (*d*, *J* = 282.0 Hz, 1F); ESI-HRMS, *m*/*z*: Calcd for $C_{16}H_{15}F_2N_2O$ [M + H]$^+$, 289.1147, found: 289.1136.

(±)-3-(2,2-Difluoro-1-hydroxy)ethyl-7-methoxy-2-phenyl-imidazo[1,2-a]pyridine (**3ad**), white solid (52 mg, 85%); m.p. 201–203 °C; ^{1}H-NMR (400 MHz, $CDCl_3$:CD_3OD = 7:1), δ_H, ppm: 3.86 (*s*, 3H), 5.27–5.32 (*m*, 1H), 5.99–6.28 (*m*, 1H), 6.78–6.81 (*m*, 1H), 7.31–7.42 (*m*, 3H), 7.49–7.59 (*m*, 2H), 8.46 (*d*, *J* = 4.8 Hz, 1H); ^{13}C-NMR (100 MHz, $CDCl_3$:CD_3OD = 7:1), δ_C, ppm: 55.5, 66.2 (*t*, *J* = 25.0 Hz), 93.7, 107.3, 114.3, 115.4 (*t*, *J* = 245.0 Hz), 127.8, 128.1, 128.4, 128.7, 133.3, 145.0, 147.3, 158.8; ^{19}F NMR (376 MHz, $CDCl_3$:CD_3OD = 7:1), δ_F, ppm: −123.98 (*d*, *J* = 282.0 Hz, 1F), −125.18 (*d*, *J* = 282.0 Hz, 1F); ESI-HRMS, *m*/*z*: Calcd for $C_{16}H_{15}F_2N_2O_2$ [M + H]$^+$, 305.1096, found: 305.1081.

(±)-6-Chloro-3-(2,2-difluoro-1-hydroxy)ethyl-2-phenyl-imidazo[1,2-a]pyridine (**3ae**), white solid (57 mg, 92%); m.p. 215–217 °C; ^{1}H-NMR (400 MHz, $CDCl_3$:CD_3OD = 7:1), δ_H, ppm: 5.30–5.38 (*m*, 1H), 6.16 (*td*, *J* = 55.6, 3.2 Hz, 1H), 7.18–7.22 (*m*, 1H), 7.36–7.42 (*m*, 3H), 7.46 (*d*, *J* = 9.2 Hz, 1H), 7.51–7.54 (*m*, 2H), 8.72 (*s*, 1H); ^{13}C-NMR (100 MHz, $CDCl_3$:CD_3OD = 7:1), δ_C, ppm: 66.3 (*t*, *J* = 25.0 Hz), 115.3 (*t*, *J* = 245.0 Hz), 116.2, 116.9, 120.5, 125.3, 127.2,

128.5, 128.6, 128.8, 132.8, 144.0, 146.3; ^{19}F NMR (376 MHz, $CDCl_3$:CD_3OD = 7:1), δ_F, ppm: −123.99 (*d*, *J* = 282.0 Hz, 1F), −124.97 (*d*, *J* = 282.0 Hz, 1F); ESI-HRMS, *m*/*z*: Calcd for $C_{15}H_{12}ClF_2N_2O$ [M + H]$^+$, 309.0601, found: 309.0616.

(±)-7-Chloro-3-(2,2-difluoro-1-hydroxy)ethyl-7-chloro-2-phenyl-imidazo[1,2-a]pyridine (**3af**), white solid (56mg, 91%); m.p. 219–221 °C; ^{1}H-NMR (400 MHz, $CDCl_3$:CD_3OD = 7:1), δ_H, ppm: 5.34–5.41 (*m*, 1H), 6.16 (*t*, *J* = 55.6 Hz, 1H), 6.83 (*d*, *J* = 7.2 Hz, 1H), 7.33–7.47 (*m*, 3H); 7.59 (*d*, *J* = 4.4 Hz, 2H), 8.62 (*d*, *J* = 7.2 Hz, 1H); ^{13}C-NMR (100 MHz, DMSO-*d6*), δ_C, ppm: 65.8 (*t*, *J* = 24.0 Hz), 113.8, 116.0, 116.3 (*t*, *J* = 243.0 Hz), 116.8, 128.8, 128.9, 129.0, 129.2, 131.3, 133.8, 145.2, 146.0; ^{19}F NMR (376 MHz, $CDCl_3$:CD_3OD = 7:1), δ_F, ppm: −124.13 (*d*, *J* = 285.8 Hz, 1F), −125.03 (*d*, *d*, *J* = 285.8 Hz, 1F); ESI-HRMS, *m*/*z*: Calcd for $C_{15}H_{12}ClF_2N_2O$ [M + H]$^+$, 309.0601, found: 309.0616.

(±)-8-Chloro-3-(2,2-difluoro-1-hydroxy)ethyl-8-chloro-2-phenyl-imidazo[1,2-a]pyridine (**3ag**), white solid (55 mg, 89%); m.p. 233–235 °C; ^{1}H-NMR (400 MHz, $CDCl_3$:CD_3OD = 7:1), δ_H, ppm: 5.32–5.39 (*m*, 1H), 6.08 (*td*, *J* = 55.6, 3.2 Hz, 1H), 6.78–6.82 (*m*, 1H), 7.34–7.45 (*m*, 4H), 7.62 (*d*, *J* = 6.4 Hz, 2H), 8.64 (*d*, *J* = 7.2 Hz, 1H); ^{13}C-NMR (100 MHz, DMSO-d6), δ_C, ppm: 65.9 (*t*, *J* = 25.0 Hz), 112.5, 116.2 (*t*, *J* = 243.0 Hz), 118.4, 121.8, 125.2, 127.2, 128.9, 129.2, 129.2, 133.8, 142.5, 145.6; ^{19}F NMR (376 MHz, $CDCl_3$:CD_3OD = 7:1), δ_F, ppm: −124.06 (*d*, *J* = 282.0 Hz, 1F), −125.11 (*d*, *J* = 282.0 Hz, 1F); ESI-HRMS, *m*/*z*: Calcd for $C_{15}H_{12}ClF_2N_2O$ [M + H]$^+$, 309.0601, found: 309.0616.

(±)-8-Bromo-3-(2,2-difluoro-1-hydroxy)ethyl-2-phenyl-imidazo[1,2-a]pyridine (**3ah**), white solid (58 mg, 83%); m.p. 259–261 °C; ^{1}H-NMR (400 MHz, $CDCl_3$:CD_3OD = 7:1), δ_H, ppm: 5.32–5.38 (*m*, 1H), 6.01–6.30 (*m*, 1H), 6.73–6.77 (*m*, 1H), 7.41–7.46 (*m*, 3H), 7.55 (*d*, *J* = 6.8 Hz, 1H), 7.64 (*d*, *J* = 6.8 Hz, 2H), 8.69 (*d*, *J* = 6.0 Hz, 1H); ^{13}C-NMR (100 MHz, DMSO-d6), δ_C, ppm: 66.0 (*t*, *J* = 25.0 Hz), 110.6, 113.0, 116.2 (*t*, *J* = 243.0 Hz), 118.4, 127.6, 128.6, 128.8, 129.1, 129.2, 133.8, 143.1, 145.6; ^{19}F NMR (376 MHz, $CDCl_3$:CD_3OD = 7:1), δ_F, ppm: −123.98 (*d*, *J* = 282.0 Hz, 1F), −125.16 (*d*, *J* = 282.0 Hz, 1F); ESI-HRMS, *m*/*z*: Calcd for $C_{15}H_{12}BrF_2N_2O$ [M + H]$^+$, 353.0096, found: 353.0107.

(±)-3-(2,2-Difluoro-1-hydroxy)ethyl-2-phenyl-6-(trifluoromethyl)-imidazo[1,2-a]pyridine (**3ai**), white solid (61 mg, 89%); m.p. 193–195 °C; ^{1}H-NMR (400 MHz, $CDCl_3$:CD_3OD = 7:1), δ_H, ppm: 5.35–5.43 (*m*, 1H), 6.17 (*td*, *J* = 55.6, 2.8 Hz, 1H), 7.33–7.39 (*m*, 4H), 7.48–7.51 (*m*, 2H), 7.59 (*d*, *J* = 9.8 Hz, 1H), 9.02 (*s*, 1H); ^{13}C-NMR (100 MHz, $CDCl_3$:CD_3OD = 7:1), δ_C, ppm: 66.1 (*t*, *J* = 25.0 Hz), 115.5 (*t*, *J* = 245.0 Hz), 116.5 (*q*, *J* = 34.0 Hz), 116.9, 117.3, 121.6–121.8 (*m*), 123.5 (*q*, *J* = 270.0 Hz), 126.8 (*q*, *J* = 5.0 Hz), 128.6, 128.7, 128.7, 132.4, 145.4, 147.1; ^{19}F NMR (376 MHz, $CDCl_3$:CD_3OD = 7:1), δ_F, ppm: −62.05, −123.79 (*d*, *J* = 282.0 Hz, 1F), −125.09 (*d*, *J* = 282.0 Hz, 1F); ESI-HRMS, *m*/*z*: Calcd for $C_{16}H_{12}F_5N_2O$ [M + H]$^+$, 343.0864, found: 343.0857.

(±)-3-(2,2-Difluoro-1-hydroxy)ethyl-2-phenyl-7-(trifluoromethyl)-imidazo[1,2-a]pyridine (**3aj**), white solid (60 mg, 87%); m.p. 206–208 °C; ^{1}H-NMR (400 MHz, CD_3OD), δ_H, ppm: 5.37–5.45 (*m*, 1H), 6.36 (*td*, *J* = 55.6, 3.6 Hz, 1H), 7.18 (*dd*, *J* = 6.4, 3.2 Hz, 1H), 7.44–7.54 (*m*, 3H), 7.66–7.69 (*m*, 2H); 7.93 (*s*, 1H), 8.96 (*d*, *J* = 7.2 Hz, 1H); ^{13}C-NMR (CD_3OD), δ_C, ppm: 66.1 (*t*, *J* = 25.0 Hz), 107.6–107.7 (*m*), 113.8 (*q*, *J* = 4.0 Hz), 115.8 (*t*, *J* = 245.0 Hz), 118.0 (*t*, *J* = 3.0 Hz), 123.4 (*q*, *J* = 272.0 Hz), 127.5 (*q*, *J* = 34.0 Hz), 128.4, 128.6, 128.8, 129.0, 132.7, 143.7, 147.3; ^{19}F NMR (376 MHz, CD_3OD), δ_F, ppm: −65.30, −125.65 (*d*, *J* = 285.8 Hz, 1F), −126.62 (*d*, *J* = 285.8 Hz, 1F); ESI-HRMS, *m*/*z*: Calcd for $C_{16}H_{12}F_5N_2O$ [M + H]$^+$, 343.0864, found: 343.0857.

(±)-6-Bromo-3-(2,2-difluoro-1-hydroxy)ethyl-7-methyl-2-phenyl-6-(trifluoro-methyl)-imidazo[1,2-a]pyridine (**3ak**), white solid (66 mg, 90%); m.p. 237–239 °C; ^{1}H-NMR (400 MHz, DMSO-d6), δ_H, ppm: 5.32–5.39 (*m*, 1H), 6.59 (*t*, *J* = 53.2 Hz, 1H), 6.86–6.88 (*m*, 1H), 7.40–7.51 (*m*, 3H), 7.54–7.70 (*m*, 3H), 8.85 (*s*, 1H); ^{13}C-NMR (100 MHz, DMSO-d6), δ_C, ppm: 22.4, 65.8 (*t*, *J* = 24.0 Hz), 110.5, 116.0, 116.3 (*t*, *J* = 243.0 Hz), 116.6, 127.4, 127.7, 129.0, 129.1, 134.0, 135.7, 144.7, 145.6; ^{19}F NMR (376 MHz, DMSO-d6), δ_F, ppm: −123.85 (*d*, *J* = 282.0 Hz, 1F), −124.80

(*d*, *J* = 282.0 Hz, 1F); ESI-HRMS, *m*/*z*: Calcd for $C_{16}H_{14}BrF_2N_2O$ [M + H]$^+$, 367.0252, found: 367.0244.

3-(2,2-Difluoroacetyl)-imidazo[1,2-a]pyridine (**4a**), white solid (15 mg, 77%); m.p. 102–104 °C; ^{1}H-NMR (400 MHz, $CDCl_3$), δ_H, ppm: 6.26 (*t*, *J* = 53.6 Hz, 1H), 7.21–7.25 (*m*, 1H), 7.63–7.68 (*m*, 1H), 7.87 (*d*, *J* = 9.2 Hz, 1H), 8.63 (*s*, 1H), 9.62 (*d*, *J* = 6.8 Hz, 1H); ^{13}C-NMR (100 MHz, $CDCl_3$), δ_C, ppm: 111.1 (*t*, *J* = 252.0 Hz), 116.2, 118.1, 119.9, 129.0, 131.0, 146.0 (*t*, *J* = 6.0 Hz), 149.7, 177.0 (*t*, *J* = 26.0 Hz); ^{19}F NMR (376 MHz, $CDCl_3$), δ, ppm: −121.40; ESI-HRMS, *m*/*z*: Calcd for $C_9H_7F_2N_2O$ [M + H]$^+$, 197.0521, found: 197.0539.

3-(2,2-Difluoroacetyl)-2-phenyl-imidazo[1,2-a]pyridine (**4b**), white solid (22 mg, 79%); m.p. 121–123 °C; ^{1}H-NMR (400 MHz, $CDCl_3$), δ_H, ppm: 5.91 (*t*, *J* = 53.2 Hz, 1H), 7.23–7.26 (*m*, 1H), 7.52–7.58 (*m*, 3H), 7.62–7.65 (*m*, 2H), 7.66–7.72 (*m*, 1H), 7.85 (*d*, *J* = 8.8 Hz, 1H), 9.78 (*d*, *J* = 6.8 Hz, 1H); ^{13}C-NMR (100 MHz, $CDCl_3$), δ_C, ppm: 106.0 (*t*, *J* = 244.0 Hz), 116.2, 117.7, 119.0, 128.8, 129.4, 129.7, 130.2, 131.6, 133.7, 148.5, 157.5, 177.2 (*t*, *J* = 25.0 Hz); ^{19}F NMR (376 MHz, $CDCl_3$), δ_F, ppm: −124.48; ESI-HRMS, *m*/*z*: Calcd for $C_{15}H_{11}F_2N_2O$ [M + H]$^+$, 273.0834, found: 273.0830.

4. Conclusions

In conclusion, we have developed a facile and efficient method for the synthesis of C3-difluoromethyl carbinol-containing imidazo[1,2-*a*]pyridines via HFIP-promoted direct C(sp^2)-H hydroxydifluoromethylation. A small library of difluoromethylated carbinols were prepared at room temperature in good to high yields by the practical green method. This HFIP-promoted strategy exhibited some definite benefits, such as being transition metal- and oxidant-free and having wide substrate generality, excellent functional group tolerance and mild reaction conditions. In addition, gram-scale and synthetic transformation experiments have also been demonstrated. Therefore, this simple and green synthesis strategy might be attractive for the further design and rapid synthesis of potentially bioactive fluorinated heterocyclic derivatives with multifunctional groups.

Supplementary Materials: The following supporting information can be downloaded at: https://www.mdpi.com/article/10.3390/molecules28227522/s1, which contains details on the crystallographic information parameters (Table S1), experimental procedure for compounds **1b–1c**, experimental procedure for compounds **1z–1ac**, ^{1}H, ^{13}C and ^{19}F NMR spectra for all compounds **3a–3ak** and **4a–4b**. References [24,37] are cited in the supplementary materials.

Author Contributions: Conceptualization, K.Y. and Z.C.; methodology, J.G., X.G. and Z.L.; formal analysis, L.W.; data curation, L.W. and Z.C.; writing—original draft preparation, J.G. and K.Y.; writing—review and editing, K.Y. and Z.L.; project administration, K.Y.; funding acquisition, Z.C. and K.Y. All authors have read and agreed to the published version of the manuscript.

Funding: This research was supported by the National Natural Science Foundation of China (No. 81960883), the Natural Science Foundation of Jiangxi Province (No. 20224BAB203010), Scientific Research Fund of Jiangxi Provincial Education Department (No. GJJ201504) and Scientific Research Project of Gannan Medical University (No. YB201903).

Institutional Review Board Statement: Not applicable.

Informed Consent Statement: Not applicable.

Data Availability Statement: All data supporting the findings of this study are available within the paper and within its Supplementary Materials published online.

Conflicts of Interest: The authors declare no conflict of interest.

References

1. Zafrani, Y.; Sod-Moriah, G.; Yeffet, D.; Berliner, A.; Amir, D.; Marciano, D.; Elias, S.; Katalan, S.; Ashkenazi, N.; Madmon, M.; et al. CF_2H, a functional group-dependent hydrogen-bond donor: Is it a more or less lipophilic bioisostere of OH, SH and CH_3? *J. Med. Chem.* **2019**, *62*, 5628–5637. [CrossRef] [PubMed]

2. Gillis, E.P.; Eastman, K.J.; Hill, M.D.; Donnelly, D.J.; Meanwell, N.A. Applications of Fluorine in Medicinal Chemistry. *J. Med. Chem.* **2015**, *58*, 8315–8359. [CrossRef] [PubMed]
3. Liu, J.; Yu, D.; Yang, Y.; You, H.; Sun, M.; Wang, Y.; Shen, X.; Liu, Z.Q. Free-radical-promoted Dehydrogenative coupling of polyfluorinated alcohol with quinone, chromone, and coumarin. *Org. Lett.* **2020**, *22*, 4844–4847. [CrossRef] [PubMed]
4. Zafrani, Y.; Yeffet, D.; Sod-Moriah, G.; Berliner, A.; Amir, D.; Marciano, D.; Gershonov, E.; Saphier, S. Difluoromethyl bioisostere: Examining the "lipophilic hydrogen bond donor" concept. *J. Med. Chem.* **2017**, *60*, 797–804. [CrossRef]
5. Selmi-Higashi, E.; Zhang, J.L.; Cambeiro, X.C.; Arseniyadis, S. Synthesis of α-difluoromethyl aryl ketones through a photoredox difluoromethylation of enol silanes. *Org. Lett.* **2021**, *23*, 4239–4243. [CrossRef]
6. Chen, X.Y.; Liu, B.; Pei, C.C.; Li, J.Y.; Zou, D.P.; Wu, Y.J.; Wu, Y.S. Visible-light-induced radical difluoromethylation/cyclization of unactivated alkenes: Access to CF_2H-substituted quinazolinones. *Org. Lett.* **2021**, *23*, 7787–7791. [CrossRef]
7. Zhang, W.; Xiang, X.X.; Chen, J.Y.; Yang, C.; Pan, Y.L.; Cheng, J.P.; Meng, Q.B.; Li, X. Direct C-H difluoromethylation of heterocycles via organic photoredox catalysis. *Nat. Commun.* **2020**, *11*, 638. [CrossRef]
8. Hertel, L.W.; Boder, G.B.; Kroin, J.S.; Rinzel, S.M.; Poore, G.A.; Todd, G.C.; Grindey, G.B. Evaluation of the antitumor activity of gemcitabine (2′,2′-Difluoro-2′-deoxycytidine). *Cancer Res.* **1990**, *50*, 4417–4422.
9. Rudd, M.T.; Bennett, D.J.; Wai, J.; Meng, Z. Homobispiperidinyl Derivatives as Liver X Receptor Beta Agonists, Compositions, and Their Use. Patent WO2017095758A1, 8 June 2017.
10. Tanaka, Y.; Kikuchi, F.; Yamamoto, T.; Nakamura, M.; Takami, K.; Murakami, M.; Daini, M.; Wada, Y.; Kakegawa, K.; Kasahara, T.; et al. Heterocyclic Compound for Gaucher Disease. Patent WO2021020363, 4 February 2021.
11. Wacker, D.A.; Nara, S.J.; Cheruku, S.; Sarkunam, K.; Jaipuri, F.A.; Thangavel, S.; Narayan, R.; Bandreddy, S.R.; Jogi, S.; Kathi, P.K. Substituted Bicyclic Compounds as Farnesoid X Receptor Modulators and Their Preparation. Patent WO2020168143, 20 August 2020.
12. Boisnard, S.; El-Ahmad, Y.; Fett, E.; Halley, F.; Nicolai, E.; Tabart, M.; Terrier, C.; Vivet, B. Preparation of Novel Substituted 6,7-Dihydro-5H-benzo[7]annulene Compounds as Inhibitors and Degraders of Estrogen Receptors. Patent WO2021063967, 8 April 2021.
13. Zhang, Y.X.; Yan, W.T.; Wang, Y.K.; Weng, Z.Q. Copper-catalyzed synthesis of indol-3-yl α-(difluoromethyl)-α-(trifluoromethyl) carbinols: Construction of difluoromethylated sp^3 carbon centers. *Org. Lett.* **2017**, *19*, 5478–5481. [CrossRef]
14. Liu, B.; Chen, X.Y.; Pei, C.C.; Li, J.Y.; Zou, D.P.; Wu, Y.J.; Wu, Y.S. Ruthenium-catalyzed ortho-C–H hydroxyfluoroalkylation of arenes with fluorinated alcohols. *J. Org. Chem.* **2022**, *87*, 14364–14373. [CrossRef]
15. Devi, N.; Singh, D.; K. Rawal, R.; Bariwal, J.; Singh, V. Medicinal attributes of imidazo[1,2-*a*]pyridine derivatives: An update. *Curr. Top. Med. Chem.* **2016**, *16*, 2963–2994. [CrossRef] [PubMed]
16. Abe, Y.; Kayakiri, H.; Satoh, S.; Inoue, T.; Sawada, Y.; Inamura, N.; Asano, M.; Aramori, I.; Hatori, C.; Sawai, H.; et al. A Novel Class of Orally Active Non-Peptide Bradykinin B2 Receptor Antagonists. 4. Discovery of Novel Frameworks Mimicking the Active Conformation. *J. Med. Chem.* **1998**, *41*, 4587–4598. [CrossRef] [PubMed]
17. Okubo, T.; Yoshikawa, R.; Chaki, S.; Okuyama, S.; Nakazato, A. Design, synthesis and structure-affinity relationships of aryloxyanilide derivatives as novel peripheral benzodiazepine receptor ligands. *Bioorg. Med. Chem.* **2004**, *12*, 423–438. [CrossRef] [PubMed]
18. Shukla, N.M.; Salunke, D.B.; Yoo, E.; Mutz, C.A.; Balakrishna, R.; David, S.A. Antibacterial activities of Groebke–Blackburn–Bienaymé-derived imidazo[1,2-*a*]pyridin-3-amines. *Biorg. Med. Chem.* **2012**, *20*, 5850–5863. [CrossRef]
19. Feng, S.; Hong, D.; Wang, B.; Zheng, X.; Miao, K.; Wang, L.; Yun, H.; Gao, L.; Zhao, S.; Shen, H.C. Discovery of Imidazopyridine Derivatives as Highly Potent Respiratory Syncytial Virus Fusion Inhibitors. *ACS Med. Chem. Lett.* **2015**, *6*, 359–362. [CrossRef]
20. Ismail, M.A.; Arafa, R.K.; Wenzler, T.; Brun, R.; Tanious, F.A.; Wilson, W.D.; Boykin, D.W. Synthesis and antiprotozoal activity of novel bis-benzamidino imidazo[1,2-*a*]pyridines and 5,6,7,8-tetrahydro-imidazo[1,2-*a*]pyridines. *Biorg. Med. Chem.* **2008**, *16*, 683–691. [CrossRef]
21. Kim, O.; Jeong, Y.; Lee, H.; Hong, S.-S.; Hong, S. Design and synthesis of imidazopyridine analogues as inhibitors of phosphoinositide 3-kinase signaling and angiogenesis. *J. Med. Chem.* **2011**, *54*, 2455–2466. [CrossRef]
22. Ma, C.H.; Chen, M.; Feng, Z.W.; Zhang, Y.; Wang, J.; Jiang, Y.-Q.; Yu, B. Functionalization of imidazo[1,2-*a*]pyridines via radical reactions. *New J. Chem.* **2021**, *45*, 9302–9314. [CrossRef]
23. Ji, X.-M.; Wei, L.; Chen, F.; Tang, R.-Y. Direct trifluoromethylation of imidazoheterocycles in a recyclable medium at room temperature. *RSC Adv.* **2015**, *38*, 29766–29773. [CrossRef]
24. Nipate, D.S.; Jaspal, S.; Shinde, V.N.; Rangan, K.; Kumar, A. TEMPO-mediated cross-dehydrogenative coupling of indoles and imidazo[1,2-*a*]pyridines with fluorinated alcohols. *Org. Lett.* **2021**, *23*, 1373–1377. [CrossRef]
25. Lefebvre, Q.; Hoffmann, N.; Rueping, M. Photoorganocatalysed and visible light photoredox catalysed trifluoromethylation of olefins and (hetero)aromatics in batch and continuous flow. *Chem. Commun.* **2016**, *52*, 2493–2496. [CrossRef] [PubMed]
26. Tfilla, P.; Baeza, A.; Najera, C. Fluorinated Alcohols As Promoters for the Metal-Free Direct Substitution Reaction of Allylic Alcohols with Nitrogenated, Silylated, and Carbon Nucleophiles. *J. Org. Chem.* **2012**, *77*, 7344–7354.
27. Pérez, J.M.; Maquilón, C.; Ramón, D.J.; Baeza, A. Hexafluoroisopropanol-Promoted Metal-Free Allylation of Silyl Enol Ethers with Allylic Alcohols. *Asian J. Org. Chem.* **2017**, *6*, 1440–1444. [CrossRef]

28. Yang, J.G.; Gui, J.; Mu, M.M.; Liu, S.M.; Li, J.S.; Ren, J.; Wang, Z.M. Synthesis of difluoromethylated carbinols via a HFIP-promoted hydroxy-difluoromethylation of aniline, indole, and pyrrole derivatives with difluoroacetaldehyde ethyl hemiacetal. *J. Org. Chem.* **2023**, *88*, 4790–4798. [CrossRef] [PubMed]
29. Motiwala, H.F.; Armaly, A.M.; Cacioppo, J.G.; Coombs, T.C.; Koehn, K.R.K.; Norwood, V.M., IV; Aubé, J. HFIP in organic synthesis. *Chem. Rev.* **2022**, *122*, 12544–12747. [CrossRef]
30. Chen, Y.X.; Wang, Y.R.; Zhong, R.; Li, J.S. HFIP promoted C3 alkylation of lawsone and 4-hydroxycoumarin with alcohols by dehydrative cross-coupling. *J. Org. Chem.* **2020**, *85*, 10638–10647. [CrossRef]
31. Yang, J.; Liu, S.; Hong, P.; Li, J.; Wang, Z.; Ren, J. Synthesis of 2,2-Difluoro-3-hydroxy-1,4-diketones via an HFIP-Catalyzed Mukaiyama Aldol Reaction of Glyoxal Monohydrates with Difluoroenoxysilanes. *J. Org. Chem.* **2022**, *87*, 1144–1153. [CrossRef]
32. CCDC: 2295549 (for **3b**) Contain the Supplementary Crystallographic Data for This Paper. These Data Are Provided Free of Charge from the Cambridge Crystallographic Data Centre. Available online: www.ccdc.cam.ac.uk/data_request/cif (accessed on 11 October 2023).
33. Li, J.; Xi, W.; Liu, S.; Ruan, C.; Zheng, X.; Yang, J.; Wang, L.; Wang, Z. HFIP-catalyzed diflfluoroalkylation of propargylic alcohols to access tetrasubstituted diflfluoroalkyl allenes. *Org. Lett.* **2021**, *23*, 7264–7269. [CrossRef]
34. Li, G.-X.; Qu, J. Friedel-Crafts alkylation of arenes with epoxides promoted by fluorinated alcohols or water. *Chem. Commun.* **2010**, *46*, 2653–2655. [CrossRef]
35. Yang, J.; Liu, S.; Gui, J.; Xiong, D.; Li, J.; Wang, Z.; Ren, J. HFIP-Promoted selective hydroxyalkylation of aniline derivatives with arylglyoxal hydrates. *J. Org. Chem.* **2022**, *87*, 6352–6361. [CrossRef]
36. Obermayer, D.; Znidar, D.; Glotz, G.; Stadler, A.; Dallinger, D.; Kappe, C.O. Design and performance validation of a conductively heated sealed-vessel reactor for organic Synthesis. *J. Org. Chem.* **2016**, *81*, 11788–11801. [CrossRef] [PubMed]
37. Ghosh, P.; Ganguly, B.; Kar, B.; Dwivedi, S.; Das, S. Green procedure for highly efficient, rapid synthesis of imidazo[1,2-*a*]pyridine and its late stage functionalization. *Synth. Commun.* **2018**, *48*, 1076. [CrossRef]

molecules

Review

Quinolin-4-ones: Methods of Synthesis and Application in Medicine

Katarzyna Gach-Janczak , Justyna Piekielna-Ciesielska, Jakub Waśkiewicz, Kamil Krakowiak, Karol Wtorek and Anna Janecka *

Department of Biomolecular Chemistry, Medical University of Lodz, Mazowiecka 6/8, 92-215 Lodz, Poland; katarzyna.gach@umed.lodz.pl (K.G.-J.); justyna.piekielna@umed.lodz.pl (J.P.-C.); jakub.waskiewicz@student.umed.lodz.pl (J.W.); kamil.krakowiak@student.umed.lodz.pl (K.K.); karol.wtorek@umed.lodz.pl (K.W.)
* Correspondence: anna.janecka@umed.lodz.pl

Abstract: Quinolinones, also called quinolones, are a group of heterocyclic compounds with a broad spectrum of biological activities. These compounds occur naturally in plants and microorganisms but can also be obtained synthetically. The first synthesis of quinolinones took place at the end of the 19th century, and the most recent methods were published just a few years ago. They allow for obtaining an unlimited number of analogs differing in biological properties. In this review, we described the plethora of methods leading to quinolin-4-ones. Several of these compounds have been used as antibiotics for over four decades, but recently, their antiproliferative effects have been of particular interest to researchers. This review summarizes the experimental progress made in the synthetic development of various routes leading to quinoline-4-ones and presents an overview of the structures, their evolution, and their relation to activity.

Keywords: chemical synthesis; antibacterial agents; antiviral agents; anticancer activity; structure–activity relationship

1. Introduction

Heterocycles containing a nitrogen atom in their structure are a very important group of organic compounds that have found numerous applications in organic synthesis, medicinal chemistry, and everyday life. Alkaloids, antibiotics, vitamins, hormones, and a large number of drugs and dyes contain azaheterocyclic ring systems. The properties of azaheterocyclic compounds depend on the specific structure of the heterocyclic ring and the type of functional groups present in the molecule. These compounds often exhibit biological activities such as antibacterial, antifungal, or anticancer [1].

An interesting group of azaheterocyclic compounds is quinolinones (also called quinolones or azaflavones in some sources). They contain a quinoline ring system linked to a carbonyl group and can be divided into quinolin-2(1*H*)-ones and quinolin-4(1*H*)-ones (Figure 1).

Quinolinones are relatively rare in nature. The quinolin-4-one skeleton can be found in plants of the *Rutaceae*, *Rubiaceae*, and *Astraceae* families and in microorganisms [2]. Some of the naturally occurring quinolin-4-ones with interesting activities isolated from plants are shown in Figure 2.

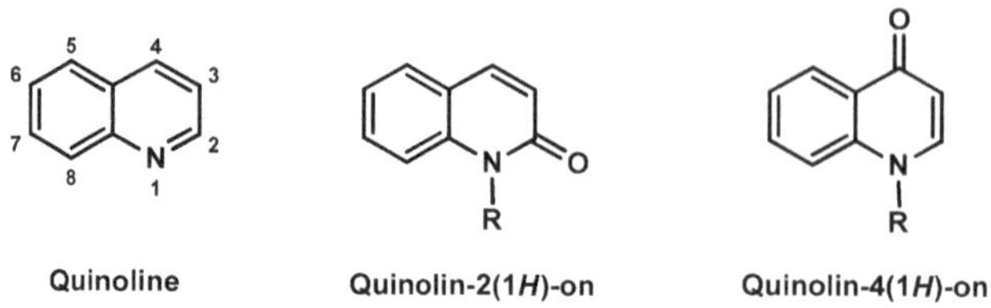

Figure 1. The structure of quinoline and quinolinones.

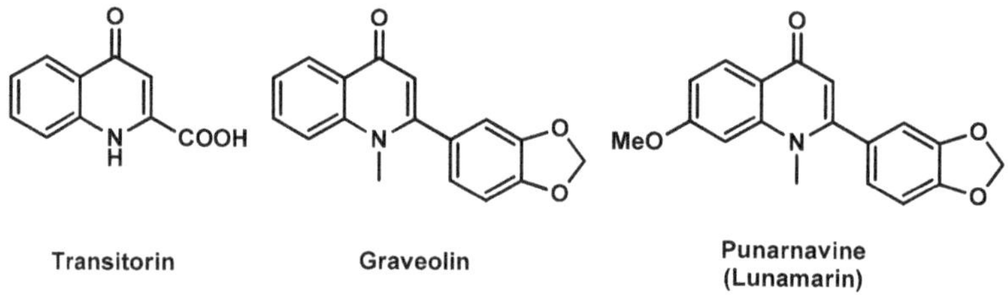

Figure 2. Quinolin-4-ones of plant origin with interesting biological properties.

Transitorin inhibits the growth of bacteria such as *Staphylococcus aureus*, *Pseudomonas aeruginosa*, and Enterobacter cloacae [3]. **Graveolin** has versatile pharmacological activity, including antifungal [4] and antimicrobial [5] properties. Graveolin has also been shown to have cytotoxic activity against A375 cutaneous melanoma cells [6]. Synthetic derivatives of graveolin inhibited the process of angiogenesis during the development of malignant lesions [7].

A derivative of graveolin with a methoxy group in position 7 is **punarnavine** (also called lunamarin). This compound was isolated from the roots of *Boerhaavia diffusa* (family *Nyctaginaceae*), a plant used in Indian folk medicine. Punarnavine was shown to possess various properties, including hepatoprotective [8], immunomodulatory [9], anti-stress [10], and anti-oxidative [11]. Recently, attention has also been drawn to the anticancer effects of punarnavine. In low, non-toxic concentrations, this compound induces apoptosis in B16F-10 mouse melanoma cells [12]. In studies using a mouse model of breast tumor 4T1, which mimics stage four human breast cancer, punarnavine was shown to inhibit primary tumor growth and reduce the progression of metastasis to lymph nodes, liver, and lung. In addition, punarnavine is non-genotoxic [13] and inhibits angiogenesis [14], making it a good candidate for an anticancer drug.

Among quinolin-4-ones of therapeutic importance isolated from microorganisms, the most interesting is **intervenolin** (Figure 3), found in aerobic Gram-positive bacteria of the genus *Nocardia* (family *Nocardiaceae*) [15]. This alkaloid selectively inhibited the proliferation of human gastric cancer cells MKN-74 and MKN-7 and colon cancer cells HTC-15.

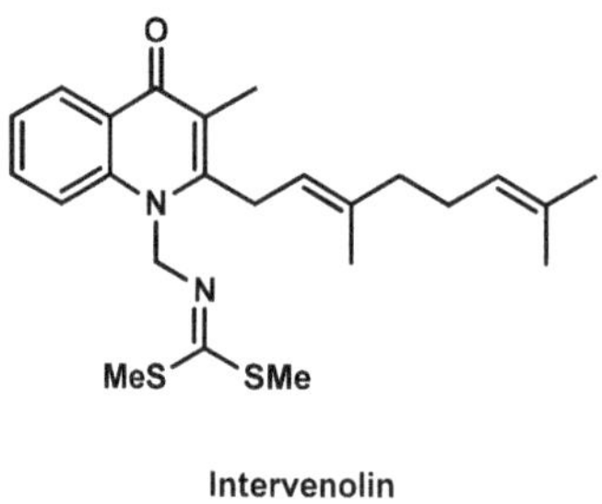

Figure 3. The structure of intervenolin, a quinolin-4-one isolated from microorganisms.

Two unique substituents can be distinguished in the structure of this compound: a bis(methylthio)methyleneaminomethyl group on the nitrogen atom and a geranyl chain on the C2, which both have a strong effect on the biological activity of intervenolin [16].

Despite their interesting biological activities, none of the quinoline-4-ones found in nature has been recommended as a drug so far. However, due to the large number of potential functionalizations at different positions, quinoline-4-ones are attractive precursors for the synthesis of biologically important structures, some of which have already become approved medications. Here, we focused on showing various routes leading to synthetic quinoline-4-ones and discussing their medical uses.

2. Chemical Methods of Synthesis of Quinolin-4-ones

Over the years, many methods have been developed for the synthesis of the quinolin-4-one core. Due to the high possibility of functionalization of the molecule, different strategies and starting materials have been used (Figure 4). These compounds can be obtained with classical methods starting from aniline derivatives, for example, via the Gould-Jacobs (route A) or Conrad-Limpach (route B), as well as starting from methyl anthranilate via Biere-Seelen synthesis (route C) or using Dieckmann condensation (route D). Less common 3-substituted quinolin-4-ones were obtained by the Snieckus reaction from an anthranilic acid amide (route E) or the Camps reaction from *N*-(2-acylaryl)amides (route F). Modern methods include transformations catalyzed with transition metals (route G) or *N*-heterocyclic carbenes (route H). Recently, a method has also been developed for the synthesis of quinolin-4-ones from isatoic anhydride via a decarboxylating cyclization reaction (route I).

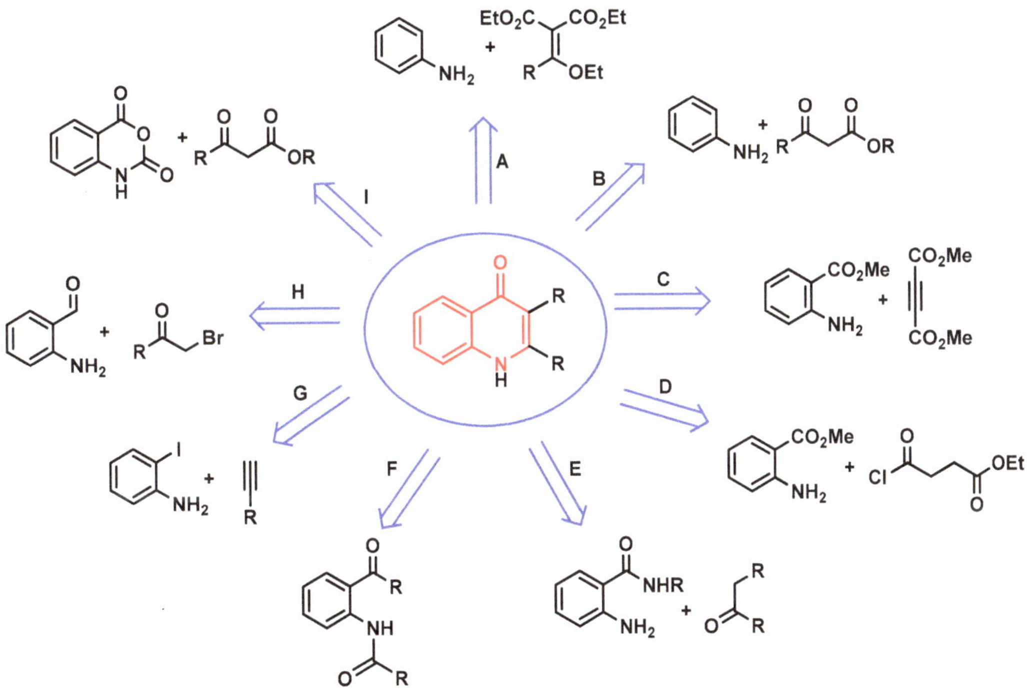

Figure 4. Quinolin-4-one retrosynthesis pathways.

Some other important reviews in the field of quinolinone synthesis should be mentioned here. Rao et al. [17] described quinolinone synthesis by electrochemically and photochemically driven methods. Dine et al. [18] provided protocols for metal-free reactions and transition metal-catalyzed procedures. Shen et al. [19] summarized the experimental progress made in the synthetic development of various catalytic routes to preparing quinoline-4-ones and their analogs.

2.1. Gould–Jacobs Method (Route A)

The Gould–Jacobs reaction [20] is an important thermal cyclization method leading to the quinolin-4-one backbone. The original approach, proposed in 1939 (Scheme 1), in the first step, involves condensation of aniline **1** and diethyl ethoxymethylidenedimalonate **2**. The resulting diester **3**, upon further heating, is readily cyclized through a ketene intermediate compound **4**, usually under drastic conditions, with the formation of 4-hydroxy-3-ethoxycarbonylquinoline **5**. Hydrolysis of the ester **5** to a carboxylic acid **6** and subsequent decarboxylation yields the corresponding quinolin-4-one **7**. This method allows for the synthesis of many commercially available drugs whose structure is based on the quinoline-4-one backbone, such as the antibiotics nalidixic acid, oxolinic acid, and norfloxacin.

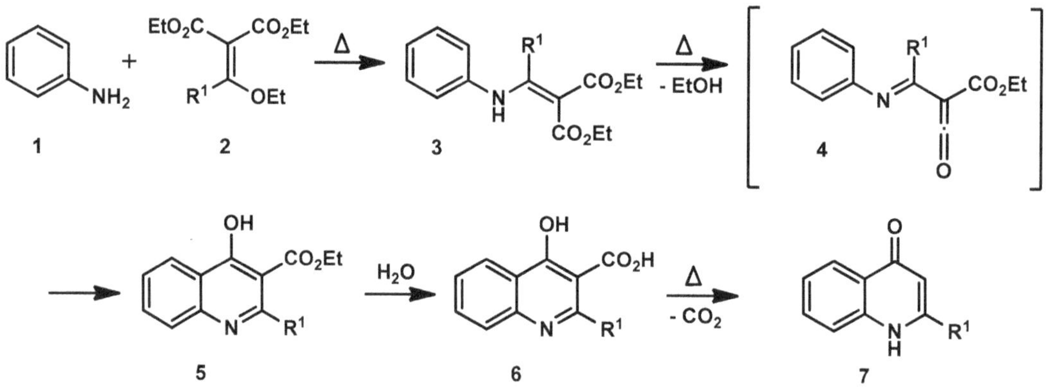

Scheme 1. Gould–Jacobs reaction.

An important aspect is also the regioselectivity of the Gould–Jacobs reaction, which is generally controlled by both steric and electronic factors [21]. Cyclization can occur at either of the two *ortho* positions of aniline, and therefore, when asymmetrically substituted anilines are used, a mixture of products is often obtained.

This classical approach requires a multi-step reaction sequence generally occurring in low overall yields, and temperatures higher than 250 °C are necessary for the cyclization step, which may lead to product decomposition and undesirable side reactions [22].

2.2. Conrad–Limpach Method (Pathway B)

The reaction pathway shown in Scheme 2 was developed by Max Conrad and Leonhard Limpach in 1887 [23] and is still successively used [24]. The method involves the condensation of aniline **1** with acetyl acetic ester **8**, with the formation of iminoester **9**, which is in equilibrium with the iminoenol form **10**. The iminoenol **10** then undergoes an intramolecular hetero-Diels–Alder reaction to form hemiketal **11**, which, after alcohol elimination and keto-enol tautomerization, yields quinolin-4-one **7**. Since in this method, the high-energy iminoenol **10** must be indirectly formed, and temporary dearomatization (hemiketal **11**) occurs, the synthesis requires rather drastic conditions (high temperature, strong acid as catalyst). Thus, in early work, cyclization was carried out by heating the Schiff base to high temperatures (200–250 °C) without using a solvent, but the reaction yields were very moderate (less than 30%). When an inert, high-boiling solvent such as mineral oil, diphenyl ether, or Dowtherm A was used, the cyclization yields increased to 95% in many cases [25].

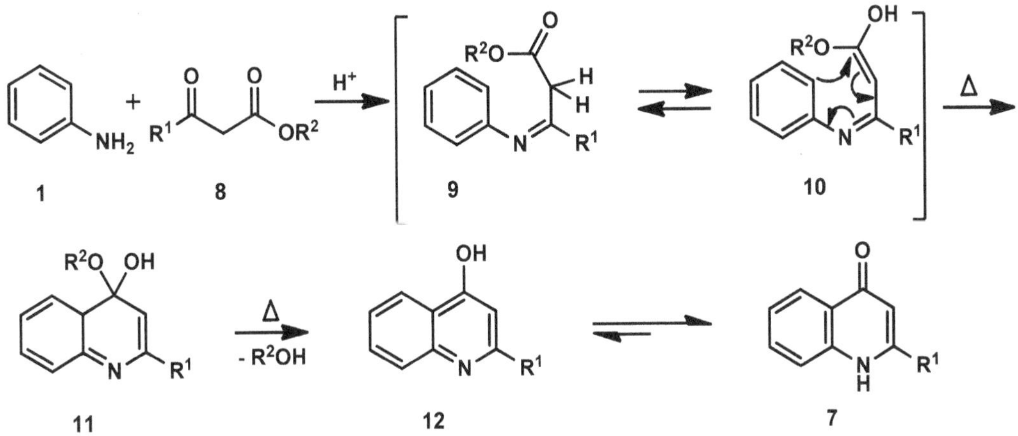

Scheme 2. Conrad–Limpach reaction.

As with the Gould–Jacobs reaction, the presented method has its limitations related to conducting the reaction at very high temperatures. Mineral oil, diphenyl ether, and Dowtherm A, commonly used as solvents, are expensive and difficult to remove from the reaction mixture [26].

2.3. Biere-Seelen's Synthesis (Route C)

Biere and Seelen's approach [27] to the synthesis of quinoline-4-ones was developed in 1979 (Scheme 3). The procedure begins with Michael's addition of dimethyl acetylenedicarboxylate **14** to methyl anthranilate **13**, leading to enaminoester **15**. This ester then undergoes cyclization to quinolin-4-one in the presence of a strong base with the formation of diester **16**. Base hydrolysis using aqueous sodium hydroxide solution occurs regioselectively at position 2 to give ester **17**, while hydrolysis carried out in 2-propanol at boiling temperature leads to the formation of dicarboxylic acid **18**. The final step is the thermal decarboxylation of ester **17** (method a) to produce a mixture of quinolin-3-carboxylic acid **19** and its methyl ester, which are easily separated by column chromatography (14% yield). In contrast, this reaction carried out with the addition of sodium hydroxide solution (method b) yielded only acid **19** with a yield of 92%. The authors also carried out copper-catalyzed thermal decarboxylation of both monoester **17** and dicarboxylic acid **18** (method c), which yielded acid **19** in 52% yield.

The cycloacylation of aniline derivatives to quinoline-4-ones in the presence of Eaton's reagent was also described [28]. This high-yielding methodology is applicable to a wide variety of functionalized anilines and requires milder conditions than those traditionally employed.

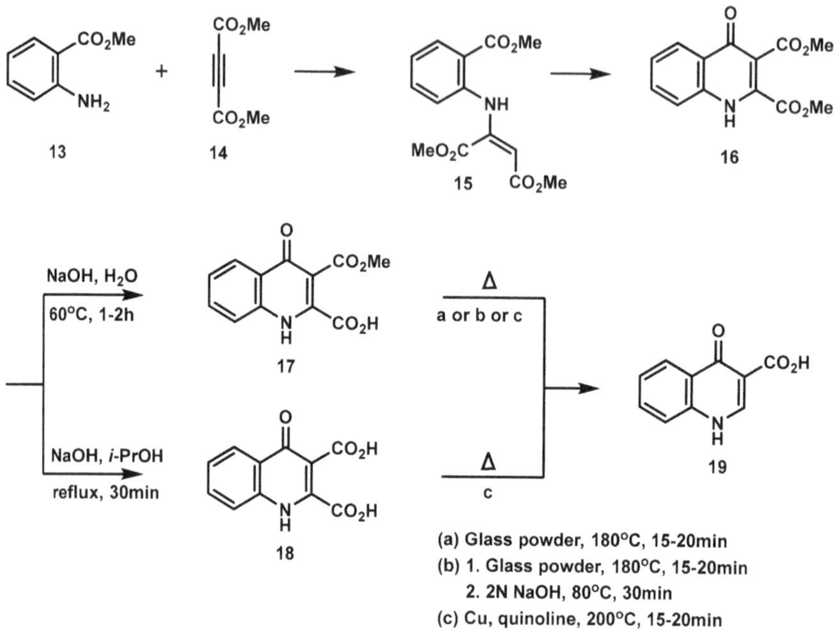

Scheme 3. Biere-Seelen's reaction.

2.4. Dieckmann's Condensation (Route D)

Dieckmann's condensation [29,30] is an intramolecular reaction that diesters undergo with the formation of cyclic β-ketoesters. As a result of the reaction of methyl anthranilate **13** with methyl acrylate **20**, a diester **21** can yield, which undergoes intramolecular cyclization in the presence of sodium hydride to give dihydroquinolinone **22** (Scheme 4). Subsequent oxidation with chloranil leads to quinolin-4-one **23**.

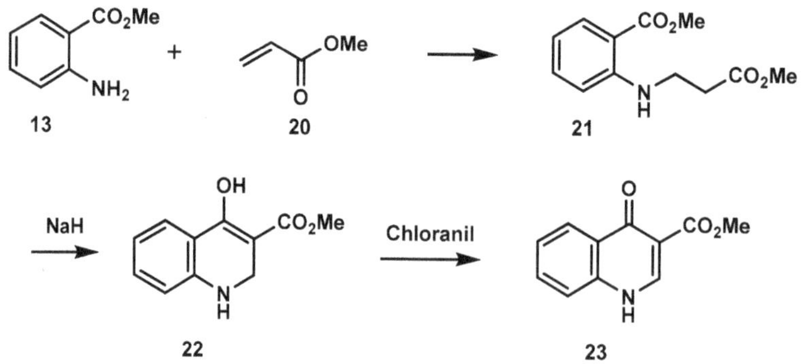

Scheme 4. Dieckmann's reaction.

2.5. Snieckus' Synthesis (Route E)

One of the few methods for the synthesis of 3-substituted quinolin-4-ones was developed by the Victor Snieckus group (Scheme 5) [31]. They proposed a regioselective sequence of *ortho*-metallation reactions with subsequent amination, which is a modification of the Niementovsky reaction, allowing for the obtaining of γ-hydroxyquinoline derivatives. Condensation of an anthranilic acid amide **24** with a ketone **25** gave imine **26**, which was difficult to purify, so it was used directly for the subsequent cyclization reaction. Treatment of imine **26** with a strong base (LDA) led to quinolin-4-one **27**. The Snieckus' method appears to be an interesting route for the synthesis of 3-substituted quinolin-4-ones due to the availability of substituted anthranilic acid amides and the relatively mild conditions of the base-promoted reaction.

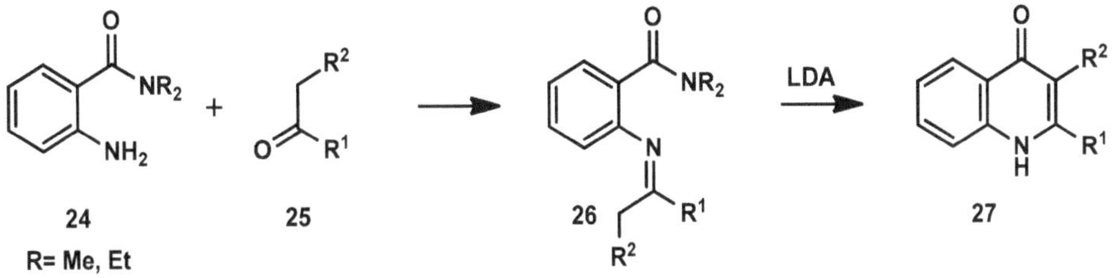

Scheme 5. Snieckus' reaction.

2.6. Cyclization of N-(2-Acylaryl)amides by Camps' Method (Route F)

In 1899, Rudolf Camps [32,33] presented the synthesis of quinolin-2-ones and quinolin-4-ones by base-catalyzed intramolecular cyclization of *N*-(2-acylaryl)amides **28**. Depending on the structure of the substrate and the conditions used, the reaction leads to the formation of quinolin-4-one **27** (a) or quinolin-2-one **29** (b) (Scheme 6).

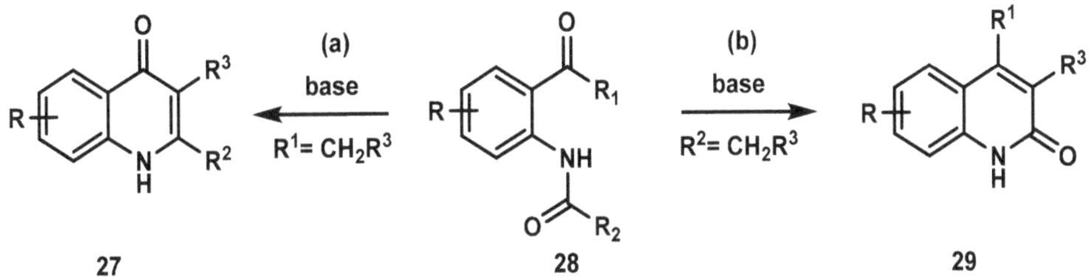

Scheme 6. Camps' cyclization reaction.

The mechanism of the Camps' reaction (Scheme 7) involves intramolecular aldol condensation of *N*-(2-acylaryl)amides **28**, which are capable of enolization. When a strong base, such as sodium hydroxide, is used, one of the protons of the CH_2 group next to the carbonyl is deprotonated, and the formed enolate ion **30** attacks the carbonyl group of the amide function. The formed alcohol **31** then undergoes β-elimination to give quinolin-4-one **7**. The synthesis of quinolin-2-one occurs by deprotonating one of the protons of the CH_2 group of the amide with a weaker base. The formed enolate ion **32** attacks the carbonyl group of the ketone, and the formed alcohol **33** undergoes elimination to give quinolin-2-one **29**.

The Camps' precursor (**28**) can be easily obtained, for example, by the condensation of 2-aminoacetophenone with acid chloride or the Friedel-Crafts acylation of anilides. In 2007, Buchwald's group [34] developed a rapid two-step synthesis of 2-aryl- and 2-vinylquinolin-4-ones from *N*-(2-acylaryl)amides **28** using the Camps' cyclization method (Scheme 8). The precursor **28** was obtained by the copper-catalyzed amidation of 2-halogenoacetophenone **34**. A series of 2-aryl-4-quinolinones **7** was obtained in very good yields (72–97%).

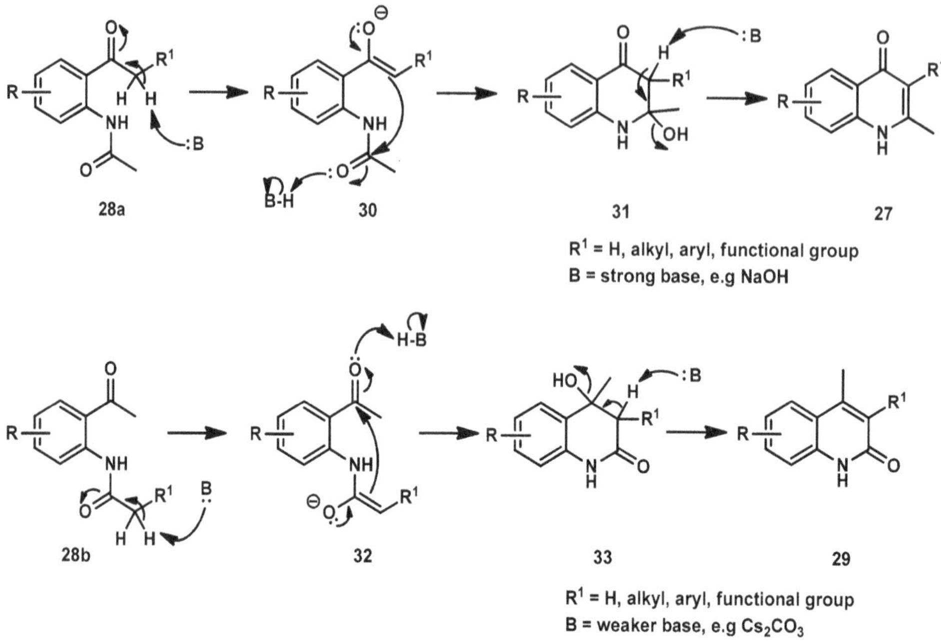

Scheme 7. Mechanisms of Camps' cyclization.

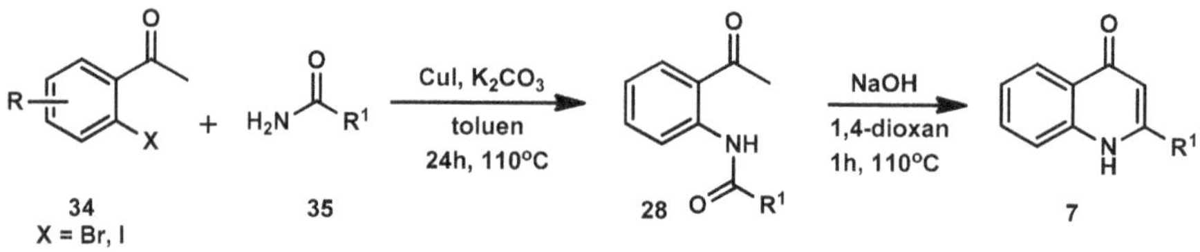

Scheme 8. Synthesis of quinoline-4-ones, according to Buchwald.

The authors also described how the type of base used affects the regioselectivity of the reaction. It was shown that *N*-(2-acetylphenyl)but-2-enamide (**36**) has the ability to cyclize to 2-vinylquinolin-4-one **37** and 4-methyl-3-vinylquinolin-2-one **38** (Scheme 9). In the presence of a stronger base (NaOH), deprotonation occurs at the α position of the ketone, followed by intramolecular aldol condensation with the formation of quinolin-4-one **37**. However, the use of a weaker base (Cs_2CO_3) gave quinolin-2-one **38** as the major product via γ deprotonation of the amide. In both cases, trace amounts of the second isomer were formed, which could be easily separated by column chromatography.

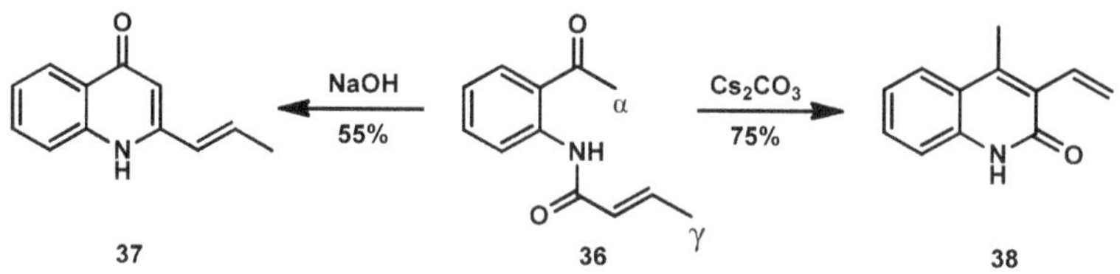

Scheme 9. Cyclization of *N*-(2-acetylphenyl)but-2-enamide (**36**) to quinolin-4-one and quinolin-2-one.

2.7. Palladium-Catalyzed Carbonylation Reaction (Route G)

Palladium-catalyzed carbonylation reactions are commonly used in the synthesis of quinolin-4-ones [19,35]. These reactions typically use carbon monoxide as the source of the carbonyl group. The coupling reaction of 2-iodoaniline **39** and terminal acetylenes **40** under CO to synthesize quinolin-4-ones was first described by Torii's group [36]. The reaction used high-pressure CO gas and $PdCl_2(PPh_3)_2$ or $PdCl_2(dppf)Cl_2$ as a catalyst. The methodology was later improved by Genelot [37,38], who introduced milder reaction conditions.

In recent years, alternative sources of CO, such as molybdenum or tungsten carbonyls, have been sought as potential replacements for gaseous CO. One example is the use of molybdenum hexacarbonyl [$Mo(CO)_6$] in the synthesis of quinolin-4-ones from 2-iodoaniline **39** and terminal acetylenes **40** [39]. This paper describes two protocols that yielded a wide range of 2-substituted quinolin-4-ones **7** with good yields. The first protocol allows the synthesis of the expected compounds as early as after 20 min of microwave (MW) heating (Scheme 10, Method A), while the second procedure is a two-step approach in a single vessel, in which the reaction is carried out at room temperature and functional groups sensitive to $Mo(CO)_6$, such as nitro groups, may be present (Scheme 10, Method B).

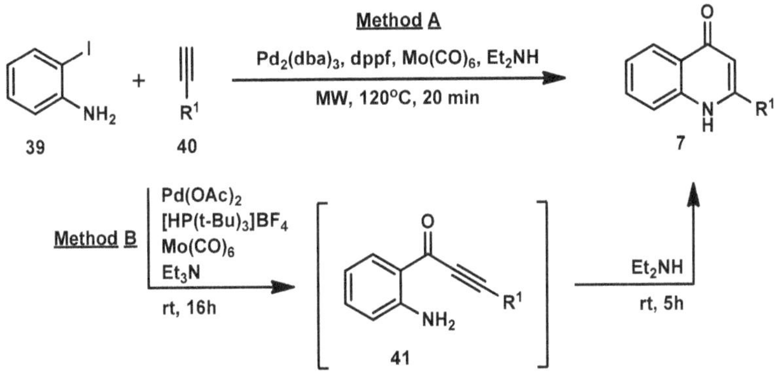

Scheme 10. Palladium-catalyzed carbonylation reaction.

The reaction mechanism involves Sonogashira carbonylation, resulting in the formation of alkyne **41**, which then undergoes cyclization upon the addition of diethylamine (Scheme 11) [40]. Molybdenum hexacarbonyl decomposes under heating, releasing CO, which then participates in the carbonylation step. As a result of the oxidative addition of the catalyst Pd^0 to 2-iodoaniline **39**, complex A is formed. In the next step, there is an insertion-1,1 of CO into the palladium–carbon bond of complex A, making complex B susceptible to attack by the nucleophile alkyne **40**. As a result, intermediate complex C is formed, and the carbonylation product **41** is formed by β-elimination and restoration of catalytically active forms of Pd^0. The addition of diethylamine to the resulting alkyne **53** leads to the formation of intermediate compound D, which undergoes cyclization to give the desired 2-substituted quinolin-4-one **7**.

The use of molybdenum hexacarbonyl as a source of CO has many advantages over traditional methods using gaseous CO. $Mo(CO)_6$ is solid, making it easier to use and transport this reactant. In addition, the lower risk of explosion improves safety and comfort. The reagent is relatively inexpensive and readily available, making it an attractive alternative to gaseous CO in large-scale industrial processes.

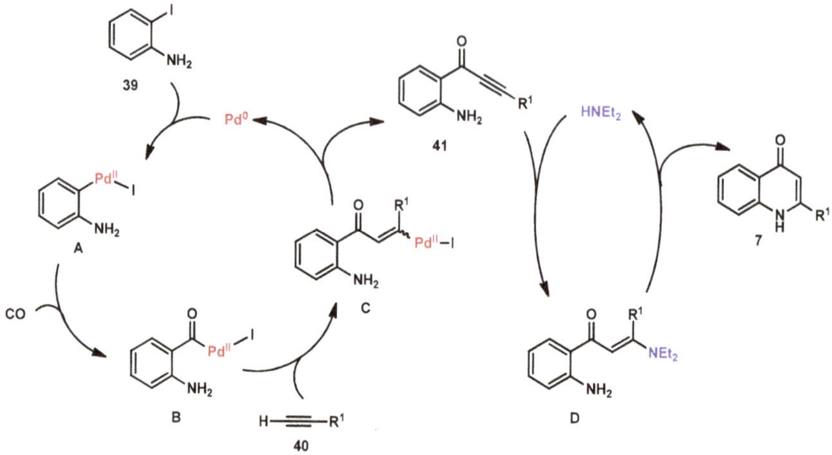

Scheme 11. Mechanism of palladium-catalyzed carbonylation reaction.

2.8. *Environmentally Friendly Syntheses of Quinolin-4-ones*

As described above, classical methods are very useful for developing an infinite variety of quinoline-4-one analogs. The drawbacks of these procedures are the use of hazardous chemicals, high temperatures, organic solvents, and long reaction times. All these factors have an impact on the environment and the economy. It is believed that the development of so-called green syntheses, which are alternative technologies seeking replacements for hazardous substances and trying to reduce waste and conserve energy, is the future of organic chemistry. Also, in the field of heterocyclic compounds, efforts are made to develop new, environmentally friendly procedures [41].

2.8.1. *N*-Heterocyclic Carbenes-Catalyzed Synthesis of Quinolin-4-ones (Route H)

N-heterocyclic carbenes (NHCs) are electron-rich chemical entities that exhibit a nucleophilic character [42]. They play an important role in modern organic chemistry. They are mainly used in reactions catalyzed by organometallic compounds and in organocatalysis. NHC-catalyzed synthesis of quinolin-4-ones has been described, among others, by Rai and co-workers [43]. The authors developed a method to obtain products in high yields by dehydrative acetylation of 2 aminobenzaldehyde **42** with α-halogenketone **43** and subsequent intramolecular heterocyclization (Scheme 12). Due to the mild reaction conditions, simplicity, and lack of by-products, this approach is an interesting alternative to already existing procedures for obtaining quinolin-4-ones **7**.

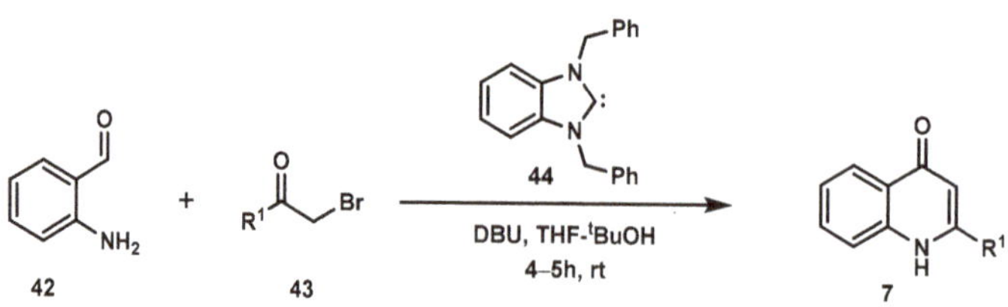

Scheme 12. NHC-catalyzed synthesis of quinoline-4-ones.

Scheme 13 shows the proposed reaction mechanism. The nucleophilic attack of the carbene **44** on the carbonyl carbon of the aldehyde **42** leads to the formation of the intermediate compound **45**. In the second step, a proton transfer occurs in **45** to produce homoenolate **46**. Nucleophilic homoenolate **46** attacks bromide **43** to give adduct **47**, which undergoes intramolecular heterocyclization. The resulting adduct **48** undergoes dehydration, yielding product **7**. The increased stability of the conjugated quinolin-4-one **7** double bonds is probably the driving force of the dehydration reaction.

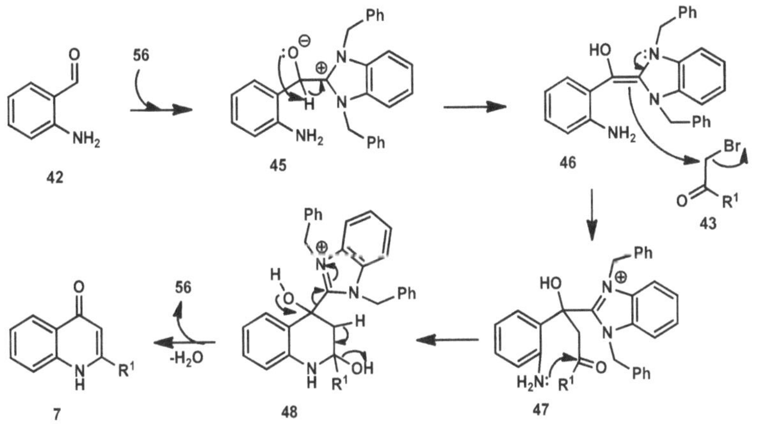

Scheme 13. Mechanism of NHC-catalyzed synthesis of quinolin-4-ones.

2.8.2. Decarboxylating Cyclization (Route I)

Recently, Sun and co-workers [44] developed an environmentally friendly decarboxylating cyclization procedure to synthesize quinolin-4-ones (Scheme 14). The starting materials are inexpensive and readily available isatoic anhydrides **49** and 1,3 dicarbonyl compounds **8**. The base generates a carbanion from compound **8**, which attacks the carbonyl group of anhydride **49** to give intermediate **50** with the simultaneous release of CO_2. Following intramolecular cyclization and dehydration, a suitably substituted quinolin-4-one **52** is formed. The reaction was carried out in water at 80 °C. The reaction was also performed using isatin **53** as a substrate in dimethyl sulfoxide (DMSO). The added oxidant, *tert*-butyl hydroperoxide, converted isatin **53** to isatoic anhydride **49**, and the subsequent addition of 1,3-dicarbonyl compound **8** gave the desired quinolin-4-one **52**. Thanks to the mild reaction conditions, the synthesis is environmentally friendly, and the only by-products are carbon dioxide and water. No toxic transition metals are used in the reaction, while a wide range of products can also be obtained on a large scale.

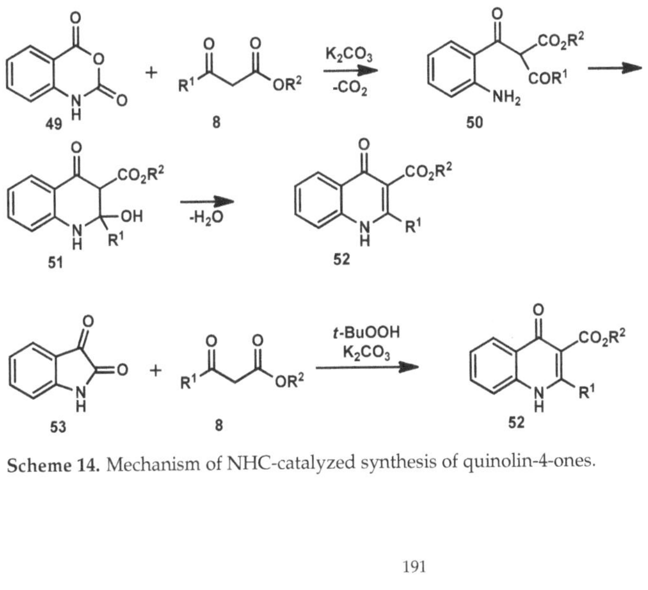

Scheme 14. Mechanism of NHC-catalyzed synthesis of quinolin-4-ones.

3. Quinolin-4-ones as Drugs

Quinolin-4-ones and their derivatives are important organic compounds because of their wide range of biological activities, which makes them attractive therapeutic agents [45]. Seventeen compounds containing a quinolin-4-one motif have been marketed to date, according to the Drug Bank Online database [46]. The majority of these drugs are antibiotics from the fluoroquinolinone group (norfloxacin, ciprofloxacin, oflokxacin, perflofloxacin, lomefloxacin, levofloxacin, grepafloxacin, sparfloxacin, moxifloxacin, besifloxacin, rosoxacin, delafloxacin, finafloxacin, gatifloxacin, but also antiviral drugs (elvitegravir) and those used to treat asthma (nedocromil) and cystic fibrosis (ivacaftor).

The quinolinone scaffold is also incorporated into compounds that display anti-malarial activity, such as endochin and its derivatives and analogs [47,48]. However, reports on the anti-malarial properties of quinolinones, compared to their antibacterial properties, are relatively limited and will not be discussed in this review.

3.1. Antibacterial Agents

Quinolones are one of the most important families of antimicrobial agents, representing fully synthetic compounds, as opposed to antibiotics, which are produced by microorganisms [2].

3.1.1. Four Generations of Quinolinones

First-generation quinoline-4-ones, nalidixic acid (technically a naphthyridone) first synthesized by Lesher [49], and oxolinic acid [50] developed in Japan in the 1970s (Figure 5) were mainly used in the treatment of urinary tract infections, but their use was limited due to the narrow spectrum of action (some Gram-negative bacteria), short half-life, and a number of side effects, and nowadays these agents are not recommended for therapy [51].

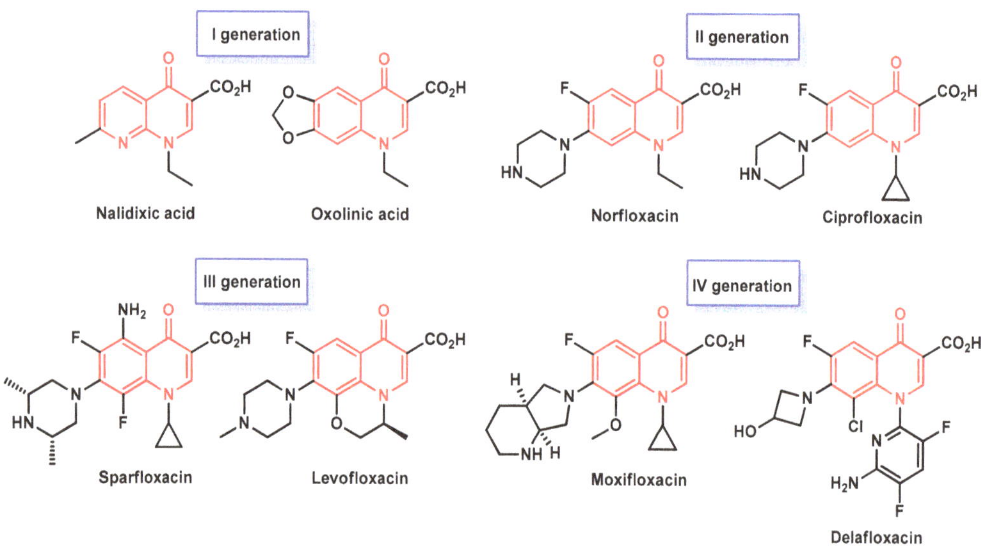

Figure 5. Examples of the I-st, II-nd, III-rd, and IV-th generation quinoline-4-ones.

Soon, new analogs with better properties, modified primarily at the nitrogen atom and at the C-6, C-7, and C-8 positions, were synthesized. In the late 1970s, there was a breakthrough in the development of quinolinone-based drugs with the introduction of the fluorine atom into the quinolin-4-one skeleton in position 6 [52–54]. This modification significantly improved the spectrum of activity and yielded the second-generation quinoline-4-ones. The first approved drug from this group was norfloxacin, which showed enhanced antimicrobial activity against Gram-negative and some Gram-positive bacteria and reduced toxicity [55–57] (Figure 5).

The third-generation quinoline-4-ones have expanded activity against Gram-positive and Gram-negative bacteria. The fourth-generation quinolinones retain the Gram-negative and Gram-positive antimicrobial activity and inhibit anaerobic bacteria [58].

Almost all of the second-, third-, and fourth-generation quinoline-4-ones possess a fluorine atom at position 6, with the exception of the most recent compounds from the fourth generation. Although the fluorine at C6 significantly improves activity, current research is focused on its removal because it is related to genotoxicity. The reduction in potency caused by its removal can be compensated for by using alternative substituents at other positions [59].

Some important antibacterials from the group of fluoroquinolin-4-ones are norfloxacin, ciprofloxacin, levofloxacin, and moxifloxacin.

Norfloxacin is not only the first synthesized fluoroquinolin-4-one derivative but is also the first quinolinone drug approved for use that contains a piperazine group. Norfloxacin is highly active against most Gram-positive and Gram-negative bacteria. It was patented in 1979 by Kyorin [60] and four years later approved for clinical use.

Another example of a second-generation quinolin-4-one and the best-known antibacterial of the fluoroquinolinone group is **ciprofloxacin**. It is a drug used for a wide range of bacterial infections when other antibiotics have failed. In the early 2000s, ciprofloxacin was also used to treat patients after inhalation exposure to anthrax bacillus after bioterrorist attacks in the United States. Ciprofloxacin was first synthesized by Klaus Grohe and patented by Bayer in 1983 [61].

Ofloxacin is a racemic second-generation quinolin-4-one. Its (*S*)-isomer, levofloxacin (a third-generation quinolin-4-one), is the first known optically active fluoroquinolinone. Both compounds were developed by Daiichi Pharmaceuticals in 1985 and 1993, respectively, and both have broad-spectrum antimicrobial activity. Levofloxacin additionally acts against penicillin-resistant *Streptococcus pneumoniae*, the pneumonia germ responsible for the development of respiratory tract infections, sinusitis, pneumonia, meningitis, and sepsis.

The important fourth-generation quinoline-4-one is **moxifloxacin**. It was developed by Bayer AG (Leverkusen, Germany) in 1989. It is used in the treatment of a wide range of infections caused by Gram-positive and Gram-negative bacteria and other microorganisms (*Chlamydia trachomatis*). Like other fluoroquinolinones, moxifloxacin easily penetrates into tissues and body fluids. It is used to treat many infections, including cellulitis, anthrax, and abdominal and respiratory tract infections. Drugs containing moxifloxacin were introduced to the market in 1999 and are currently sold in more than 80 countries around the world.

3.1.2. Mechanism of Antimicrobial Action of Quinolinones

The mechanism of action of fluoroquinolin-4-ones is interference with nucleic acid synthesis. They inhibit bacterial DNA synthesis by disrupting the enzymes topoisomerase IV and DNA gyrase, resulting in rapid bacterial death [62,63]. As a general rule, Gram-negative bacterial activity correlates with the inhibition of DNA gyrase, and Gram-positive activity corresponds with the inhibition of DNA type IV topoisomerase. First- and second-generation fluoroquinolinones bind either to DNA gyrase or to DNA topoisomerase IV, depending on the bacteria and the drug, whereas newer fluoroquinolin-4-ones may target both of these enzymes with equal affinity in any organism [2].

The broad spectrum of activity and excellent tissue penetration make fluoroquinolin-4-ones potent antibacterial drugs, but an increasing number of resistant pathogens becomes a factor limiting their use [51,64–66].

3.1.3. Structure–Antibacterial Activity of Fluoroquinolin-4-ones

Many analyses have been carried out to determine the relationship between the structure and antibacterial activity of fluoroquinolin-4-ones [52,59] (Figure 6). Substituents placed on the N1 nitrogen atom are essential for this activity. The cyclopropyl group is considered the best choice (e.g., ciprofloxacin, moxifloxacin). Antibacterial activity of fluoroquinolinones substituted at N1 may result from the steric hindrance of the substituent and also from other spatial effects that affect the molecule's interaction with the DNA gyrase. The carboxyl group on the C3 and the ketone group in position 4 facilitate penetration of the drug into the bacterial cell. In the C5 position, the introduction of small substituents, such as an amine group (e.g., sparfloxacin) or a methyl group (e.g., grepafloxacin), increases activity, especially against Gram-positive bacteria. In the case of larger substituents, the antimicrobial activity drops drastically. The fluorine atom in position 6 is necessary for strong activity. Substitution at the C7 enhances antimicrobial activity. It is the most adaptable site for modifications. Various groups are permitted, but the optimal substituents are 5- and 6-membered *N*-heterocycles [53]. The introduction of a pyrrolidine ring into this position (e.g., moxifloxacin) results in increased activity against Gram-positive bacteria, while the introduction of piperazine (e.g., ciprofloxacin) is against Gram-negative bacteria. Further substitution of both ring types with an alkyl group increases the solubility of the molecule in water [54]. However, a substituent at the C7 position can be responsible for central nervous system (CNS) side effects and for interactions with non-steroidal anti-inflammatory drugs (NSAIDs) [51]. Fluoroquinolinones with pyrrolidine or piperazine rings containing alkyl groups (e.g., grepafloxacin) cause minimal interactions with other drugs and a lower risk of CNS disorders compared to non-alkylated structures (e.g., ciprofloxacin). Substituents on the C8 improve efficacy against anaerobic bacteria. Although the presence of a halogen at this position (e.g., sparfloxacin) is associated with an expanded spectrum of action against anaerobic bacteria, improved oral absorption, and water solubility, it is, unfortunately, also a major cause of phototoxicity. The presence of a methoxy group at the C8 prevents the phototoxicity of fluoroquinolin-4-ones and may reduce the risk of drug resistance development [65].

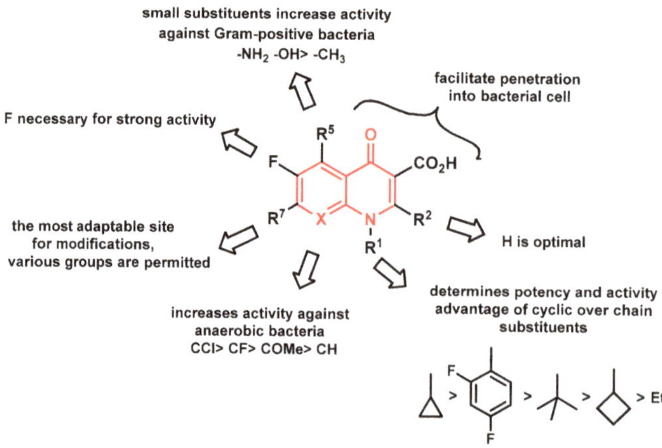

Figure 6. Relationship between structure and antibacterial activity of fluoroquinolin-4-ones.

3.2. Cystic Fibrosis Drugs

In addition to broad antimicrobial activity, quinolin-4-ones also exhibit other biological properties. **Ivacaftor** (Figure 7) is a drug designed to treat patients with cystic fibrosis. Cystic fibrosis is an incurable inherited disease caused by mutations in the CFTR gene. This gene provides instructions for making a protein called the CF transmembrane conductance regulator, which functions as a channel across the membrane of cells that produce mucus, sweat, saliva, tears, and digestive enzymes. Mutations in the CFTR gene cause clinical symptoms in many areas of the body: the lungs, sinuses, gastrointestinal tract, and reproductive tract.

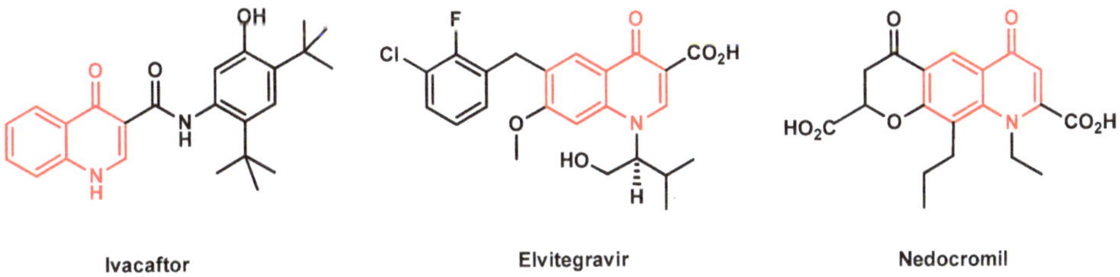

Figure 7. Structure of quinoline-4-one drugs approved for the treatment of cystic fibrosis, HIV, and asthma.

There are an estimated 100,000 people with cystic fibrosis worldwide. The median survival 50 years ago was 10 years; with today's medications, patients live up to 50 years. In 2012, Vertex Pharmaceuticals launched a quinolin-4-one derivative, ivacaftor, approved by the FDA to treat cystic fibrosis. This is the first drug that treats the underlying cause rather than the symptoms of the disease. Ivacaftor improves the transport of chloride ions through the ion channels by binding to the channels directly and inducing a non-conventional mode of gating, which in turn increases the probability that the defective channel is open [67–69].

3.3. Antiviral Drugs

Elvitegravir (Figure 7) was the first quinoline-4-one drug developed for the treatment of HIV infection. Its mechanism of action involves the inhibition of HIV integrase, the enzyme essential for the virus to integrate its genetic code into the host's DNA. Elvitegravir was developed by the pharmaceutical company Gilead Sciences, which licensed the drug from Japan Tobacco in 2008, and it was approved by the FDA for clinical use in 2014 [70–73].

3.4. Antiallergic Drugs

Nedocromil (Figure 7) is a quinolinone-based medication whose action is to prevent shortness of breath, wheezing, and other breathing problems caused by asthma. Nedocromil acts as a mast cell stabilizer, inhibits the degranulation of mast cells, and prevents the release of histamine and tryptase, preventing the synthesis of prostaglandins and leukotrienes [74].

4. Anticancer Activity of Quinolin-4-ones

According to the World Health Organization (WHO), cancer is one of the most common causes of death before the age of 70, and it is a major global health problem [75]. The search for new anticancer drug candidates is one of the very important goals of medicinal chemistry.

The high costs of obtaining new drugs encourage the search for anticancer agents among already available drugs or their derivatives [76,77]. Quinolin-4-ones possess many desirable features, such as low toxicity, favorable physicochemical properties, a pharmacokinetic profile, and a low incidence of resistance development. The simplicity of the synthesis of these compounds and the large number of potential functionalization sites have made them attractive precursors in the search for new anticancer drug candidates. Many antibiotics from the fluoroquinolinone group, which have been known for years, are now being investigated for their potential anticancer activity [78–81].

4.1. Quinolin-4-one Antibiotics with Anticancer Potential

Ciprofloxacin has been shown to induce apoptosis in a variety of cancer cell lines, including lung or breast cancer cells [82–85]. Various ciprofloxacin derivatives with anticancer properties are described in detail by Nowakowska et al. [86].

Lomefloxacin shows high cytotoxic activity against COLO829 melanoma cells [87,88].

Norfloxacin analogs were potent against the HeLa cell line with 100% inhibition of cell viability [89].

Levofloxacin significantly altered gene expression in a direction favoring anticancer activity. It seems to be an attractive candidate for breast cancer treatment. It inhibited proliferation and induced apoptosis in a panel of breast cancer cell lines while sparing normal breast cells [90,91].

In 2003, China's State Food and Drug Administration (SFDA) approved **7-chloroquinolin-4-on** (**54**) for the treatment of breast cancer and non-small-cell lung cancer (NSCLC). Its antitumor activity involves blocking DNA synthesis and damaging the tumor cell's DNA matrix. However, the usefulness of this drug is limited due to its poor solubility. Zhou's group [92] synthesized derivatives of this compound, obtaining a series of five 3-amido-7-chloroquinolin-4-ones **55a–e** with an amide side chain in position 3 and alkyl substituents on the nitrogen atom (Figure 8).

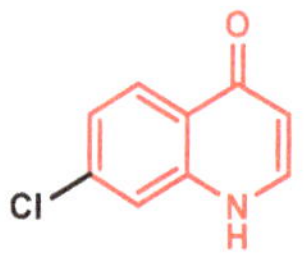

54

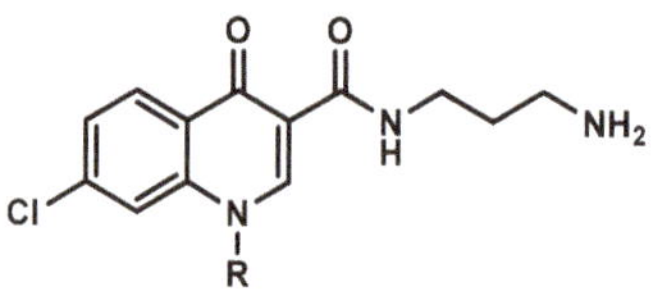

55a–e

a R= C_2H_5
b R= $CH_2CH{=}CH_2$
c R= n-C_4H_9
d R= CH_2Ph
e R= $(CH_2)_3Ph$

Figure 8. Structure of 7-chloroquinolin-4-one and its 3-amido derivatives **55a–e**.

These compounds exhibited a more potent antiproliferative effect against a panel of human tumor cell lines than the lead structure **54**. The best results were obtained for compound **55e**, which showed the most potent antiproliferative activity and exhibited selective cytotoxicity against HepG2 cells.

In the search for more efficient anticancer candidates, the hybridization of various quinolinones with other anticancer pharmacophores has also been investigated [93]. However, up to date, none of the quinoline-4-ones has been approved in Europe or the US for the treatment of cancer.

4.2. Mode of Action of Quinolinones as Anticancer Agents

Data indicate that fluoroquinol-4-ones inhibit various proteins and enzymes involved in cancer cell growth, such as topoisomerase II, protein kinases, phosphoinositide 3-kinases

(PI3K), and histone deacetylase (HDAC) [94–98]. Such inhibition leads to cell death by apoptosis and cell cycle arrest in the S and G2/M phases [99]. Furthermore, dysfunction of mitochondria, likely due to the similarity between mitochondrial and bacterial cell structures, has been shown as a mechanism of action [97]. The effects of fluoroquinolinones on migration and metastasis have also been reported. In particular, they could suppress matrix metalloproteinase 9 (MMP9) production [100–103]. Additionally, some quinoline-4-ones were found to be potent inhibitors of tubulin polymerization with high selectivity against tumor cells. 2-Phenylquinolin-4-one displayed an antiproliferative effect in several cancer types through inhibition of tubulin polymerization [104]. Fluoroquinolinones can also hamper cancer growth by indirect effects via modulation of the immune response [105].

4.3. Structure–Anticancer Activity of Quinoline-4-ones

Structure–activity relationship studies performed in many laboratories all over the world allowed us to determine which positions in the quinoline ring and which substituents are the most important for anticancer activity [106,107].

Figure 9 summarizes the effects of various substituents and their positions on the biological activity of quinoline-4-ones. A substitution on the nitrogen atom, as well as the presence of a carbonyl group at position 4, are essential for overall potency [78,81,92]. A cyclopropyl group on N1 was found to increase the activity compared to the ethyl group. Substituted phenyl or thiazole rings in this position also have a beneficial effect [81]. Alkyl groups in position 2 turned out to be more advantageous for antineoplastic activity than aryl groups [108]. It is believed that the substituent in position 3 should be coplanar with the quinoline ring [109]. The nature of substituents at C5 can affect the molecule's penetration into cells and binding to the target. The optimal substituent at position 6 is a fluorine atom [86,89]. The substituent at the C7 carbon is responsible for direct interaction with topoisomerase II in a complex with DNA. The introduction of aromatic rings at this position improves antitumor properties, as does the methoxy group at the C8 carbon atom [89].

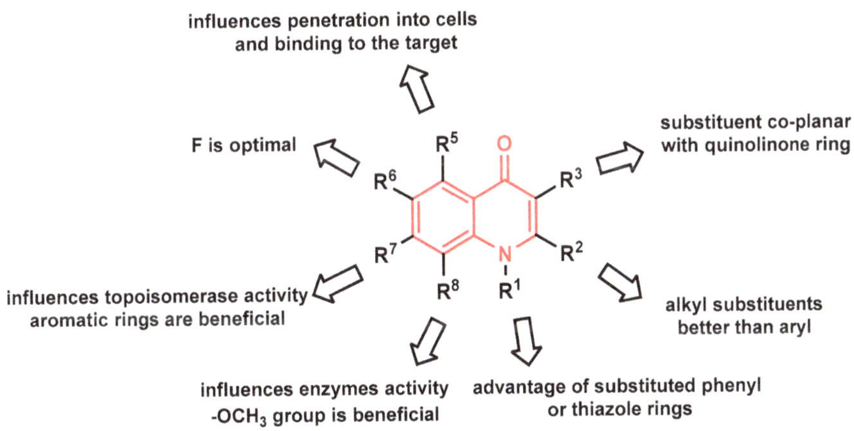

Figure 9. Relationship between structure and anticancer activity of quinolin-4-ones.

5. Conclusions

The scaffold of quinolinones is found in nature and has been used for the development of new, biologically active substances. Among many possible isomeric forms, the most common are quinolin-4-ones, which are present in several natural structures but mostly in fully synthetic compounds. Several quinoline-4-ones have been approved as antimicrobial agents and turned out to be invaluable in treating various bacterial infections.

More recently, it has been demonstrated that some quinoline-4-ones exhibit cytotoxic activity against various cancer cell lines and show excellent bioavailability as compared to the current therapeutic drugs for cancer treatment. These properties make them promising candidates for new anticancer agents. The high costs of designing, synthesizing, and marketing new medications encourage the search for anticancer candidates among already available drugs. Therefore, numerous synthetic modifications of existing and approved for clinical use quinoline-4-ones are being conducted, providing more detailed structure–activity relationship studies that may lead to more active moieties. Advances in developing computer-aided models to predict desired features of drugs may be invaluable in designing new generations of quinoline-4-ones. The future directions should be focused on increasing efficacy and reducing undesired side effects and off-target interactions.

Author Contributions: Conceptualization, A.J. and K.G.-J.; methodology, K.G.-J., J.P.-C. and A.J.; writing—original draft preparation, K.G.-J., J.P.-C., J.W., K.K., K.W. and A.J.; writing—review and editing, A.J. and K.G.-J.; visualization, K.W. All authors have read and agreed to the published version of the manuscript.

Funding: This work was supported by the Medical University of Lodz (grant No. 503/1-156-02/503-11-001).

Institutional Review Board Statement: Not applicable.

Informed Consent Statement: Not applicable.

Data Availability Statement: Not applicable.

Conflicts of Interest: The authors declare no conflicts of interest.

References

1. Shiro, T.; Fukaya, T.; Tobe, M. The chemistry and biological activity of heterocycle-fused quinolinone derivatives: A review. *Eur. J. Med. Chem.* **2015**, *97*, 397–408. [CrossRef] [PubMed]
2. Millanao, A.R.; Mora, A.Y.; Villagra, N.A.; Bucarey, S.A.; Hidalgo, A.A. Biological effects of quinolones: A family of broad-spectrum antimicrobial agents. *Molecules* **2021**, *26*, 7153. [CrossRef] [PubMed]
3. Al-Khalil, S.; Alkofahi, A.; El-Eisawi, D.; Al-Shabib, A. Transtorine, a new quinoline alkaloid from *Ephedra transitoria*. *J. Nat. Prod.* **1998**, *61*, 262–263. [CrossRef] [PubMed]
4. Oliva, A.; Meepagala, K.M.; Wedge, D.E.; Harries, D.; Hale, A.L.; Aliotta, G.; Duke, S.O. Natural fungicides from *Ruta graveolens* L. leaves, including a new quinolone alkaloid. *J. Agric. Food Chem.* **2003**, *51*, 890–896. [CrossRef]
5. Ivanova, A.; Mikhova, B.; Najdenski, H.; Tsvetkova, I.; Kostova, I. Antimicrobial and cytotoxic activity of *Ruta graveolens*. *Fitoterapia* **2005**, *76*, 344–347. [CrossRef]
6. Ghosh, S.; Bishayee, K.; Khuda-Bukhsh, A.R. Graveoline isolated from ethanolic extract of *Ruta graveolens* triggers apoptosis and autophagy in skin melanoma cells: A novel apoptosis-independent autophagic signaling pathway. *Phytother. Res.* **2014**, *28*, 1153–1162. [CrossRef]
7. An, Z.-Y.; Yan, Y.-Y.; Peng, D.; Ou, T.-M.; Tan, J.-H.; Huang, S.-L.; An, L.-K.; Gu, L.-Q.; Huang, Z.-S. Synthesis and evaluation of graveoline and graveolinine derivatives with potent anti-angiogenesis activities. *Eur. J. Med. Chem.* **2010**, *45*, 3895–3903. [CrossRef]
8. Rawat, A.K.; Mehrotra, S.; Tripathi, S.C.; Shome, U. Hepatoprotective activity of *Boerhaavia diffusa* L. roots-a popular Indian ethnomedicine. *J. Ethnopharmacol.* **1997**, *56*, 61–66. [CrossRef]
9. Manu, K.A.; Kuttan, G. Immunomodulatory activities of punarnavine, an alkaloid from *Boerhaavia diffusa*. *Immunopharmacol. Immunotoxicol.* **2009**, *31*, 377–387. [CrossRef]
10. Dhingra, D.; Valecha, R. Punarnavine, an alkaloid isolated from ethanolic extract of *Boerhaavia diffusa* Linn. reverses depression-like behaviour in mice subjected to chronic unpredictable mild stress. *Indian J. Exp. Biol.* **2014**, *52*, 799–807.
11. Olaleye, M.T.; Akinmoladun, A.C.; Ogunboye, A.A.; Akindahunsi, A.A. Antioxidant activity and hepatoprotective property of leaf extracts of *Boerhaavia diffusa* Linn against acetaminophen-induced liver damage in rats. *Food Chem. Toxicol.* **2010**, *48*, 2200–2205. [CrossRef] [PubMed]
12. Manu, K.A.; Kuttan, G. Punarnavine induces apoptosis in B16F-10 melanoma cells by inhibiting NF-kappaB signaling. *Asian Pac. J. Cancer Prev.* **2009**, *10*, 1031–1038. [PubMed]

13. Aher, V.; Chattopadhyay, P.; Goyary, D.; Veer, V. Evaluation of the genotoxic and antigenotoxic potential of the alkaloid punarnavine from *Boerhaavia diffusa*. *Planta Med.* **2013**, *79*, 939–945. [CrossRef] [PubMed]
14. Saraswati, S.; Alhaider, A.A.; Agrawal, S.S. Punarnavine, an alkaloid from *Boerhaavia diffusa* exhibits anti-angiogenic activity via downregulation of VEGF in vitro and in vivo. *Chem. Biol. Interact.* **2013**, *206*, 204–213. [CrossRef]
15. Kawada, M. Small molecules modulating tumor-stromal cell interactions: New candidates for anti-tumor drugs. *J. Antibiot.* **2016**, *69*, 411–414. [CrossRef]
16. Abe, H.; Kawada, M.; Inoue, H.; Ohba, S.; Masuda, T.; Hayashi, C.; Igarashi, M.; Nomoto, A.; Watanabe, T.; Shibasaki, M. Structure-activity relationship study of intervenolin derivatives: Synthesis, antitumor, and anti-*Helicobacter pylori* activities. *Tetrahedron* **2013**, *69*, 7608–7617. [CrossRef]
17. Rao, B.; Sreenivasulu, C.; Kishore, D.R.; Satyanarayana, G. Trending strategies for the synthesis of quinolinones and isoquinolinones. *Tetrahedron* **2022**, *127*, 133093. [CrossRef]
18. Dine, I.; Mulugeta, E.; Melaku, Y.; Belete, M. Recent advances in the synthesis of pharmaceutically active 4-quinolone and its analogues: A review. *RSC Adv.* **2023**, *13*, 8657–8682. [CrossRef]
19. Shen, C.; Wang, A.; Xu, J.; An, Z.; Loh, K.Y.; Zhang, P.; Liu, X. Recent advances in the catalytic synthesis of 4-quinolones. *Chem* **2019**, *5*, 1059–1107. [CrossRef]
20. Gould, R.G.; Jacobs, W.A. The synthesis of certain substtituted quinolines and 5,6-benzoquinolines. *J. Am. Chem. Soc.* **1939**, *61*, 2890–2895. [CrossRef]
21. Morgentin, R.; Pasquet, G.; Boutron, P.; Jung, F.; Lamorlette, M.; Maudet, M.; Ple, P. Strategic studies in the synthese of novel 6,7-substiuted quinolones and 7- or 6-substitued 1,6- and 1,7-naphthyridones. *Tetrahedron* **2008**, *64*, 2772–2782. [CrossRef]
22. Smith, R.B.; Faki, H.; Leslie, R. Limitations of the Jacobs-Gould reaction using microwave irradiation. *Synth. Commun.* **2011**, *41*, 1492–1499. [CrossRef]
23. Conrad, M.; Limpach, L. Synthesen von Chinolinderivaten mittelst Acetessigester. *Ber. Dtsch. Chem. Ges.* **1887**, *20*, 944–948. [CrossRef]
24. Pou, S.; Dodean, R.A.; Frueh, L.; Liebman, K.M.; Gallagher, R.T.; Jin, H.; Jacobs, R.T.; Nilsen, A.; Stuart, D.R.; Doggett, J.S.; et al. A new scalable synthesis of ELQ-300, ELQ-316, and other antiparasitic quinolones. *Org. Process Res. Dev.* **2021**, *25*, 1841–1852. [CrossRef]
25. Brouet, J.C.; Gu, S.; Peet, N.P.; Williams, J.D. Survey of solvents for the Conrad–Limpach synthesis of 4-hydroxyquinolones. *Synth. Commun.* **2009**, *39*, 1563–1569. [CrossRef]
26. Shilabin, A.G.; Dzhekieva, L.; Misra, P.; Jayaram, B.; Pratt, R.F. 4-Quinolones as noncovalent inhibitors of high molecular mass penicillin-binding proteins. *ACS Med. Chem. Lett.* **2012**, *9*, 592–595. [CrossRef]
27. Biere, H.; Seelen, W.A. general synthesis of 4-oxo-1, 4-dihydropyridinecarboxylic acid derivatives. *Just. Lieb. Annal. Chem.* **1976**, *11*, 1972–1981. [CrossRef]
28. Zewge, D.; Chen, C.Y.; Deer, C.; Dormer, P.G.; Hughes, D.L. A mild and efficient synthesis of 4-quinolones and quinolone heterocycles. *J. Org. Chem.* **2007**, *72*, 4276–4279. [CrossRef]
29. Schaefer, J.P.; Bloomfield, J.J. The Dieckmann condensation (Including the Thorpe-Ziegler condensation). *Org. React.* **1967**, *15*, 1–203.
30. DeGraffenreid, M.R.; Bennett, S.; Caille, S.; Gonzalez-Lopez de Turiso, F.; Hungate, R.W.; Julian, L.D.; Kaizerman, J.A.; McMinn, D.L.; Sun, D.; Yan, X.; et al. An efficient and scalable one-pot double Michael addition-Dieckmann condensation for the synthesis of 4,4-disubstituted cyclohexane β-keto esters. *J. Org. Chem.* **2007**, *72*, 7455–7458. [CrossRef]
31. Chongau, R.J.; Siddiqui, M.A.; Snieckus, V. A modified von Niementowski quinoline synthesis from anthranilamides. *Tetrahedron Lett.* **1986**, *27*, 5323–5326. [CrossRef]
32. Camps, R. Synthese von α-und y-Oxychinolinen. *Chem. Ber.* **1899**, *32*, 3228–3234. [CrossRef]
33. Fisyuk, A.; Kostyuchenko, A.S.; Goncharov, D.S. Camps Reaction and Related Cyclizations. *Russ. J. Org. Chem.* **2020**, *56*, 1863–1892. [CrossRef]
34. Jones, C.P.; Anderson, K.W.; Buchwald, S.L. Sequential Cu-catalyzed amidation-base-mediated Camps cyclization: A two-step synthesis of 2-aryl-4-quinolones from halophenones. *J. Org. Chem.* **2007**, *72*, 7968–7973. [CrossRef]
35. Silva, V.L.M.; Silva, A.M.S. Palladium-catalyzed synthesis and transformation of quinolones. *Molecules* **2019**, *24*, 228. [CrossRef]
36. Torii, S.; Okumoto, H.; Xu, L.H.; Sadakane, M.; Shostakovsky, M.V.; Ponomaryov, A.B.; Kalinin, V.N. Syntheses of chromones and quinolones via Pd-catalyzed carbonylation of o-iodophenols and anilines in the presence of acetylenes. *Tetrahedron* **1993**, *49*, 6773–6784. [CrossRef]
37. Genelot, M.; Bendjeriou, A.; Dufaud, V.; Djakovitch, L. Optimised procedures for the one-pot selective syntheses of indoxyls and 4-quinolones by a carbonylative Sonogashira/cyclisation sequence. *Appl. Catal.* **2009**, *369*, 125–132. [CrossRef]
38. Genelot, M.; Dufaud, V.; Djakovitch, L. Heterogeneous metallo-organocatalysis for the selective one-pot synthesis of 2-benzylidene-indoxyl and 2-phenyl-4-quinolone. *Tetrahedron* **2011**, *67*, 976–981. [CrossRef]

39. Åkerbladh, L.; Nordeman, P.; Wejdemar, M.; Odell, L.R.; Larhed, M. Synthesis of 4-quinolones via a carbonylative Sonogasira cross-coupling using molybdenum hexacarbonyl as a CO source. *J. Org. Chem.* **2015**, *80*, 1464–1471. [CrossRef]
40. Liu, J.; Chen, J.; Xia, C. A simple and efficient recyclable phosphine-free catalytic system for alkoxycarbonylation and carbonylative Sonogashira coupling reactions of aryl iodides. *J. Catal.* **2008**, *253*, 50–56. [CrossRef]
41. Kumaraswamy, B.; Hemalatha, K.; Pal, R.; Matada, G.S.P.; Hosamani, K.R.; Aayishamma, I.; Aishwarya, N.V.S.S. An insight into sustainable and green chemistry approaches for the synthesis of quinoline derivatives as anticancer agents. *Eur. J. Med. Chem.* **2024**, *275*, 116561. [CrossRef] [PubMed]
42. Bellotti, P.; Koy, M.; Hopkinson, M.N.; Glorius, F. Recent advances in the chemistry and applications of *N*-heterocyclic carbenes. *Nat. Rev. Chem.* **2021**, *5*, 711–725. [CrossRef] [PubMed]
43. Rai, V.K.; Verma, F.; Sahu, G.P.; Singh, M.; Rai, A. One-pot Allan-Robinson/Friedländer route to chromen-/quinolin-4-ones through the domino acetylative cyclisation of 2-hydroxy-/2-aminobenzaldehydes. *Eur. J. Org. Chem.* **2018**, *48*, 537–544. [CrossRef]
44. Ma, Y.; Zhu, Y.; Zhang, D.; Meng, Y.; Tang, T.; Wang, K.; Ma, J.; Wang, J.; Sun, P. Eco-friendly decarboxylative cyclization in water: Practical access to the anti-malarial 4-quinolones. *Green Chem.* **2019**, *21*, 478–482. [CrossRef]
45. Dube, P.S.; Legoabe, L.J.; Fan, R.M. Quinolone: A versatile therapeutic compound class. *Mol. Divers.* **2022**, *27*, 1501–1526. [CrossRef]
46. Available online: https://go.drugbank.com/structures/search/small_molecule_drugs/structure#results (accessed on 1 December 2024).
47. Beteck, R.M.; Smit, F.J.; Haynes, R.K.; N'Da, D.D. Recent progress in the development of anti-malarial quinolones. *Malar. J.* **2014**, *13*, 339. [CrossRef]
48. Fan, Y.L.; Cheng, X.W.; Wu, J.B.; Liu, M.; Zhang, F.Z.; Xu, Z.; Feng, L.S. Antiplasmodial and antimalarial activities of quinolone derivatives: An overview. *Eur. J. Med. Chem.* **2018**, *146*, 1–14. [CrossRef]
49. Lesher, G.Y.; Froelich, E.J.; Gruett, M.D.; Bailey, J.H.; Brundage, R.P. 1,8-Naphthyridine derivatives. A new class of chemotherapeutic agents. *J. Med. Pharm. Chem.* **1962**, *5*, 1063–1065. [CrossRef]
50. Madsen, P.O.; Rhodes, P.R. Oxolinic acid, a new chemotherapeutic agent in the treatment of urinary tract infection. *J. Urol.* **1971**, *105*, 870–872. [CrossRef]
51. Tang, K.; Zhao, H. Quinolone Antibiotics: Resistance and Therapy. *Infect. Drug Resist.* **2023**, *16*, 811–820. [CrossRef]
52. Gootz, T.D.; Brighty, K.E. Fluoroquinolone antibacterials: SAR mechanism of action, resistance, and clinical aspects. *Med. Res. Rev.* **1996**, *16*, 433–486. [CrossRef]
53. Koga, H.; Itoh, A.; Murayama, S.; Suzue, S.; Irikura, T. Structure-activity relationships of antibacterial 6,7- and 7,8-disubstituted 1-alkyl-1,4-dihydro-4-oxoquinoline-3-carboxylic acids. *J. Med. Chem.* **1980**, *23*, 1358–1363. [CrossRef] [PubMed]
54. Domagala, J.M. Structure-activity and structure-side-effect relationships for the quinolone antibacterials. *J. Antimicrob. Chemother.* **1994**, *33*, 685–706, Erratum in *J. Antimicrob. Chemother.* **1994**, *34*, 851. [CrossRef] [PubMed]
55. Andriole, C.L.; Andriole, V.T. Are all quinolones created equal? *Medicguide Infect. Dis.* **2002**, *21*, 1–5.
56. Khanna, A.; Kumar, N.; Rana, R.; Jyoti; Sharma, A.; Muskan; Kaur, H.; Bedi, P.M.S. Fluoroquinolones tackling antimicrobial resistance: Rational design, mechanistic insights and comparative analysis of norfloxacin vs ciprofloxacin derivatives. *Bioorg. Chem.* **2024**, *153*, 107773. [CrossRef]
57. Sharma, P.C.; Jain, A.; Jain, S. Fluoroquinolone antibacterials: A review on chemistry, microbiology and therapeutic prospects. *Acta Pol. Pharm.* **2009**, *66*, 587–604.
58. King, D.E.; Malone, R.; Lilley, S.H. New classification and update on the quinolone antibiotics. *Am. Fam. Physician* **2000**, *61*, 2741–2748.
59. Pham, T.D.M.; Ziora, Z.M.; Blaskovich, M.A.T. Quinolone antibiotics. *Med. Chem. Comm.* **2019**, *10*, 1719–1739. [CrossRef]
60. Irikura, T. Piperazinyl Derivatives of Quinoline Carboxylic Acids. U.S. Patent US4146719A, 27 March 1979.
61. Zerbes, R.; Naab, P.; Franckowiak, G.; Diehl, H. One-Pot Process for the Preparation of 3-Quinolonecarboxylic Acid Derivatives. U.S. Patent US5639886A, 17 June 1997.
62. Aldred, K.; Kerns, R.J.; Osheroff, N. Mechanism of quinolone action and resistance. *Biochemistry* **2014**, *53*, 1565–1574. [CrossRef]
63. Spencer, A.C.; Panda, S.S. DNA gyrase as a target for quinolones. *Biomedicines* **2023**, *11*, 371. [CrossRef]
64. Correia, S.; Poeta, P.; Hébraud, M.; Capelo, J.L.; Igrejas, G. Mechanisms of quinolone action and resistance: Where do we stand? *J. Med. Microbiol.* **2017**, *66*, 551–559. [CrossRef] [PubMed]
65. Bush, N.; Diez-Santos, I.; Abbott, L.R.; Maxwell, A. Quinolones: Mechanism, lethality and their contributions to antibiotic resistance. *Molecules* **2020**, *25*, 5662. [CrossRef] [PubMed]
66. Zhang, G.F.; Zhang, S.; Pan, P.; Liu, X.; Feng, L.S. 4-Quinolone derivatives and their activities against Gram positive pathogens. *Eur. J. Med. Chem.* **2018**, *143*, 710–723. [CrossRef] [PubMed]
67. Van Goor, F.; Hadida, S.; Grootenhuis, P.D.; Burton, B.; Cao, D.; Neuberger, T.; Turnbull, A.; Singh, A.; Joubran, J.; Hazlewood, A.; et al. Rescue of CF airway epithelial cell function in vitro by a CFTR potentiator, VX-770. *Proc. Nat. Acad. Sci. USA* **2009**, *106*, 18825–18830. [CrossRef]

68. Sloane, P.A.; Rowe, S.M. Cystic fibrosis transmembrane conductance regulator protein repair as a therapeutic strategy in cystic fibrosis. *Curr. Opin. Pulm. Med.* **2010**, *16*, 591–597. [CrossRef]
69. Hadida, S.; Van Goor, F.; Zhou, J.; Arumugam, V.; McCartney, J.; Hazlewood, A.; Decker, C.; Negulescu, P.; Grootenhuis, P.D.J. Discovery of *N*-(2,4-di-tert-butyl-5-hydroxyphenyl)-4-oxo-1,4-dihydroquinoline-3-carboxamide (VX-770, ivacaftor), a potent and orally bioavailable CFTR potentiator. *J. Med. Chem.* **2014**, *57*, 9776–9795. [CrossRef]
70. Sato, M.; Motomura, T.; Aramaki, H.; Matsuda, T.; Yamashita, M.; Ito, Y.; Kawakami, H.; Matsuzaki, Y.; Watanabe, W.; Yamataka, K.; et al. Novel HIV-1 integrase inhibitors derived from quinolone antibiotics. *J. Med. Chem.* **2006**, *49*, 1506–1508. [CrossRef]
71. Shimura, K.; Kodama, E.; Sakagami, Y.; Matsuzaki, Y.; Watanabe, W.; Yamataka, K.; Watanabe, Y.; Ohata, Y.; Doi, S.; Sato, M.; et al. Broad antiretroviral activity and resistance profile of the novel human immunodeficiency virus integrase inhibitor elvitegravir (JTK-303/GS-9137). *J. Virol.* **2008**, *82*, 764–774. [CrossRef]
72. Unger, N.R.; Worley, M.V.; Kisgen, J.J.; Sherman, E.M.; Childs-Kean, L.M. Elvitegravir for the treatment of HIV. *Expert Opin. Pharmacother.* **2016**, *17*, 2359–2370. [CrossRef]
73. Ohata, Y.; Tomonaga, M.; Watanabe, Y.; Tomura, K.; Kimura, K.; Akaki, T.; Adachi, K.; Kodama, E.N.; Matsuzaki, Y.; Hayashi, H. Antiviral Activity and Resistance Profile of the Novel HIV-1 Non-Catalytic Site Integrase Inhibitor JTP-0157602. *J. Virol.* **2022**, *96*, e0184321. [CrossRef]
74. Ben-Eli, H.; Solomon, A. Topical antihistamines, mast cell stabilizers, and dual-action agents in ocular allergy: Current trends. *Curr. Opin. Allergy Clin. Immunol.* **2018**, *18*, 411–416. [CrossRef] [PubMed]
75. Siegel, R.L.; Miller, K.D.; Fuchs, H.E.; Jemal, A. Cancer Statistics, 2022. *CA Cancer J. Clin.* **2022**, *72*, 7–33. [CrossRef] [PubMed]
76. Paul, M.; Gafter-Gvili, A.; Fraser, A.; Leibovici, L. The anti-cancer effects of quinolone antibiotics? *Eur. J. Clin. Microbiol. Infect. Dis.* **2007**, *26*, 825–831. [CrossRef] [PubMed]
77. Yadav, V.; Talwar, P. Repositioning of fluoroquinolones from antibiotic to anti-cancer agents: An underestimated truth. *Biomed. Pharmacother.* **2019**, *111*, 934–946. [CrossRef]
78. Batalha, P.N.; Vieira de Souza, M.C.; Peña-Cabrera, E.; Cruz, D.C.; da Costa Santos Boechat, F. Quinolones in the search for new anticancer agents. *Curr. Pharm. Des.* **2016**, *22*, 6009–6020. [CrossRef]
79. Gao, Y.; Shang, Q.; Li, W.; Guo, W.; Stojadinovic, A.; Mannion, C.; Man, Y.G.; Chen, T. Antibiotics for cancer treatment: A double-edged sword. *J. Cancer* **2020**, *11*, 5135–5149. [CrossRef]
80. Abdel-Aal, M.A.A.; Abdel-Aziz, S.A.; Shaykoon, M.S.A.; Abuo-Rahma, G.E.A. Towards anticancer fluoroquinolones: A review article. *Arch. Pharm.* **2019**, *352*, e1800376. [CrossRef]
81. Bisacchi, G.; Hale, M. A "Double-Edged" Scaffold: Antitumor power within the antibacterial quinolone. *Curr. Med. Chem.* **2016**, *23*, 520–577. [CrossRef]
82. Kloskowski, T.; Gurtowska, N.; Nowak, M.; Joachimiak, R.; Bajek, A.; Olkowska, J.; Drewa, T. The influence of ciprofloxacin on viability of A549, HepG2, A375.S2, B16 and C6 cell lines in vitro. *Acta Pol. Pharm.* **2011**, *68*, 859–865. [CrossRef]
83. Beberok, A.; Wrześniok, D.; Rok, J.; Rzepka, Z.; Respondek, M.; Buszman, E. Ciprofloxacin triggers the apoptosis of human triple-negative breast cancer MDA-MB-231 cells via the p53/Bax/Bcl-2 signaling pathway. *Int. J. Oncol.* **2018**, *52*, 1727–1737. [CrossRef]
84. Beberok, A.; Wrześniok, D.; Minecka, A.; Rok, J.; Delijewski, M.; Rzepka, Z.; Respondek, M.; Buszman, E. Ciprofloxacin-mediated induction of S-phase cell cycle arrest and apoptosis in COLO829 melanoma cells. *Pharmacol. Rep.* **2018**, *70*, 6–13. [CrossRef] [PubMed]
85. Kloskowski, T.; Szeliski, K.; Fekner, Z.; Rasmus, M.; Dąbrowski, P.; Wolska, A.; Siedlecka, N.; Adamowicz, J.; Drewa, T.; Pokrywczyńska, M. Ciprofloxacin and levofloxacin as potential drugs in genitourinary cancer treatment-the effect of dose-response on 2D and 3D cell cultures. *Int. J. Mol. Sci.* **2021**, *22*, 11970. [CrossRef] [PubMed]
86. Nowakowska, J.; Radomska, D.; Czarnomysy, R.; Marciniec, K. Recent development of fluoroquinolone derivatives as anticancer agents. *Molecules* **2024**, *29*, 3538. [CrossRef] [PubMed]
87. Beberok, A.; Wrześniok, D.; Szlachta, M.; Rok, J.; Rzepka, Z.; Respondek, M.; Buszman, E. Lomefloxacin induces oxidative stress and apoptosis in COLO829 melanoma cells. *Int. J. Mol. Sci.* **2017**, *18*, 2194. [CrossRef]
88. Beberok, A.; Rzepka, Z.; Rok, J.; Banach, K.; Wrześniok, D. UV radiation enhances lomefloxacin-mediated cytotoxic, growth-inhibitory and pro-apoptotic effect in human melanoma cells through excessive reactive oxygen species generation. *Int. J. Mol. Sci.* **2020**, *21*, 8937. [CrossRef]
89. Qurban, F.; Shahzad, S.A.; Khaskheli, M.S.; Khan, S.U.; Khan, S.A.; Rauf, W.; Islam, S.; Mannan, A. Design, synthesis and evaluation of novel norfloxacin analogs as potent anticancer and antioxidant agents. *Future Med. Chem.* **2024**, *16*, 1777–1789. [CrossRef]
90. He, X.; Yao, Q.; Hall, D.D.; Song, Z.; Fan, D.; You, Y.; Lian, W.; Zhou, Z.; Duan, L.; Chen, B. Levofloxacin exerts broad-spectrum anticancer activity via regulation of THBS1, LAPTM5, SRD5A3, MFAP5 and P4HA1. *Anticancer Drugs* **2022**, *33*, e235–e246. [CrossRef]

91. Yu, M.; Li, R.; Zhang, J. Repositioning of antibiotic levofloxacin as a mitochondrial biogenesis inhibitor to target breast cancer. *Biochem. Biophys. Res. Commun.* **2016**, *471*, 639–645. [CrossRef]
92. Zhou, P.; Huang, L.; Zhou, J.; Jiang, B.; Zhao, Y.; Deng, X.; Zhao, Q.; Li, F. Discovery of novel 4(1H)-quinolone derivatives as potential antiproliferative and apoptosis inducing agents. *Bioorg. Med. Chem. Lett.* **2017**, *27*, 4185–4189. [CrossRef]
93. Gao, F.; Zhang, X.; Wang, T.; Xiao, J. Quinolone hybrids and their anti-cancer activities: An overview. *Eur. J. Med. Chem.* **2019**, *165*, 59–79. [CrossRef]
94. Willmore, E.; De Caux, S.; Sunter, N.J.; Tilby, M.J.; Jackson, G.H.; Austin, C.A.; Durkacz, B.W. A novel DNA-dependent protein kinase inhibitor, NU7026, potentiates the cytotoxicity of topoisomerase II poisons used in the treatment of leukemia. *Blood* **2004**, *103*, 4659–4665. [CrossRef] [PubMed]
95. Wohlkonig, A.; Chan, P.F.; Fosberry, A.P.; Homes, P.; Huang, J.; Kranz, M.; Leydon, V.R.; Miles, T.J.; Pearson, N.D.; Perera, R.L.; et al. Structural basis of quinolone inhibition of type IIA topoisomerases and target-mediated resistance. *Nat. Struct. Mol. Biol.* **2010**, *17*, 1152–1153. [CrossRef] [PubMed]
96. Golub, A.G.; Yakovenko, O.Y.; Bdzhola, V.G.; Sapelkin, V.M.; Zien, P.; Yarmoluk, S.M. Evaluation of 3-carboxy-4(1H)-quinolones as inhibitors of human protein kinase CK2. *J. Med. Chem.* **2006**, *49*, 6443–6450. [CrossRef] [PubMed]
97. Wendorff, T.J.; Schmidt, B.H.; Heslop, P.; Austin, C.A.; Berger, J.M. The structure of DNA-bound human topoisomerase II alpha: Conformational mechanisms for coordinating inter-subunit interactions with DNA cleavage. *J. Mol. Biol.* **2012**, *424*, 109–124. [CrossRef]
98. Sha, S.; Han, H.W.; Gao, F.; Liu, T.B.; Li, Z.; Zu, C.; Zhong, W.Q.; Zhu, H.L. Discovery of fluoroquinolone derivatives as potent, selective inhibitors of PI3K. *MedChemComm* **2015**, *6*, 2029–2035. [CrossRef]
99. Seo, K.W.; Holt, R.; Jung, Y.S.; Rodriguez, C.O.; Chen, X.; Rebhun, R.B. Fluoroquinolone-mediated inhibition of cell growth, S-G2/M cell cycle arrest, and apoptosis in canine osteosarcoma cell lines. *PLoS ONE* **2012**, *7*, e42960. [CrossRef]
100. Lawrence, J.W.; Darkin-Rattray, S.; Xie, F.; Neims, A.H.; Rowe, T.C. 4-Quinolones cause a selective loss of mitochondrial DNA from mouse L1210 leukemia cells. *J. Cell. Biochem.* **1993**, *51*, 165–174. [CrossRef]
101. Huang, C.-Y.; Yang, J.-L.; Chen, J.-J.; Tai, S.-B.; Yeh, Y.-H.; Liu, P.-F.; Lin, M.-W.; Chung, C.-L.; Chen, C.-L. Fluoroquinolones suppress Tgf-β and Pma-induced Mmp-9 production in cancer cells: Implications in repurposing quinolone antibiotics for cancer treatment. *Int. J. Mol. Sci.* **2021**, *22*, 11602. [CrossRef]
102. Kan, J.-Y.; Hsu, Y.-L.; Chen, Y.-H.; Chen, T.-C.; Wang, J.-Y.; Kuo, P.-L. Gemifloxacin, a fluoroquinolone antimicrobial drug, inhibits migration and invasion of human colon cancer cells. *BioMed Res. Int.* **2013**, *2013*, 159786. [CrossRef]
103. Michot, J.M.; Seral, C.; Van Bambeke, F.; Mingeot-Leclercq, M.P.; Tulkens, P.M. Influence of efflux transporters on the accumulation and efflux of four quinolones (ciprofloxacin, levofloxacin, garenoxacin, and moxifloxacin) in J774 macrophages. *Antimicrob. Agents Chemother.* **2005**, *49*, 2429–2437. [CrossRef]
104. Chen, Y.C.; Lu, P.H.; Pan, S.L.; Teng, C.M.; Kuo, S.C.; Lin, T.P.; Ho, Y.F.; Huang, Y.C.; Guh, J.H. Quinolone analogue inhibits tubulin polymerization and induces apoptosis via Cdk1-involved signaling pathways. *Biochem. Pharmacol.* **2007**, *74*, 10–19. [CrossRef] [PubMed]
105. Assar, S.; Nosratabadi, R.; Khorramdel Azad, H.; Masoumi, J.; Mohamadi, M.; Hassanshahi, G. A review of immunomodulatory effects of fluoroquinolones. *Immunol. Investig.* **2021**, *50*, 1007–1026. [CrossRef] [PubMed]
106. Azzman, N.; Anwar, S.; Syazani Mohamed, W.A.; Ahemad, N. Quinolone derivatives as anticancer agents: Importance in medicinal chemistry. *Curr. Top. Med. Chem.* **2024**, *24*, 1134–1157. [CrossRef] [PubMed]
107. Dhiman, P.; Arora, N.; Thanikachalam, P.V.; Monga, V. Recent advances in the synthetic and medicinal perspective of quinolones: A review. *Bioorg. Chem.* **2019**, *92*, 103291. [CrossRef]
108. Jaskulska, A.; Gach-Janczak, K.; Drogosz-Stachowicz, J.; Janecki, T.; Janecka, A.E. Synthesis and anticancer aroperties of aew 3-methylidene-1-sulfonyl-2,3-dihydroquinolin-4(1*H*)-ones. *Molecules* **2022**, *27*, 3597. [CrossRef]
109. Hryhoriv, H.; Kovalenko, S.M.; Georgiyants, M.; Sidorenko, L.; Georgiyants, V. A Comprehensive review on chemical synthesis and chemotherapeutic potential of 3-heteroaryl fluoroquinolone hybrids. *Antibiotics* **2023**, *12*, 625. [CrossRef]

molecules

MDPI

Review

2-Azidobenzaldehyde-Based [4+2] Annulation for the Synthesis of Quinoline Derivatives

Xiaofeng Zhang [1,2,*,†], Miao Liu [1,3,†], Weiqi Qiu [1,4] and Wei Zhang [1,*]

1 Center for Green Chemistry and Department of Chemistry, University of Massachusetts Boston, 100 Morrissey Blvd, Boston, MA 02125, USA; liumiaomarcus@gmail.com (M.L.); qiuwb@bc.edu (W.Q.)
2 Department of Medicinal Chemistry, Cerevel Therapeutics, Cambridge, MA 02141, USA
3 Department of Mechanical Engineering, University of Wisconsin Milwaukee, Milwaukee, WI 53211, USA
4 Department of Chemistry, Boston College, 2609 Beacon Street, Chestnut Hill, MA 20467, USA
* Correspondence: xfxiaofengzhang@gmail.com (X.Z.); wei2.zhang@umb.edu (W.Z.); Tel.: +1-617-287-6147 (W.Z.)
† These authors contributed equally to this work.

Abstract: Quinoline is a privileged heterocyclic ring which can be found in many drug molecules and bioactive compounds. The development of synthetic methods for making quinoline derivatives continuously attracts the interest of organic and medicinal chemists. This paper highlights 2-azidobenzaldehyde-based [4+2] annulation for the synthesis of quinoline derivatives including fused and spiro-quinolines, quinoline-4-ols, 4-aminoquinolines, and related compounds.

Keywords: quinolines; azidobenzaldehyde; [4+2] annulation; heterocyclic; bioactive

1. Introduction

Quinoline is a privileged scaffold for bioactive molecules [1–6]. Shown in Figure 1 are some quinoline-containing drug molecules, including Mefloquine (antimalarial) [7–10], Brequinar (anticancer) [11–13], Pitavastatin (cholesterol-lowering) [14–16], Plaquenil (antimalarial) [17,18], Ciprofloxacin (antibacterial) [19–21], and Lenvatinib (anticancer) [22–24]. Other than these commercial drugs, a great number of quinoline derivatives have been reported or are under development as druggable molecules for anticancer [25,26], antibacterial [27–29], antifungal [30–32], antiviral [33–36], antimalarial [37–39], antioxidant [40–42], anti-inflammatory [43–45], CNS effect [46,47], cardiovascular, anticonvulsant, analgesic, and anthelmintic research [48–51].

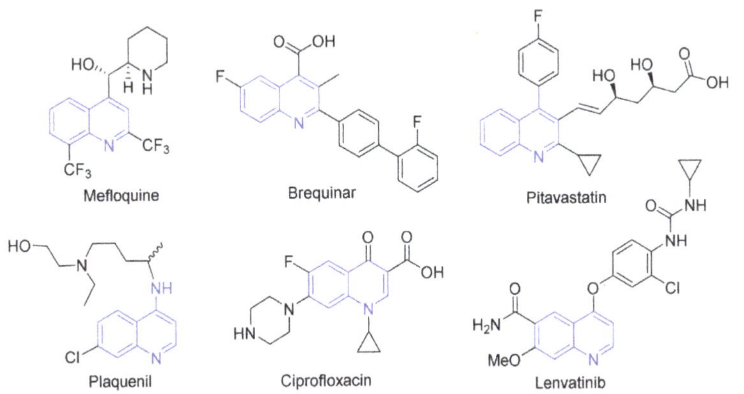

Figure 1. Representative structures of quinoline drugs.

The synthesis of quinolines has continuously attracted the interest of organic and medicinal chemists. Over the years, many methods including name reactions such as Skraup, Doebner–von Miller, Friedlander, Ptzinger, Conrad–Limpach, and Combes syntheses [52,53] have been developed for making quinolines. Alam and Patel have reviewed the general synthetic methods [54,55], which included new methods such as cascade reactions [56] and metal-free one-pot synthesis [57,58].

Shown in Scheme 1 are three general approaches to assembling the quinoline ring. (**I**) [3+3] Annulation. The early developed methods such as the Skraup, Doebner–von Miller, Conrad–Limpach, Gould–Jacobs, and Combes syntheses are in this category. The drawback of this approach is that it has a low regioselectivity for the synthesis of multi-substituted quinolines [52,54,57]. (**II**) [4+2] Annulation. There are two ways to put together the pyridine ring according to [4+2] annulation (**II-a** and **II-b**). The first one has low regioselectivity and a limited substrate scope [52,55,56]. The second one is more popular and uses readily available substrates to give products with a good regioselectivity. Reactions such as the Friedlander and Ptzinger reactions are in this category [52–55,58]. (**III**) Cyclization. In theory, there are four different ways to form the pyridine ring of quinolines (**III-a** to **III-d**) according to cyclization. But they have not been fully developed due to the difficult reaction process and the limited availability of the substrates [59,60].

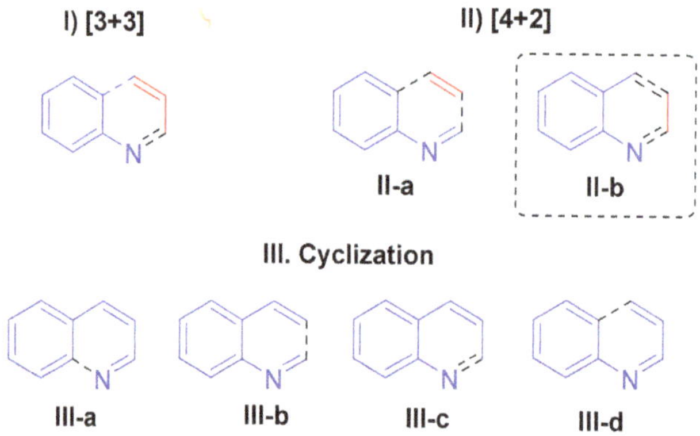

Scheme 1. Common synthetic strategies for making quinolines.

Presented in this paper are azidobenzaldehyde-based [4+2] annulation reactions (**II-b** type) for the synthesis of substituted quinolines [61]. For example, [4+2] annulation could be accomplished using a direct Diels–Alder reaction or via sequential condensation and cyclization reactions. In addition to quinolines, versatile 2-azidobenzaldehydes **1** could be used for the synthesis of other heterocyclic rings such as indoles [62,63], 3,4-dihydroquinazolines [64,65], triazolobenzodiazepines [66], 2*H*-indazoles [67], benzoxazepinones [68], and quinolines. Meanwhile, 2-azidobenzaldehydes could be easily prepared from commercially available 2-halobenzaldehydes, 2-aminobenzaldehydes, and 2-nitrobenzaldehydes [62,64,66–68]. Over 30 papers on the 2-azidobenzaldehyde-based synthesis of quinolines could be found in the literature and are summarized in this paper.

2. Results

2.1. Synthesis of 4-Unsubstituted Quinolines

There are many 4-nonsubstituted quinoline drugs and drug candidates (Figure 2), such as Bedaquiline for the treatment of active tuberculosis [69,70], Indacaterol for the treatment of chronic obstructive pulmonary disease [71,72], Brexpiprazole for the treatment of schizophrenia [73], Mardepodect for the treatment of schizophrenia [74], and Acridine carboxamide for the treatment of cancer [75].

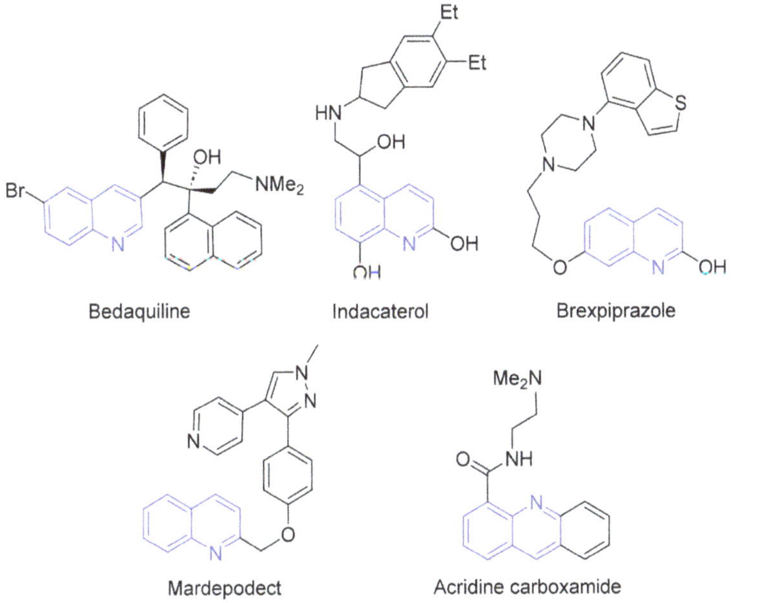

Figure 2. Examples of 4-unsubstituted quinoline-based marketed drugs and clinical candidates.

Utilizing the Diels–Alder reaction of benzisoxazole [76–78] as a key step, Zhang and co-workers developed a one-pot reaction involving the denitrogenation of 2-azidobenzaldehyde **1**, the formation of benzisoxazole **4**, [4+2] cycloaddition with fumarate ester **2**, and, finally, dehydrative aromatization of intermediate **5** for the synthesis of quinolinedicarboxylates **3** (Scheme 2) [79].

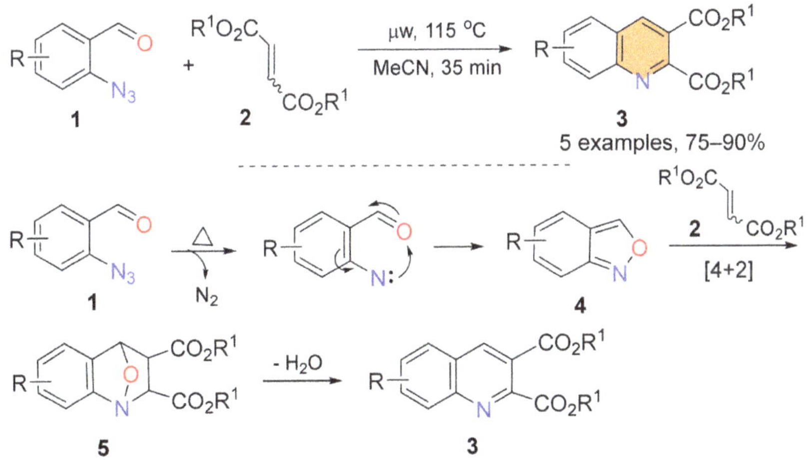

Scheme 2. Cascade reactions for quinolinedicarboxylates **3**.

Zeng and Chen reported the Cu-catalyzed synthesis of 3-NO_2 quinolines **7**. The addition of nitroalkene **6** to the azide group of **1** affords **8**, which then undergoes cyclization to form **9**. Proton transfer and the release of N_2 gas from **9** affords **10**, which leads to the formation of 3-NO_2 quinolines **7** after dehydration (Scheme 3) [80].

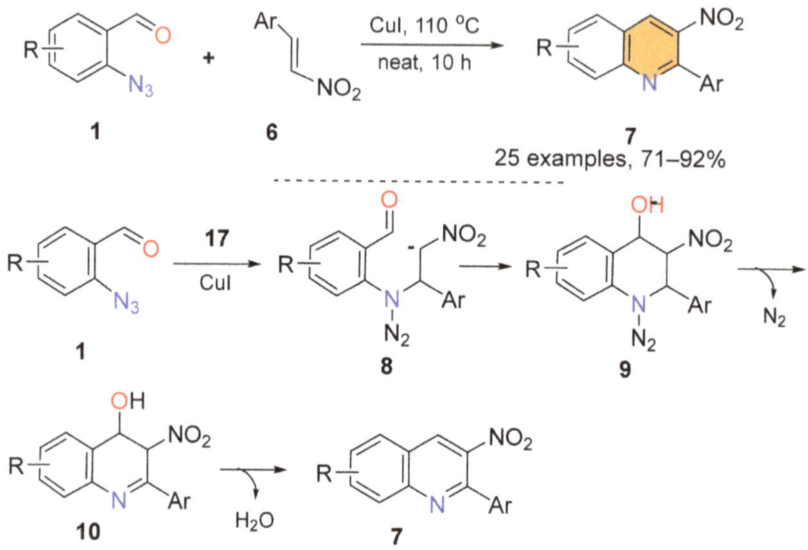

Scheme 3. Cu-catalyzed cascade reaction for 3-NO_2 quinolines **7**.

The intramolecular aza-Wittig reaction can be used for making *N*-heterocyclics [81,82]. Zhang and He reported the synthesis of 2,3-substituted quinolines **14** through a sequence involving the Staudinger, Knoevenagel, and aza-Wittig reactions [83,84]. The first intermediates **11** generated according to the Staudinger reaction of 2-azidobenzaldehydes **1** and PPh_3 undergo the Knoevenagel reaction with different carbonyl compounds **12a–c** to generate the corresponding intermediates **13a–c**. The intramolecular aza-Wittig reaction releasing $Ph_3P{=}O$ gives 2,3-substituted quinoline products **14a–c** (Scheme 4).

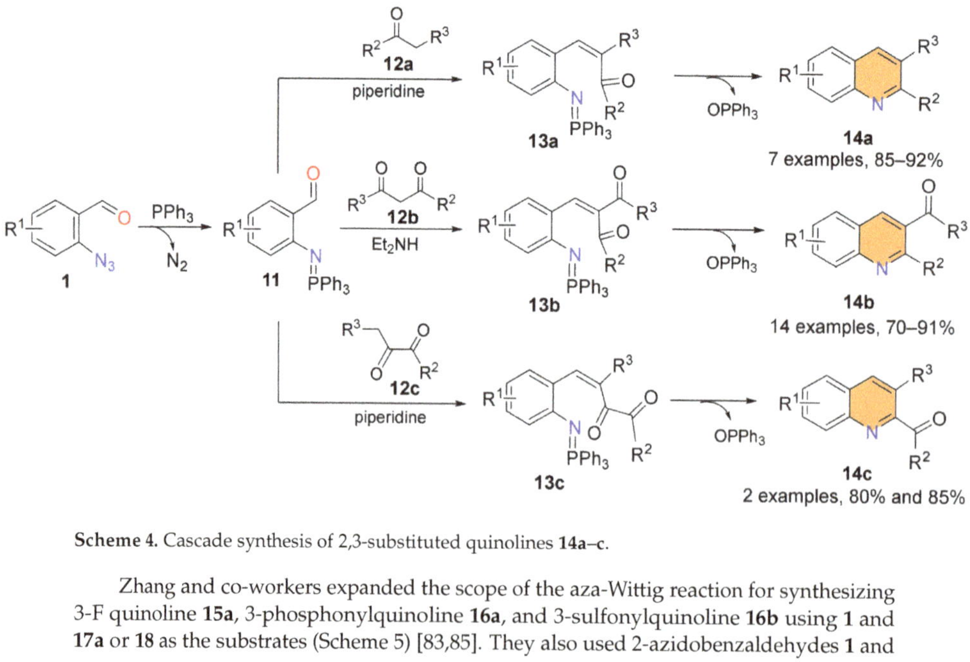

Scheme 4. Cascade synthesis of 2,3-substituted quinolines **14a–c**.

Zhang and co-workers expanded the scope of the aza-Wittig reaction for synthesizing 3-F quinoline **15a**, 3-phosphonylquinoline **16a**, and 3-sulfonylquinoline **16b** using **1** and **17a** or **18** as the substrates (Scheme 5) [83,85]. They also used 2-azidobenzaldehydes **1** and

1-substituted propan-2-ones **19** in the synthesis of bioactive 2-substituted quinolines such as HTLV-1 inhibitors **20a** and **20b**, antileishmanial agent **20c**, and CF_3- and CHF_2-substituted quinolines **20d** and **20e**, shown in Scheme 6 [86–88].

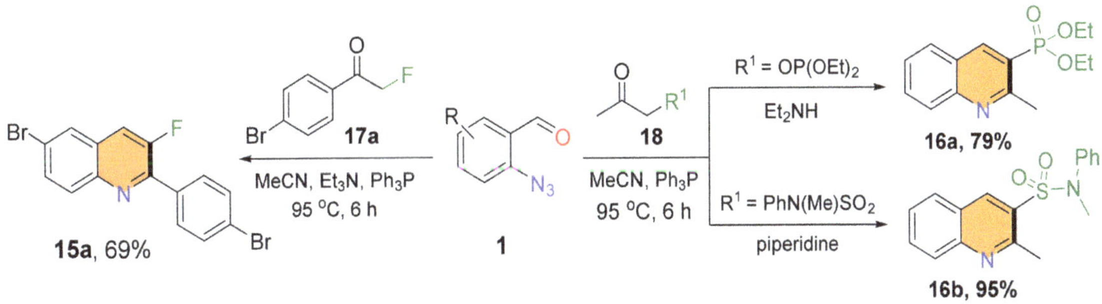

Scheme 5. Synthesis of 3-F quinolines **15a**, 3-phosphonylquinolines **16a**, and 3-sulfonylquinoline **16b**.

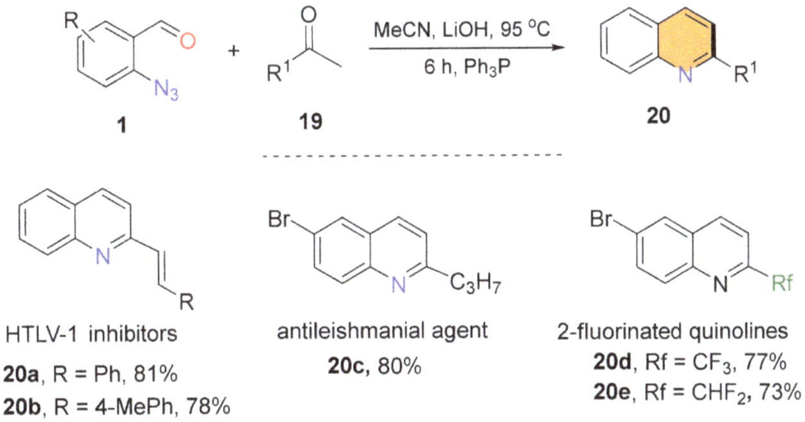

Scheme 6. Synthesis of bioactive 2-substituted quinolines **20**.

Due to the importance of F groups in drug molecules [89–93], the Zhang group introduced a reaction between compounds **1** and **21** for making 2-fluorinated quinolines **22a–f**, bearing CF_2, CF_3, and C_2F_5 at 70–87% yields (Scheme 7). These compounds are related to histone acetyltransferase (HAT) inhibitors [94–96]. Shi and co-workers reported the synthesis of 2,3-substituted quinolines **25** via a condensation/Michael/Staudinger/aza-Wittig/aromatization reaction (Scheme 8) [97]. The *ortho*-Azido-β-nitro-styrenes **23**, synthesized by the condensation of 2-azidobenzaldehydes **1** and $MeNO_2$, undergo the Michael addition with ketones **24** to form **26**, which then reacts with PPh_3 according to the Staudinger reaction to afford intermediates **27**. The intramolecular aza-Wittig reaction of **27** gives **28**, followed by the aromatic removal of $MeNO_2$ to afford **25** (Scheme 8).

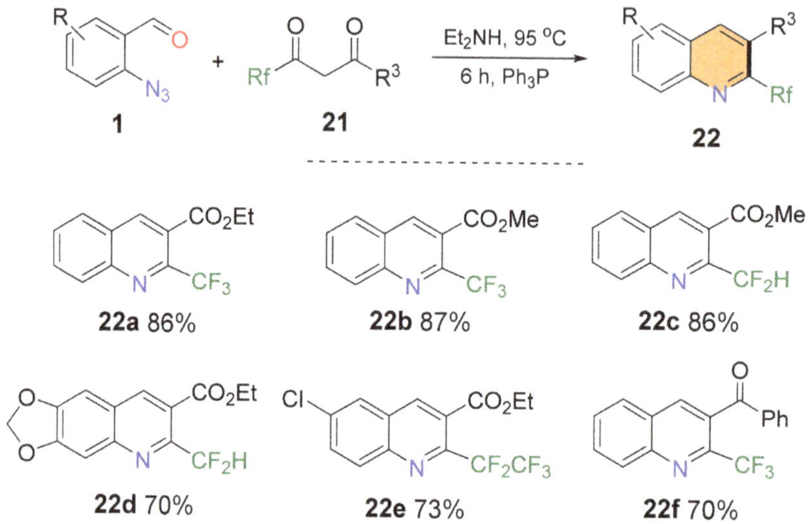

Scheme 7. Direct synthesis of 2-fluorinated quinolines **22**.

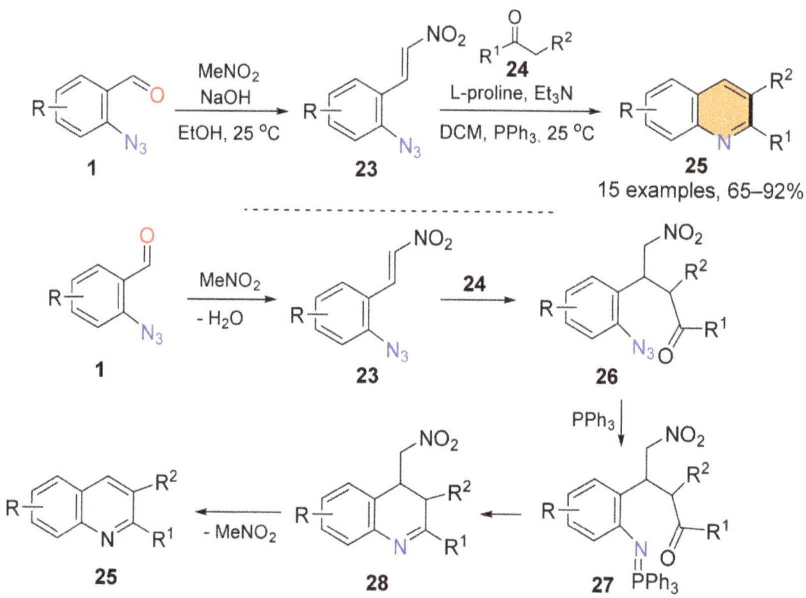

Scheme 8. One-pot synthesis of 2,3-substituted quinolines **25**.

In addition to aza-Wittig cyclization, for making quinolines, Chassaing and co-workers reported intramolecular imine formation of aldehyde and aniline for the synthesis of 2-substituted quinolines **31**. The reaction process involves the formation of *ortho*-azidocinnamoyl compounds **30** via the condensation of 2-azidobenzaldehydes **1** and ketones **29**. The irradiation of aryl azides **30** generates nitrenes **32**, which are then converted into amino intermediates **33** in protic media to give products **31** after imine cyclization (Scheme 9) [98].

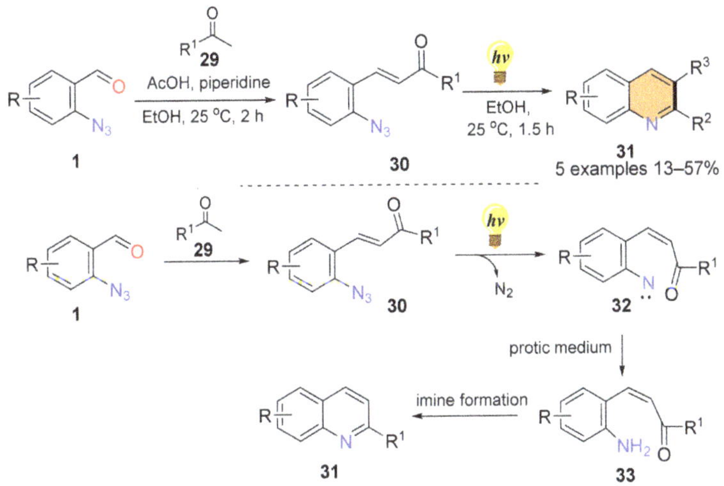

Scheme 9. Photosynthesis of 2-substituted quinolines **31**.

Reddy and co-workers reported a cyclization reaction of azidophenyl propargyl alcohol **34** for making 2,3-substituted quinolines **35** (Scheme 10) [99]. The azidophenyl propargyl alcohol **34**, generated according to the Favorskii reaction of 2-azidobenzaldehydes **1**, undergoes the Ni-catalyzed addition of its R^1 group to the propargylic system to form intermediates **36**, followed by E/Z isomerization to afford intermediate **37**. The stabilized azido group enables the subsequent denitrogenative cyclization into cyclic intermediates **38**. The denickelation of **38** under the photo conditions gives intermediate **39**, followed by dehydrative aromatization to afford 2,3-substituted quinolines **35**.

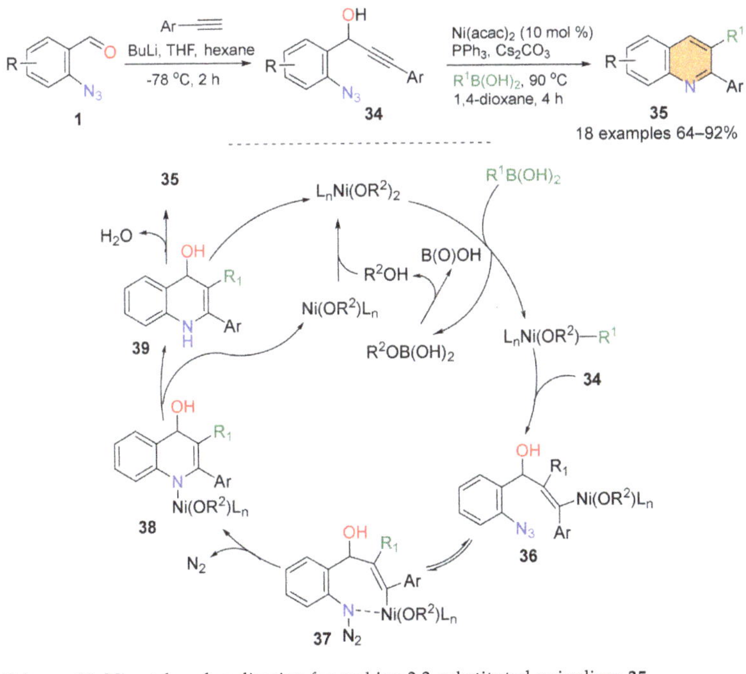

Scheme 10. Ni-catalyzed cyclization for making 2,3-substituted quinolines **35**.

The Yamamoto group utilized *o*–propargyl arylazides **40** derived from the dehydration of compound **34** for the synthesis of multi-substituted quinolines **41**–**44** (Scheme 11). The electrophilic cyclization of **40** in the presence of catalytic amounts of $AuCl_3/AgNTf_2$ affords products **41**. The method was applied to the synthesis of 3-Br/I quinolines **42** in the presence of I_2, Br_2, or NIS [100]. Zhou and co-workers extended the scope of electrophilic cyclization induced by pseudohalogen $(SeCN)_2$ for the synthesis of quinolylselenocyanates **43** [101]. Wang and Quan reported the synthesis of SCF_3-substituted quinolines **44** through radical cyclization of *o*–propargyl arylazides **40** [102].

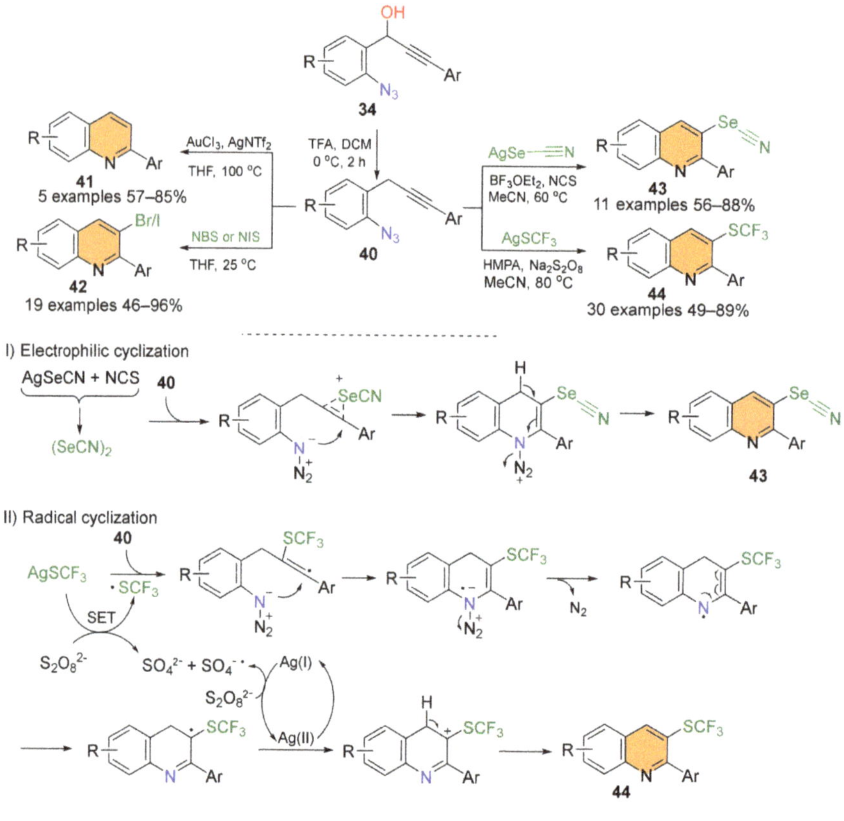

Scheme 11. Alkyne–azide cyclization for making quinolines **41**–**44**.

The ring-opening of aziridines is a useful tool for the construction of N-heterocycles via cycloaddition or cyclization reactions [103–106]. Wan and co-workers utilized ring-opening and the cyclization of aziridines **45** for the synthesis of 3-substituted quinolines **46** (Scheme 12) [107]. The TfOH-promoted denitrogenation of the azide group of **1** gave intermediate **47** and then **48** after deprotonation. The addition of 48 to aziridines **45** forms **49**, followed by the cleavage of TsNH2 to afford **50**. Acid-promoted cyclization of **50** gives **51**, which is converted into 3-aryl quinoline **46** after aromatization (Scheme 12).

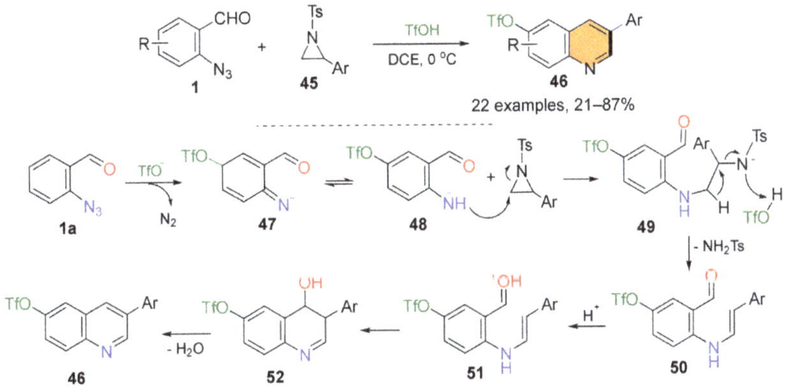

Scheme 12. Ring-opening and cyclization of aziridines for the synthesis of quinolines **46**.

2.2. *Synthesis of Polycyclic Quinolines*

Polycyclic quinolines can be found in a number of commercial drugs and other bioactive compounds, such as the natural alkaloid Mappicine [108], the poly ADP ribose polymerase (PARP) inhibitor for cancer Talazoparib [109], the antibacterial drug Levofloxacin [110], the natural product Flindersine [111], and the immune response modifier Imiquimod (Figure 3) [112]. The cyclization of 2-azidobenzaldehydes can be integrated into multistep synthesis, one-pot synthesis, or a multi-component reaction (MCR) to make *N*-heterocycles [63,68,113,114]. The aza-Wittig cyclization-based synthesis of polycyclic quinolines is covered in this section. Shown in Scheme 13 are two pathways for the synthesis of cyclohexane-fused quinolines **55** (ee up to 98%) and **57** involving Knoevenage/aza-Wittig/dehydration reactions and Mannich/aza-Wittig/deamination sequences, respectively [115,116].

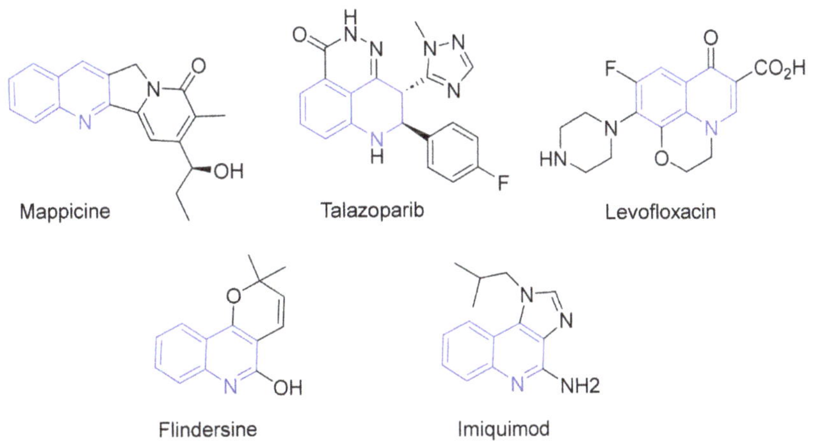

Figure 3. Polycyclic quinoline-based marked drugs and natural products.

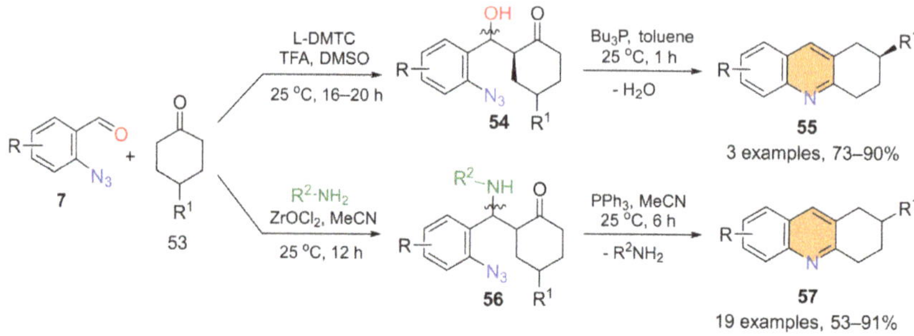

Scheme 13. One-pot synthesis of cyclohexane-fused quinolines **55** and **57**.

A three-component reaction (3-CR)-initiated synthesis of furan-fused quinolines was reported by Ding and co-workers (Scheme 14). Compounds **60** generated from the condensation of 2-azidobenzaldehydes **1**, dialkyl acetylenedicarboxylate **58**, and isocyanides **59** are used for Staudinger and aza-Wittig cyclization to make furan-fused quinolines **61** [117]. He's group reported sequential 3-CR/Staudinger/aza-Wittig cyclization/dehydroaromatization reaction processes for making furan-fused quinolines **65** (Scheme 14) [118].

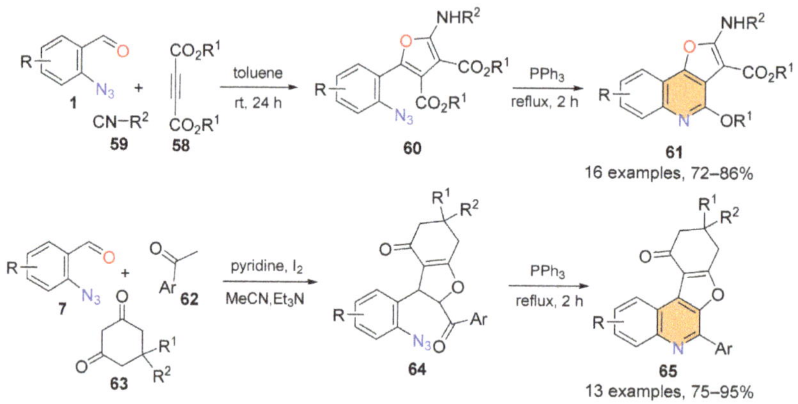

Scheme 14. MCR-initiated synthesis of furan-fused quinolines **61** and **65**.

He and Bharate's groups independently reported the reaction of aldehydes, 1,3-dicarbonyl compounds **66**, amines **67**, and nitroalkanes to make pyrrole-fused quinolines **69** (Scheme 15) [119,120]. He and co-workers also reported a 3-CR/Staudinger/aza-Wittig process for making pyrrole-fused quinolines **73** using **1**, acetyl compounds **70**, and TosMIC **71** as the starting materials (Scheme 15) [121]. Interestingly, an example of making cyclopropa[*c*]indeno [1,2-*b*]quinolines **77** was developed according to the 3-CR/Staudinger/aza-Wittig sequence. The products have a highly condensed ring system, including cyclopropane (Scheme 16) [122].

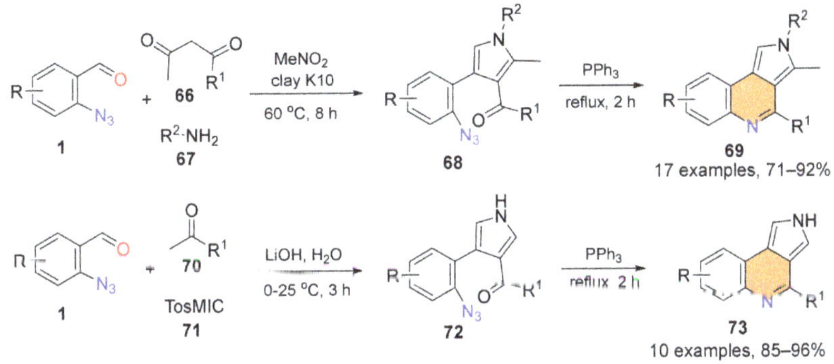

Scheme 15. MCR-initiated synthesis of pyrrole-fused quinolines **69** and **73**.

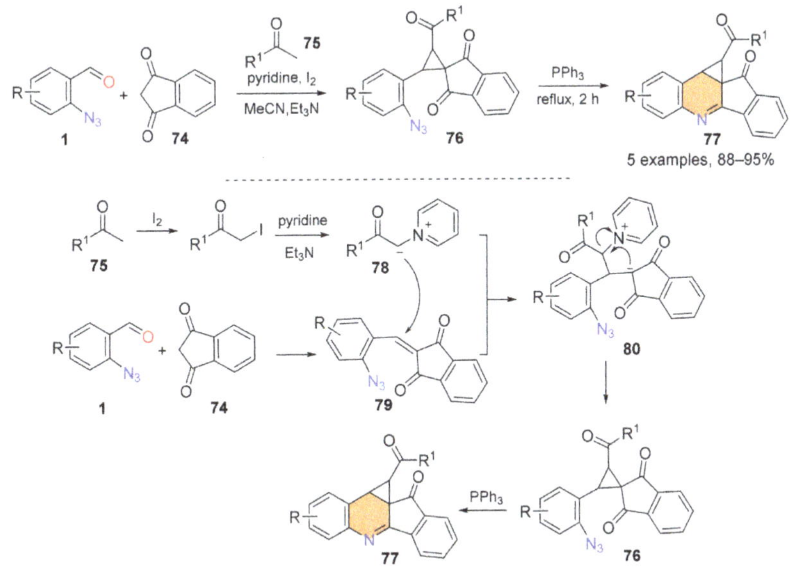

Scheme 16. MCR-initiated synthesis of highly condensed quinolines **77**.

Other than aza-Wittig cyclization for the construction of quinolines, Zhang and co-workers reported cascade denitrogenation/aza-Diels–Alder/dehydrative aromatization reactions for the synthesis of pyrrolidine-2,5-dione-fused quinolines using 2-azidobenzaldehydes **1** and *N*-substituted maleimides **81** as the starting materials (Scheme 17) [79]. This method (Scheme 17a) is more efficient than stepwise synthesis for the synthesis of **82a** (Scheme 17b) [123]. In another case, Li and Wang's group utilized compounds **1** and 3-aza-1,6-enynes **87** as the substrates for making tetrahydrobenzo[*b*][1,8]naphthyridines **88** (Scheme 18) [124]. The reactions involved aza-Diels–Alder cycloaddition of intermediates **90** derived from the Au-catalyzed reaction of enynes **87** to form **91**, followed by dehydrative aromatization to give products **88**.

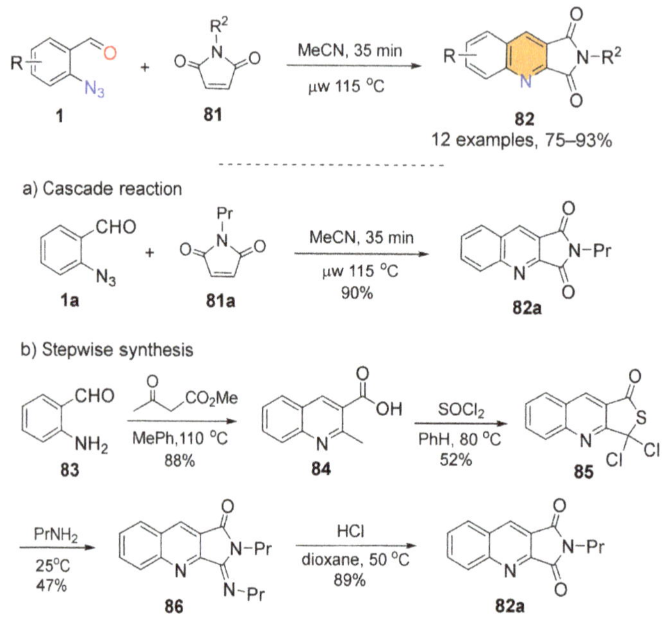

Scheme 17. Cascade reactions for pyrroloquinolinediones **82**.

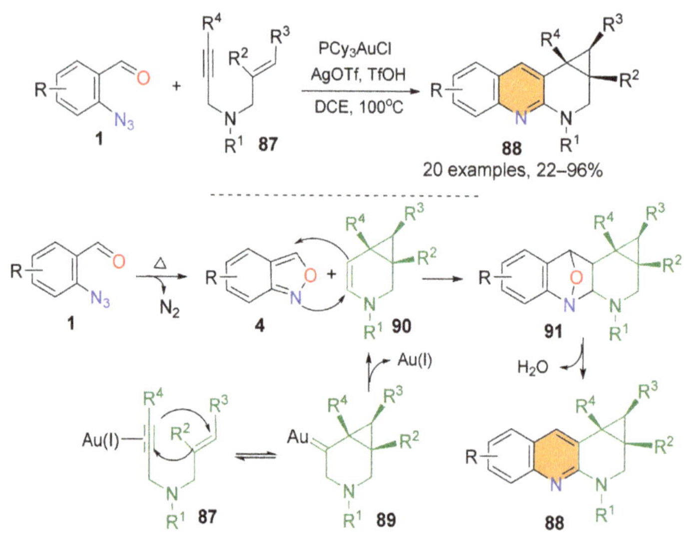

Scheme 18. Cascade reactions for tetrahydrobenzo[*b*][1,8]naphthyridines **88**.

Gharpure and co-workers introduced Lewis-acid-promoted cascade reactions involving Friedel–Crafts/alkyne indol-2-yl cation cyclization/vinyl cation trapping for the synthesis of pyrrolizino-quinolines **93** [125]. Benzyl alcohol and TMSOTf promote the Friedel–Crafts reaction of **1** and indoles **92** to form **93**, which undergoes alcohol elimination and cyclization to afford vinyl cations **96**. The trapping of the vinyl cations with the azide group generates pyrrolizino-quinolines **93** (Scheme 19) [125]. Alves and co-workers introduced sequential [3+2] cycloaddition and condensation reactions for making triazole-fused quinolines **98** (Scheme 20) [126]. The 1,3-dipolar cycloaddition of **1** and 1,3-dicarbonyl compounds **97** leads to the formation of 1,2,3-triazoles **99**. The intramolecular condensation of enolates **100** generated from the DBU deprotonation of **99** affords **101** and then triazole-fused quinolines **98** after dehydration.

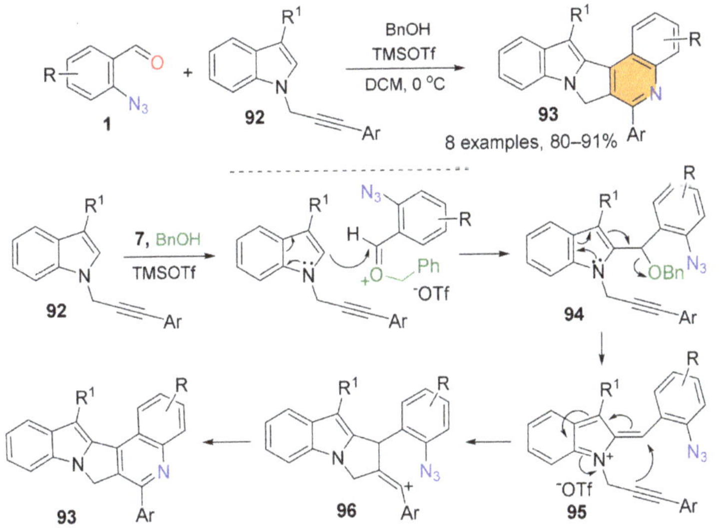

Scheme 19. Cascade synthesis of pyrrolizino-quinolines **93**.

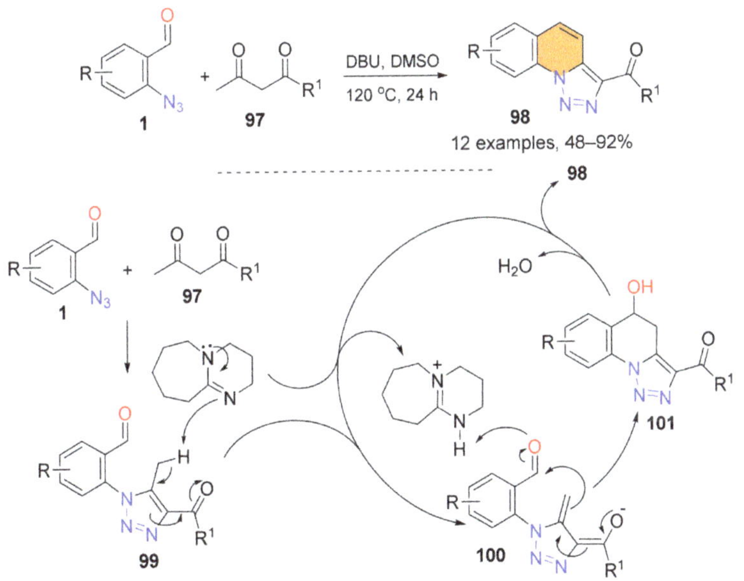

Scheme 20. Direct synthesis of [1–3] triazolo[1,5-a] quinolines **98**.

Alajarin and co-workers reported the cyclization of ketenimines for making spiro-quionlines **106** and **106′** using azides **102** and **102′** bearing five- and six-membered cyclic acetal functions (such as 1,3-dioxolane, 1,3-dithiolane, 1,3-dioxane, and 1,3-dithiane) as the starting materials (Scheme 21) [127,128]. The treatment of azides **102** or **102′** with PPh_3 affords iminophosphorane **103** or **103′**, followed by aza-Wittig reactions with disubstituted ketenes giving ketenimines **104** or **104′** and then *ortho*-azaxylylenes **105** or **105′** after deprotonation. The cyclization of **105** or **105′** gives the products spiroquinolines **106** or **106′**.

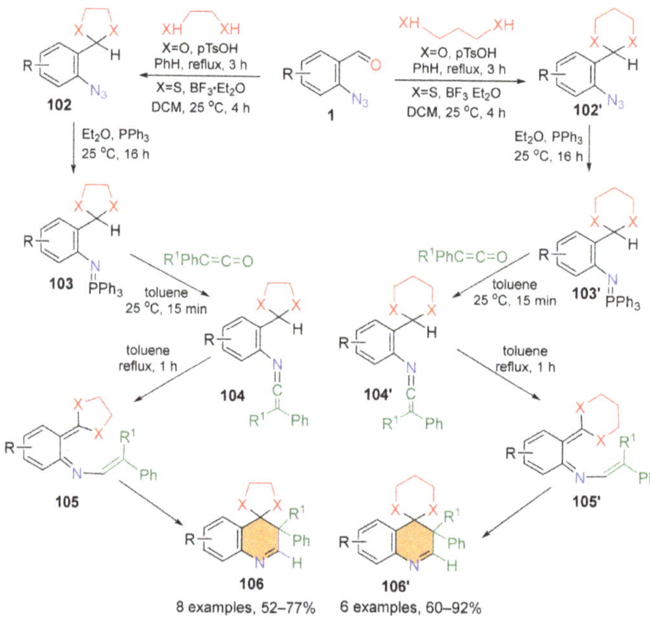

Scheme 21. Tandem synthesis of spiroquinolines **106** and **106′**.

2.3. *Synthesis of 4-Hydroxyquinoline Derivatives*

There are some drugs and natural products that have 4-hydroxyquinoline moiety, such as the furoquinoline alkaloid Skimmianine, with anticancer and anti-inflammatory effects [129,130]; Cabozantinib, used for the treatment of medullary thyroid cancer [131]; Tasquinimod as an immunomodulator for the treatment of blood cancers [132]; AMG-208 as a clinical trial drug for cancer [133]; Anlotinib, with antineoplastic and anti-angiogenic activities [134,135]; and Delafloxacin as a fluoroquinolone antibiotic for the treatment of acute bacterial skin and skin structure infections (Figure 4) [136].

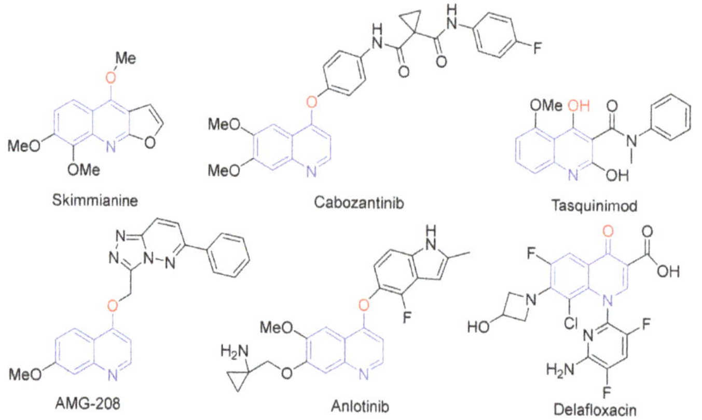

Figure 4. Examples of 4-hydroxy and related quinolines as drug molecules.

Other than aniline-based [3+3] annulation [137] and [4+2] annulation [138] reactions for making 4-hydroxyquinolines, 2-azidobenzaldehydes have also been employed for the [4+2] annulation-based synthesis of 4-alkoxy quinolines. Gharpure and coworkers reported the reactions of **1** with hydroxyalkynes for the synthesis of different kinds of 4-alkoxy quinolines, **107**, **108**, and **109** (Scheme 22) [139–141]. The synthesis first led to the formation of oxonium ions **110** followed by intramolecular [4+2] cycloaddition and aromatization to give 4-alkoxy quinolines **109**, with a new seven-membered heterocyclic ring (Scheme 22).

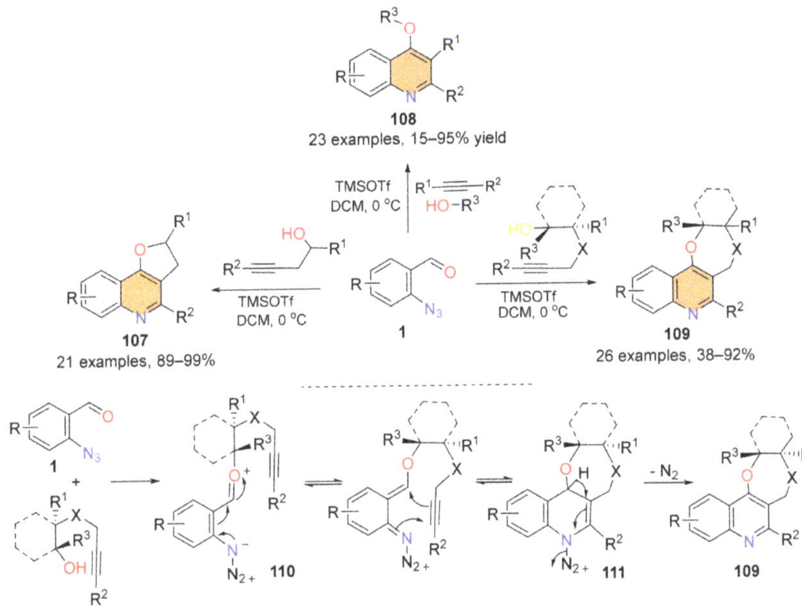

Scheme 22. Cascade synthesis of 4-alkoxy quinolines **107**, **108**, and **109**.

Metal-catalyzed 6-*endo*-dig azide–yne cyclization of azide-tethered alkynes **112** has been developed for the synthesis of multi-substituted 4-alkoxy quinolines **113–117** (Scheme 23) [142–144]. For example, Xu and coworkers synthesized **115** and **116** via the $AgSbF_6$-catalyzed cyclization of azide-tethered alkynes. The cyclization of Ag(I)-activated alkyne-tethered azides **112** affords **118**, followed by N_2 gas release from α-imino silver carbenes **119** and then reaction with R^1X to afford halonium zwitterions **120**. After a concerted rearrangement of **120**, product **115** or **116** is obtained.

Zhang and coworkers reported the direct synthesis of quinolin-4-ols, which are structurally related to histone acetyltransferase (HAT) inhibitors [145] and the key intermediates for making HMG-CoA reductase inhibitors [146]. The reaction of **1** and α-fluoro-β-ketoesters **121** or **123** via sequential aldol, aza-Wittig, and dehydrofluorinative aromatization reactions gives quinolin-4-ols **122** or **124** (Scheme 24) [83]. Green chemistry metrics analysis for this one-pot synthesis and two reported multi-step syntheses [138] of HAT inhibitor **144a** was conducted [79,82,147] to validate synthetic efficiency and low waste generation (Scheme 25).

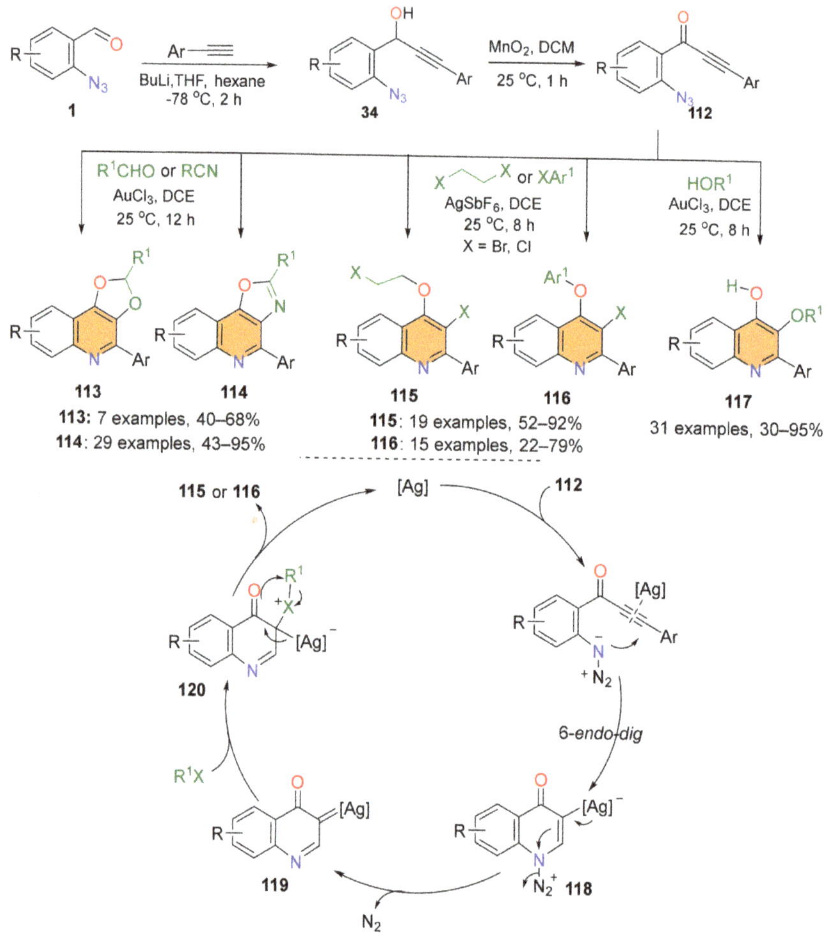

Scheme 23. Metal-catalyzed synthesis of diverse 4-alkoxy quinolines **113–117**.

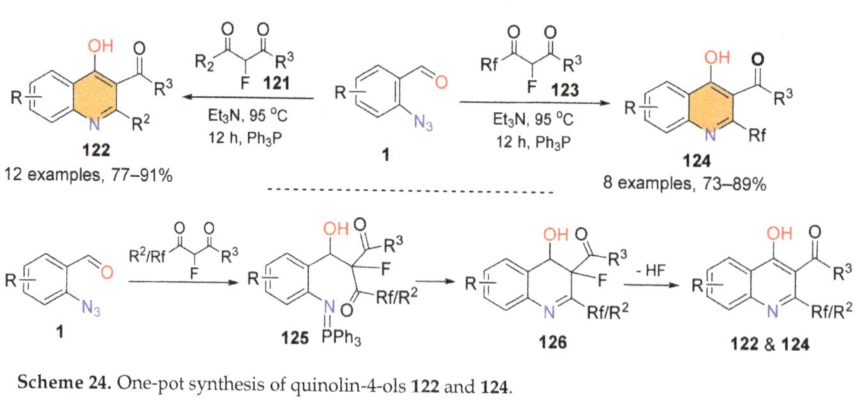

Scheme 24. One-pot synthesis of quinolin-4-ols **122** and **124**.

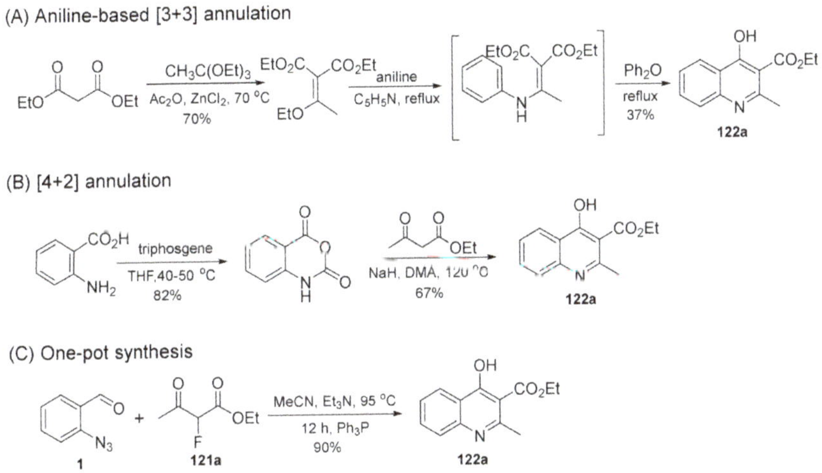

Scheme 25. Three methods for making **122a**.

2.4. Synthesis of 4-Amino-Quinolines

Shown in Figure 5 are some 4-amino-quinoline drugs and drug candidates such as Chloroquine, Amodiaquine, Bosutinib, Amsacrine, and Dovitinib for the treatment of malarial [148–150], leukemia [151,152], and cancer diseases [153]. Common methods in the literature for the synthesis of 4-aminoquinolines are aniline-based multi-step reactions [96,154–156]. Sharada and coworkers employed azides **1** and amines **127** for the synthesis of 4-aminoquinolines **129** through the aza-Diels–Alder reaction of intermediates **128** with dimethylacetylenedicarboxylate (DMAD) (Scheme 26) [157]. Shown in Scheme 27 are two examples of quinoline synthesis using 2-azidobenzaldehyde-based [4+2] cycloadditions. In the first case of using fumarate esters as the dienophiles, quinolines **3** are generated from dehydrative aromatization involving C-O bond cleavage. In the second case of using DMAD as a dienophile for cycloaddition with **128**, 4-aminoquinolines **129** result from the aromatization via the N-N bond cleavage of **130** [79,157].

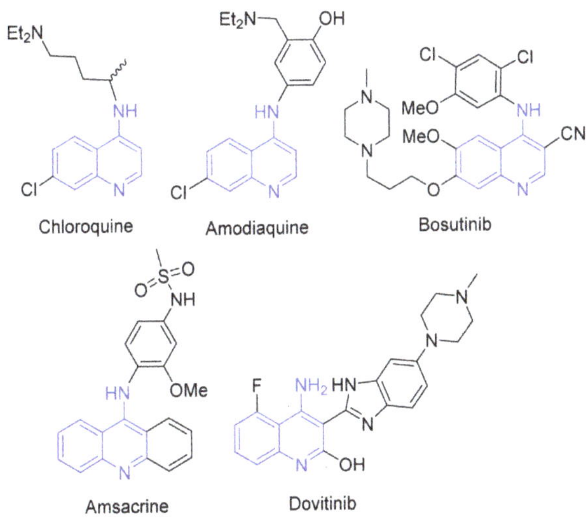

Figure 5. 4-Aminoquinoline-based drugs and drug candidates.

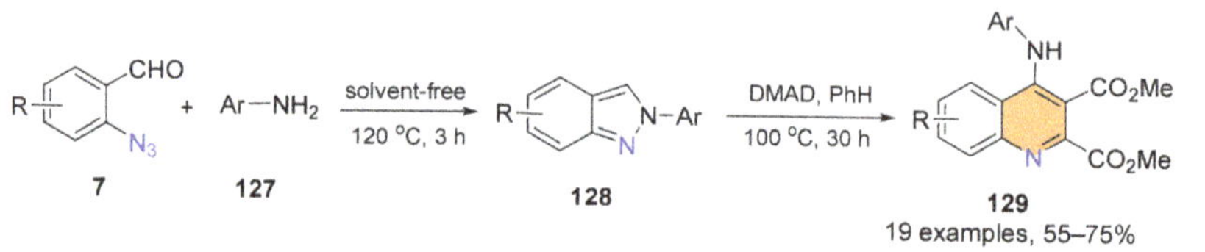

Scheme 26. One-pot synthesis of 4-aminoquinolines **129**.

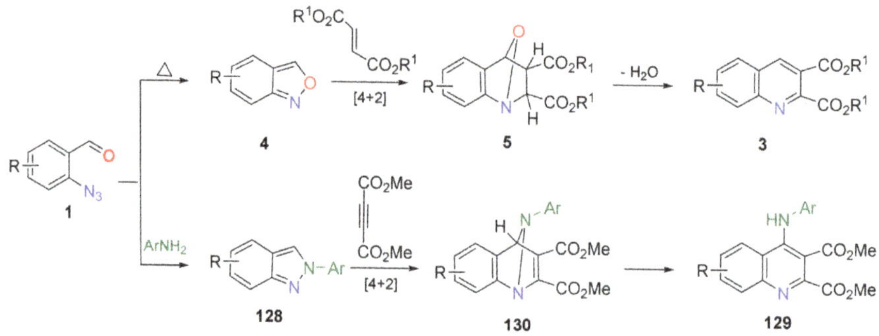

Scheme 27. Comparison of [4+2] cycloaddition for making compounds **3** and **129**.

Zhang and co-workers reported a three-component reaction of azides **1**, α-fluoro-β-ketoesters **121**, and amines **131** or **133** for the synthesis of 4-aminoquinolines **132** and **134** involving Mannich, aza-Wittig and dehydrofluorinative aromatization reactions (Scheme 28) [158]. They also applied this method to the synthesis of 2-CF_3 quinolines **137a** and **137b** with a pyrrolidine or a piperidine at the 4-position (Scheme 29).

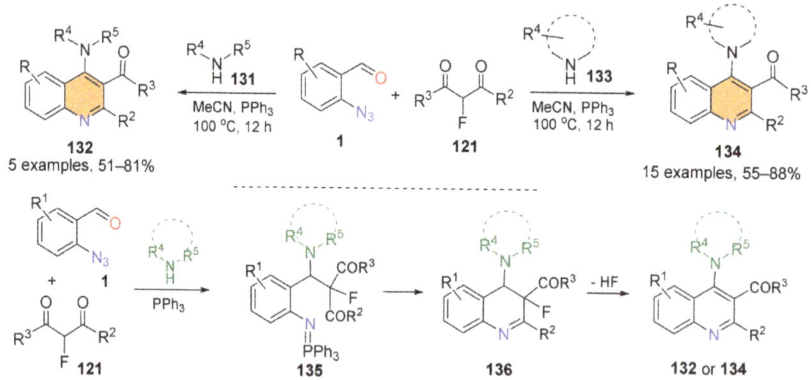

Scheme 28. Three-component reaction to make 4-aminoquinolines **132** and **134**.

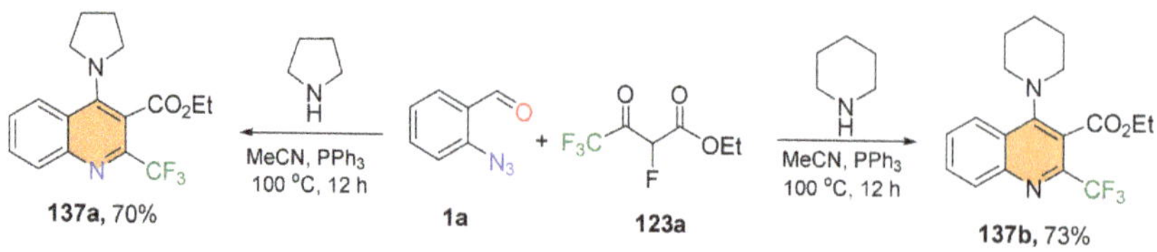

Scheme 29. One-pot synthesis of CF_3-containing 4-aminoquinolines **137a** and **137b**.

3. Conclusions

Presented in this paper are 2-azidobenzaldehyde-initiated reactions for the synthesis of diverse quinoline compounds, including fused, spiro-, and polycyclic quinolines, as well as substituted quinolines such as quinoline-4-ols and 4-aminoquinolines. These biologically significant quinoline compounds could be well utilized in medicinal chemistry programs for drug discovery. Also, 2-azidobenzaldehyde-initiated synthesis can be developed as one-pot stepwise synthesis or a multicomponent reaction for operation simplicity, process efficiency, and economizing in terms of steps, pots, and atoms. Some of the synthetic methods presented in this paper provide novel pathways for making quinolines which could also be used for the synthesis of other heterocyclic compounds.

Author Contributions: M.L., W.Q. and X.Z., literature search and original manuscript writing; W.Z., revision and supervision. All authors have read and agreed to the published version of the manuscript.

Funding: This research received no external funding.

Conflicts of Interest: X.Z. is employee of the Cerevel Therapeutics, other authors declare no conflicts of interest.

References

1. Chung, P.Y.; Bian, Z.X.H.; Pun, Y.D.; Chan, A.S.; Chui, C.H.; Tang, J.C.; Lam, K.H. Recent advances in research of natural and synthetic bioactive quinolines. *Future Med. Chem.* **2015**, *7*, 947–967. [CrossRef] [PubMed]
2. Kumar, S.; Bawa, S.; Gupta, H. Biological activities of quinoline derivatives. *Mini Rev. Med. Chem.* **2009**, *9*, 1648–1654. [CrossRef] [PubMed]
3. Broch, S.; Henon, H.; Debaud, L.A.; Fogeron, M.L.; Bonnefoy-Berard, N.; Anizon, F.; Moreau, P. Synthesis and biological activities of new di- and trimeric quinoline derivatives. *Bioorg. Med. Chem.* **2010**, *18*, 7132–7143. [CrossRef] [PubMed]
4. Dib, M.; Ouchetto, H.; Ouchetto, K.; Hafid, A.; Khouili, M. Recent Developments of Quinoline Derivatives and their Potential Biological Activities. *Curr. Org. Synth.* **2021**, *18*, 248–269. [CrossRef] [PubMed]
5. Chu, X.M.; Wang, C.; Liu, W.; Liang, L.L.; Gong, K.K.; Zhao, C.Y.; Sun, K.L. Quinoline and quinolone dimers and their biological activities: An overview. *Eur. Med. Chem.* **2019**, *161*, 101–117. [CrossRef] [PubMed]
6. Matada, B.S.; Pattanashettar, R.; Yernale, N.G. A comprehensive review on the biological interest of quinoline and its derivatives. *Bioorg. Med. Chem.* **2021**, *32*, 115973. [CrossRef] [PubMed]
7. Martins, A.C.; Paoliello, M.M.B.; Docea, A.O.; Santamaria, A.; Tinkov, A.A.; Skalny, A.V.; Aschner, M. Review of the mechanism underlying mefloquine-induced neurotoxicity. *Crit. Rev. Toxicol.* **2021**, *51*, 209–216. [CrossRef] [PubMed]
8. Kucharski, D.J.; Jaszczak, M.K.; Boratynski, P.J. A Review of Modifications of Quinoline Antimalarials: Mefloquine and (hydroxy)Chloroquine. *Molecules* **2022**, *27*, 1003. [CrossRef]
9. Dos Santos, M.C.; Scaini, J.L.R.; Lopes, M.V.C.; Rodrigues, B.G.; Silva, N.O.; Borges, C.R.L.; Dos Santos, S.C.; Dos Santos Machado, K.; Werhli, A.V.; da Silva, P.E.A.; et al. Mefloquine synergism with anti-tuberculosis drugs and correlation to membrane effects: Biologic, spectroscopic and molecular dynamics simulations studies. *Bioorg. Chem.* **2021**, *110*, 104786. [CrossRef]
10. Cifci, B.; Yildiz, Y.; Altin, E.; Habibi, H.; Kocer, B.; Dizbay, M. Successful treatment of HIV-associated progressive multifocal leukoencephalopathy (PML) with mirtazapine, mefloquine, and IVIG combination therapy: A case report. *J. Neurovirol.* **2023**, *29*, 111–115. [CrossRef]
11. Dodion, P.F.; Wagener, T.; Stoter, G.; Drozd, A.; Lev, L.M.; Skovsgaard, T.; Renard, J.; Cavalli, F. Phase II trial with Brequinar (DUP-785, NSC 368390) in patients with metastatic colorectal cancer: A study of the Early Clinical Trials Group of the EORTC. *Ann. Oncol.* **1990**, *1*, 79–80. [CrossRef] [PubMed]
12. Andersen, P.I.; Krpina, K.; Ianevski, A.; Shtaida, N.; Yang, E.; Jo, J.; Koit, S.; Tenson, T.; Hukkanen, V.; Anthonsen, M.W.; et al. Novel Antiviral Activities of Obatoclax, Emetine, Niclosamide, Brequinar, and Homoharringtonine. *Viruses* **2019**, *11*, 964. [CrossRef] [PubMed]
13. Horvat, N.K.; Lesinski, G.B. Bring on the brequinar: An approach to enforce the differentiation of myeloid-derived suppressor cells. *J. Clin. Investig.* **2022**, *132*, e158661. [CrossRef] [PubMed]
14. Sadeq, A.; Elnour, A.A.; Farah, F.H.; Ramadan, A.; Baraka, M.A.; Don, J.; Amoodi, A.A.; Sam, K.G.; Mazrouei, N.A.; Alkaabi, M. A Systematic Review of Randomized Clinical Trials on the Efficacy and Safety of Pitavastatin. *Curr. Rev. Clin. Exp. Pharmacol.* **2023**, *18*, 120–147. [CrossRef] [PubMed]
15. Katanasaka, Y.; Hirano, S.; Sunagawa, Y.; Miyazaki, Y.; Sato, H.; Funamoto, M.; Shimizu, K.; Shimizu, S.; Sari, N.; Hasegawa, K.; et al. Clinically Administered Doses of Pitavastatin and Rosuvastatin. *Int. Heart J.* **2021**, *62*, 1379–1386. [CrossRef] [PubMed]
16. Gowripattapu, S.; Sathis Kumar, D.; Selvamuthukumar, S. Formulation and Statistical Evaluation of Tablets Containing Pitavastatin- Self Nano Emulsifying Drug Delivery Systems. *Curr. Drug Deliv.* **2023**, *20*, 414–432. [CrossRef] [PubMed]

17. Weisinger, H.S.; Pesudovs, K.; Collin, H.B. Management of patients undergoing hydroxychloroquine (Plaquenil) therapy. *Clin. Exp. Optom.* **2000**, *83*, 32–36. [CrossRef]
18. Veld, A.E.I.; Grievink, H.W.; van der Plas, J.L.; Eveleens Maarse, B.C.; van Kraaij, S.J.W.; Woutman, T.D.; Schoonakker, M.; Klarenbeek, N.B.; de Kam, M.L.; Kamerling, I.M.C.; et al. Immunosuppression by hydroxychloroquine: Mechanistic proof in in vitro experiments but limited systemic activity in a randomized placebo-controlled clinical pharmacology study. *Immunol. Res.* **2023**, *71*, 617–627. [CrossRef]
19. Chen, Z.; Wu, J.; Wang, W.; Tang, X.; Zhou, L.; Lv, Y.; Zheng, Y. Investigation of the Pathogenic Mechanism of Ciprofloxacin in Aortic Aneurysm and Dissection by an Integrated Proteomics and Network Pharmacology Strategy. *J. Clin. Med.* **2023**, *12*, 1270. [CrossRef]
20. Loupias, P.; Laumaille, P.; Morandat, S.; Mondange, L.; Guillier, S.; El Kirat, K.; Da Nascimento, S.; Biot, F.; Taudon, N.; Dassonville-Klimpt, A.; et al. Synthesis and study of new siderophore analog-ciprofloxacin conjugates with antibiotic activities against *Pseudomonas aeruginosa* and *Burkholderia* spp. *Eur. J. Med. Chem.* **2023**, *245*, 114921. [CrossRef]
21. Yuan, L.; Wu, H.; Wang, J.; Zhou, M.; Zhang, L.; Xiang, J.; Liao, Q.; Luo, L.; Qian, M.; Zhang, D. Pharmacokinetics, withdrawal time, and dietary risk assessment of enrofloxacin and its metabolite ciprofloxacin, and sulfachloropyridazine-trimethoprim in Taihe black-boned silky fowls. *J. Food Sci.* **2023**, *88*, 1743–1752. [CrossRef] [PubMed]
22. Bo, Z.; Chen, B.; Zhao, Z.; He, Q.; Mao, Y.; Yang, Y.; Yao, F.; Yang, Y.; Chen, Z.; Yang, J.; et al. Prediction of Response to Lenvatinib Monotherapy for Unresectable Hepatocellular Carcinoma by Machine Learning Radiomics: A Multicenter Cohort Study. *Clin. Cancer Res.* **2023**, *29*, 1730–1740. [CrossRef] [PubMed]
23. Peng, Z.; Fan, W.; Zhu, B.; Wang, G.; Sun, J.; Xiao, C.; Huang, F.; Tang, R.; Cheng, Y.; Huang, Z.; et al. Lenvatinib Combined with Transarterial Chemoembolization as First-Line Treatment for Advanced Hepatocellular Carcinoma: A Phase III, Randomized Clinical Trial (LAUNCH). *J. Clin. Oncol.* **2023**, *41*, 117–127. [CrossRef] [PubMed]
24. Wang, X.H.; Liu, C.J.; Wen, H.Q.; Duan, X.H.; Jiao, Y.Q.; Liu, Y.J.; Chen, M.S.; Zhu, K.S.; Mao, X.H.; Zhou, Q.F. Effectiveness of lenvatinib plus immune checkpoint inhibitors in primary advanced hepatocellular carcinoma beyond oligometastasis. *Clin. Transl. Med.* **2023**, *13*, e1214. [CrossRef] [PubMed]
25. Afzal, O.; Kumar, S.; Haider, M.R.; Ali, M.R.; Kumar, R.; Jaggi, M.; Bawa, S. A review on anticancer potential of bioactive heterocycle quinoline. *Eur. J. Med. Chem.* **2015**, *97*, 871–910. [CrossRef] [PubMed]
26. Arafa, R.K.; Hegazy, G.H.; Piazza, G.A.; Abadi, A.H. Synthesis and in vitro antiproliferative effect of novel quinoline-based potential anticancer agents. *Eur. J. Med. Chem.* **2013**, *63*, 826–832. [CrossRef] [PubMed]
27. Fu, H.G.; Li, Z.W.; Hu, X.X.; Si, S.Y.; You, X.F.; Tang, S.; Wang, Y.X.; Song, D.Q. Synthesis and Biological Evaluation of Quinoline Derivatives as a Novel Class of Broad-Spectrum Antibacterial Agents. *Molecules* **2019**, *24*, 548. [CrossRef]
28. Dorababu, A. Recent update on antibacterial and antifungal activity of quinoline scaffolds. *Arch. Pharm.* **2021**, *354*, e2000232. [CrossRef]
29. Jin, G.; Xiao, F.; Li, Z.; Qi, X.; Zhao, L.; Sun, X. Design, Synthesis, and Dual Evaluation of Quinoline and Quinolinium Iodide Salt Derivatives as Potential Anticancer and Antibacterial Agents. *ChemMedChem* **2020**, *15*, 600–609. [CrossRef]
30. Chen, Y.J.; Ma, K.Y.; Du, S.S.; Zhang, Z.J.; Wu, T.L.; Sun, Y.; Liu, Y.Q.; Yin, X.D.; Zhou, R.; Yan, Y.F.; et al. Antifungal Exploration of Quinoline Derivatives against Phytopathogenic Fungi Inspired by Quinine Alkaloids. *J. Agric. Food Chem.* **2021**, *69*, 12156–12170. [CrossRef]
31. Cretton, S.; Dorsaz, S.; Azzollini, A.; Favre-Godal, Q.; Marcourt, L.; Ebrahimi, S.N.; Voinesco, F.; Michellod, E.; Sanglard, D.; Gindro, K.; et al. Antifungal Quinoline Alkaloids from Waltheria indica. *J. Nat. Prod.* **2016**, *79*, 300–307. [CrossRef] [PubMed]
32. Yang, G.Z.; Zhu, J.K.; Yin, X.D.; Yan, Y.F.; Wang, Y.L.; Shang, X.F.; Liu, Y.Q.; Zhao, Z.M.; Peng, J.W.; Liu, H. Design, Synthesis, and Antifungal Evaluation of Novel Quinoline Derivatives Inspired from Natural Quinine Alkaloids. *J. Agric. Food Chem.* **2019**, *67*, 11340–11353. [CrossRef] [PubMed]
33. Das, P.; Deng, X.; Zhang, L.; Roth, M.G.; Fontoura, B.M.; Phillips, M.A.; De Brabander, J.K. SAR-Based Optimization of a 4-Quinoline Carboxylic Acid Analogue with Potent Antiviral Activity. *ACS Med. Chem. Lett.* **2013**, *4*, 517–521. [CrossRef] [PubMed]
34. De la Guardia, C.; Stephens, D.E.; Dang, H.T.; Quijada, M.; Larionov, O.V.; Lleonart, R. Antiviral Activity of Novel Quinoline Derivatives against Dengue Virus Serotype 2. *Molecules* **2018**, *23*, 672. [CrossRef] [PubMed]
35. Seliem, I.A.; Panda, S.S.; Girgis, A.S.; Moatasim, Y.; Kandeil, A.; Mostafa, A.; Ali, M.A.; Nossier, E.S.; Rasslan, F.; Srour, A.M.; et al. New quinoline-triazole conjugates: Synthesis, and antiviral properties against SARS-CoV-2. *Bioorg. Chem.* **2021**, *114*, 105117. [CrossRef] [PubMed]
36. Wang, M.; Zhang, G.; Zhao, J.; Cheng, N.; Wang, Y.; Fu, Y.; Zheng, Y.; Wang, J.; Zhu, M.; Cen, S.; et al. Synthesis and antiviral activity of a series of novel quinoline derivatives as anti-RSV or anti-IAV agents. *Eur. J. Med. Chem.* **2021**, *214*, 113208. [CrossRef] [PubMed]
37. Vinindwa, B.; Dziwornu, G.A.; Masamba, W. Synthesis and Evaluation of Chalcone-Quinoline Based Molecular Hybrids as Potential Anti-Malarial Agents. *Molecules* **2021**, *26*, 4093. [CrossRef]
38. Roy, D.; Anas, M.; Manhas, A.; Saha, S.; Kumar, N.; Panda, G. Synthesis, biological evaluation, Structure—Activity relationship studies of quinoline-imidazole derivatives as potent antimalarial agents. *Bioorg. Chem.* **2022**, *121*, 105671. [CrossRef]
39. Jesu Jaya Sudan, R.; Lesitha Jeeva Kumari, J.; Iniyavan, P.; Sarveswari, S.; Vijayakumar, V. Evaluation of xanthene-appended quinoline hybrids as potential leads against antimalarial drug targets. *Mol. Divers.* **2023**, *27*, 709–727. [CrossRef]

40. Orhan Puskullu, M.; Tekiner, B.; Suzen, S. Recent Studies of Antioxidant Quinoline Derivatives. *Mini Rev. Med. Chem.* **2013**, *13*, 365–372. [CrossRef]
41. Parameswaran, K.; Sivaguru, P.; Lalitha, A. Synthesis of novel bis(pyrimido [5,4-c]quinoline-2,4(1H,3H)-dione) and its derivatives: Evaluation of their antioxidant properties. *Bioorg. Med. Chem. Lett.* **2013**, *23*, 3873–3878. [CrossRef] [PubMed]
42. Ismaili, L.; Nadaradjane, A.; Nicod, L.; Guyon, C.; Xicluna, A.; Robert, J.F.; Refouvelet, B. Synthesis and antioxidant activity evaluation of new hexahydropyrimido [5,4-c]quinoline-2,5-diones and 2-thioxohexahydropyrimido [5,4-c]quinoline-5-ones obtained by Biginelli reaction in two steps. *Eur. J. Med. Chem.* **2008**, *43*, 1270–1275. [CrossRef] [PubMed]
43. Tseng, C.H.; Tung, C.W.; Wu, C.H.; Tzeng, C.C.; Chen, Y.H.; Hwang, T.L.; Chen, Y.L. Discovery of Indeno [1,2-c]quinoline Derivatives as Potent Dual Antituberculosis and Anti-Inflammatory Agents. *Molecules* **2017**, *22*, 1001. [CrossRef] [PubMed]
44. El-Feky, S.A.; Abd El-Samii, Z.K.; Osman, N.A.; Lashine, J.; Kamel, M.A.; Thabet, H. Synthesis, molecular docking and anti-inflammatory screening of novel quinoline incorporated pyrazole derivatives using the Pfitzinger reaction II. *Bioorg. Chem.* **2015**, *58*, 104–116. [CrossRef] [PubMed]
45. Chia, E.W.; Pearce, A.N.; Berridge, M.V.; Larsen, L.; Perry, N.B.; Sansom, C.E.; Godfrey, C.A.; Hanton, L.R.; Lu, G.L.; Walton, M.; et al. Synthesis and anti-inflammatory structure–activity relationships of thiazine–quinoline–quinones: Inhibitors of the neutrophil respiratory burst in a model of acute gouty arthritis. *Bioorg. Med. Chem.* **2008**, *16*, 9432–9442. [CrossRef] [PubMed]
46. Bewley, B.R.; Spearing, P.K.; Weiner, R.L.; Luscombe, V.B.; Zhan, X.; Chang, S.; Cho, H.P.; Rodriguez, A.L.; Niswender, C.M.; Conn, P.J.; et al. Discovery of a novel, CNS penetrant M4 PAM chemotype based on a 6-fluoro-4-(piperidin-1-yl)quinoline-3-carbonitrile core. *Bioorg. Med. Chem. Lett.* **2017**, *27*, 4274–4279. [CrossRef] [PubMed]
47. Zajdel, P.; Marciniec, K.; Maslankiewicz, A.; Satala, G.; Duszynska, B.; Bojarski, A.J.; Partyka, A.; Jastrzebska-Wiesek, M.; Wrobel, D.; Wesolowska, A.; et al. Quinoline- and isoquinoline-sulfonamide derivatives of LCAP as potent CNS multi-receptor—5-HT1A/5-HT2A/5-HT7 and D2/D3/D4—Agents: The synthesis and pharmacological evaluation. *Bioorg. Med. Chem.* **2012**, *20*, 1545–1556. [CrossRef]
48. Kumar, S.; Kaushik, D.; Bawa, S.; Khan, S.A. Design, Synthesis and Screening of Quinoline-Incorporated Thiadiazole as a Potential Anticonvulsant. *Chem. Biol. Drug Des.* **2012**, *79*, 104–111. [CrossRef]
49. Muruganantham, N.; Sivakumar, R.; Anbalagan, N.; Gunasekaran, V.; Leonard, J.T. Synthesis, Anticonvulsant and Antihypertensive Activities of 8-Substituted Quinoline Derivatives. *Biol. Pharm. Bull.* **2004**, *27*, 1683–1687. [CrossRef]
50. Shanley, H.T.; Taki, A.C.; Byrne, J.J.; Jabbar, A.; Wells, T.N.C.; Samby, K.; Boag, P.R.; Nguyen, N.; Sleebs, B.E.; Gasser, R.B. A High-Throughput Phenotypic Screen of the 'Pandemic Response Box' Identifies a Quinoline Derivative with Significant Anthelmintic Activity. *Pharmaceuticals* **2022**, *15*, 257. [CrossRef]
51. Williams, A.; Villamor, L.; Fussell, J.; Loveless, R.; Smeyne, D.; Philp, J.; Shaikh, A.; Sittaramane, V. Discovery of Quinoline-Derived Trifluoromethyl Alcohols as Antiepileptic and Analgesic Agents That Block Sodium Channels. *ChemMedChem* **2022**, *17*, e202100547. [CrossRef] [PubMed]
52. Kouznetsov, V.V.; Mendez, L.Y.V.; Gomez, C.M.M. Recent Progress in the Synthesis of Quinolines. *Curr. Org. Chem.* **2005**, *9*, 141–161. [CrossRef]
53. Weyesa, A.; Mulugeta, E. Recent advances in the synthesis of biologically and pharmaceutically active quinoline and its analogues: A review. *RSC Adv.* **2020**, *10*, 20784–20793. [CrossRef] [PubMed]
54. Nainwal, L.M.; Tasneem, S.; Akhtar, W.; Verma, G.; Khan, M.F.; Parvez, S.; Shaquiquzzaman, M.; Akhter, M.; Alam, M.M. Green recipes to quinoline: A review. *Eur. J. Med. Chem.* **2019**, *164*, 121–170. [CrossRef] [PubMed]
55. Prajapati, S.M.; Patel, K.D.; Vekariya, R.H.; Panchal, S.N.; Patel, H.D. Recent advances in the synthesis of quinolines: A review. *RSC Adv.* **2014**, *4*, 24463–24476. [CrossRef]
56. Barluenga, J.; Rodriguez, F.; Fananas, F.J. Recent Advances in the Synthesis of Indole and Quinoline Derivatives through Cascade Reactions. *Chem. Asian J.* **2009**, *4*, 1036–1048. [CrossRef] [PubMed]
57. Ramann, G.A.; Cowen, B.J. Recent Advances in Metal-Free Quinoline Synthesis. *Molecules* **2016**, *21*, 986. [CrossRef]
58. Bharate, J.B.; Vishwakarma, R.A.; Bharate, J.B. Metal-free domino one-pot protocols for quinoline synthesis. *RSC Adv.* **2015**, *5*, 42020–42053. [CrossRef]
59. Teja, C.; Khan, F.R.N. Radical Transformations towards the Synthesis of Quinoline: A Review. *Chem. Asian J.* **2020**, *15*, 4153–4167. [CrossRef]
60. Vessally, E.; Edjlali, L.; Hosseinian, A.; Bekhradnia, A.; Esrafili, M.D. Novel routes to quinoline derivatives from *N*-propargylamines. *RSC Adv.* **2016**, *6*, 49730–49746. [CrossRef]
61. Doraghi, F.; Karimian, S.; Aledavoud, S.P.; Amini, A.; Larijani, B.; Mahdavi, M. 2-azidobenzaldehyde: A versatile scaffold for the generation of *N*-heterocyclic compounds. *J. Mol. Struct.* **2023**, *1294*, 136510. [CrossRef]
62. Das, S.K.; Roy, S.; Khatua, H.; Chattopadhyay, B. Ir-Catalyzed Intramolecular Transannulation/C(sp2)–H Amination of 1,2,3,4-Tetrazoles by Electrocyclization. *J. Am. Chem. Soc.* **2018**, *140*, 8429–8433. [CrossRef] [PubMed]
63. Roy, S.; Das, S.K.; Khatua, H.; Das, S.; Singh, K.N.; Chattopadhyay, B. Iron-Catalyzed Radical Activation Mechanism for Denitrogenative Rearrangement Over C(sp3)–H Amination. *Angew. Chem. Int. Ed.* **2021**, *60*, 8772–8780. [CrossRef] [PubMed]
64. Ansari, A.J.; Joshi, G.; Yadav, U.P.; Maurya, A.K.; Agnihotri, V.K.; Kalra, S.; Kumar, R.; Singh, S.; Sawant, D.M. Exploration of Pd-catalysed four-component tandem reaction for one-pot assembly of pyrazolo [1,5-c]quinazolines as potential EGFR inhibitors. *Bioorg. Chem.* **2019**, *93*, 103314. [CrossRef] [PubMed]

65. Zhao, L.; Yang, M.L.; Liu, M.; Ding, M.W. New efficient synthesis of polysubstituted 3,4-dihydroquinazolines and 4H-3,1-benzothiazines through a Passerini/Staudinger/aza-Wittig/addition/nucleophilic substitution sequence. *Beilstein J. Org. Chem.* **2022**, *18*, 286–292. [CrossRef] [PubMed]
66. Vroemans, R.; Bamba, F.; Winters, J.; Thomas, J.; Jacobs, J.; Van Meervelt, L.; John, J.; Dehaen, W. Sequential Ugi reaction/base-induced ring closing/IAAC protocol toward triazolobenzodiazepine-fused diketopiperazines and hydantoins. *Beilstein J. Org. Chem.* **2018**, *14*, 626–633. [CrossRef] [PubMed]
67. Vidyacharan, S.; Murugan, A.; Sharada, D.S. C(sp2)–H Functionalization of 2H-Indazoles at C3-Position via Palladium(II)-Catalyzed Isocyanide Insertion Strategy Leading to Diverse Heterocycles. *J. Org. Chem.* **2016**, *81*, 2837–2848. [CrossRef]
68. De Moliner, F.; Bigatti, M.; De Rosa, C.; Banfi, L.; Riva, R.; Basso, A. Synthesis of triazolo-fused benzoxazepines and benzoxazepinones via Passerini reactions followed by 1,3-dipolar cycloadditions. *Mol. Divers.* **2014**, *18*, 473–482. [CrossRef]
69. Singh, B. Bedaquiline in Drug-Resistant Tuberculosis: A Mini-Review. *Curr. Mol. Pharmacol.* **2023**, *16*, 243–253. [CrossRef]
70. Conradie, F.; Bagdasaryan, T.R.; Borisov, S.; Howell, P.; Mikiashvili, L.; Ngubane, N.; Samoilova, A.; Skornykova, S.; Tudor, E.; Variava, E.; et al. Bedaquiline–Pretomanid–Linezolid Regimens for Drug-Resistant Tuberculosis. *N. Engl. J. Med.* **2022**, *387*, 810–823. [CrossRef]
71. Frampton, J.E. QVA149 (Indacaterol/Glycopyrronium Fixed-Dose Combination): A Review of Its Use in Patients with Chronic Obstructive Pulmonary Disease. *Drugs* **2014**, *74*, 465–488. [CrossRef]
72. Cazzola, M.; Bardaro, F.; Stirpe, E. The role of indacaterol for chronic obstructive pulmonary disease (COPD). *J. Thorac. Dis.* **2013**, *5*, 559–566. [CrossRef] [PubMed]
73. Umezaki, Y.; Motomura, H.; Egashira, R.; Toyofuku, A.; Naito, T. A Case of Oral Cenesthopathy Treated with the Combination of Brexpiprazole and Mirtazapine. *Clin. Neuropharmacol.* **2023**, *46*, 123–125. [CrossRef] [PubMed]
74. Shapovalov, V.; Kopanitsa, L.; Pruteanu, L.L.; Ladds, G.; Bailey, D.S. Transcriptomics-Based Phenotypic Screening Supports Drug Discovery in Human Glioblastoma Cells. *Cancers* **2021**, *13*, 3780. [CrossRef] [PubMed]
75. Murray, V.; Chen, J.K.; Galea, A.M. The Potential of Acridine Carboxamide Pt Complexes as Anti-Cancer Agents: A Review. *Anti-Cancer Agent. Med. Chem.* **2014**, *14*, 695–705. [CrossRef] [PubMed]
76. Nair, V.; Menon, R.S.; Biju, A.T.; Abhilash, K.G. 1,2-Benzoquinones in Diels–Alder reactions, dipolar cycloadditions, nucleophilic additions, multicomponent reactions and more. *Chem. Soc. Rev.* **2012**, *41*, 1050–1059. [CrossRef] [PubMed]
77. Bharate, S.B.; Mudududdla, R.; Bharate, J.B.; Battini, N.; Battula, S.; Yadav, R.R.; Singh, B.; Vishwakarma, R.A. Tandem one-pot synthesis of flavans by recyclable silica–$HClO_4$ catalyzed Knoevenagel condensation and [4+2]-Diels–Alder cycloaddition. *Org. Biomol. Chem.* **2012**, *10*, 5143–5150. [CrossRef] [PubMed]
78. Houk, K.N.; Liu, F.; Yang, Z.; Seeman, J.I. Evolution of the Diels–Alder Reaction Mechanism since the 1930s: Woodward, Houk with Woodward, and the Influence of Computational Chemistry on Understanding Cycloadditions. *Angew. Chem.* **2021**, *60*, 12660–12681. [CrossRef]
79. Zhang, X.F.; Dhawan, G.; Muthengi, A.; Liu, S.; Wang, W.; Legris, M.; Zhang, W. One-pot and catalyst-free synthesis of pyrroloquinolinediones and quinolinedicarboxylates. *Green Chem.* **2017**, *19*, 3851–3855. [CrossRef]
80. Zheng, L.; Zeng, Z.G.; Yan, Q.; Jia, F.C.; Jia, L.; Chen, Y. Copper-Catalyzed Synthesis of 3-NO2 Quinolines from o-Azidobenzaldehyde and Nitro-olefins and its Application in the Concise Synthesis of Quindolines. *Adv. Synth. Catal.* **2018**, *360*, 4037–4042. [CrossRef]
81. Ma, X.; Zhang, X.; Awad, J.M.; Xie, G.; Qiu, W.; Muriph, R.E. Sequential decarboxylative [3+2] cycloaddition and Staudinger/aza-Wittig reactions for diastereoselective synthesis of tetrahydro-pyrroloquinazolines and tetrahedro-pyrrolobenzodiazepines. *Tetrahedron Lett.* **2020**, *61*, 151392. [CrossRef]
82. Ma, X.; Zhang, X.F.; Awad, J.M.; Xie, G.; Qiu, W.; Zhang, W. One-pot synthesis of tetrahydro-pyrrolobenzodiazepines and tetrahydro-pyrrolobenzodiazepinones through sequential 1,3-dipolar cycloaddition/N-alkylation (N-acylation)/Staudinger/aza-Wittig reactions. *Green Chem.* **2019**, *21*, 4489–4494. [CrossRef]
83. Zhang, X.F.; Ma, X.; Qiu, W.; Evans, J.; Zhnag, W. Cascade Knoevenagel and aza-Wittig reactions for the synthesis of substituted quinolines and quinolin-4-ols. *Green Chem.* **2019**, *21*, 349–354. [CrossRef]
84. Qu, F.; He, P.; Hu, R.; Cheng, X.; Wang, S.; Wu, J. Efficient Synthesis of Quinolines via a Knoevenagel/Staudinger/aza-Wittig Sequence. *Synth. Commun.* **2015**, *45*, 2802–2809. [CrossRef]
85. Malkova, K.; Bubyrev, A.; Kalinin, S.; Dar'in, D. Facile access to 3-sulfonylquinolines via Knoevenagel condensation/aza-Wittig reaction cascade involving ortho-azidobenzaldehydes and β-ketosulfonamides and sulfones. *Beilstein J. Org. Chem.* **2023**, *19*, 800–807. [CrossRef] [PubMed]
86. Fakhfakh, M.A.; Fournet, A.; Prina, E.; Mouscadet, J.F.; Franck, X.; Hocquemiller, R.; Figadere, B. Synthesis and biological evaluation of substituted quinolines: Potential treatment of protozoal and retroviral co-infections. *Bioorg. Med. Chem.* **2003**, *11*, 5013–5023. [CrossRef] [PubMed]
87. Gopinath, V.S.; Pinjari, J.; Dere, R.T.; Verma, A.; Vishwakarma, P.; Shivahare, R.; Moger, M.; Kumar Goud, P.S.; Ramanathan, V.; Bose, P.; et al. Design, synthesis and biological evaluation of 2-substituted quinolines as potential antileishmanial agents. *Eur. J. Med. Chem.* **2013**, *69*, 527–536. [CrossRef]
88. Gopaul, K.; Shintre, S.A.; Koorbanally, N.A. A Review on the Synthesis and Anti-cancer Activity of 2-substituted Quinolines. *Anti-Cancer Agent. Med. Chem.* **2015**, *15*, 631–646. [CrossRef]
89. Swallow, S. Chapter Two—Fluorine in Medicinal Chemistry. *Progress Med. Chem.* **2015**, *54*, 65–133. [CrossRef]

90. Ramachandran, P.V. Welcome to 'Fluorine in Medicinal Chemistry'. *Future Med. Chem.* **2009**, *1*, 771–772. [CrossRef]
91. Purser, S.; Moore, P.R.; Swallow, S.; Gouverneur, V. Fluorine in medicinal chemistry. *Chem. Soc. Rev.* **2008**, *37*, 320–330. [CrossRef] [PubMed]
92. Gillis, E.P.; Eastman, K.J.; Hill, M.D.; Donnelly, D.J.; Meanwell, N.A. Applications of Fluorine in Medicinal Chemistry. *J. Med. Chem.* **2015**, *58*, 8315–8359. [CrossRef] [PubMed]
93. Bohm, H.J.; Banner, D.; Bendels, S.; Kansy, M.; Kuhn, B.; Muller, K.; Obst-Sander, U.; Stahl, M. Fluorine in Medicinal Chemistry. *ChemBioChem* **2004**, *5*, 637–643. [CrossRef] [PubMed]
94. Zhou, Y.; Wang, J.; Gu, Z.; Wang, S.; Zhu, W.; Acena, J.L.; Soloshonok, V.A.; Izawa, K.; Liu, H. Next Generation of Fluorine-Containing Pharmaceuticals, Compounds Currently in Phase II–III Clinical Trials of Major Pharmaceutical Companies: New Structural Trends and Therapeutic Areas. *Chem. Rev.* **2016**, *116*, 422–518. [CrossRef] [PubMed]
95. Mai, A.; Rotili, D.; Tarantino, D.; Ornaghi, P.; Tosi, F.; Vicidomini, C.; Sbardella, G.; Nebbioso, A.; Miceli, M.; Altucci, L.; et al. Small-Molecule Inhibitors of Histone Acetyltransferase Activity: Identification and Biological Properties. *J. Med. Chem.* **2006**, *49*, 6897–6907. [CrossRef] [PubMed]
96. Goncalves, V.; Brannigan, J.A.; Whalley, D.; Ansell, K.H.; Saxty, B.; Holder, A.A.; Wilkinson, A.J.; Tate, E.W.; Leatherbarrow, R.J. Discovery of Plasmodium vivaxN-Myristoyltransferase Inhibitors: Screening, Synthesis, and Structural Characterization of their Binding Mode. *J. Med. Chem.* **2012**, *55*, 3578–3582. [CrossRef]
97. Yu, Z.; Zheng, H.; Yuan, W.; Tang, Z.; Zhang, A.; Shi, D. An unexpected one-pot synthesis of multi-substituted quinolines via a cascade reaction of Michael/Staudinger/aza-Wittig/aromatization of ortho-azido-β-nitro-styrenes with various carbonyl compounds. *Tetrahedron* **2013**, *69*, 8137–8141. [CrossRef]
98. Chaabouni, S.; Pinkerton, N.M.; Legrand, S.; Abid, S.; Galaup, C.; Chassaing, S.; Koivisto, J.; Hirvonen, J.; Kostiainen, M.A.; Bimbo, L.M. Photochemistry of ortho-azidocinnamoyl derivatives: Facile and modular synthesis of 2-acylated indoles and 2-substituted quinolines under solvent control. *Synlett* **2017**, *28*, 2614–2618. [CrossRef]
99. Kumar, G.R.; Kumar, R.; Rajesh, M.; Reddy, M.S. A nickel-catalyzed anti-carbometallative cyclization of alkyne–azides with organoboronic acids: Synthesis of 2,3-diarylquinolines. *Chem. Commun.* **2018**, *54*, 759–762. [CrossRef]
100. Huo, Z.; Gridnev, I.D.; Yamamoto, Y. A Method for the Synthesis of Substituted Quinolines via Electrophilic Cyclization of 1-Azido-2-(2-propynyl)benzene. *J. Org. Chem.* **2010**, *75*, 1266–1270. [CrossRef]
101. Wang, J.; Wang, Y.Z.; Liu, Y.; Yan, X.; Yan, Y.; Chao, S.; Shang, X.; Ni, T.; Zhou, P. Synthesis of Isoquinolylselenocyanates and Quinolylselenocyanates via Electrophilic Selenocyanogen Cyclization Induced by Pseudohalogen $(SeCN)_2$ Generated in situ. *Adv. Synth. Catal.* **2022**, *364*, 187–192. [CrossRef]
102. Qiu, Y.F.; Niu, Y.J.; Wei, X.; Cao, B.Q.; Wang, X.C.; Quan, Z.J. AgSCF3/Na2S2O8-Promoted Trifluoromethylthiolation/Cyclization of o-Propargyl Arylazides/o-Alkynyl Benzylazides: Synthesis of SCF3-Substituted Quinolines and Isoquinolines. *J. Org. Chem.* **2019**, *84*, 4165–4178. [CrossRef] [PubMed]
103. Chai, Z.; Zhu, Y.M.; Yang, P.J.; Wang, S.; Wang, S.; Liu, Z.; Yang, G. Copper(I)-Catalyzed Kinetic Resolution of N-Sulfonylaziridines with Indoles: Efficient Construction of Pyrroloindolines. *J. Am. Chem. Soc.* **2015**, *137*, 10088–10091. [CrossRef] [PubMed]
104. Kang, B.; Miller, A.W.; Goyal, S.; Nguyen, S.T. $Sc(OTf)_3$-catalyzed condensation of 2-alkyl-N-tosylaziridine with aldehydes or ketones: An efficient synthesis of 5-alkyl-1,3-oxazolidines. *Chem. Commun.* **2009**, *26*, 3928–3930. [CrossRef] [PubMed]
105. Ungureanu, I.; Klotz, P.; Mann, A. Phenylaziridine as a Masked 1,3 Dipole in Reactions with Nonactivated Alkenes. *Angew. Chem.* **2000**, *39*, 4615–4617. [CrossRef]
106. Wani, I.A.; Sayyad, M.; Ghorai, M.K. Domino ring-opening cyclization (DROC) of activated aziridines and epoxides with nitrones via dual-catalysis "on water". *Chem. Commun.* **2017**, *53*, 4386–4389. [CrossRef]
107. Yi, R.; Li, X.; Wan, B. Ring-opening and cyclization of aziridines with aryl azides: Metal-free synthesis of 6-(triflyloxy)quinolines. *Org. Chem. Front.* **2018**, *5*, 3488–3493. [CrossRef]
108. Lin, C.H.; Tsai, M.R.; Wang, Y.S.; Chang, N.C. An Efficient Approach to 3,4-Disubstituted Pyridin-2-ones. Formal Synthesis of Mappicine Ketone. *J. Org. Chem.* **2003**, *68*, 5688–5691. [CrossRef]
109. Wainberg, Z.A.; Singh, A.S.; Konecny, G.E.; McCann, K.E.; Hecht, J.R.; Goldman, J.; Chmielowski, B.; Finn, R.S.; O'Brien, N.; Von Euw, E.; et al. Preclinical and Clinical Trial Results Using Talazoparib and Low-Dose Chemotherapy. *Clin. Cancer Res.* **2023**, *29*, 40–49. [CrossRef]
110. Wronska, N.; Majoral, J.P.; Appelhans, D.; Bryszewska, M.; Lisowska, K. Synergistic Effects of Anionic/Cationic Dendrimers and Levofloxacin on Antibacterial Activities. *Molecules* **2019**, *24*, 2894. [CrossRef]
111. Gentili, P.L.; Ortica, F.; Romani, A.; Favaro, G. Effects of Proximity on the Relaxation Dynamics of Flindersine and 6(5H)-Phenanthridinone. *J. Phys. Chem. A* **2007**, *111*, 193–200. [CrossRef] [PubMed]
112. Papakostas, D.; Stockfleth, E. Topical treatment of basal cell carcinoma with the immune response modifier imiquimod. *Future Oncol.* **2015**, *11*, 2985–2990. [CrossRef] [PubMed]
113. Xiong, J.; He, H.T.; Yang, H.Y.; Zeng, Z.G.; Zhong, C.R.; Shi, H.; Ouyang, M.L.; Tao, Y.Y.; Pang, Y.L.; Zhang, Y.H.; et al. Synthesis of 4-Tetrazolyl-Substituted 3,4-Dihydroquinazoline Derivatives with Anticancer Activity via a One-Pot Sequential Ugi-Azide/Palladium-Catalyzed Azide-Isocyanide Cross-Coupling/Cyclization Reaction. *J. Org. Chem.* **2022**, *87*, 9488–9496. [CrossRef] [PubMed]
114. Kumar, R.; Arigela, R.K.; Samala, S.; Kundu, B. Diversity Oriented Synthesis of Indoloazepinobenzimidazole and Benzimidazotriazolobenzodiazepine from N1-Alkyne-1,2-diamines. *Chem. A Eur. J.* **2015**, *21*, 18828–18833. [CrossRef] [PubMed]

115. Ramachary, D.B.; Shruthi, K.S. A Brønsted Acid-Amino Acid as a Synergistic Catalyst for Asymmetric List-Lerner-Barbas Aldol Reactions. *J. Org. Chem.* **2016**, *81*, 2405–2419. [CrossRef] [PubMed]
116. Zhang, W.; Zheng, W.; Zuo, G.; Li, X.; Zhang, X.F.; Zhang, W. One-pot Mannich/aza-Wittig/deaminative aromatization reactions for the synthesis of 1,2,3,4-tetrahydroacridines and cyclohepta[b]quinolines. *New J. Chem.* **2023**, *47*, 220–223. [CrossRef]
117. Sun, M.; Lu, Y.; Zhao, L.; Ding, M. One-pot and divergent synthesis of furo [3,2-c]quinolines and quinazolin-4(3H)-ones via sequential isocyanide-based three-component/Staudinger/aza-Wittig reaction. *Tetrahedron* **2021**, *80*, 131868. [CrossRef]
118. Wei, J.; Nie, B.; Peng, R.; Cheng, X.; Wang, S. A Facile Synthesis of Benzofuro [2,3-c]quinolines via a Multicomponent Reaction and Staudinger–Aza-Wittig–Dehydroaromatization Sequence. *Synlett* **2016**, *27*, 626–630. [CrossRef]
119. Bharate, J.B.; Sharma, R.; Aravinda, S.; Gupta, V.K.; Bharate, S.B.; Vishwakarma, R.A. Montmorillonite clay catalyzed synthesis of functionalized pyrroles through domino four-component coupling of amines, aldehydes, 1,3-dicarbonyl compounds and nitroalkanes. *RSC Adv.* **2013**, *3*, 21736–21742. [CrossRef]
120. Qu, F.; Hu, R.; Gao, L.; Wu, J.; Cheng, X.H.; Wang, S.; He, P. New and Efficient Synthesis of 2,3,4-Trisubstituted 2H-Pyrrolo [3,4-c]quinolines via a MCR/Staudinger/Aza-Wittig Sequence. *Synthesis* **2015**, *47*, 3701–3710. [CrossRef]
121. Shi, Y.; Liao, L.; He, P.; Hu, Y.; Cheng, H.; Wang, S.; Wu, J. Synthesis of 2H-pyrrolo [3,4-C]quinoline via an Aldol/Van Leusen/Staudinger/aza-Wittig sequence. *Synth. Commun.* **2016**, *46*, 1357–1363. [CrossRef]
122. Nie, B.; Wu, L.; Hu, R.; Sun, Y.; Wu, J.; He, P.; Huang, N. Synthesis of cyclopropa[c]indeno [1,2-b]quinolines through a MCR/Staudinger/aza-Wittig sequence. *Synth. Commun.* **2017**, *47*, 1368–1374. [CrossRef]
123. Van Es, T.; Staskun, B. Chlorine- and Sulphur-substituted Pyrrolo [3, 4-b]quinolines and Related Derivatives arising from the Aminolysis of 3, 3, 9-Trichlorothieno [3, 4-b]quinolin-1(3H)-one. *S. Afr. J. Chem.* **2003**, *56*, 40–46.
124. Yi, R.; Li, X.; Wan, B. Merging Gold Catalysis and Brønsted Acid Catalysis for the Synthesis of Tetrahydrobenzo[b][1,8]naphthyridines. *Adv. Synth. Catal.* **2018**, *360*, 875–880. [CrossRef]
125. Gharpure, S.J.; Shelke, Y.G. Lewis Acid Mediated Cascade Friedel–Craft/Alkyne Indol-2-yl Cation Cyclization/Vinyl Cation Trapping for the Synthesis of N-Fused Indole Derivatives. *Org. Lett.* **2017**, *19*, 5406–5409. [CrossRef] [PubMed]
126. Costa, G.P.; Bach, M.F.; Moraes, M.C.; Barcellos, T.; Lenardao, E.J.; Silva, M.S.; Alve, D. Sequential Organocatalytic Synthesis of [1,2,3]Triazolo [1,5-a]quinolines. *Adv. Synth. Catal.* **2020**, *362*, 5044–5055. [CrossRef]
127. Alajarin, M.; Bonillo, B.; Ortin, M.M.; Sanchez-Andrada, P.; Vidal, A. Hydricity-Promoted [1,5]-H Shifts in Acetalic Ketenimines and Carbodiimides. *Org. Lett.* **2006**, *8*, 5645–5648. [CrossRef]
128. Alajarin, M.; Bonillo, B.; Ortin, M.M.; Sanchez-Andrada, P.; Vidal, A. Tandem 1,5-Hydride Shift/6π Electrocyclization of Ketenimines and Carbodiimides Substituted with Cyclic Acetal and Dithioacetal Functions: Experiments and Computations. *Eur. J. Org. Chem.* **2011**, *2011*, 1896–1913. [CrossRef]
129. Ratheesh, M.; Sindhu, G.; Helen, A. Anti-inflammatory effect of quinoline alkaloid skimmianine isolated from *Ruta graveolens* L. *Inflamm. Res.* **2013**, *62*, 367–376. [CrossRef]
130. Zuo, Y.; Pu, J.; Chen, G.; Shen, W.; Wang, B. Study on the activity and mechanism of skimmianine against human non-small cell lung cancer. *Nat. Prod. Res.* **2019**, *33*, 759–762. [CrossRef]
131. Wang, L.H.; Wehland, M.; Wise, P.M.; Infanger, M.; Grimm, D.; Kreissl, M.C. Cabozantinib, Vandetanib, Pralsetinib and Selpercatinib as Treatment for Progressed Medullary Thyroid Cancer with a Main Focus on Hypertension as Adverse Effect. *Int. J. Mol. Sci.* **2023**, *24*, 2312. [CrossRef] [PubMed]
132. Shen, L.; Pili, R. Tasquinimod targets suppressive myeloid cells in the tumor microenvironment. *Oncoimmunology* **2019**, *8*, e1072672. [CrossRef] [PubMed]
133. Hong, D.S.; Rosen, P.; Lockhart, A.C.; Fu, S.; Janku, F.; Kurzrock, R.; Khan, R.; Amore, B.; Caudillo, I.; Deng, H.; et al. A first-in-human study of AMG 208, an oral MET inhibitor, in adult patients with advanced solid tumors. *Oncotarget* **2015**, *6*, 18693–18706. [CrossRef] [PubMed]
134. Yuan, S.; Peng, L.; Liu, Y.; Till, B.G.; Yan, X.; Zhang, J.; Zhu, L.; Wang, H.; Zhang, S.; Li, H.; et al. Low-dose anlotinib confers improved survival in combination with immune checkpoint inhibitor in advanced non-small cell lung cancer patients. *Cancer Immunol. Immunother.* **2023**, *72*, 437–448. [CrossRef] [PubMed]
135. Zhao, R.; Zou, W.; Zhao, X. Treatment of neurofibromatosis type II with anlotinib: A case report and literature review. *Anti-Cancer Drugs* **2023**, *34*, 1065–1068. [CrossRef] [PubMed]
136. He, R.; Lin, F.; Yu, B.; Qiu, J.; Zheng, L. The efficacy and adverse events of delafloxacin in the treatment of acute bacterial infections: A systematic review and meta-analysis of randomized controlled trials. *Front. Pharmacol.* **2022**, *13*, 975578. [CrossRef] [PubMed]
137. Gould, R.G.; Jacobs, W.A. The Synthesis of Certain Substituted Quinolines and 5,6-Benzoquinolines. *J. Am. Chem. Soc.* **1939**, *10*, 2890–2895. [CrossRef]
138. Jentsch, N.G.; Hume, J.D.; Crull, E.B.; Beauti, S.M.; Pham, A.H.; Pigza, J.A.; Kessl, J.J.; Donahue, M.G. Quinolines from the cyclocondensation of isatoic anhydride with ethyl acetoacetate: Preparation of ethyl 4-hydroxy-2-methylquinoline-3-carboxylate and derivatives. *Beilstein J. Org. Chem.* **2018**, *14*, 2529–2536. [CrossRef]
139. Gharpure, S.J.; Nanda, S.K.; Adate, P.A.; Shelke, Y.G. Lewis Acid Promoted Oxonium Ion Driven Carboamination of Alkynes for the Synthesis of 4-Alkoxy Quinolines. *J. Org. Chem.* **2017**, *82*, 2067–2080. [CrossRef]
140. Hande, P.E.; Mishra, M.; Ali, F.; Kapoor, S.; Datta, A.; Gharpure, S.J. Design and Expeditious Synthesis of Quinoline-Pyrene-Based Ratiometric Fluorescent Probes for Targeting Lysosomal pH. *ChemBioChem* **2020**, *21*, 1492–1498. [CrossRef]

141. Gharpure, S.J.; Nanda, S.K.; Fartade, D.J. Formal [4+2] Cycloaddition of o-Aza-Quinone Methide for the Synthesis of 1,4-Heterocycle-Fused Quinolines. *Adv. Synth. Catal.* **2021**, *363*, 2562–2567. [CrossRef]
142. Su, H.; Bao, M.; Pei, C.; Hu, W.H.; Qiu, L.H.; Xu, X.F. Gold-catalyzed dual annulation of azide-tethered alkynes with nitriles: Expeditious synthesis of oxazolo [4,5-c]quinolines. *Org. Chem. Front.* **2019**, *6*, 2404–2409. [CrossRef]
143. Su, H.; Bao, M.; Huang, J.J.; Qiu, L.H.; Xu, X.F. Silver-Catalyzed Carbocyclization of Azide-Tethered Alkynes: Expeditious Synthesis of Polysubstituted Quinolines. *Adv. Synth. Catal.* **2019**, *361*, 826–831. [CrossRef]
144. Huang, J.; Su, H.; Bao, M.; Qiu, L.; Zhang, Y.; Xu, X. Gold(iii)-catalyzed azide-yne cyclization/O–H insertion cascade reaction for the expeditious construction of 3-alkoxy-4-quinolinone frameworks. *Org. Biomol. Chem.* **2020**, *18*, 3888–3892. [CrossRef]
145. Lenoci, A.; Tomassi, S.; Conte, M.; Benedetti, R.; Rodriguez, V.; Carradori, S.; Secci, D.; Castellano, S.; Sbardella, G.; Filetici, P.; et al. Quinoline-Based p300 Histone Acetyltransferase Inhibitors with Pro-apoptotic Activity in Human Leukemia U937 Cells. *ChemMedChem* **2014**, *9*, 542–548. [CrossRef] [PubMed]
146. Zhao, S.; Zhou, W.; Liu, J. Synthesis and HMG-CoA reductase inhibition of 2-cyclopropyl-4-thiophenyl-quinoline mevalonolactones. *Bioorg. Med. Chem.* **2009**, *17*, 7915–7923. [CrossRef] [PubMed]
147. Ma, X.; Qiu, W.; Liu, L.; Zhang, X.; Awad, J.M.; Evans, J.; Zhang, W. Synthesis of tetrahydropyrrolothiazoles through one-pot and four-component N,S-acetalation and decarboxylative [3+2] cycloaddition. *Green Synth. Catal.* **2021**, *2*, 74–77. [CrossRef]
148. Ward, S.A. Mechanisms of chloroquine resistance in malarial chemotherapy. *Trends Pharmacol. Sci.* **1988**, *9*, 241–246. [CrossRef]
149. Homewood, C.A.; Warhurst, D.C.; Peters, W.; Baggaley, V.C. Lysosomes, pH and the Anti-malarial Action of Chloroquine. *Nature* **1972**, *235*, 50–52. [CrossRef]
150. Kim, T.H.; Kim, H.K.; Hwang, E.S. Novel anti-adipogenic activity of anti-malarial amodiaquine through suppression of PPARγ activity. *Arch. Pharm. Res.* **2017**, *40*, 1336–1343. [CrossRef]
151. Brummendorf, T.H.; Cortes, J.E.; Milojkovic, D.; Gambacorti-Passerini, C.; Clark, R.E.; le Coutre, P.; Garcia-Gutierrez, V.; Chuah, C.; Kota, V.; Lipton, J.H.; et al. Bosutinib versus imatinib for newly diagnosed chronic phase chronic myeloid leukemia: Final results from the BFORE trial. *Leukemia* **2022**, *36*, 1825–1833. [CrossRef] [PubMed]
152. Mezger, K.; Ebert, S.; Muhle, H.E.; Stadt, U.Z.; Borkhardt, A.; Dilloo, D.; Faber, J.; Feuchtinger, T.; Imschweiler, T.; Jorch, N.; et al. Amsacrine combined with etoposide and methylprednisolone is a feasible and safe component in first-line intensified treatment of pediatric patients with high-risk acute lymphoblastic leukemia in CoALL08-09 trial. *Pediatr. Blood Cancer* **2022**, *69*, e29997. [CrossRef] [PubMed]
153. Rizzo, A.; Dalia Ricci, A.; Brandi, G. Neoadjuvant Dovitinib in Early- and Intermediate-Stage Hepatocellular Carcinoma. *Oncologyst* **2022**, *27*, e976. [CrossRef] [PubMed]
154. Thomas, K.D.; Adhikari, A.V.; Shetty, N.S. Design, synthesis and antimicrobial activities of some new quinoline derivatives carrying 1,2,3-triazole moiety. *Eur. J. Med. Chem.* **2010**, *45*, 3803–3810. [CrossRef] [PubMed]
155. Lin, A.J.; Loo, T.L. Synthesis and antitumor activity of halogen-substituted 4-(3,3-dimethyl-1-triazeno)quinolines. *J. Med. Chem.* **1978**, *21*, 268–272. [CrossRef] [PubMed]
156. Lavrard, H.; Larini, P.; Popowycz, F. Superacidic Cyclization of Activated Anthranilonitriles into 2-Unsubstituted-4-aminoquinolines. *Org. Lett.* **2017**, *19*, 4203–4206. [CrossRef] [PubMed]
157. Vidyacharan, S.; Sagar, A.; Sharada, D.S. A new route for the synthesis of highly substituted 4-aminoquinoline drug like molecules via aza hetero–Diels–Alder reaction. *Org. Biomol. Chem.* **2015**, *13*, 7614–7618. [CrossRef]
158. Zhang, X.F.; Ma, X.M.; Qiu, W.; Award, J.M.; Evans, J.; Zhang, W. One-Pot Mannich, Aza-Wittig and Dehydrofluorinative Aromatization Reactions for Direct Synthesis of 2,3-Disubstituted 4-Aminoquinolines. *Adv. Synth. Catal.* **2020**, *362*, 5513–5517. [CrossRef]

MDPI AG
Grosspeteranlage 5
4052 Basel
Switzerland
Tel.: +41 61 683 77 34

Molecules Editorial Office
E-mail: molecules@mdpi.com
www.mdpi.com/journal/molecules

www.ingramcontent.com/pod-product-compliance
Lightning Source LLC
Chambersburg PA
CBHW040835250926
51850CB00007BA/1322
9783725869961